THRESHOLD LOGIC

THRESHOLD LOGIC

P. M. LEWIS II

Information Scientist
General Electric Research and
Development Center

C. L. COATES

Professor of Electrical Engineering
University of Texas

John Wiley & Sons, Inc. New York · London · Sydney

 Library of Congress Catalog Card Number: 66–27896. Printed in the United States of America.

PREFACE

The rapid growth of the information processing field over the past twenty years has stemmed from two related, yet separate factors: the conception of more complex and sophisticated computer systems and the ability to build such systems economically and reliably. It is with the second, namely the design of digital systems, that this book deals.

Most digital computers are currently made of basic building blocks such as AND, OR, NAND, or NOR gates. These are embodiments of the basic Boolean algebra connectives and are, in some sense, perfectly natural choices. Not only are they conveniently related to the Boolean expressions, but they are relatively easy to implement physically using components with fairly wide tolerances. However, since new circuit techniques now promise the inexpensive mass fabrication of stable, reliable components, there is renewed interest in using a logically more powerful basic building block—the threshold gate. Although this type of gate requires tighter tolerances than more conventional gates, it promises a substantial reduction in the number of gates, the total number of components, and the number of interconnections.

Because of their increased logical complexity, the efficient use of threshold gates requires the development of a new switching theory. Among other names, this has been called threshold logic, linear input logic, and majority logic. The first published work in this field was that of McNaughton in 1957. Since then a rather extensive research effort has resulted in contributions by many individuals at various universities and industrial laboratories.

At present the state of threshold logic is roughly comparable to the more conventional switching theory. Many interesting and challenging problems remain, but enough has been learned about the theory so that real digital systems can be effectively designed. We feel, therefore, that a book on this subject is timely.

We cover the entire area of threshold logic in this book. Since the ultimate aim of any switching theory is synthesis, we have tried to maintain this perspective in our presentation. We have attempted to organize the material so that it will be useful to both the researcher and to the design engineer who wants to use threshold gates. With this in mind the concept of the function tree is used as a unifying framework for much of the theoretical development. This allows the presentation of single and multiple gate synthesis problems to be combined. We believe that this approach will enable both the researcher and the design engineer to attain new insight. Finally, we have tried to give the reader an "intuitive feel" for the theory. Many practical problems can be solved without the "general algorithm" by using intuition and skill based upon the experience of applying the theorems to simple examples. We hope that our organization of the material will help in this direction.

For the designer interested mainly in synthesizing threshold gate networks rather than in the details of the theory, we suggest the following sections as a first reading: Chapter 1; Chapter 2; Chapter 3, Sections 1 through 7 and 11; Chapter 4, Sections 1 through 3, 9, and 10; Chapter 5; Chapter 6; and Chapter 7. All starred sections or portions of sections can be omitted. Most of the theorems in these sections, and throughout the book, are followed by examples illustrating their use in synthesis. This often allows the reader to skip over the detailed proof of the theorem on a first reading.

For the student and researcher, a considerably more detailed exposition of the theory is also included. The first four chapters develop the general framework on which the theory is based. Chapter 5 is the culmination of the function tree approach to single and multiple gate synthesis. Chapter 6 specializes these concepts to majority gates and symmetrical functions. Chapter 7 develops some pertinent properties based upon duality concepts and gives some additional synthesis procedures. These first seven chapters constitute the elementary parts of the theory. Chapter 8 gives some advanced uses of the function tree including a test for single-gate realizability. Chapter 9 discusses general methods for solving simultaneous inequalities with emphasis on linear programming. Chapter 10 considers adaptive realization procedures such as those used by the Perceptron, Adaline, etc.

Although the book deals with the advanced as well as the elementary aspects of threshold logic, the only mathematical prerequisites are the elements of Boolean algebra such as might be obtained in a first course in switching theory. A certain amount of mathematical maturity might be helpful, however, in following the detailed proofs and arguments.

The development of threshold logic is the result of the research contributions of many individuals. Although these contributors are too numerous to list individually here, they are referenced throughout the text and in the Bibliography. We wish to thank especially Dr. R. E. Stearns, who helped substantially in the preparation of the chapter on linear programming. Also we would like to thank the management of the General Electric Research and Development Center and in particular to Drs. J. L. Lawson and R. L. Shuey, who established and maintain the free and stimulating environment in which our research was performed and who provided the time and opportunity to write this book. The assistance also of the Bureau of Engineering Research and the University Research Institute of the University of Texas is gratefully acknowledged.

P. M. LEWIS II
C. L. COATES

August 1966

CONTENTS

Chapter 1

INTRODUCTORY CONCEPTS

Digital computers, digital communication and control systems, and many other digital systems contain networks whose input and output variables can have only two different values, that is, are binary-valued. The analysis and synthesis of such *digital* or *logical networks* is the subject matter of the areas of Switching Theory and Logical Design.

More specifically, these areas deal with two general problems: the characterization of the input-output relations of digital networks (using, for example, Boolean algebra), and the synthesis of digital networks from their input-output relations using certain specified logic elements as building blocks. Conventionally, these building blocks have included such elements as relay contacts, AND gates, OR gates, and NOR gates.

All of these conventional elements share two very useful properties: they are easily constructed from standard components, and their input-output relations are simple and easily described in Boolean algebra—the same language that is used to describe the input-output relations of the network to be synthesized. These two properties combine to give simplicity and elegance to the theory and practice based on these elements.

There is no reason, however, for the building blocks to be limited to such logically simple elements. In fact, the need for digital networks that are larger, faster, more reliable, and more compact, together with the emergence of technologies that allow mass fabrication of stable reliable components, increases the desirability that the basic gate perform a more complex logical operation. In this way the number of gates that is needed to realize a given function can be materially reduced. This is of particular significance if the circuit complexity of the gate is not appreciably increased over that of a more conventional gate.

The threshold gate is a class of logical elements that has these properties. Threshold Logic is the branch of Switching Theory that deals with the use of this gate as the basic building block for the synthesis of digital networks. This new gate offers some potentially significant advantages compared with conventional gates, but at the same time there are additional complexities involved in its theory and application.

1.1 DEFINITION OF THE THRESHOLD GATE

The input-output relation of a digital network is expressed as a *logical* or *switching function*, which is a binary-valued function of binary-valued variables. There are n independent binary variables, $x_1,x_2,\ldots,x_n$, each of which can take on the values 0 or 1. (Sometimes binary variables that take on other values are used, but they are indicated by different letters.) Switching functions are denoted by $F(x_1,x_2,\ldots,x_n)$ and can take on the values 0 or 1.

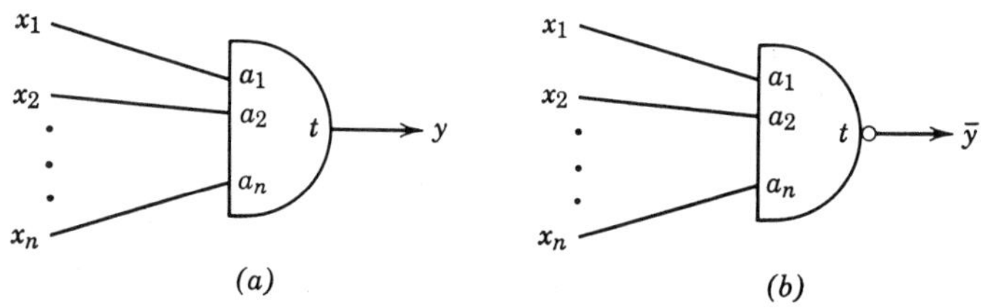

Fig. 1.1. Diagrammatic representation of a T gate. (*a*) Normal T gate. (*b*) Inverting T gate.

A model of a *threshold gate* or *T gate* is represented diagrammatically as in Fig. 1.1. It has n binary-valued inputs and a single binary-valued output. The inputs are shown as the n independent variables $x_1,x_2,\ldots,x_n$. The output is denoted by y and can take on the values 0 or 1.

The internal parameters of a T gate are a *threshold* t and *weights* or *coefficients* $a_1,a_2,\ldots,a_n$ where t and each a_j, ($j = 1,2,\ldots,n$), can be any fixed positive or negative real number. Each coefficient a_j is associated with a particular input x_j. The model operates as follows.

For each combination of values of the input variables $x_1,x_2,\ldots,x_n$, the corresponding value of the output y is

$$\begin{aligned} y &= 1 \quad \text{if} \quad \sum_{j=1}^{n} a_j x_j \geq t \\ y &= 0 \quad \text{if} \quad \sum_{j=1}^{n} a_j x_j < t \end{aligned} \tag{1.1}$$

where the sum and product operations are the usual arithmetic ones. The notation

$$y = \left\langle \sum_{j=1}^{n} a_j x_j \right\rangle_t \tag{1.2}$$

is used to represent Eq. 1.1.

Equations 1.1 and 1.2 characterize the T gate in terms of a linear arithmetic function that involves the internal parameters. This function is represented by f and is called the *separating function* for the gate; that is,

$$f(x_1,x_2,\ldots,x_n) = \sum_{j=1}^{n} a_j x_j$$

It is permissible, therefore, to write

$$y = \langle f(x_1,x_2,\ldots,x_n) \rangle_t \tag{1.3}$$

Although the T gate defined by Eqs. 1.1 to 1.3 has its inputs connected to the independent variables $x_1,x_2,\ldots,x_n$, this is not essential to the definition. In general, a T gate may have m inputs, some of which are connected to independent variables and some to the outputs of other T gates. In the case where the m inputs are connected to the outputs of m other T gates denoted by $y_1,y_2,\ldots,y_m$, Eq. 1.2 becomes

$$y = \left\langle \sum_{j=1}^{m} a_j y_j \right\rangle_t$$

and the law of operation is again defined by Eq. 1.1 with $n = m$ and $x_j = y_j$.

The T gate of Eq. 1.3 yields a binary-valued output for each combination of binary-valued inputs; therefore the output y is a switching function of the input variables. This switching function can be represented in a number of ways. Equation 1.3 is the *separating-function representation* in which the value of the function is expressed in terms of the independent variables x_j and the parameters a_j and t. Another often-used representation is the *Boolean-function representation*:

$$y = F(x_1,x_2,\ldots,x_n) \tag{1.4}$$

in which the value of the function is expressed in terms of only the independent variables x_j and the Boolean† operators, $\dotplus$, $\cdot$, and $\bar{\ }$.

The actual form of the representation is significant since it may imply a realization of the function. Equation 1.3, for example, implies a T-gate realization while Eq. 1.4 may imply a realization with AND and OR gates. In fact, translating from one representation of a switching function to

† The symbol $\dotplus$ denotes the Boolean addition operator. The symbol $+$ is used for normal arithmetic addition.

another is usually the key to the analysis and synthesis problems. The translation from the representation in terms of the separating function (Eq. 1.3) to the Boolean-function representation (Eq. 1.4) is straightforward. This is the analysis problem and is illustrated by Example 1.1. Translating from the Boolean-function to the separating-function representation is a more involved procedure. Its solution implies the synthesis of the T gate from its input-output relation. As such it is called the "one-gate realization problem."

To clarify the relationship between the notation and the realization, we adopt the convention that each formula for the separating function implies a particular T-gate realization. Specifically

$$\langle f(x_1,x_2,\ldots,x_n)\rangle_t = \left\langle \sum_{j=1}^{n} a_j x_j \right\rangle_t$$

implies the gate with threshold t, input variables x_j, and corresponding weights a_j.

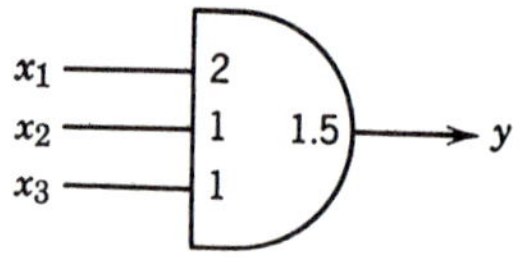

Fig. 1.2. T gate of Example 1.1.

Example 1.1. Consider the T gate shown in Fig. 1.2. The output in terms of the separating function is

$$y = \langle 2x_1 + x_2 + x_3 \rangle_{1\cdot5}$$

The output as a Boolean function of the input variables is

$$y = x_1\bar{x}_2\bar{x}_3 \dotplus x_1x_2\bar{x}_3 \dotplus x_1\bar{x}_2x_3 \dotplus \bar{x}_1x_2x_3 \dotplus x_1x_2x_3 = x_1 \dotplus x_2x_3$$

This follows immediately from Table 1.1 (Section 2.4 gives a computationally simpler way to solve this type of problem).

Table 1.1. Tabular Representation of $y = \langle 2x_1 + x_2 + x_3 \rangle_{1\cdot5}$

x_1, x_2, x_3	$2x_1 + x_2 + x_3$	$y = \langle 2x_1 + x_2 + x_3 \rangle_{1\cdot5}$
0 0 0	0	0
0 0 1	1	0
0 1 0	1	0
0 1 1	2	1
1 0 0	2	1
1 0 1	3	1
1 1 0	3	1
1 1 1	4	1

There are, of course, many separating-function representations of this same Boolean function. One other example is

$$y = \langle 6x_1 + 4x_2 + 3x_3 \rangle_5$$

as can be verified by a similar table.

1.2 ADVANTAGES OF THRESHOLD LOGIC

It follows from their definition that T gates are more powerful logical elements than are AND and OR gates. In fact, a T gate can be considered a generalization of an AND or OR gate. A T gate for which all weights have unit value and the threshold value is $\frac{1}{2}$ is an n-input OR gate. The same T gate, but with threshold $n - \frac{1}{2}$, is an n-input AND gate. Thus AND and OR gates are special T gates. Generalizations result by allowing the weights and the threshold to have different values. In a similar sense, the complementing T gate is a generalization of the NAND and NOR gates.

Because of the wide range of possible weight and threshold values, the size and complexity of the class of logical functions that can be realized by a single T gate are large when compared with those of the conventional elements. As a result, the number of gates, the number of inputs, and the total number of physical components needed to realize a particular logical function are, on the average, considerably smaller for T gates than for conventional gates. Actually, many functions that require a number of AND and OR gates can be realized by a single T gate.

An example of this is the class of functions $M_k(x_1,x_2,\ldots,x_n)$ which have unit value if and only if k or more of their inputs have unit value. For such functions

$$M_k(x_1,x_2,\ldots,x_n) = \langle x_1 + x_2 + \cdots + x_n \rangle_{k-1/2}$$

Specifically, consider $M_2(x_1,x_2,\ldots,x_7)$, which has unit value when 2 or more of the 7 input variables have unit value. Such a function can be realized by a single T gate with 7 inputs. To realize the same function with AND-OR logic requires 12 gates with a total of 38 inputs. This is illustrated in Fig. 1.3.

Although $M_k(x_1,x_2,\ldots,x_n)$ type functions are familiar examples, they represent a very small portion of the logical functions of n variables that can be realized by a single T gate. On the other hand, there are many functions that require more than one T gate for their realization. These, however, can always be realized by a network of T gates, and the number required is never larger than the number of AND, OR, and INVERTER gates that would be needed.

To justify this statement, we have to show only that AND, OR, and INVERTER gates are all special T gates. This has already been demonstrated for AND and OR gates and is certainly also true for INVERTER gates. For example, the single-input T gate with weight -1 and threshold $-\frac{1}{2}$ is an INVERTER.

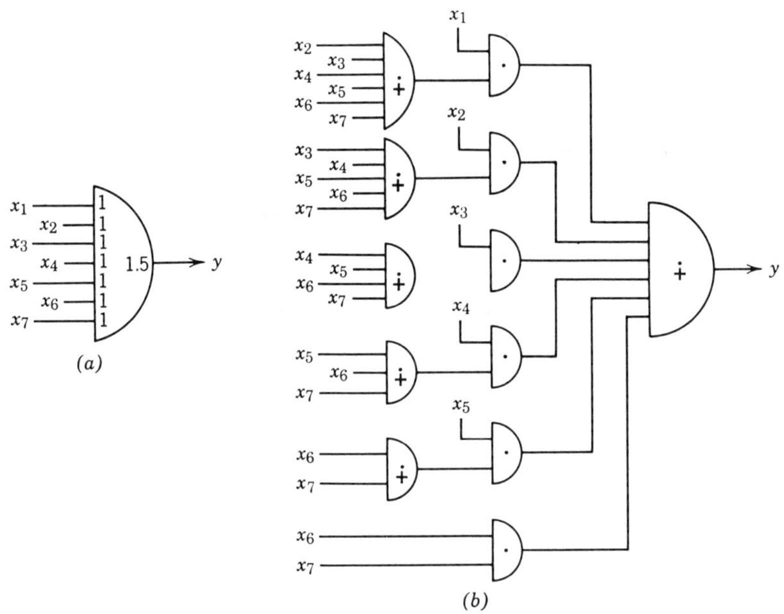

$$y = x_1(x_2 \dotplus x_3 \dotplus x_4 \dotplus x_5 \dotplus x_6 \dotplus x_7) \dotplus x_2(x_3 \dotplus x_4 \dotplus x_5 \dotplus x_6 \dotplus x_7) \dotplus x_3(x_4 \dotplus x_5 \dotplus x_6 \dotplus x_7) \dotplus x_4(x_5 \dotplus x_6 \dotplus x_7) \dotplus x_5(x_6 \dotplus x_7) \dotplus x_6x_7$$

Fig. 1.3. Realization of the logical function $y = M_2(x_1,x_2,\ldots,x_7)$ using (*a*) one T gate and (*b*) AND and OR gates.

Thus T gates are general logical elements; their use results, among other things, in a considerable saving in numbers of gates, components, and inputs. This is their principal advantage. However, to obtain this advantage requires that some potential difficulties be overcome.

First, there is the problem of synthesizing the given Boolean function. For this synthesis, it is necessary to convert the Boolean representation to a representation in terms of separating functions. From looking at a Boolean function we cannot easily tell how to represent it as an interconnection of threshold gates. In particular, most of the elaborate machinery of Boolean algebra is not useful in T gate synthesis.

The second problem concerns the sensitivity of the T gate to variations in weights, threshold, and logical variables. In any physical realization of a threshold gate, these parameters will vary from their nominal values, and if this variation is too large the gate will no longer operate correctly. Furthermore, the allowable variation for which the gate will still operate correctly depends on the nominal values of the weights and threshold. For example, it is intuitively obvious and will be proved later that an OR gate with two unit weights and a threshold value of $\frac{1}{2}$ has larger allowable variations than an AND gate with 15 unit weights and a threshold value of $14\frac{1}{2}$.

It will become clear that some of the advantages of gate reduction are bought at the expense of tighter tolerances on components. Thus, in evaluating the advantages of T gates, and indeed throughout the theory and practice of Threshold Logic, the concepts of sensitivity and tolerance play a significant role.

1.3 THRESHOLD OR LINEARLY SEPARABLE FUNCTIONS

Geometric Interpretation

For many purposes it is convenient to define an abstract space in which many of the quantities of interest can be plotted.

Switching space is an n-dimensional Euclidean space where each coordinate axis corresponds to one of the independent variables. Each combination of values that the independent variables can assume plots as a point in the space. The 2^n such points, $p_0, p_1, \ldots, p_{2^n-1}$, lie on the vertices of a unit hypercube (n cube) in the n-dimensional space. A given logical function F assigns to each p ($p = p_0, p_1, \ldots, p_{2^n-1}$) a corresponding value $F(p)$ which is either 1 or 0. In the case where "don't-care" conditions exist, F is defined on only a subset of the points; the remaining points are the "don't-care" points. Let P_0 and P_1 be the subsets of vertices of the n cube for which $F(p)$ has values 0 and 1 respectively.

Now consider the sets P_0 and P_1 defined by any logical function F that represents the output y of a single T gate, where

$$y = F(x_1, x_2, \ldots, x_n) = \left\langle \sum_{j=1}^{n} a_j x_j \right\rangle_t$$

From the rule of operation (Eq. 1.1), the P_0 vertices of the n cube are separated from the P_1 vertices by the hyperplane whose equation is

$$\sum_{j=1}^{n} a_j z_j = t \tag{1.5}$$

where the domain of z_j ($j = 1,2,\ldots,n$) is the real line. This hyperplane includes the points that satisfy

$$\sum_{j=1}^{n} a_j x_j = t$$

for $x_j = 0, 1$. Figure 1.4a shows such a separating hyperplane for a two-variable function.

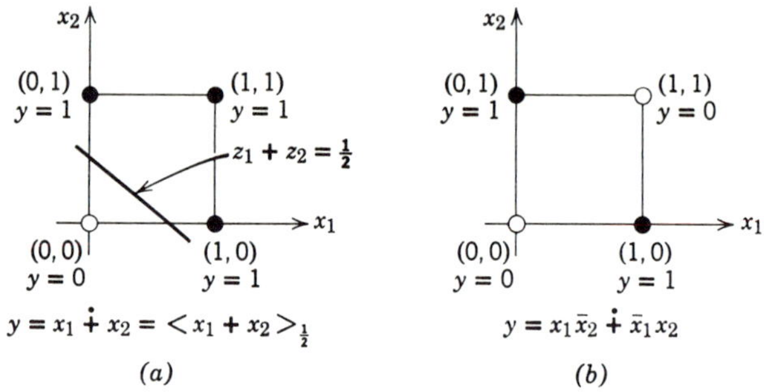

Fig. 1.4. (*a*) Threshold function and (*b*) nonthreshold function.

The class of logical functions that can be realized by a single T gate is called the *linearly separable* (ℓ.s.) or the *threshold function* class. It follows from the previous discussion that members of this class can be characterized as follows:

Theorem 1.1. A logical function F is of the linearly separable class if and only if there exists a hyperplane that separates the P_0 vertices of the n cube from the P_1 vertices. If the hyperplane is represented by Eq. 1.5, then the output of a T gate that realizes F is

$$y = F(x_1,x_2,\ldots,x_n) = \left\langle \sum_{j=1}^{n} a_j x_j \right\rangle_t \tag{1.6}$$

or, what is the same thing,

$$y = F(p) = \langle f(p) \rangle_t$$

Not every partition of the 2^n vertices of the n cube into subsets P_0 and P_1 can be separated by a hyperplane; therefore, not every logical function can be realized by a single T gate. An example of such a function is

$$y = x_1\bar{x}_2 \dotplus \bar{x}_1 x_2 \tag{1.7}$$

It is evident from Fig. 1.4b that no two-dimensional hyperplane exists which separates the P_0 from the P_1 vertices.

This geometrical model is quite useful for answering certain questions. However, it has the disadvantage of being nonvisual for more than three dimensions. The following section defines a different interpretation that does not have this limitation.

The Map Interpretation

Each T gate specifies a separating function f and a logical function F both of which are defined on the vertices of the n cube. For each vertex p for which F is defined, there is associated the ordered triplet $[p,f(p),F(p)]$ where $f(p)$ is some real number and where $F(p) = 0$ or 1. The set $\{[p,f(p),F(p)]\}$ of ordered triplets is called the *map* of F by f.

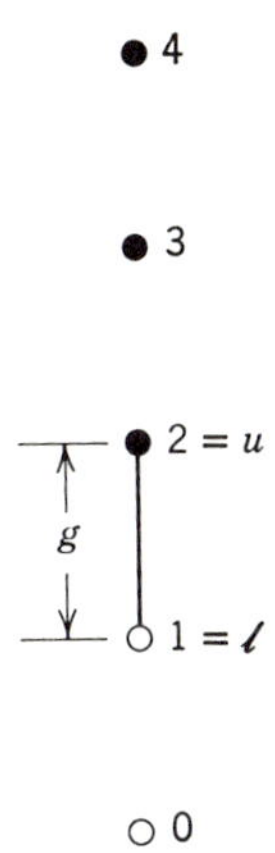

Fig. 1.5. Visualization of the map of Example 1.1.

A convenient visualization of the map is shown in Fig. 1.5 for the functions of Example 1.1 where

$$F(x_1,x_2,x_3) = x_1 \dotplus x_2x_3$$

and

$$f(x_1,x_2,x_3) = 2x_1 + x_2 + x_3$$

For each p, a point is located on the real line at location $f(p)$. This point is represented by ○ or ● according as $F(p) = 0$ or 1, respectively. Note that the visualization includes no information concerning p, nor is there any notation made if two or more p's have the same $f(p)$ value. Essentially the map is the result of the separating function f mapping the vertices of the unit cube in n space onto the real line. Each $f(p)$ is called the *image* of p.

A map is divided into disjoint subsets, called the *zero* and *unit* parts, according to whether $F(p) = 0$ or 1. Let u denote the smallest $f(p)$ in the unit part. If the unit part is empty, for example, when $F(p) = 0$ for all p, then let $u = \infty$. Let ℓ denote the largest value of $f(p)$ in the zero part. If the zero part is empty, for example, when $F(p) = 1$ for all p, then let $\ell = -\infty$. A map is separated if $u > \ell$. In such cases the terminology "*F is separated by f*" or "*f realizes F*" is used. In the map of Fig. 1.5, $u = 2$ and $\ell = 1$; hence it is separated.

Of particular importance to a separated map is the set of all numbers y such that $\ell < y \leq u$. This set defines the half-closed interval on the real line that is called the *gap*. It is denoted by $u{:}\ell$ where u and ℓ are the *upper* and *lower* boundaries respectively. Note that u is a point of the gap but that ℓ is not. The *length* g of the gap is defined as $u - \ell$.

Reference to Eq. 1.1 and to the above description of the map shows that the class of linearly separable functions can be characterized as follows.

> **Theorem 1.2.** A logical function F is of the linearly separable class if and only if there exists a linear arithmetic function of the form
>
> $$f(x_1,x_2,\ldots,x_n) = \sum_{j=1}^{n} a_j x_j$$
>
> that yields a separated map for F.

If y is the output of a T gate, then from Theorem 1.2 it follows that, for each p

$$\begin{aligned} f(p) &\geq u \quad \text{if and only if} \quad y = 1 \\ f(p) &\leq \ell \quad \text{if and only if} \quad y = 0 \end{aligned} \tag{1.8}$$

We represent Eq. 1.8 as

$$y = \langle f(p) \rangle_{u:\ell} \tag{1.9}$$

Comparing Eq. 1.9 with Eq. 1.1, it is evident that if F is separated by f, then the T gate that realizes F has its threshold t within the gap; that is, $\ell < t \leq u$. As a result, we often replace the t in the diagrammatic representation of Fig. 1.1 by the gap $u{:}\ell$.

1.4 T-GATE REALIZATIONS OF SPECIAL FUNCTIONS

Logical functions which, for all p, have unit value or have zero value are called *constant functions* and are denoted by $y \equiv 1$ and $y \equiv 0$. The following are important properties of constant functions.

> **Theorem 1.3.** The unit and zero constant functions are linearly separable. Moreover, if f is an arbitrary linear-arithmetic function, then it separates the unit function with gap $\sigma_n{:}-\infty$ and the zero function with

gap $\infty:\sigma_p$ where σ_n and σ_p are the minimum and maximum values of f, respectively. That is, for all p,

$$1 \equiv y = \langle f(p) \rangle_{\sigma_n:-\infty} \qquad \text{and} \qquad 0 \equiv y = \langle f(p) \rangle_{\infty:\sigma_p} \tag{1.10}$$

The validity of Theorem 1.3 follows immediately from the observation that in each case the gap is selected so that all real line images of the n cube vertices due to the f mapping lie above the gap for $y \equiv 1$ and below the gap for $y \equiv 0$. This is illustrated in Fig. 1.6.

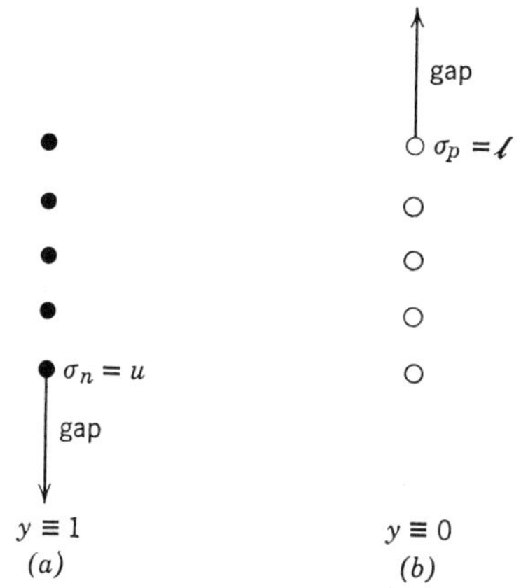

Fig. 1.6. Maps of constant functions due to an arbitrary separating function f.

Another pair of Boolean functions that are linearly separable and can be realized by an easily described class of separating functions are the logical OR and the logical AND of n variables. The results are set forth in the following theorem.

Theorem 1.4. The logical functions $y = x_1 \dotplus x_2 \dotplus \cdots \dotplus x_n$ and $y = x_1x_2 \ldots x_n$ are linearly separable and are separated by an arbitrary linear-arithmetic function $f(x_1,x_2,\ldots,x_n)$ if and only if all coefficients are positive. The gaps for the OR and AND functions are $a_n:0$ and $\sigma:(\sigma - a_n)$ respectively, where $\sigma = \sum_{j=1}^{n} a_j$ and $a_n \leq a_j$ $(j = 1,2,\ldots,n)$. In other words,

$$x_1 \dotplus x_2 \dotplus \cdots \dotplus x_n = \Big\langle \sum_{j=1}^{n} a_j x_j \Big\rangle_{a_n:0} \tag{1.11}$$

$$x_1x_2 \ldots x_n = \Big\langle \sum_{j=1}^{n} a_j x_j \Big\rangle_{\sigma:(\sigma - a_n)}$$

The validity follows from the observation that the map of the OR function has only the image $f(p_0) = 0$ below the gap while the AND function has all images below the gap except $f(p_{2^n-1}) = \sigma$, where $p_0 = 00\ldots0$ and $p_{2^n-1} = 11\ldots1$. This is illustrated in Fig. 1.7.

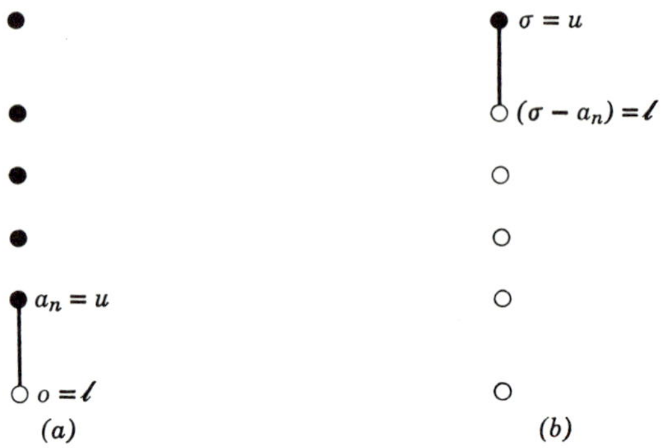

Fig. 1.7. Maps for the OR and AND functions due to a separating function with positive coefficients. (*a*) Map for the OR function. (*b*) Map for the AND function.

1.5 THRESHOLD GATE CIRCUITS

The principal advantage of the T gate is that it can be constructed from about the same number of physical components as more conventional logic gates, yet it has a much greater logic capability. At this point it is appropriate to consider briefly a few of the physical implementations.

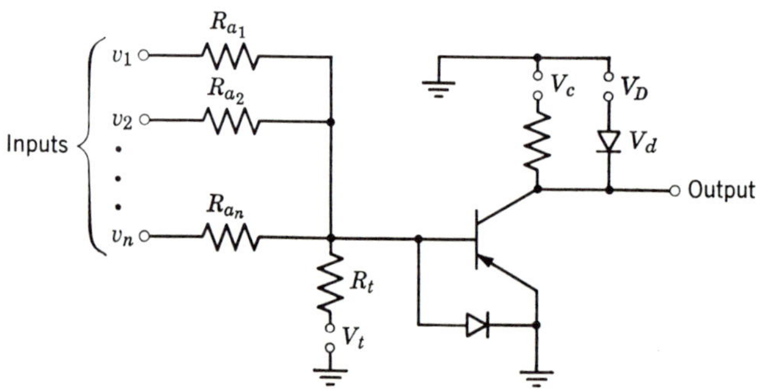

Fig. 1.8. A resistor-transistor complementing T gate circuit.

Figure 1.8 shows one type of resistor-transistor complementing T gate (Rowe).† Each input variable x_j is represented by the voltage variable v_j. Voltage values V_1 and V_0 represent the unit and zero values, respectively, of the logical variables. The resistors R_{a_j} determine the weights (only

† Names in parentheses correspond to references in the bibliography.

positive weights are available) while the threshold is determined by resistor R_t, voltage source V_t, and the characteristic of the transistor. The input circuit is a resistance-adding network that computes the weighted sum of the inputs. The weighted sum for those input combinations for which the logical value of the output (complemented) is unity is not sufficient to cause the transistor to conduct. Hence the output voltage is V_1 and consists of the sum of the clamp supply voltage V_D and the clamp diode voltage V_d. For those input combinations for which the logical value of the output is zero, the weighted sum is sufficient to cause the transistor to conduct and to reach a saturation condition. The output has value V_0, which is the collector-emitter saturation voltage and is usually a few tenths of a volt.

Figure 1.9 shows a magnetic core T-gate circuit (Karnaugh). Each input variable x_j is represented by the current variable i_j, and the corresponding

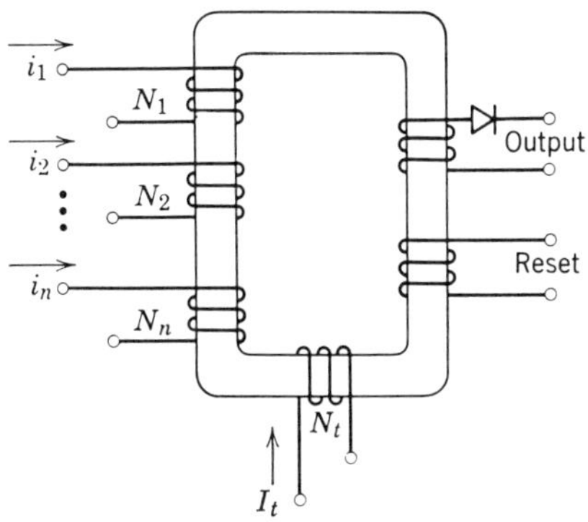

Fig. 1.9. A magnetic core T-gate circuit.

weight is represented by the number of turns N_j. Both positive and negative integer weight values are possible depending upon the relative direction of the winding. Current values of I_1 and I_0 represent the unit and zero values respectively of the logical variables, and usually $I_0 \approx 0$. The threshold is represented by a separate current I_t that is applied to a winding of N_t turns. The circuit operation occurs in two phases. Initially the magnetic core is in negative saturation. During the first or setting phase, the input currents are applied and the weighted sum of the inputs

$$\sum_{j=1}^{n} N_j i_j + N_t I_t$$

is sufficient to drive the core to positive saturation if and only if the particular input combination yields unit logical output. During the second or readout phase, the input currents are removed and a pulse, sufficient to drive the core to negative saturation, is applied to the reset winding. If, at the end of the setting phase, the core is in positive saturation, a pulse will appear on the output winding when the core is reset, and this corresponds to unit logical output. If the core is in negative saturation at the end of the setting phase, no pulse will appear during reset and this corresponds to zero logical output.

A third example of a T-gate circuit is the parametron (Goto) shown in Fig. 1.10. Each input variable x_j is represented by the current variable i_j, and the corresponding weight is represented by the number of turns N_j.

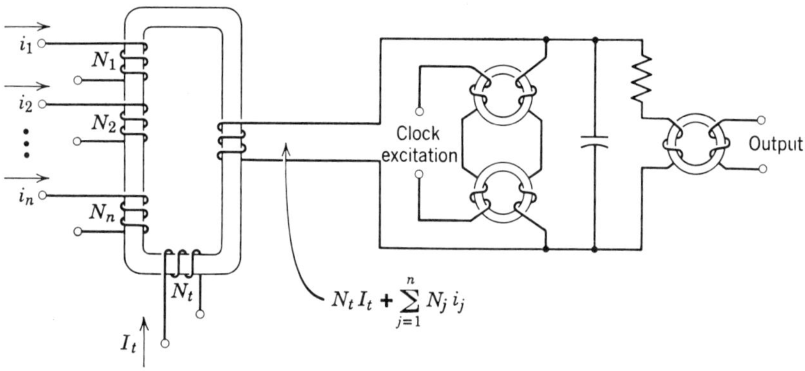

Fig. 1.10. A parametron T-gate circuit.

Again, both positive and negative integer weight values are possible, depending upon the relative direction of the winding. Each i_j is a sinusoidal current of fixed amplitude and fixed frequency ω with two possible phase values (called positive and negative) that are 180° apart. These represent the unit and zero values of the logical variable. The threshold is represented by a separate winding to which is applied a separate input sinusoid. The magnetic circuit determines the weighted linear sum of the input currents and the threshold, and applies the resultant signal to the resonant circuit of an oscillator which is tuned to the frequency ω. The circuit operates as follows:

Before the start of each clock phase, the oscillator is dormant, that is, not oscillating. At the beginning of the clock phase a clock excitation is applied to a nonlinear inductance that is a part of the oscillator tuned circuit. The clock excitation, which consists of a d–c component and a sinusoid of frequency 2ω, causes the oscillator to generate a signal of frequency ω. The phase of this oscillation is the same as the phase of the linear weighted sum from the input magnetic circuit. As a result the phase of the oscillator will be positive or negative according to whether the phase of the resultant input is positive or negative.

1.6 THRESHOLD GATE SENSITIVITY MODELS

A factor of particular significance in the consideration of Threshold Logic is the sensitivity of the T gate to signal and component fluctuations. From a consideration of the map it is clear that fluctuations of the circuit signals and components will cause the image points of the map to vary with respect to each other and will, therefore, cause the gap to vary. Moreover, fluctuations in the threshold circuit cause the threshold to change location within the gap. As long as these fluctuations neither eliminate the gap nor cause the threshold to lie outside the gap, the gate will continue to give the correct output for each input combination. Thus the gate has a certain allowable tolerance or sensitivity to component and signal variations. We wish to investigate this tolerance and determine how it depends upon the gate parameters.

The relationship between sensitivity and gate parameters depends, to a large extent, on the physical realization of the T gate. Such factors as how the logical variables are to be realized and exactly what parameters are allowed to vary have a significant effect on the form of the result.

Thus, to some extent, every different detailed physical realization of a T gate will have detailed differences in its sensitivity analysis. However, most circuit realizations are approximations, in some sense, to one of a very few *T gate models* such as that described by Eq. 1.1. Analysis of these models yields certain *inherent sensitivity* factors which are intrinsic to the model and which will, at least approximately, also apply to the particular realization. There may be some additional *circuit sensitivity* factors which depend on the details of the implementing circuit, but they are beyond the scope of the present treatment.

At this point we consider the sensitivity of two T gate models. The model described by Eq. 1.1 is called the *Zero-One Model* and is approximated by the resistor-transistor circuit and the magnetic core circuit for T gates.

The second model, called the *Minus-Plus-One Model*, is defined similarly except that the possible values of the binary variables are 1 and -1 rather than 1 and 0 respectively. It is approximated by the parametron T gate circuit. For most considerations no distinction between the two models is necessary. With regard to sensitivity, however, it is necessary to consider them separately. Moreover, it is convenient at this time to restrict our attention to only the *definite* T gates of each model. A *definite* T gate is one for which all coefficients or weights are of the same sign; that is, its weights are either all positive or all negative.

A more general treatment† of sensitivity is given in Appendix A. There the logical variables can be represented by arbitrary quantities V_1 and V_0, and the weights can be of mixed signs.

1.7 SENSITIVITY: DEFINITE ZERO-ONE MODEL

This analysis is a "worst-case" approach to sensitivity. It postulates that tolerances have been assigned to the parameters of the model and that the gate must operate error free for all variations within these tolerances. This means that the T gate realizes the same switching function when fluctuations are absent as when they are present within the prescribed amounts.

Let $\mathscr{A}$, $\mathscr{T}$, and $\mathscr{X}$ denote the percentage fluctuation tolerances on the weights, threshold, and input values respectively. Although the zero value of the input variables does not vary, it is permissible to assume that the percentage variation is $\mathscr{X}$. All tolerances are defined so that (1) the gate will operate error-free for all possible combinations of weight variations of percentage less than $\mathscr{A}$, all possible threshold variations of percentage less than $\mathscr{T}$, and all possible combinations of input variations of percentage less than $\mathscr{X}$; (2) for some combination of weight variations of percentage $\mathscr{A}$, some combination of threshold variations of percentage $\mathscr{T}$, and some combination of input value variations of percentage $\mathscr{X}$, there is an input combination for which the gate output is in error. Under these conditions $\mathscr{A}$, $\mathscr{T}$, and $\mathscr{X}$ are called *allowable fluctuation tolerances* and the operation is called *error-free*. We are interested in determining how particular values of $\mathscr{A}$, $\mathscr{T}$, and $\mathscr{X}$ restrict the parameters of a T gate so that it operates error-free.

According to the first part of Eq. 1.1, all input combinations for which $y = 1$ are such that the linear sum of the weighted inputs exceeds the threshold. For these points the worst-case fluctuations are those that make the linear weighted sum as small as possible and that make the threshold as large as possible. When these worst-case fluctuations are included, the inequality that must be satisfied for error-free operation, when all coefficients are positive, is

$$\sum_{j=1}^{n} a_j(1 - \mathscr{A})x_j(1 - \mathscr{X}) \geq t(1 + \mathscr{T})$$

or

$$(1 - \mathscr{A})(1 - \mathscr{X}) \sum_{j=1}^{n} a_j x_j \geq t(1 + \mathscr{T})$$

† See also Muroga, 4.

This inequality must be satisfied for all input combinations such that $y = 1$. The most important combinations in this regard are those for which

$$\sum_{j=1}^{n} a_j x_j = u < \infty$$

since this is the smallest value such that $y = 1$. (The restriction $u < \infty$ excludes only the constant function $y \equiv 0$.) Using this information, the above condition for error-free operation becomes

$$(1 - \mathscr{A})(1 - \mathscr{X})u \geq t(1 + \mathscr{T}) \tag{1.12}$$

From the second inequality of Eq. 1.1, the worst-case fluctuations cause the linear weighted sum to be as large as possible and the threshold to be as small as possible. When the worst-case fluctuations less than $\mathscr{A}$, $\mathscr{X}$, and $\mathscr{T}$ are included, the inequality that must be satisfied for error-free operation, when all coefficients are positive, is

$$\sum_{j=1}^{n} a_j(1 + \mathscr{A})x_j(1 + \mathscr{X}) \leq t(1 - \mathscr{T})$$

or

$$(1 + \mathscr{A})(1 + \mathscr{X}) \sum_{j=1}^{n} a_j x_j \leq t(1 - \mathscr{T})$$

This relation must be satisfied for all input combinations such that $y = 0$. The most important combinations in this regard are those for which

$$\sum_{j=1}^{n} a_j x_j = \ell > -\infty$$

since this is the largest value such that $y = 0$. (The restriction $\ell > -\infty$ excludes only the constant function $y \equiv 1$.) The above condition for error-free operation then becomes

$$(1 + \mathscr{A})(1 + \mathscr{X})\,\ell \leq t(1 - \mathscr{T}) \tag{1.13}$$

When all coefficients are negative, the inequalities of Eqs. 1.12 and 1.13 become the inequalities of Eqs. 1.14 and 1.15 respectively:

$$(1 + \mathscr{A})(1 + \mathscr{X})u \geq t(1 - \mathscr{T}) \tag{1.14}$$

$$(1 - \mathscr{A})(1 - \mathscr{X})\ell \leq t(1 + \mathscr{T}) \tag{1.15}$$

Using these results, the following constraint must be satisfied by a T gate in order to achieve error-free operation when parameter-fluctuation tolerances $\mathscr{A}$, $\mathscr{X}$, and $\mathscr{T}$ are specified.

Theorem 1.5. The following inequality must be satisfied by any Definite Zero-One Model T Gate to operate error free when the allowable parameter fluctuations are less than $\mathscr{A}$, $\mathscr{X}$, and $\mathscr{T}$ and when u and ℓ are finite:

$$\left|\frac{u-\ell}{u+\ell}\right| \geq \frac{\mathscr{A}+\mathscr{X}+\mathscr{T}+\mathscr{A}\mathscr{X}\mathscr{T}}{1+\mathscr{A}\mathscr{X}+\mathscr{A}\mathscr{T}+\mathscr{X}\mathscr{T}} \tag{1.16}$$

Proof. From Eqs. 1.12 and 1.13 it follows that for error-free operation, when all coefficients are positive,

$$\frac{(1-\mathscr{A})(1-\mathscr{X})u}{1+\mathscr{T}} \geq t \geq \frac{(1+\mathscr{A})(1+\mathscr{X})\ell}{1-\mathscr{T}}$$

or

$$\frac{(1-\mathscr{A})(1-\mathscr{X})u}{1+\mathscr{T}} \geq \frac{(1+\mathscr{A})(1+\mathscr{X})\ell}{1-\mathscr{T}}$$

Multiplying both sides of the inequality by $(1+\mathscr{T})(1-\mathscr{T})$ and combining terms gives

$$\frac{u-\ell}{u+\ell} = \left|\frac{u-\ell}{u+\ell}\right| \geq \frac{\mathscr{A}+\mathscr{X}+\mathscr{T}+\mathscr{A}\mathscr{X}\mathscr{T}}{1+\mathscr{A}\mathscr{X}+\mathscr{A}\mathscr{T}+\mathscr{X}\mathscr{T}}$$

and this establishes the result when all coefficients are positive. A similar procedure starting with Eqs. 1.14 and 1.15 yields the result

$$\frac{u-\ell}{-(u+\ell)} = \left|\frac{u-\ell}{u+\ell}\right| \geq \frac{\mathscr{A}+\mathscr{X}+\mathscr{T}+\mathscr{A}\mathscr{X}\mathscr{T}}{1+\mathscr{A}\mathscr{X}+\mathscr{A}\mathscr{T}+\mathscr{X}\mathscr{T}}$$

when all coefficients are negative. This completes the proof. ∎

Theorem 1.5 shows that parameter tolerances constrain the gate by bounding the allowable values of $|(u-\ell)/(u+\ell)|$. This quantity is called the *relative gap*. If the relative gap is specified, then the allowable fluctuation tolerances are constrained by Theorem 1.5. It shows that the larger the gap length and the closer it is located to the origin of the map, the larger is the relative gap and hence the larger are the allowable fluctuation tolerances.

Corollary 1.5.1. The minimum relative gap G^* for error-free operation of a Definite Zero-One Model T Gate with finite u and ℓ and with specified allowable fluctuation tolerances $\mathscr{A}$, $\mathscr{T}$, and $\mathscr{X}$ is

$$\min\left|\frac{u-\ell}{u+\ell}\right| \equiv G^* = \frac{\mathscr{A}+\mathscr{X}+\mathscr{T}+\mathscr{A}\mathscr{X}\mathscr{T}}{1+\mathscr{A}\mathscr{X}+\mathscr{A}\mathscr{T}+\mathscr{X}\mathscr{T}}$$

Proof. Theorem 1.5 shows that the set of possible relative gaps for error-free operation has as its lower limit point the value G^*. Moreover, errors always occur for this value; hence it is not a member of the set.

Therefore, from the proof of Theorem 1.5 and the definitions of $\mathscr{A}$, $\mathscr{X}$, and $\mathscr{T}$, the result follows immediately.|

The relative gap of Eq. 1.16 is not necessarily the limit for error-free operation for every Definite Zero-One T gate. Similarly a Definite Zero-One T gate that satisfies the inequality of Theorem 1.5 is not necessarily error-free. To accomplish both, requires that the threshold value be properly selected. For each T gate with specified $\mathscr{A}$, $\mathscr{X}$, and $\mathscr{T}$, there is a necessary and sufficient value for the threshold, called the *optimum threshold* and denoted by T^*, such that Eq. 1.16 defines the limit of error-free operation. This same threshold value assures error-free operation when Theorem 1.5 is satisfied. This is set forth in the following corollary.

Corollary 1.5.2. The optimum threshold value T^* for a Definite Zero-One Model T Gate with finite u and ℓ so that the relative gap of Corollary 1.5.1 is the limit of error-free operation for specified $\mathscr{A}$, $\mathscr{T}$, and $\mathscr{X}$ is

$$T^* = \frac{2u\ell}{u + \ell} \cdot \frac{1 + \mathscr{A}\mathscr{X} + \mathscr{A}\mathscr{T} + \mathscr{X}\mathscr{T}}{1 - \mathscr{T}^2}$$

Moreover, with this threshold value, Theorem 1.5 becomes a necessary and sufficient condition for error-free operation.

Proof. Equations 1.12 and 1.13 define values of t so that T gates with positive coefficients operate error-free. The limit of error-free operation occurs when the inequalities of Eqs. 1.12 and 1.13 are equalities. Then each worst-case fluctuated gap boundary

$$(1 - \mathscr{A})(1 - \mathscr{X})u \qquad \text{and} \qquad (1 + \mathscr{A})(1 + \mathscr{X})\ell$$

has the same value as the corresponding worst-case fluctuated threshold $t(1 + \mathscr{T})$ and $t(1 - \mathscr{T})$, respectively. Solving each of these equations for t gives

$$t = \frac{(1 - \mathscr{A})(1 - \mathscr{X})u}{1 + \mathscr{T}}$$

$$t = \frac{(1 + \mathscr{A})(1 + \mathscr{X})\ell}{1 - \mathscr{T}}$$

Adding these together and collecting terms gives

$$2t = \frac{(1 + \mathscr{A}\mathscr{X} + \mathscr{A}\mathscr{T} + \mathscr{X}\mathscr{T})(u + \ell) + (\mathscr{A} + \mathscr{X} + \mathscr{T} + \mathscr{A}\mathscr{X}\mathscr{T})(\ell - u)}{1 - \mathscr{T}^2}$$

Solving Eq. 1.16 for the term $(\mathscr{A} + \mathscr{X} + \mathscr{T} + \mathscr{A}\mathscr{X}\mathscr{T})$ and substituting into the above equation gives the desired threshold value for the positive coefficient case, after a simplification that recognizes $|(u - \ell)/(u + \ell)|$

$= (u - \ell)/(u + \ell)$ for this case. The validity of the threshold value is established for the case of all negative coefficients by a similar procedure. It begins with the equalities from Eqs. 1.14 and 1.15 and in the simplification recognizes that $|(u - \ell)/(u + \ell)| = -(u - \ell)/(u + \ell)$.

That the optimum threshold makes Theorem 1.5 a necessary and sufficient condition for error-free operation is justified as follows. For this threshold and for all relative gaps greater than that of Eq. 1.16, it is evident from the above argument that the inequalities of Eqs. 1.12 and 1.13 or Eqs. 1.14 and 1.15 are satisfied, and hence no errors result.|

The expressions for minimum relative gap of Eq. 1.16 and optimum threshold of Corollary 1.5.2 can be simplified when $\mathscr{T}^2$, $\mathscr{A}\mathscr{X}$, $\mathscr{A}\mathscr{T}$, and $\mathscr{X}\mathscr{T} << 1$. Under these conditions, approximate expressions for these quantities become

$$G^* \approx \mathscr{A} + \mathscr{X} + \mathscr{T} \tag{1.17}$$

$$T^* \approx \frac{2u\ell}{u + \ell} \tag{1.18}$$

This shows that the minimum relative gap can be considered as the "available tolerance" which can be divided, as desired, among $\mathscr{A}$, $\mathscr{T}$, and $\mathscr{X}$. The optimum threshold is independent of this division and depends only upon the values of u and ℓ that satisfy Eq. 1.17.

Another special case of interest occurs when $\mathscr{A} = \mathscr{X} = 0$. Under these conditions the minimum relative gap and the optimum threshold become

$$G^* = \mathscr{T}$$

$$T^* = \tfrac{1}{2}(u + \ell)$$

1.8 SENSITIVITY: DEFINITE MINUS-PLUS-ONE MODEL

The logical input variables of a Minus-Plus-One Model T Gate are denoted by ζ_j $(j = 1,2,\ldots,n)$ and have possible values -1, 1 corresponding to x_j values of 0, 1, respectively. The output variable is denoted by w.

For each combination of values of the input variables $\zeta_1,\zeta_2,\ldots,\zeta_n$, the corresponding value of output w is

$$\begin{aligned} w &= 1 \qquad \text{if} \quad \sum_{j=1}^{n} a_j\zeta_j \geq t \\ w &= -1 \qquad \text{if} \quad \sum_{j=1}^{n} a_j\zeta_j < t \end{aligned} \tag{1.19}$$

The value of ζ_j is related to the corresponding value of x_j by the relation

$$\zeta_j = x_j - \bar{x}_j \tag{1.20}$$

Using this expression, the conditions of Eq. 1.19 can be expressed in terms of the x_j variables as

$$w = 1 \qquad \text{if} \quad \sum_{j=1}^{n} a_j x_j - \sum_{j=1}^{n} a_j \bar{x}_j \geq t$$

$$w = -1 \qquad \text{if} \quad \sum_{j=1}^{n} a_j x_j - \sum_{j=1}^{n} a_j \bar{x}_j < t$$

When worst-case fluctuations less than $\mathscr{A}$, $\mathscr{T}$, and $\mathscr{X}$ are included, and when t and all coefficients are positive, the conditions that must be satisfied, for the error-free realization of a given function, are:

$w = 1$ implies

$$(1 - \mathscr{A})(1 - \mathscr{X}) \sum_{j=1}^{n} a_j x_j - (1 + \mathscr{A})(1 + \mathscr{X}) \sum_{j=1}^{n} a_j \bar{x}_j \geq t(1 + \mathscr{T})$$

$w = -1$ implies

$$(1 + \mathscr{A})(1 + \mathscr{X}) \sum_{j=1}^{n} a_j x_j - (1 - \mathscr{A})(1 - \mathscr{X}) \sum_{j=1}^{n} a_j \bar{x}_j \leq t(1 - \mathscr{T})$$

Completing the indicated multiplication, rearranging the terms, and using the relation $(x_j + \bar{x}_j) = 1$, these inequalities become

$w = 1$ implies

$$(1 + \mathscr{A}\mathscr{X}) \sum_{j=1}^{n} a_j(x_j - \bar{x}_j) - (\mathscr{A} + \mathscr{X}) \sum_{j=1}^{n} a_j \geq t(1 + \mathscr{T})$$

(1.21)

$w = -1$ implies

$$(1 + \mathscr{A}\mathscr{X}) \sum_{j=1}^{n} a_j(x_j - \bar{x}_j) + (\mathscr{A} + \mathscr{X}) \sum_{j=1}^{n} a_j \leq t(1 - \mathscr{T})$$

The above conditions must be satisfied for all input combinations. If for given values of $\mathscr{A}$, $\mathscr{X}$, and $\mathscr{T}$, they are satisfied by those input combinations for which

$$\sum_{j=1}^{n} a_j(x_j - \bar{x}_j) = u \qquad (w = 1) \tag{1.22}$$

and

$$\sum_{j=1}^{n} a_j(x_j - \bar{x}_j) = \ell \qquad (w = -1)$$

then they are satisfied for all input combinations. Therefore, combining Eq. 1.21 and Eq. 1.22 and letting $\sigma = \sum_{j=1}^{n} a_j$, the conditions that must be satisfied for error-free operation, when t and all coefficients are positive, become

$$\begin{aligned}(1 + \mathscr{A}\mathscr{X})u - (\mathscr{A} + \mathscr{X})\sigma &\geq t(1 + \mathscr{T})\\ (1 + \mathscr{A}\mathscr{X})\ell + (\mathscr{A} + \mathscr{X})\sigma &\leq t(1 - \mathscr{T})\end{aligned} \tag{1.23}$$

Similarly, the corresponding conditions for error-free operation when t is negative and all coefficients are positive are found to be

$$\begin{aligned}(1 + \mathscr{A}\mathscr{X})u - (\mathscr{A} + \mathscr{X})\sigma &\geq t(1 - \mathscr{T})\\ (1 + \mathscr{A}\mathscr{X})\ell + (\mathscr{A} + \mathscr{X})\sigma &\leq t(1 + \mathscr{T})\end{aligned}$$

The conditions for error-free operation of this model are obtained from these conditions by methods that parallel those used in Section 1.7. The particular results are set forth in the following theorem and corollary.

Theorem 1.6. The following inequality must be satisfied by any Definite Minus-Plus-One Model T Gate to operate error-free when the allowable parameter fluctuations are less than $\mathscr{A}$, $\mathscr{X}$, and $\mathscr{T}$ and when u and ℓ are finite:

$$\mathscr{T}\left|\frac{u + \ell}{u - \ell}\right| + \frac{\mathscr{A} + \mathscr{X}}{1 + \mathscr{A}\mathscr{X}}\left|\frac{2\sigma}{u - \ell}\right| \leq 1$$

Moreover, this inequality is the limiting expression for error-free operation.

Corollary 1.6.1. The optimum threshold value T^* for a Definite Minus-Plus-One Model T Gate with finite u and ℓ so that Theorem 1.6 defines the limit of error-free operation for specified $\mathscr{A}$, $\mathscr{X}$, and $\mathscr{T}$ is

$$T^* = \frac{u + \ell}{2}(1 + \mathscr{A}\mathscr{X})$$

Moreover, with this threshold value, Theorem 1.6 becomes a necessary and sufficient condition for error-free operation.

Thus, for the Definite Minus-Plus-One Model, the quantity $|\sigma/(u - \ell)|$ enters as a sensitivity parameter in addition to the relative gap $|(u - \ell)/(u + \ell)|$. For the definite models, $|\sigma|$ is called the *fan-in* of the gate and $|\sigma/(u - \ell)|$ is called the *normalized fan-in*. The model will operate error free with components of larger tolerances as the normalized fan-in is decreased and the relative gap is increased. For those realizations in which the gap length $(u - \ell)$ is normalized at unity, the previous statement corresponds to decreasing both $|\sigma|$ and $|u|$.

A case of special interest occurs when $u = -\ell$. Under these conditions

$T^* = 0$, and thus the necessary and sufficient condition for error-free operation from Theorem 1.6 becomes

$$\left|\frac{u - \ell}{2\sigma}\right| \geq \frac{\mathscr{A} + \mathscr{X}}{1 + \mathscr{A}\mathscr{X}}$$

Under the additional restrictions that the fan-in is an odd number and that all weights have unit value, we have a special T gate called a *majority gate.* It is readily apparent that the output of a majority gate with positive weights is unity if and only if more than half of the input variables have unit value. For the Definite Minus-Plus-One Model majority gate with positive weights, the gap is $u{:}\ell = 1{:}-1$, $T^* = 0$, and the condition for error-free operation is

$$\frac{1}{\sigma} \geq \frac{\mathscr{A} + \mathscr{X}}{1 + \mathscr{A}\mathscr{X}}$$

Thus only fan-in is restricted by the allowable tolerances.

It is shown in Appendix A that these two parameters, relative gap and normalized fan-in, also characterize a somewhat more general T-gate model: namely, the General Definite Model. In this model the coefficients all have the same sign, but the physical quantities that represent the logical variables can have arbitrary values either positive or negative. Both models considered in this chapter are special cases of the General Definite Model. Thus relative gap and normalized fan-in are the appropriate sensitivity parameters for a rather wide class of models. For this reason, they are of primary importance in our study of reliability.

In general the sensitivity parameters enter into the theory in two ways. First, there is the problem of *maximally reliable realizations.* For single-gate realizations, for example, this implies finding that single-gate realization which has maximum relative gap and/or minimum normalized fan-in. Similarly, for multigate realizations, we maximize the reliability of the least reliable gate.

The second problem is that of *realizations within prescribed tolerances.* Here we assume that certain allowable parameter fluctuations have been specified. These, then, imply corresponding restrictions on relative gap and/or normalized fan-in. Naturally, the tighter these restrictions, the more gates may be required in the realization. Our goal is to find any realization such that each gate satisfies the specified restrictions, and if possible one that requires the fewest number of gates.

Table 1.2 shows how the relative gap restricts the allowable tolerances for both models when

$$\mathscr{A} = \mathscr{X} = \mathscr{T}$$

and when fan-in $\sigma = 10$ and gap length $g = 1$.

Table 1.2. Table of allowable values of $\mathscr{A}$, $\mathscr{X}$, and $\mathscr{T}$ for Error-Free Operation when $\mathscr{A} = \mathscr{X} = \mathscr{T}$, $\sigma = 10$, and $g = 1$

	$\mathscr{A} = \mathscr{X} = \mathscr{T}$	
u	Zero-One Model	Minus-Plus-One Model
	%	%
1	33.3	2.44
2	11.1	2.32
3	6.7	2.22
4	4.8	2.12
5	3.7	2.04
6	3.0	1.96

1.9 CHARACTERIZATION OF LINEAR SEPARABILITY BY SIMULTANEOUS INEQUALITIES

One of the problems that will be considered in detail is that of determining whether an arbitrarily specified logical function F is of the linearly separable class. The method of simultaneous inequalities is a natural approach to this problem.

Let F represent a logical function defined on the vertices p, $(p = p_0, p_1, \ldots, p_{2^n-1})$ of the n cube and let the coordinates of vertex p be $(x_1{}^p, x_2{}^p, \ldots, x_n{}^p)$. If F is a linearly separable function, then according to Eq. 1.8 there is a set of real numbers $(a_1, a_2, \ldots, a_n, u, \ell)$ such that, for each p,

$$\sum_{i=1}^{n} a_i x_i{}^p \geq u \qquad \text{if } F(p) = 1$$

$$\sum_{i=1}^{n} a_i x_i{}^p \leq \ell \qquad \text{if } F(p) = 0 \tag{1.24}$$

$$u > \ell$$

Therefore, F defines a set of $(2^n + 1)$ inequalities in $n + 2$ unknowns $(a_1, a_2, \ldots, a_n, u, \ell)$; this is called the R_0 *set*. If don't-care points exist, then fewer inequalities are needed (see Section 4.10). Since the inequalities are homogeneous, any convenient normalization can be used to reduce the number of unknowns by one. For example, one of the unknowns can be arbitrarily selected or the gap length, $u - \ell$, can be assigned unit value. These results are summarized as follows.

Theorem 1.7. An arbitrarily selected logical function F is of the linearly separable class if and only if there exists a set of real numbers $(a_1,a_2,\ldots,a_n, u, \ell)$ that satisfy the inequalities of Eq. 1.24. Moreover, any such set of numbers defines the weights and the gap of a T Gate that realizes F.

Example 1.2. For the logical function $y = F(x_1,x_2,x_3) = x_1 \dotplus x_2x_3$, the table of combinations and the associated linear inequalities are given in Table 1.3. It is easily verified that these inequalities are consistent (that is, have a solution) and that one solution is the set ($a_1 = 2$, $a_2 = a_3 = 1$, $u = 2$, $\ell = 1$).

Table 1.3.

x_1 x_2 x_3	y	Inequality
0 0 0	0	$0 \leq \ell$
0 0 1	0	$a_3 \leq \ell$
0 1 0	0	$a_2 \leq \ell$
0 1 1	1	$a_2 + a_3 \geq u$
1 0 0	1	$a_1 \geq u$
1 0 1	1	$a_1 + a_3 \geq u$
1 1 0	1	$a_1 + a_2 \geq u$
1 1 1	1	$a_1 + a_2 + a_3 \geq u$
		$u > \ell$

Frequently it can be established that a logical function is not linearly separable by selecting only certain of the inequalities, based upon the form of the function, and by showing that they define an inconsistency so that a solution to Eq. 1.24 is impossible. Although this is a heuristic procedure, it is illustrated by Example 1.3.

Example 1.3. Consider the logical function

$$y = F(x_1,x_2,x_3,x_4) = x_1x_2 \dotplus x_3x_4$$

Since

$$y = F(1,1,0,0) = 1 \quad \text{and} \quad y = F(0,0,1,1) = 1$$

then two of the inequalities that must be satisfied are

$$a_1 + a_2 \geq u \quad \text{and} \quad a_3 + a_4 \geq u$$

In each inequality one of the weights must be greater than or equal to $u/2$. Let a_1 and a_3 be such weights. Clearly, then,

$$a_1 + a_3 \geq u$$

However, since $F(1,0,1,0) = 0$, it follows that

$$a_1 + a_3 \leq \ell$$

These two inequalities imply that $\ell \geq u$, which violates one of the inequalities of Eq. 1.24.

The difficulty with the simultaneous inequality approach to threshold logic is twofold. First, it does not appear to be suited to other than one-gate realization problems. Second, for other than small values of n, determining the consistency and a solution to a set of $(2^n + 1)$ simultaneous inequalities is a formidable task. One of the ways to circumvent this difficulty, as we shall see later, is to reduce the number of inequalities by means of threshold function properties. This approach is only partially successful and, even then, still requires solving a set of inequalities simultaneously.

Basically there are two methods for solving a set of simultaneous inequalities: (1) the successive elimination of variables and (2) linear programming. Both of these are described in some detail in Chapter 9. At this time, we discuss them only in broad outline and describe their advantages and limitations.

The successive-elimination-of-variables method is similar to the successive elimination of variables for simultaneous equations. Suitably selected inequalities are added or subtracted in pairs to eliminate one variable at a time, until inequalities with only one variable remain. Usually this variable is constrained between two bounds within which the value of the variable can be chosen arbitrarily. This value when substituted into other inequalities defines bounds on a second variable within which it can be chosen arbitrarily, and so on. The resultant set of values is always a solution to the original inequalities. Moreover, this method defines, sequentially, the set of all possible solutions.

As this method is carried out, the number of inequalities that must be considered simultaneously at each stage may be considerably larger than the original number. This is in marked contrast to simultaneous equations wherein the number decreases successively. For this reason the method is usually limited to cases where the number of inequalities is small.

Linear programming is a bounded iterative procedure for determining a solution to a consistent set of linear inequalities and simultaneously maximizing or minimizing some linear function of the variables. With regard to the threshold-gate problem, for example, the normalized fan-in can be minimized or the relative gap can be maximized. This is certainly a desirable feature. In addition, there are many efficient computer programs available for solving linear programming problems; this is another desirable feature.

However, there are a number of drawbacks to using linear programming as a basis for threshold logic. Even for the single-gate realization problem discussed here, linear programming often requires considerably more computations than the alternative methods to be discussed later. Moreover, in the multigate case, there is at present no practical linear programming algorithm for multigate synthesis with controlled sensitivity. (For some first attempts at such a theory, see the references in Chapter 9.)

Although the linear programming approach has some interesting features, its disadvantages make it unsuitable for the purposes of this book. Our main goal is to develop a theory that is applicable to both the single- and the multigate cases, and that is computationally simple and utilizes the function in algebraic form rather than as a set of simultaneous inequalities.

Chapter 2

ELEMENTARY PROPERTIES

2.1 INTRODUCTION

Our principal goal in threshold logic is the development of general analysis and synthesis procedures for T-gate networks, taking into account, when necessary, the sensitivity characteristics of the gates. Before proceeding with this general study, we present in this chapter a number of very simple and elementary properties of T-gate realizations. Most of these properties are obvious once stated, and require little or no proof, but they are often quite useful in analysis, synthesis, and particularly in simplification of T-gate networks.

The notation for single-gate networks was defined in Chapter 1. It was demonstrated that a logical function of the linearly separable class has a Boolean-function representation $y = F(x_1, x_2, \ldots, x_n)$ and a representation in terms of separating function

$$y = \langle f(x_1, x_2, \ldots, x_n) \rangle_{u:\ell}$$

The separating-function representation implies the single T-gate realization.

For multigate networks, the notation is quite similar. It was also shown in Chapter 1 that an arbitrary logical function can be realized by an interconnection of T gates. The final gate in such a network will, in general, have inputs $y_1, y_2, \ldots, y_m$ and output y. The output can be represented either as a Boolean function or as a separating function of the input variables:

$$y = F(y_1, y_2, \ldots, y_m) = \langle a_1 y_1 + a_2 y_2 + \cdots + a_m y_m \rangle_{u:\ell} \qquad (2.1)$$

Each y_j may be the output of a T gate that can be expressed as

$$y_j = F_j(y_{1j}, y_{2j}, \ldots, y_{mj}) = \langle a_{1j} y_{1j} + a_{2j} y_{2j} + \cdots + a_{mj} y_{mj} \rangle_{u_j:\ell_j}$$

and so on until all are functions of the independent variables $x_1,x_2,\ldots,x_n$. A specific example of the separating-function representation is

$$y = \langle a_1\langle a_1'x_1 + a_2'x_2 + a_3'\langle a_1''x_1 + a_2''x_2\rangle_{u'':\ell''}\rangle_{u_1':\ell_1'} + a_2\langle a_2'x_2 + a_3'x_3 + a_4'x_4\rangle_{u_2':\ell_2'} + a_3x_3 + a_4x_4\rangle_{u:\ell} \qquad (2.2)$$

As in the one-gate case, each separating-function representation implies a T-gate realization. Hence we call such a representation a *realization* of the logical function. The realizations of Eqs. 2.1 and 2.2. are shown in Figs. 2.1a and 2.1b respectively.

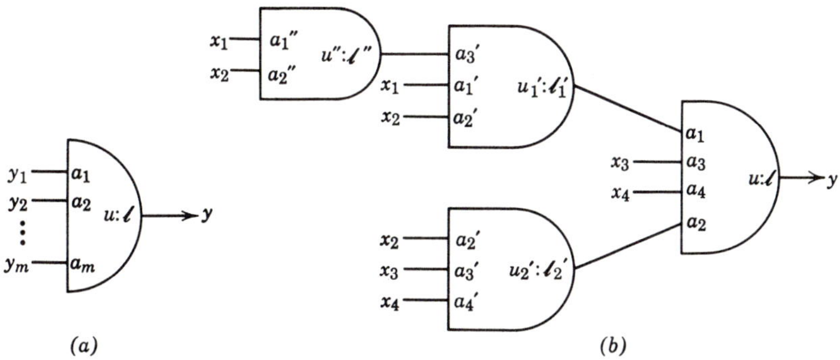

Fig. 2.1. T-gate realizations implied by separating-function representations of logical functions. (*a*) Realization of Eq. 2.1. (*b*) Realization of Eq. 2.2.

Since each y_j is some function of the independent variables, the value for an arbitrarily selected y_j is not necessarily independent of the values of the other y_j. For example, if $y_1 = x_1\bar{x}_2$ and $y_2 = \bar{x}_1x_2$, then $y_1 = 1$ implies $y_2 = 0$, and the combination $y_1 = y_2 = 1$ is not attainable.

All of the possible 2^m combinations of values of $y_1,y_2,\ldots,y_m$ can be obtained only if the value of each y_j is independent of the values of the other y_j variables. This is the case when each y_j is the independent variable x_j, or when it is a function of $x_1,x_2,\ldots,x_m$, that is not restrictive. For example, when $y_1 = x_1x_2$, $y_2 = x_3x_4$, and so on.

In general, the properties of T gates are valid for both the input variables $y_1,y_2,\ldots,y_m$, and the independent variables $x_1,x_2,\ldots,x_n$. In fact y_j variables are more general since they include the possibility that $y_j = x_j$ for $j = 1,2,\ldots,n = m$. The only important distinction is that the y_j notation may not have all of the possible 2^m combination of values attainable. When this restriction is important, it is explicitly set forth.

2.2 SIMPLE T-GATE REALIZATIONS OF LOGICAL FUNCTIONS

An arbitrary logical function can always be realized by a number of simple T-gate networks. These realizations are not necessarily efficient with regard to the number of gates required, but they are useful in many cases. For example, if a given switching function F can be expressed as the logical sum of two or more functions, $F_1,F_2,\ldots,F_m$; that is,

$$y = F(y_1,y_2,\ldots,y_m) = y_1 \dotplus y_2 \dotplus \cdots \dotplus y_m$$

where

$$y_j = F_j(x_1,x_2,\ldots,x_n)$$

then an obvious realization of F with input variables y is

$$y = \langle y_1 + y_2 + \cdots + y_m \rangle_{1:0}$$

which is an OR gate realization.

Similarly, if F can be expressed as a logical product

$$y = y_1 y_2 \ldots y_m$$

then an AND gate realization is

$$y = \langle y_1 + y_2 + \cdots + y_m \rangle_{m:(m-1)}$$

As a slightly more complicated example, consider

$$y = y_1 \dotplus y_2 y_3 y_4$$

This can be realized by a combination of the AND and OR realizations. First we let

$$y_0 = y_2 y_3 y_4$$

and obtain the realization

$$y_0 = y_2 y_3 y_4 = \langle y_2 + y_3 + y_4 \rangle_{3:2}$$

Then the final realization is

$$y = y_1 \dotplus y_0 = \langle 3y_1 + y_2 + y_3 + y_4 \rangle_{3:2}$$

To see that this is, in fact, a realization for y, observe that:

(1) For $y_1 = 1$, the value of $a_{y_1} = 3$, is sufficient to exceed the threshold and hence the value of the function is 1

(2) For $y_1 = 0$, the realization is just that for y_0, and hence is also correct

The idea behind these examples can be expressed in the following theorem.

Theorem 2.1. Let

$$y_0 = F_0(y_2,y_3,\ldots,y_m) = \langle f_0(y_2,y_3,\ldots,y_m)\rangle_{u_0:\ell_0}$$

and let σ_{p_0} and σ_{n_0} denote the largest and smallest possible values of $f_0(y_2,y_3,\ldots,y_m)$ respectively.

A. If $y = y_1 \dotplus y_0$, then

$$y = \langle a_1y_1 + f_0(y_2,y_3,\ldots,y_m)\rangle_{u_0:\ell_0}$$

for any $a_1 \geq u_0 - \sigma_{n_0}$.

B. If $y = y_1y_0$, then

$$y = \langle a_1y_1 + f_0(y_2,y_3,\ldots,y_m)\rangle_{(u_0+a_1):(\ell_0+a_1)}$$

for any $a_1 \geq \sigma_{p_0} - \ell_0$.

Proof

A. We consider separately the two cases $y_1 = 1$ and $y_1 = 0$. For $y_1 = 1$ it follows that $y = 1$ and for $y_1 = 0$ that $y = y_0$. Informally, it can be seen from Fig. 2.2 that when $y_1 = 1$, the weight

$$a_1 = u_0 - \sigma_{n_0}$$

is large enough so that

$$a_1 + f_0$$

will equal or exceed u_0, even though f_0 has its smallest value of σ_{n_0}. For $y_1 = 0$, the variable y_1 has no effect and f_0 separates y_0 with gap $u_0{:}\ell_0$. More precisely: Let $y_1 = 1$. Then $y = y_1 \dotplus y_0 = 1$. Since $a_1 \geq u_0 - \sigma_{n_0}$, then

$$a_1y_1 + f_0(y_2,y_3,\ldots,y_m) \geq a_1 + \sigma_{n_0} \geq (u_0 - \sigma_{n_0}) + \sigma_{n_0} = u_0$$

where the latter is the upper gap boundary. Therefore, for $y_1 = 1$,

$$y = \langle a_1y_1 + f_0(y_2,y_3,\ldots,y_m)\rangle_{u_0:\ell_0} = 1$$

Let $y_1 = 0$. Then $y = y_1 \dotplus y_0 = y_0$ and

$$\langle a_1y_1 + f_0(y_2,y_3,\ldots,y_m)\rangle_{u_0:\ell_0} = y_0$$

B. If $y = y_1y_0 = 1$, then $y_1 = 1$ and $y_0 = 1$. Under this condition

$$a_1y_1 + f_0(y_2,y_3,\ldots,y_m) \geq a_1 + u_0$$

where the latter is the upper gap boundary. Therefore

$$y = \langle a_1y_1 + f_0(y_2,y_3,\ldots,y_m)\rangle_{(u_0+a_1):(\ell_0+a_1)} = 1$$

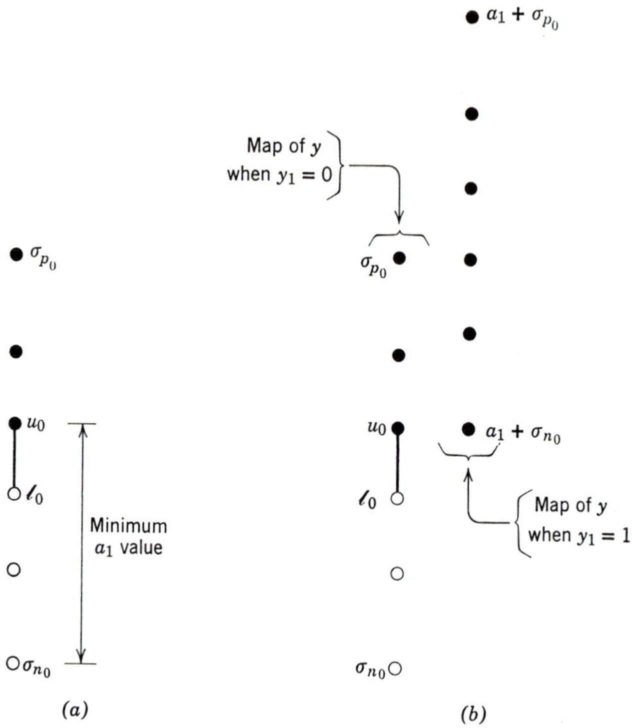

Fig. 2.2. Illustration of proof of Theorem 2.1A. (*a*) Map of arbitrary $y_0 = F_0(y_2,y_3, \ldots,y_m)$. (*b*) Map of $y = y_1 \dotplus F_0(y_2,y_3,\ldots,y_m)$ divided into two parts.

If $y = y_1y_0 = 0$, then $y_1 = 0$ or $y_0 = 0$. Since y_1 and y_0 cannot both be unity and since $a_1 \geq \sigma_{p_0} - \ell_0$, it follows that

$$a_1y_1 + f_0(y_2,y_3,\ldots,y_m) \leq \max\{a_1,\sigma_{p_0}\} \leq \max\{a_1,a_1 + \ell_0\} = a_1 + \ell_0$$

where the latter is the lower gap boundary. Therefore

$$y = \langle a_1y_1 + f_0(y_2,y_3,\ldots,y_m)\rangle_{(u_0+a_1):(\ell_0+a_1)} = 0$$

This argument is illustrated by Fig. 2.3.▎

The values of a_1 of Theorem 2.1 are usually selected for parts A and B as $(u_0 - \sigma_{n_0})$ and $(\sigma_{p_0} - \ell_0)$ respectively. These are the minimum values of a_1 for which the relative gap has the largest value that can be obtained with the given separating function. Larger values of a_1 do not increase the relative gap but do increase the fan-in. (For T gates with both positive and negative coefficients, the fan-in is $\sigma_{p_0} - \sigma_{n_0}$.)

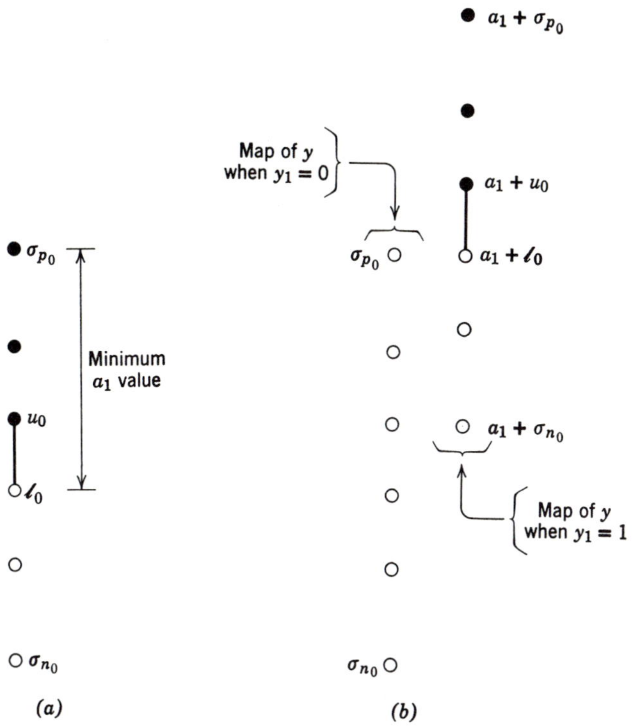

Fig. 2.3. Illustration of proof of Theorem 2.1B. (*a*) Map of arbitrary $y_0 = F_0(y_2,y_3,\ldots,y_m)$. (*b*) Map of $y = y_1F_0(y_2,y_3,\ldots,y_m)$ divided into two parts.

Example 2.1. The following are logical functions and a corresponding realization (separating function and gap) that is obtained by the successive application of Theorem 2.1.

(1) $y = x_1 \dotplus x_2x_3\ldots x_n = \langle (n-1)x_1 + x_2 + x_3 + \cdots + x_n \rangle_{(n-1):(n-2)}$

(2) $y = x_1(x_2 \dotplus x_3 \dotplus \cdots \dotplus x_n) = \langle (n-1)x_1 + x_2 + x_3 + \cdots + x_n \rangle_{n:(n-1)}$

(3) $y = x_1 \dotplus x_2(x_3 \dotplus x_4(x_5 \dotplus x_6))$
$= \langle 8x_1 + 5x_2 + 3x_3 + 2x_4 + x_5 + x_6 \rangle_{8:7}$

(4) $y = x_1(x_2x_3 \dotplus x_4x_5) = x_1(y_0 \dotplus x_4x_5) = \langle 3x_1 + 2y_0 + x_4 + x_5 \rangle_{5:4}$

But

$$y_0 = x_2x_3 = \langle x_2 + x_3 \rangle_{2:1}$$

Hence

$$y = \langle 3x_1 + 2\langle x_2 + x_3 \rangle_{2:1} + x_4 + x_5 \rangle_{5:4}$$

Theorem 2.1 can be used to realize any arbitrary logical function by a network of T gates; however, the number of gates that are used may be far from minimal. For example, when the logical function

$$y = x_1x_2 \dotplus x_1x_3 \dotplus x_2x_3$$

is realized using Theorem 2.1 a network of two T gates is required. However, it is readily verified that this function (the majority function of three variables) is linearly separable and has a realization

$$y = \langle x_1 + x_2 + x_3 \rangle_{2:1}$$

It is evident that the majority function can be factored as

$$y = x_1x_2 \dotplus (x_1 \dotplus x_2)x_3$$

of which the general form is

$$y = y_1y_2 \dotplus (y_1 \dotplus y_2)F_0(y_3,y_4,\ldots,y_m)$$

where the function F_0 is modified by two variables y_1 and y_2.

This situation is a combination of both parts of Theorem 2.1 into one function by the introduction of the additional variable y_1. Note that

$$y = y_2 \dotplus F_0(y_3,y_4,\ldots,y_m)$$

when $y_1 = 1$ and

$$y = y_2F_0(y_3,y_4,\ldots,y_m)$$

when $y_1 = 0$. The following theorem is a general statement of this situation.

Theorem 2.2. Let

$$y_0 = F_0(y_3,y_4,\ldots,y_m) = \langle f_0(y_3,y_4,\ldots,y_m)\rangle_{u_0:\ell_0}$$

and let σ_{p_0} and σ_{n_0} denote the largest and smallest possible values of $f_0(y_3,y_4,\ldots,y_m)$ respectively. If

$$y = y_1y_2 \dotplus (y_1 \dotplus y_2)y_0$$

then

$$y = \langle ay_1 + ay_2 + f_0(y_3,y_4,\ldots,y_m)\rangle_{(u_0+a):(\ell_0+a)}$$

for all

$$a \geq \max\{(u_0 - \sigma_{n_0}), (\sigma_{p_0} - \ell)\}$$

Proof. Let $y_1 = y_2 = 1$. Then

$$y = y_1y_2 \dotplus (y_1 \dotplus y_2)y_0 = 1$$

Since $a \geq u_0 - \sigma_{n_0}$, then

$$\begin{aligned} ay_1 + ay_2 + f_0(y_3,y_4,\ldots,y_m) &\geq 2a + \sigma_{n_0} \\ &\geq u_0 - \sigma_{n_0} + a + \sigma_{n_0} = u_0 + a \end{aligned}$$

where the latter is the upper gap boundary. Therefore, when $y_1 = y_2 = 1$,

$$y = \langle ay_1 + ay_2 + f_0(y_3,y_4,\ldots,y_m)\rangle_{(u_0+a):(\ell_0+a)} = 1$$

Let $y_1 = y_2 = 0$. Then

$$y = y_1y_2 \dotplus (y_1 \dotplus y_2)y_0 = 0$$

Since $a \geq \sigma_{p_0} - \ell_0$, then

$$ay_1 + ay_2 + f_0(y_3,y_4,\ldots,y_m) \leq \sigma_{p_0} \leq a + \ell_0$$

where the latter is the lower gap boundary. Therefore when $y_1 = y_2 = 0$,

$$y = \langle ay_1 + ay_2 + f_0(y_3,y_4,\ldots,y_m)\rangle_{(u_0+a):(\ell_0+a)} = 0$$

Let $y_1 = \bar{y}_2 = 1$, or $y_1 = \bar{y}_2 = 0$. Then the validity follows from Theorem 2.1B and the proof is completed.|

Example 2.2. Consider the logical function

$$y = x_1x_2 \dotplus x_1x_3 \dotplus x_2x_3 \dotplus x_1x_4x_5 \dotplus x_2x_4x_5$$

which can be factored as

$$y = x_1x_2 \dotplus (x_1 \dotplus x_2)y_0$$

where

$$y_0 = x_3 \dotplus x_4x_5 = \langle 2x_3 + x_4 + x_5\rangle_{2:1}$$

from Theorem 2.1. Using Theorem 2.2, we see that a must be selected so that

$$a \geq \max\{(u_0 - \sigma_{n_0}), (\sigma_{p_0} - \ell_0)\} = \max\{(2-0), (4-1)\} = 3$$

Choosing $a = 3$, we get

$$y = \langle 3x_1 + 3x_2 + 2x_3 + x_4 + x_5\rangle_{5:4}$$

Example 2.3. Consider the logical function

$$y = y_1y_2 \dotplus (y_1 \dotplus y_2)y_0$$

where

$$y_1 = x_1(x_2 \dotplus x_3 \dotplus x_4) \qquad y_2 = x_1 \dotplus x_2(x_3 \dotplus x_4)$$

and

$$y_0 = x_2 \dotplus x_3 \dotplus x_4x_5$$

From Theorem 2.1,

$$y_1 = \langle 3x_1 + x_2 + x_3 + x_4\rangle_{4:3} \qquad y_2 = \langle 3x_1 + 2x_2 + x_3 + x_4\rangle_{3:2}$$

and

$$y_0 = \langle 2x_2 + 2x_3 + x_4 + x_5 \rangle_{2:1}$$

By Theorem 2.2,

$$y = \langle 5\langle 3x_1 + x_2 + x_3 + x_5 \rangle_{4:3} + 5\langle 3x_1 + 2x_2 + x_3 + x_4 \rangle_{3:2} + 2x_2 + 2x_3 + x_4 + x_5 \rangle_{7:6}$$

2.3 EQUIVALENT T-GATE REALIZATIONS

The previous section considered some simple procedures for realizing a logical function based upon the possibility of decomposing the Boolean representation into special forms. In this section we are concerned with means whereby a given realization can be altered without changing the logical function that is realized. Since the form of the realization is implied by the separating-function representation, in essence, we are concerned with different possible forms of the separating-function representation of a logical function. Different separating-function representations of the same logical function are called *equivalent*. Equivalent separating-function representations may yield identical maps, in which case the maps are called *equal*; or they may yield different maps, in which case the maps are called *similar*.

One way of altering the separating function is to change the coefficients and the variables in such a way that the value of the separating function for each input combination is unaltered. In such cases the gap is unchanged and the two maps are equal. A simple but useful example of this is to replace a variable y_j by the terms $1 - \bar{y}_j$ as set forth in Theorem 2.3.

Theorem 2.3. If

$$y = \langle a_1 y_1 + a_2 y_2 + \cdots + a_m y_m \rangle_{u:\ell}$$

then any $y_j, j = 1,2,\ldots,m$ can be replaced by $(1 - \bar{y}_j)$ and conversely, and the result is an equivalent representation of y, which has a map equal to the given map.

Proof. The validity follows from the observation that $y_j = 1 - \bar{y}_j$.∎

Another type of alteration changes not only the coefficients and the variables of the separating function, but also the gap. A simple but useful example is the multiplication of each coefficient by some constant b or the addition of a constant b to the separating function as set forth in Theorem 2.4. In these cases the maps are similar.

Theorem 2.4. If

$$y = \langle \sum_{j=1}^{m} a_j y_j \rangle_{u:\ell}$$

then

A.

$$y = \langle \sum_{j=1}^{m} b a_j y_j \rangle_{bu:b\ell}$$

for any real $b > 0$ and

B.

$$y = \langle b + \sum_{j=1}^{m} a_j y_j \rangle_{(u+b):(\ell+b)}$$

for any real b.

An important use for Theorems 2.3 and 2.4 is to change the sign of weights in a separating-function representation. Many T-gate circuits use resistors to represent the coefficients, in which case negative values are not admissible. When negative coefficients arise, a transformation is needed that gives an equivalent representation with only positive coefficients.

Procedure 2.1. Changing Coefficient Sign. Let

$$y = \langle \sum_{j=1}^{m} a_j y_j \rangle_{u:\ell}$$

(1) For each $a_j < 0$, replace $a_j y_j$ by $a_j - a_j \bar{y}_j$ (Theorem 2.3).

(2) Remove all constants generated by (1) and replace $u:\ell$ by $(u - \sigma_n)$: $(\ell - \sigma_n)$. (Theorem 2.4), where σ_n is the sum of the negative coefficients.

Example 2.4. Let

$$y = \langle -2x_1 + x_2 - x_3 \rangle_{-1:-2}$$

Replacing x_1 by $1 - \bar{x}_1$ and x_3 by $1 - \bar{x}_3$ according to Theorem 2.3 gives

$$y = \langle -2 + 2\bar{x}_1 + x_2 - 1 + \bar{x}_3 \rangle_{-1:-2} = \langle -3 + 2\bar{x}_1 + x_2 + \bar{x}_3 \rangle_{-1:-2}$$

Using the constant -3 to change the gap according to Theorem 2.4 gives

$$y = \langle 2\bar{x}_1 + x_2 + \bar{x}_3 \rangle_{2:1}$$

Another operation that converts one separating-function representation to a different equivalent one involves replacing two terms of the form $ay_1 + ay_2$ by the two terms $a[y_1 \dotplus y_2] + a[y_1 y_2]$. Whereas in the first case the input variables are y_1 and y_2, in the second they are $[y_1 \dotplus y_2]$ and $[y_1 y_2]$. Theorem 2.5 is a statement of the result.

Theorem 2.5. If

$$y = \langle a_1y_1 + a_2y_2 + \cdots + a_my_m\rangle_{u:\ell}$$

then any two terms of the form $a_iy_i + a_jy_j$ where $a_i = a_j$ can be replaced by the two terms

$$a_i[y_i \dotplus y_j] + a_i[y_iy_j]$$

and conversely; and the result is an equivalent representation of y which has a map equal to the given map.

Proof. The proof follows immediately from the observation that the terms $a_iy_i + a_jy_j$ in the original representation and the terms

$$a_i[y_i \dotplus y_j] + a_i[y_iy_j]$$

in the final one both contribute 0, a_i, or $2a_i$ to the sum, according as

$$y_i = y_j = 0,\ y_i \oplus y_j = 1\dagger,\ \text{or}\ y_i = y_j = 1$$

respectively.|

The principal advantage of this type of transformation is that the inputs $[y_i \dotplus y_j]$ and $[y_iy_j]$ may be simpler functions of the variables $x_1,x_2,\ldots,x_n$ than are the original inputs. This is illustrated in Example 2.5. A special case of particular interest is set forth in the following corollary.

Corollary 2.5.1. If

$$y = \langle a_1y_1 + a_2y_2 + \cdots + a_my_m\rangle_{u:\ell}$$

then any two terms of the form $a_iy_i + a_jy_j$ where $a_i = a_j$ can be replaced by the term

A. $a_i[y_i \dotplus y_j]$ if $y_iy_j \equiv 0$ for all p_j, or

B. $a_i + a_iy_iy_j$, if $[y_i \dotplus y_j] \equiv 1$ for all p_j, or conversely,

and the result is an equivalent representation of y which has a map equal to the given map.

Example 2.5. Assume that we have the realization

$$y = \langle 3x_9 + x_8 + x_7 + x_6 + x_5 + 2[\bar{x}_1x_2x_3] + 2[x_1x_2\bar{x}_3] + 2[x_1x_3]\rangle_{4:3}$$

The implied T gate network consists of four gates: one for each of the three terms and one for the final gate.

By Corollary 2.5.1,

$$2[\bar{x}_1x_2x_3] + 2[x_1x_2\bar{x}_3] + 2[x_1x_3] = 2[\bar{x}_1x_2x_3 \dotplus x_1x_2\bar{x}_3 \dotplus x_1x_3]$$

† $\oplus$ denotes the exclusive OR; that is, $y_i \oplus y_j = y_i\bar{y}_j \dotplus \bar{y}_iy_j$.

where the latter can be simplified using well-known Boolean identities to the form

$$2[x_2x_3 \dotplus x_1x_2 \dotplus x_1x_3]$$

which is the three-variable majority function and has the realization

$$2\langle x_1 + x_2 + x_3\rangle_{2:1}$$

The equivalent realization becomes

$$y = \langle 3x_9 + x_8 + x_7 + x_6 + x_5 + 2\langle x_1 + x_2 + x_3\rangle_{2:1}\rangle_{4:3}$$

Example 2.6. Assume that we have the realization

$$y = \langle 2x_1 + x_2 + x_3 + x_4 + \langle x_5 + x_6 + x_7 + x_8\rangle_{4:3}\rangle_{2:1}$$

but that sensitivity constraints restrict the realization to gaps of 3:2 or smaller. It is evident that

$$\langle x_5 + x_6 + x_7 + x_8\rangle_{4:3} = [x_5x_6x_7x_8] = \langle x_5 + x_6 + \langle x_7 + x_8\rangle_{2:1}\rangle_{3:2}$$

where the latter satisfies the gap requirements, but it does so at the expense of adding an additional gate. This is not particularly desirable.

By Theorem 2.4B, a constant can be added to the separating function and to the gap so that

$$y = \langle 2x_1 + x_2 + x_3 + x_4 + 1 + [x_5x_6x_7x_8]\rangle_{3:2}$$

Moreover, the constant can be represented by the Boolean function $1 = [x_5 \dotplus \bar{x}_5]$ so that

$$y = \langle 2x_1 + x_2 + x_3 + x_4 + [x_5 \dotplus \bar{x}_5] + [x_5x_6x_7x_8]\rangle_{3:2}$$

By the converse of Corollary 2.5.1, the term $[x_5 \dotplus \bar{x}_5]$ can be replaced by $x_5 + \bar{x}_5$, in which case

$$y = \langle 2x_1 + x_2 + x_3 + x_4 + x_5 + \bar{x}_5 + [x_5x_6x_7x_8]\rangle_{3:2}$$

By Corollary 2.5.1, the terms $\bar{x}_5 + [x_5x_6x_7x_8]$ can be replaced by the term

$$[\bar{x}_5 \dotplus x_5x_6x_7x_8] = [\bar{x}_5 \dotplus x_6x_7x_8]$$

which has the realization $\langle 3\bar{x}_5 + x_6 + x_7 + x_8\rangle_{3:2}$. The final realization is

$$y = \langle 2x_1 + x_2 + x_3 + x_4 + x_5 + \langle 3\bar{x}_5 + x_6 + x_7 + x_8\rangle_{3:2}\rangle_{3:2}$$

It requires only two gates both of which satisfy the sensitivity constraints.

A second special case of Theorem 2.5 which is of particular interest occurs when the weights of the terms that are being replaced have values equal to or greater than the upper gap boundary. This is specifically set forth in the following corollary and is illustrated in Example 2.7.

Corollary 2.5.2. If

$$y = \langle a_1y_1 + a_2y_2 + \cdots + a_my_m\rangle_{u:\ell}$$

then any two terms of the form $a_iy_i + a_jy_j$ where $a_i = a_j \geq u$ can be replaced by

$$a_i[y_i \dotplus y_j]$$

and conversely, and the result is an equivalent representation of y.

Example 2.7. Assume that we have the realization

$$y = \langle 3[x_1x_2] + 3x_3 + 2x_4 + x_5 + x_6 + x_7\rangle_{3:2}$$

but that the fan-in of 11 is too high. Using Corollary 2.5.2, the terms $3[x_1x_2] + 3x_3$ are replaced by $3[x_1x_2 \dotplus x_3]$, and the equivalent representation becomes

$$y = \langle 3\langle 2x_3 + x_1 + x_2\rangle_{2:1} + 2x_4 + x_5 + x_6 + x_7\rangle_{3:2}$$

where the inner and outer gates have fan-in values of 4 and 8 respectively.

Actually Theorem 2.5 is a special case of Theorem 2.6. Although the latter is more general, the proofs of the two are similar.

Theorem 2.6. If

$$y = \langle a_1y_1 + a_2y_2 + \cdots + a_my_m\rangle_{u:\ell}$$

then any set of terms $a_{j_1}y_{j_1} + a_{j_2}y_{j_2} + \cdots + a_{j_r}y_{j_r}$ where $a_{j_1} = a_{j_2} = \cdots = a_{j_r}$ can be replaced by the terms

$$a_{j_1}\sum_{i=1}^{r} M_i(y_{j_1},y_{j_2},\ldots,y_{j_r})$$

where $M_i(y_{j_1},y_{j_2},\ldots,y_{j_r})$ is the function which has unit value if and only if i or more of the variables $y_{j_1},y_{j_2},\ldots,y_{j_r}$ are unity. This transformation yields equivalent representations with equal maps.

Thus far† we have considered how to modify terms of a separating-function representation to obtain a different equivalent one. Now a transformation is considered that yields an equivalent separating-function representation where the number of gates is reduced. In order for this result to be applicable, the form of the original representation must be

$$y = \langle y_0 + f(y_1,y_2,\ldots,y_m)\rangle_{(\ell+1):\ell} = \langle y_0 + f(p)\rangle_{(\ell+1):\ell} \tag{2.3}$$

† The remainder of this section can be omitted on the first reading.

where each coefficient of $f(p)$ is equal to or greater than unity and where

$$y_0 = \langle f_0(y_1, y_2, \ldots, y_m)\rangle_{u_0:\ell_0} = \langle f_0(p)\rangle_{u_0:\ell_0}$$

Observe that the gap length of the final gate is unity and that the input y_0 to the final gate has unit coefficient.

It is interesting to note that if the separating-function representations for the y of Theorems 2.1 and 2.2 had been in terms of the variables y_0, y_1 and y_0, y_1, y_2 respectively, then these representations would have been of the form of Eq. 2.3. For example,

$$y = y_1 \dotplus y_0 = \langle y_0 + y_1\rangle_{1:0}$$
$$y = y_1 y_0 = \langle y_0 + y_1\rangle_{2:1}$$
$$y = y_0(y_1 \dotplus y_2) \dotplus y_1 y_2 = \langle y_0 + y_1 + y_2\rangle_{2:1}$$

In this form, therefore, the following theorem could be used to give the same final separating-function representations as were obtained by Theorems 2.1 and 2.2.

Theorem 2.7. (*J. Mezei*).† If

$$y = \langle y_0 + f(p)\rangle_{(\ell+1):\ell} = \langle\langle f_0(p)\rangle_{u_0:\ell_0} + f(p)\rangle_{(\ell+1):\ell}$$

where each coefficient of $f(p)$ is equal to or greater than unity, then

$$y = \langle f_0(p) + af(p)\rangle_{(a\ell+u_0):(a\ell+\ell_0)} \tag{2.4}$$

for any

$$a \geq \max\{(u_0 - \sigma_{n_0}), (\sigma_{p_0} - \ell_0)\}$$

where σ_{p_0} and σ_{n_0} denote the largest and smallest possible values of $f_0(p)$ respectively. The maps of the two representations are similar.

Proof. The proof is straightforward but lengthy. The basic idea, however, is easy to state. Loosely speaking, the map of f is expanded by multiplication by the constant a. This causes the interval corresponding to the original gap $(\ell + 1):\ell$ to be expanded and displaced to the interval $a(\ell + 1):a\ell$. The value of a is selected so that the gap $(a\ell + u_0):(a\ell + \ell_0)$ of the resultant representation lies within the interval $a(\ell + 1):a\ell$. This is not, however, the only restriction on the value of a. In the representation

$$\langle y_0 + f(p)\rangle_{(\ell+1):\ell}$$

y_0 contributes the amount 0 or 1 to each possible value of $y_0 + f(p)$. In the representation

$$\langle f_0(p) + af(p)\rangle_{(a\ell+u_0):(a\ell+\ell_0)}$$

† Unpublished, but quoted in Elgot.

the corresponding contributions of $f_0(p)$ to $f_0(p) + af(p)$ are

$$\sigma_{n_0} \leq f_0(p) \leq \ell_0 \quad \text{when } y_0 = 0$$

and

$$u_0 \leq f_0(p) \leq \sigma_{p_0} \text{ when } y_0 = 1$$

The value of a must be selected so that the effect of the contribution of y_0 to

$$\langle y_0 + f(p) \rangle_{(\ell+1):\ell}$$

is the same as the effect of the contribution of $f_0(p)$ to

$$\langle f_0(p) + af(p) \rangle_{(a\ell+u_0):(a\ell+\ell_0)}$$

Figure 2.4 gives a simplified illustration.

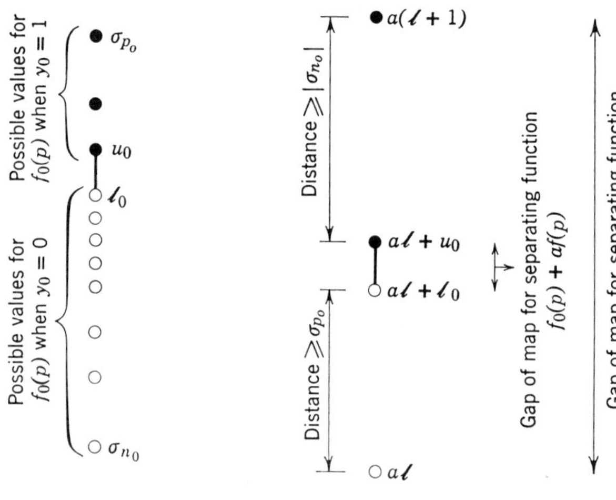

Fig. 2.4. An illustration of Theorem 2.7. (*a*) The map of $\langle f_0(p) \rangle_{u_0:\ell_0}$. (*b*) Gaps of the maps of $\langle f_0(p) + af(p) \rangle_{(a\ell+u_0):(a\ell+\ell_0)}$ and $\langle af(p) \rangle_{a(\ell+1):a\ell}$.

For the detailed proof, the possible values of p are divided into four cases: Case 1, $y = y_0 = 0$; Case 2, $y = 0$, $y_0 = 1$; Case 3, $y = 1$, $y_0 = 0$; Case 4, $y = y_0 = 1$.

CASE 1. $y = y_0 = 0$.
For each p such that $y = y_0 = 0$, it follows that

$$y_0 + f(p) = f(p) \leq \ell$$

We must show that

$$f_0(p) + af(p) \leq a\ell + \ell_0$$

Since $y_0 = 0$ implies $f_0(p) \leq \ell_0$, then

$$f_0(p) + af(p) \leq \ell_0 + a\ell$$

and the desired result is obtained.

CASE 2. $y = 0, y_0 = 1.$
For each p such that $y = 0$, $y_0 = 1$ it follows that

$$y_0 + f(p) = 1 + f(p) \leq \ell$$

or that $f(p) \leq \ell - 1$. We must show that

$$f_0(p) + af(p) \leq a\ell + \ell_0$$

Since $y_0 = 1$, then the least upper bound for $f_0(p)$ is $f_0(p) \leq \sigma_{p_0}$, in which case

$$f_0(p) + af(p) \leq \sigma_{p_0} + af(p) \leq \sigma_{p_0} + a(\ell - 1)$$

Recall that

$$a \geq \max \{(u_0 - \sigma_{n_0}), (\sigma_{p_0} - \ell_0)\}$$

which implies that $\sigma_{p_0} \leq a + \ell_0$. Using this, it follows that

$$f_0(p) + af(p) \leq \sigma_{p_0} + a(\ell - 1) \leq a + \ell_0 + a\ell - a = a\ell + \ell_0$$

and the validity of Case 2 is established.

CASE 3. $y = 1, y_0 = 0.$
For each p such that $y = 1$, $y_0 = 0$, it follows that

$$y_0 + f(p) = f(p) \geq \ell + 1$$

We must show that

$$f_0(p) + af(p) \geq a\ell + u_0$$

Since $y_0 = 0$, the lower bound for $f_0(p)$ is $f_0(p) \geq \sigma_{n_0}$, in which case

$$f_0(p) + af(p) \geq \sigma_{n_0} + af(p) \geq \sigma_{n_0} + a(\ell + 1)$$

The constraint $a \geq u_0 - \sigma_{n_0}$ implies $\sigma_{n_0} \geq u_0 - a$, in which case

$$f_0(p) + af(p) \geq \sigma_{n_0} + a(\ell + 1) \geq u_0 - a + a\ell + a = a\ell + u_0$$

and Case 3 is established.

CASE 4. $y = y_0 = 1$.
For each p such that $y = y_0 = 1$, it follows that

$$y_0 + f(p) = 1 + f(p) \geq \ell + 1$$

or that $f(p) \geq \ell$. We must show that

$$f_0(p) + af(p) \geq a\ell + u_0$$

Since $y_0 = 1$, then the lower bound for $f_0(p)$ is $f_0(p) \geq u_0$, in which case

$$f_0(p) + af(p) \geq u_0 + af(p) \geq u_0 + a\ell$$

and Case 4 is established. This completes the proof of Theorem 2.7.|

Example 2.8. Consider the realization

$$y = \langle x_1 + x_2 + x_3 + \langle 2x_4 + x_5 + x_6 \rangle_{2:1} \rangle_{3:2}$$

for the Boolean function

$$y = x_1x_2x_3 \dotplus M_2(x_1,x_2,x_3)(x_4 \dotplus x_5x_6)$$

Clearly this satisfies the conditions required of Theorem 2.7, which therefore can be applied to convert the above two-gate realization to one that requires a single gate. The constraints on a are

$$a \geq \max \{(u_0 - \sigma_{n_0}), (\sigma_{p_0} - \ell_0)\} = \max \{2,3\} = 3$$

Choosing $a = 3$, Theorem 2.7 yields the realization

$$y = \langle 3x_1 + 3x_2 + 3x_3 + 2x_4 + x_5 + x_6 \rangle_{8:7}$$

2.4 T-GATE REALIZATION OF COMPLEMENT FUNCTIONS

If F denotes an arbitrary switching function, then the complement of F is denoted by $\bar{F}$ where F and $\bar{F}$ are related as follows:

$$F(p) = 1 \quad \text{implies} \quad \bar{F}(p) = 0$$
$$F(p) = 0 \quad \text{implies} \quad \bar{F}(p) = 1$$

Moreover, if $y = F(y_1,y_2,\ldots,y_m)$, then $\bar{y} = \bar{F}(y_1,y_2,\ldots,y_m)$. In this section we are concerned with transformations that convert a T-gate realization of F to a T-gate realization of $\bar{F}$.

Consider first a linearly separable function

$$y = F(x_1,x_2,\ldots,x_n) = \left\langle \sum_{j=1}^{n} a_jx_j \right\rangle_t$$

As described in Section 1.3, the equation

$$\sum_{j=1}^{n} a_j x_j = t$$

can be interpreted in n space as a hyperplane which separates the 1 points of F from the 0 points.

To obtain a realization for

$$\bar{y} = \bar{F}(x_1, x_2, \ldots, x_n)$$

we must interchange the roles of these 1 points and 0 points. The idea is that the same plane which separated these points for F will also separate them for $\bar{F}$, except that the positive and negative sides of the plane must be interchanged. This can be done by multiplying both sides of the plane equation by -1:

$$-\sum_{j=1}^{n} a_j x_j = -t$$

Thus a realization for $\bar{F}$ is

$$\bar{y} = \bar{F}(x_1, x_2, \ldots, x_n) = \left\langle -\sum_{j=1}^{n} a_j x_j \right\rangle_{-t}$$

Although this geometric argument is appealing, the same result can be demonstrated algebraically. In fact, the algebraic proof applies to a more general situation. Consider the switching function F where

$$y = F(y_1, y_2, \ldots, y_m) = \langle f(y_1, y_2, \ldots, y_m) \rangle_{u:\ell}$$

Let $y_1^p, y_2^p, \ldots, y_m^p$ denote some combination of values of the variables $y_1, y_2, \ldots, y_m$. Then each p such that

$$F(y_1^p, y_2^p, \ldots, y_m^p) = 1 \qquad \text{implies} \quad \sum_{j=1}^{m} a_j y_j^p \geq u$$

and each p such that

$$F(y_1^p, y_2^p, \ldots, y_m^p) = 0 \qquad \text{implies} \quad \sum_{j=1}^{m} a_j y_j^p \leq \ell$$

It follows from this, therefore, that

$$\bar{F}(y_1^p, y_2^p, \ldots, y_m^p) = 0 \qquad \text{implies} \quad \sum_{j=1}^{m} a_j y_j^p \geq u \quad \text{or} \sum_{j=1}^{m} -a_j y_j^p \leq -u$$

and (2.5)

$$\bar{F}(y_1^p, y_2^p, \ldots, y_m^p) = 1 \qquad \text{implies} \quad \sum_{j=1}^{m} a_j y_j^p \leq \ell \quad \text{or} \quad \sum_{j=1}^{m} -a_j y_j^p \geq -\ell$$

Theorem 2.8. If

$$y = F(y_1,y_2,\ldots,y_m) = \langle f(y_1,y_2,\ldots,y_m)\rangle_{u:\ell} = \left\langle \sum_{j=1}^{m} a_jy_j \right\rangle_{u:\ell}$$

then

A. $\bar{y} = \bar{F}(y_1,y_2,\ldots,y_m) = \langle -f(y_1,y_2,\ldots,y_m)\rangle_{-\ell:-u}$

$$= \left\langle \sum_{j=1}^{m} -a_jy_j \right\rangle_{-\ell:-u}$$

and

B. $\bar{y} = \bar{F}(y_1,y_2,\ldots,y_m) = \langle f(\bar{y}_1,\bar{y}_2,\ldots,\bar{y}_m)\rangle_{(\sigma-\ell):(\sigma-u)}$

$$= \left\langle \sum_{j=1}^{m} a_j\bar{y}_j \right\rangle_{(\sigma-\ell):(\sigma-u)}$$

where

$$\sigma = \sum_{j=1}^{m} a_j$$

Proof. The proof of part A follows from Eqs. 2.5. Part B follows by applying Theorems 2.3 and 2.4 to each of the variables y_j. ▌

If

$$y = F(y_1,y_2,\ldots,y_m)$$

is a linearly separable function of the variables $y_1,y_2,\ldots,y_m$, then $\bar{F}(y_1,y_2,\ldots,y_m)$ is likewise linearly separable. Moreover, an interesting relationship exists between the map of

$$y = \langle f(y_1,y_2,\ldots,y_m)\rangle_{u:\ell}$$

and the maps of

$$\bar{y} = \langle -f(y_1,y_2,\ldots,y_m)\rangle_{-\ell:-u}$$

and

$$\bar{y} = \langle f(\bar{y}_1,\bar{y}_2,\ldots,\bar{y}_m)\rangle_{(\sigma-\ell):(\sigma-u)}$$

For this let $\bar{f}(y_1,y_2,\ldots,y_m)$ denote the function $f(\bar{y}_1,\bar{y}_2,\ldots,\bar{y}_m)$. It follows immediately from Theorem 2.8 that for each attainable p,

$$f(p) = \bar{f}(\bar{p}) \tag{2.6}$$

Consider the map M_1 for the separating-function representation

$$y = \langle f(y_1,y_2,\ldots,y_m)\rangle_{u:\ell}$$

We define a one-one transformation on M_1 that defines a new map M_2 as follows:

Let β be some real number. For each p let $f(p)$ of M_1 correspond to the point $2\beta - f(p)$ of M_2. If $u{:}\ell$ is the gap of M_1, then the gap of M_2 is $(2\beta - \ell){:}(2\beta - u)$.

The map M_2 is called the *β-rotation of* M_1, since for each p, if $f(p)$ of M_1 is represented as $\beta + \Delta_j$, then the corresponding point of M_2 is $\beta - \Delta_j$. The visualization of M_2 is obtained from that of M_1 by rotating M_1 through 180° about the point β. This concept is used in relating the maps of Theorem 2.8 as set forth in the following corollary and illustrated in Fig. 2.5. The validity follows immediately from Theorem 2.8 and the above discussion.

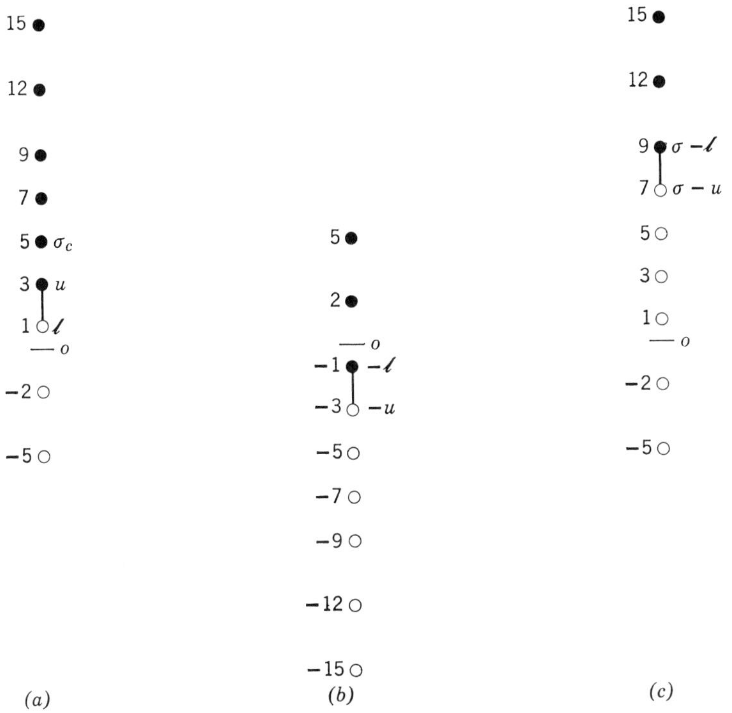

Fig. 2.5. Maps of $y = F(y_1,y_2,\ldots,y_m)$ and of $\bar{y} = \bar{F}(y_1,y_2,\ldots,y_m)$ of Corollary 2.8.1. (*a*) Map of $y = \langle f(y_1,y_2,\ldots,y_m)\rangle_{u:\ell}$. (*b*) Map of $\bar{y} = \langle -f(y_1,y_2,\ldots,y_m)\rangle_{-\ell:-u}$. (*c*) Map of $\bar{y} = \langle f(\bar{y}_1,\bar{y}_2,\ldots,\bar{y}_m)\rangle_{(\sigma-\ell):(\sigma-u)}$.

Corollary 2.8.1. The maps of

$$\bar{y} = \langle -f(y_1,y_2,\ldots,y_m)\rangle_{-\ell:-u}$$

and of

$$\bar{y} = \langle f(\bar{y}_1,\bar{y}_2,\ldots,\bar{y}_m)\rangle_{(\sigma-\ell):(\sigma-u)}$$

are 0 rotations and σ_c rotations respectively of the map of

$$y = \langle f(y_1,y_2,\ldots,y_m)\rangle_{u:\ell}$$

where $2\sigma_c = \sigma$.

If

$$y = F(y_1,y_2,\ldots,y_m)$$

is realized by a network of one or more T gates where the network inputs may or may not be the independent variables $x_1,x_2,\ldots,x_n$, then Theorem 2.8B provides a convenient means for obtaining a realization for

$$\bar{y} = \bar{F}(y_1,y_2,\ldots,y_m)$$

This is set forth in the following corollary.

Corollary 2.8.2. If F is realized by a T-gate network, then a realization of $\bar{F}$ is obtained by complementing each of the input variables to the network and replacing each gap $u_j{:}\ell_j$ by $(\sigma_j - \ell_j){:}(\sigma_j - u_j)$ where σ_j is the sum of the coefficients of the jth T gate.

Example 2.9. From Example 2.5, the final realization for

$$y = x_9[M_1(x_8,x_7,x_6,x_5) \dotplus M_2(x_1,x_2,x_3)] \dotplus M_2(x_1,x_2,x_3)M_2(x_8,x_7,x_6,x_5) \dotplus x_8x_7x_6x_5$$

is

$$y = \langle 3x_9 + x_8 + x_7 + x_6 + x_5 + 2\langle x_1 + x_2 + x_3\rangle_{2:1}\rangle_{4:3}$$

One realization for the complement function is

$$\bar{y} = \langle 3\bar{x}_9 + \bar{x}_8 + \bar{x}_7 + \bar{x}_6 + \bar{x}_5 + 2\langle \bar{x}_1 + \bar{x}_2 + \bar{x}_3\rangle_{2:1}\rangle_{6:5}$$

Both realizations are shown in Fig. 2.6.

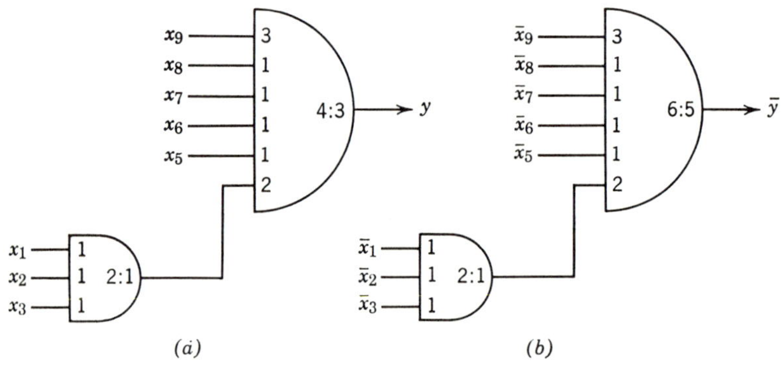

Fig. 2.6. T-gate realizations of the functions of Example 2.9. (*a*) Realization of $y = F(x_1,x_2,\ldots,x_n)$. (*b*) Realization of $\bar{y} = \bar{F}(x_1,x_2,\ldots,x_n)$.

2.5 THE COMPLEMENTING SYMMETRY CLASS OF FUNCTIONS

Let $y = F(x_1,x_2,\ldots,x_n)$ be an arbitrary switching function for which there is some T-gate network realization, for example,

$$y = \left\langle \sum_{j=1}^{n} a_j \left\langle \sum_{i=1}^{n} a_{ij}x_i \right\rangle_{u_j:\ell_j} \right\rangle_{u:\ell}$$

Clearly any independent variable can be renamed, and the resultant is a realization of the new switching function. A case of particular interest is where a variable is replaced by its complement.

Theorem 2.9. If $y = F(x_1,x_2,\ldots,x_k,\ldots, x_n)$ has some realization, for example,

$$y = \left\langle \sum_{j=1}^{n} a_j \langle a_{1j}x_1 + a_{2j}x_2 + \cdots + a_{kj}x_k + \cdots + a_{nj}x_n \rangle_{u_j:\ell_j} \right\rangle_{u:\ell}$$

then $y_k = F(x_1,x_2,\ldots,\bar{x}_k,\ldots,x_n)$ has a realization, equivalent to that for y except that x_k is replaced everywhere by $\bar{x}_k$ and $\bar{x}_k$ by x_k; in the above example,

$$y = \left\langle \sum_{j=1}^{n} a_j \langle a_{1j}x_1 + a_{2j}x_2 + \cdots + a_{kj}\bar{x}_k + \cdots + a_{nj}x_n \rangle_{u_j:\ell_j} \right\rangle_{u:\ell}$$

To each switching function, therefore, there corresponds a symmetry class of 2^n switching functions that are obtained by variable complementation. The realizations of these 2^n functions are similarly related by the complementation of corresponding variables.

If the switching function is ℓ.s., then all members of the complementing symmetry class are ℓ.s. This, together with Procedure 2.1, provides a particularly useful result. Recall that Procedure 2.1 provides a means whereby the realization of a linearly separable function can be converted to one for which all weights are positive. To make this conversion however, it is necessary to replace some of the input variables by their negative complements and to alter the gap accordingly. The result is a separating-function representation with all coefficients positive but where some of the variables are complemented. By Theorem 2.9 the complemented variables can be uncomplemented in both the Boolean and the separating-function representations, and the result will be the corresponding representations of a member of the complementing symmetry class for which all variables are uncomplemented. This establishes the following corollary.

Corollary 2.9.1. If $F(x_1,x_2,\ldots,x_n)$ is an arbitrary switching function, then in the complementing symmetry class of F there is a switching function $\phi(x_1,x_2,\ldots,x_n)$ which is linearly separable if and only if F is linearly separable and the realization of ϕ has all coefficients positive when the input variables are $x_1,x_2,\ldots,x_n$. Moreover, a realization for F can be obtained from the realization of ϕ by complementing some of the variables and changing the gap.

From this it is evident that any test and synthesis procedure for linearly separable functions need only consider the case where the weights are positive and the input variables are $x_1,x_2,\ldots,x_n$. This result will be considered further in the next chapter on linearly separable functions.

2.6 DUAL SWITCHING FUNCTIONS

To every switching function $F(x_1,x_2,\ldots,x_n)$ there corresponds the *dual function* $F^d(x_1,x_2,\ldots,x_n)$, which is defined as follows:

$$F^d(x_1,x_2,\ldots,x_n) = \bar{F}(\bar{x}_1,\bar{x}_2,\ldots,\bar{x}_n)$$

To obtain the dual of a function, both the function and the independent variables are complemented. Clearly the dual of F^d is the original function F. If $y = F(x_1,x_2,\ldots,x_n)$ then $y^d = F^d(x_1,x_2,\ldots,x_n)$. A function $F(x_1,x_2,\ldots,x_n)$ is called *self-dual* if

$$F(x_1,x_2,\ldots,x_n) = F^d(x_1,x_2,\ldots,x_n)$$

If $y = x$, where x is an independent variable, then $y^d = x$ and y is a self-dual function. A less trivial example of a self-dual function is $y = M_2(x_1,x_2,x_3)$. The following results are an immediate consequence of these definitions and of Theorems 2.8 and 2.9.

Theorem 2.10. If $y = F(x_1,x_2,\ldots,x_n)$ has some realization

$$y = \langle a_1y_1 + a_2y_2 + \cdots + a_my_m\rangle_{u:\ell}$$

where each $y_j = F_j(x_1,x_2,\ldots,x_n)$, then

$$y^d = F^d(y_1,y_2,\ldots,y_m) = \langle a_1y_1{}^d + a_2y_2{}^d + \cdots + a_my_m{}^d\rangle_{(\sigma-\ell):(\sigma-u)}$$

where

$$\sigma = \sum_{j=1}^{m} a_j$$

For the special case where F is a linearly separable function and where each $y_j = x_j$, since $y_j{}^d = x_j$, we have the following corollary.

Corollary 2.10.1. If

$$y = F(x_1,x_2,\ldots,x_n) = \langle f(x_1,x_2,\ldots,x_n)\rangle_{u:\ell} = \Big\langle \sum_{j=1}^{n} a_jx_j\Big\rangle_{u:\ell}$$

then

$$y^d = F^d(x_1,x_2,\ldots,x_n) = \langle f(x_1,x_2,\ldots,x_n)\rangle_{(\sigma-\ell):(\sigma-u)}$$

$$= \Big\langle \sum_{j=1}^{n} a_jx_j\Big\rangle_{(\sigma-\ell):(\sigma-u)}$$

In this case if F is linearly separable, then so is F^d and conversely. Moreover, the separating functions for each are identical; however, the gaps are displaced symmetrically about the center of the map σ_c as shown in Fig. 2.7. If the gap of the map of F were in the center, that is, if $\ell < \sigma_c < u$, then F would be a self-dual function. Note that the map of a self-dual function must be such that, for each p, $f(p) \neq \sigma_c$.

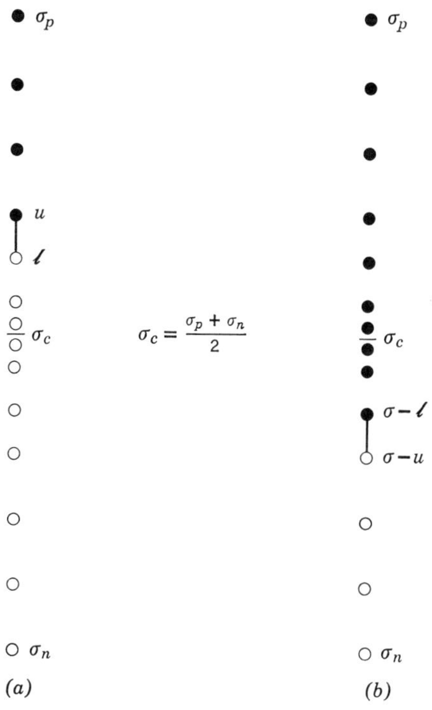

Fig. 2.7. Maps of the linearly separable functions F and F^d. (*a*) Map of $y = F(x_1,x_2,\ldots,x_n)$. (*b*) Map of $y^d = F^d(x_1,x_2,\ldots,x_n)$.

If a switching-function F is realized by a network of T gates, then applying Theorem 2.10 to every gate of the realization for F^d, we obtain the following corollary.

> ***Corollary 2.10.2.*** If F is realized by a T-gate network, then a realization for F^d is obtained by replacing each gap $u_j:\ell_j$ of each gate by the gap $(\sigma_j - \ell_j):(\sigma_j - u_j)$ where σ_j is the sum of the coefficients of the jth gate and by replacing each constant input, if any, by its complemented value.

Example 2.10. Consider the switching function of Example 2.5 for which

$$y = \langle 3x_9 + x_8 + x_7 + x_6 + x_5 + 2\langle x_1 + x_2 + x_3\rangle_{2:1}\rangle_{4:3}$$

was a realization. According to Corollary 2.10.2, a realization for the dual function is

$$y^d = \langle 3x_9 + x_8 + x_7 + x_6 + x_5 + 2\langle x_1 + x_2 + x_3\rangle_{2:1}\rangle_{6:5}$$

Both realizations are shown in Fig. 2.8. It is evident both from the realizations and from the separating-function representations of the two functions that $\langle x_1 + x_2 + x_3\rangle_{2:1}$ is a self-dual function.

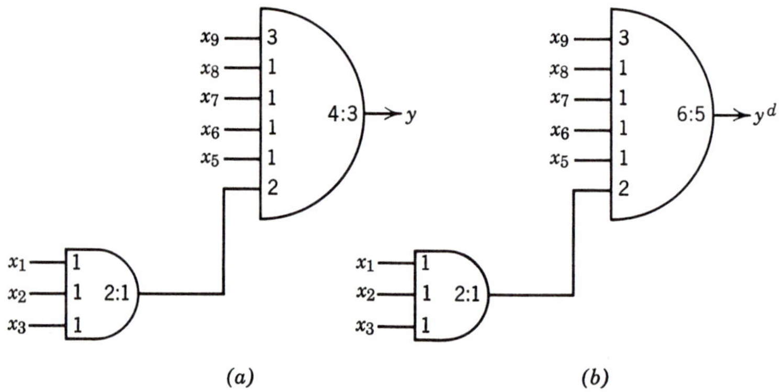

Fig. 2.8. T-gate realizations of y and y^d of Example 2.10. (*a*) Realization of y. (*b*) Realization of y^d.

The concepts of duality and dual functions play an important part in the theory of threshold logic. We return to this topic in Chapter 7.

2.7 T-GATE NETWORK ANALYSIS: AN EXAMPLE

The problems of threshold logic that are of particular interest are those of synthesis: given an arbitrarily specified switching function, how is it realized by an interconnection of T gates? This section, however, presents a simple example of analysis: given an arbitrarily specified T gate network, what is the switching function that it realizes? As usual, analysis is much simpler and straightforward than synthesis. In fact, the only problems are concerned with the formulation of systematic and computationally simple procedures.

Rather than give a general procedure which would require notation complicated beyond its importance, we present a simple example. From this the general procedure is evident.

Example 2.11. Consider the single-gate network described by

$$y = \langle 5x_1 - 4x_2 + 4x_3 + 3x_4 + 3x_5 + 2x_6 + x_7 + x_8 - x_9 \rangle_{0:-1}$$

The objective is to find the switching function realized by this gate and to express this function in a minimal sum-of-products form. The problem is complicated somewhat by the negative coefficients of x_2 and x_9, but this is easily remedied by an application of Procedure 2.1; thus

$$y = \langle 5x_1 + 4\bar{x}_2 + 4x_3 + 3x_4 + 3x_5 + 2x_6 + x_7 + x_8 + \bar{x}_9 \rangle_{5:4}$$

It is easy to indicate some of the terms of y. For example, x_3x_7 is such a term since all input combinations for which $x_3 = x_7 = 1$ have a sum of weights greater than or equal to $a_3 + a_7 = 5 = u$. The problem is to find a systematic procedure for finding all such terms.

The first step after making all the coefficients positive is to order the variables according to their coefficients, as has already been done in this example.

Now consider the variable with the largest coefficient: in this case x_1. Clearly if there are any terms at all, there will be some in which x_1 is a factor; our first task, then, is to find all terms in which x_1 is a factor. We therefore write

$$x_1($$

The left parenthesis denotes that there may be other variables multiplying x_1. We now look at the weight of x_1, which in this case is 5, and we see that x_1 by itself will exceed the threshold. Hence, the only term in which x_1 is a factor is x_1 itself. Therefore, we erase the left parenthesis and replace it by a Boolean plus sign

$$x_1 \dotplus$$

Next we consider the variable with the next largest weight, $\bar{x}_2$. Clearly if there are any additional terms, they must have $\bar{x}_2$ as a factor. We therefore set out to find all the terms with $\bar{x}_2$ as a factor and we write

$$x_1 \dotplus \bar{x}_2($$

Since the weight of $\bar{x}_2$ alone is not sufficient to equal or exceed u, if there are any such terms, they must include other variables. But x_3 is the variable with the next largest weight. If there are any terms with $\bar{x}_2$ as a factor, some of them, at least, must also have x_3 as a factor. We therefore write

$$x_1 \dotplus \bar{x}_2(x_3($$

But the weights of $\bar{x}_2$ and x_3 are sufficient to exceed u, and so we have found a new term. We replace the final parenthesis with a Boolean plus sign and continue:

$$x_1 \dotplus \bar{x}_2(x_3 \dotplus$$

We repeat this reasoning with each of the variables x_4 through $\bar{x}_9$ and obtain

$$x_1 \dotplus \bar{x}_2(x_3 \dotplus x_4 \dotplus x_5 \dotplus x_6 \dotplus x_7 \dotplus x_8 \dotplus \bar{x}_9) \dotplus$$

We have now found all the terms with $\bar{x}_2$ as a factor.

We now consider the variable with the next largest weight, x_3, and by the same reasoning obtain

$$x_1 \dotplus \bar{x}_2(x_3 \dotplus x_4 \dotplus x_5 \dotplus x_6 \dotplus x_7 \dotplus x_8 \dotplus \bar{x}_9) \\ \dotplus x_3(x_4 \dotplus x_5 \dotplus x_6 \dotplus x_7 \dotplus x_8 \dotplus \bar{x}_9)$$

We then proceed to x_4. Consider only the terms with x_4 as a factor:

$$x_4($$

Of these, we consider those terms with x_5 as a factor:

$$x_4(x_5 \dotplus$$

with x_6 as a factor:

$$x_4(x_5 \dotplus x_6 \dotplus$$

and with x_7 as a factor:

$$x_4(x_5 \dotplus x_6 \dotplus x_7($$

The weights of x_4 and x_7 are not sufficient to exceed u. We therefore continue beyond the last parenthesis adding new variables among the group we have not yet considered (in this case x_8 and $\bar{x}_9$) until u is exceeded:

$$x_4(x_5 \dotplus x_6 \dotplus x_7(x_8 \dotplus$$

and then

$$x_4(x_5 \dotplus x_6 \dotplus x_7(x_8 \dotplus \bar{x}_9) \dotplus$$

Finally, we obtain

$$x_4(x_5 \dotplus x_6 \dotplus x_7(x_8 \dotplus \bar{x}_9) \dotplus x_8\bar{x}_9)$$

The overall function is now

$$x_1 \dotplus \bar{x}_2(x_3 \dotplus x_4 \dotplus x_5 \dotplus x_6 \dotplus x_7 \dotplus x_8 \dotplus \bar{x}_9) \\ \dotplus x_3(x_4 \dotplus x_5 \dotplus x_6 \dotplus x_7 \dotplus x_8 \dotplus \bar{x}_9) \\ \dotplus x_4(x_5 \dotplus x_6 \dotplus x_7(x_8 \dotplus \bar{x}_9) \dotplus x_8\bar{x}_9) \dotplus$$

Continuing this procedure and considering x_5 and x_6, we obtain

$$\begin{aligned} x_1 &\dotplus \bar{x}_2(x_3 \dotplus x_4 \dotplus x_5 \dotplus x_6 \dotplus x_7 \dotplus x_8 \dotplus \bar{x}_9) \\ &\dotplus x_3(x_4 \dotplus x_5 \dotplus x_6 \dotplus x_7 \dotplus x_8 \dotplus \bar{x}_9) \\ &\dotplus x_4(x_5 \dotplus x_6 \dotplus x_7\,(x_8 \dotplus \bar{x}_9) \dotplus x_8\bar{x}_9) \\ &\dotplus x_5(x_6 \dotplus x_7(x_8 \dotplus \bar{x}_9) \dotplus x_8\bar{x}_9) \\ &\dotplus x_6x_7x_8\bar{x}_9 \dotplus \end{aligned}$$

If now we attempt to find the additional terms with x_7 as a factor, we get

$$x_7($$
$$x_7x_8($$
$$x_7x_8\bar{x}_9($$

We have exhausted the variables, and the sum of the weights does not equal u. Therefore there are no such terms; we erase this last product:

$$x_7x_8\bar{x}_9$$

and we are finished. To obtain the sum-of-products expression, this final expression must be multiplied out to remove the parentheses.

If the given example had been a multigate realization, the x's would have been replaced by y's but the procedure would have been the same. However, it would then be necessary to perform the same procedure on the gates realizing the y's and so on, until the Boolean expression for the y's had been obtained. This expression would then be substituted into the expression for the switching function and the result simplified.

Chapter 3

REDUCED FUNCTIONS

3.1 INTRODUCTION

With this chapter we begin a general study of T-gate networks. Our ultimate objective is the development of general, efficient synthesis procedures. As with all network theories, however, synthesis depends upon a detailed understanding of the characteristics of the basic elements from which the network is constructed. For T-gate networks this implies a detailed knowledge of single T gates and of the class of switching functions that they realize—namely, the linearly separable class. This chapter is concerned with this class of functions. Unless otherwise indicated, we assume that all are functions of the independent variables $x_1,x_2,\ldots,x_n$ and that each function assigns the value 0 or 1 to each of the 2^n combinations of values. In other words, there are no don't-care points. This is an important assumption. Most of the results of this chapter are not directly applicable to functions with don't-care points; such functions are considered in Section 4.10.

Linearly separable switching functions are characterized in part by their associated set of reduced functions. If $y = F(x_1,x_2,\ldots,x_n)$ is an arbitrary switching function, then a *reduced function of* F is obtained by assigning unit and zero values to arbitrarily selected disjoint subsets S_i and S_j of the independent variables. Either S_i or S_j can be the empty set. The resultant reduced function is denoted as

$$y(S_i = 1,S_j = 0) = F(S_i = 1,S_j = 0,x_{\alpha_1},x_{\alpha_2},\ldots,x_{\alpha_k})$$

where $x_{\alpha_1},x_{\alpha_2},\ldots,x_{\alpha_k}$ represent the remaining variables. If S_i is the single variable x_1 and S_j is the empty set, then one of the reduced functions is

$$y(x_1 = 1) = F(1,x_2,x_3,\ldots,x_n)$$

and the other is

$$y(x_1 = 0) = F(0,x_2,x_3,\ldots,x_n)$$

Consider, for example,

$$y = F(x_1,x_2,x_3,x_4) = x_1(x_2 \dotplus x_3) \dotplus x_2x_3 \dotplus \bar{x}_1x_4$$

then

$$y(x_1 = 1) = F(1,x_2,x_3,x_4) = x_2 \dotplus x_3 \dotplus x_2x_3 = x_2 \dotplus x_3$$

and

$$y(x_1 = 0) = F(0,x_2,x_3,x_4) = x_2x_3 \dotplus x_4$$

Although values are assigned to only x_1, the corresponding reduced function is not necessarily dependent upon all of the other variables. Note that $y(x_1 = 1)$ above is not a function of x_4.

In the above example, when $S_i = \{x_1\}$ and $S_j = \{x_2\}$ then

$$y(x_1 = 1,x_2 = 0) = F(1,0,x_3,x_4) = x_3$$

Similarly,

$$y(x_1 = 0,x_2 = 1) = F(0,1,x_3,x_4) = x_3 \dotplus x_4$$
$$y(x_1 = x_2 = 1) = F(1,1,x_3,x_4) = 1$$
$$y(x_1 = x_2 = 0) = F(0,0,x_3,x_4) = x_4$$

Note that the reduced function $y(x_1 = x_2 = 1)$ is the constant unit function or the function identically one for all values of the input variables.

Another important concept is that of functional containment. Let

$$y_1 = F_1(x_1,x_2,\ldots,x_n) = F_1(p)$$

and

$$y_2 = F_2(x_1,x_2,\ldots,x_n) = F_2(p)$$

denote two switching functions defined on the same n-cube. We say that F_1 *contains* F_2, and denote this by

$$F_1 \geq F_2$$

if for each p such that $F_2(p) = 1$ it follows that $F_1(p) = 1$. If P_{I_1} and P_{I_2} denote the sets of values of p such that, respectively, $F_1(p) = 1$ and $F_2(p) = 1$, then $F_1 \geq F_2$ if and only if $P_{I_1} \geq P_{I_2}$, that is, if set P_{I_1} contains set P_{I_2}. If F_1 properly contains F_2, this is denoted by

$$F_1 > F_2$$

It follows immediately from the definition that

$$F_1 \geq F_2$$

if and only if for all p

$$\text{A.}\quad F_1(p) \dotplus F_2(p) = F_1(p)$$

or

$$\text{B.}\quad F_1(p)F_2(p) = F_2(p) \tag{3.1}$$

or

$$\text{C.}\quad \bar{F}_1(p)F_2(p) \equiv 0$$

Two functions F_1 and F_2 are said to be *comparable* if one of the following relations is valid:

$$F_1 \geq F_2$$

$$F_2 \geq F_1$$

Note that this includes the relation $F_1 = F_2$. In the above example,

$$y(x_1 = 0, x_2 = 1) > y(x_1 = 1, x_2 = 0)$$

however, $y(x_1 = 1)$ and $y(x_1 = 0)$ are not comparable. The concept of functional comparability is used in the following theorem to relate two ℓ.s. functions.

Theorem 3.1. Any two ℓ.s. switching functions

$$F_1(p) = \langle f(p) \rangle_{u_1:\ell_1}$$

and

$$F_2(p) = \langle f(p) \rangle_{u_2:\ell_2}$$

that can be realized by the same separating function but with different gaps, are comparable. Furthermore, $F_1 > F_2$ if and only if $u_1 < u_2$.

Proof. Since each switching function can have the same separating function, then the image points of each p in the maps of F_1 and F_2 are identical as shown in Fig. 3.1. Now $F_1(p) = 1$ for all p such that $f(p) \geq u_1$ and $F_2(p) = 1$ for all p such that $f(p) \geq u_2$; therefore the unit parts of each map are those image points above and including the upper gap boundary. If $F_1(p) = F_2(p)$ for all p, then the unit parts of each map are the same, in which case $u_1 = u_2$. If $u_1 < u_2$, then the unit part of the map of F_1 properly contains the unit part of the map of F_2; hence $F_1 > F_2$. Conversely, $F_1 > F_2$ implies that there exists a p_α such that $F_1(p_\alpha) = 1$ and $F_2(p_\alpha) = 0$ and therefore that $u_1 < u_2$. This completes the proof.▎

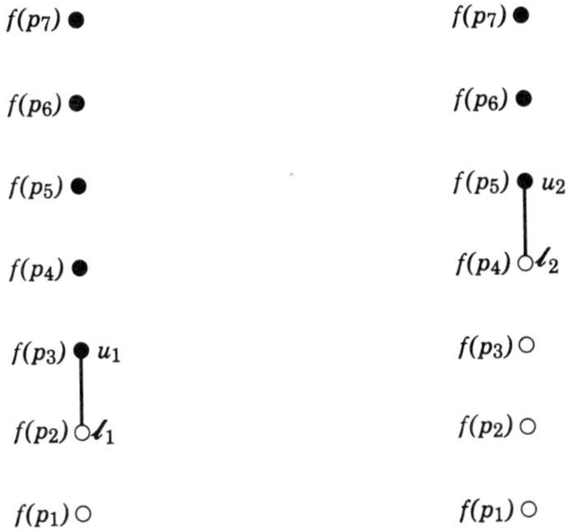

(a) (b)

Fig. 3.1. The maps of two ℓ.s. switching functions that have the same separating function. (*a*) Map of $F_1(p) = \langle f(p)\rangle_{u_1:\ell_1}$. (*b*) Map of $F_2(p) = \langle f(p)\rangle_{u_2:\ell_2}$.

Example 3.1. Let

$$y_1 = \langle 2x_1 + x_2 + x_3 + x_4 + x_5 \rangle_{2:1}$$

and

$$y_2 = \langle 2x_1 + x_2 + x_3 + x_4 + x_5 \rangle_{4:3}$$

Then according to Theorem 3.1, $y_1 > y_2$ since $u_1 < u_2$. To substantiate this, we express y_1 and y_2 as Boolean functions where

$$y_1 = x_1 \dotplus M_2(x_2,x_3,x_4,x_5)$$

and

$$y_2 = x_1 M_2(x_2,x_3,x_4,x_5) \dotplus x_2x_3x_4x_5$$

from which it is apparent that $y_1 > y_2$.

The study of ℓ.s. switching functions utilizes special forms of the Boolean representation. One of these is a *minimum sum of products* form (also known as a *minimal sum of prime implicants* and as a *disjunctive irredundant normal form*). In this form the function is represented as a sum of products (terms) of the independent variables which is irredundant in that no term or no variable in any term can be deleted without changing the function being represented. In other words, the function is a minimal sum of prime implicants.

For example

$$y_1 = x_1x_2 \dotplus \bar{x}_1x_4$$

is a minimum sum of products form while

$$y_2 = x_1x_2 \dotplus \bar{x}_1x_2\bar{x}_3$$

is not since it can be represented by

$$y_2 = x_1x_2 \dotplus x_2\bar{x}_3$$

Similarly

$$y_3 = x_1x_2\bar{x}_3 \dotplus \bar{x}_1x_2x_4 \dotplus x_2\bar{x}_3x_4$$

is not a minimum sum of products form since the last term is redundant. When this is deleted the minimum sum of products form is

$$y_3 = x_1x_2\bar{x}_3 \dotplus \bar{x}_1x_2x_4$$

The unate class of switching functions is defined in terms of the minimum sum of products (MSP) form. Specifically, a switching function is *unate* if and only if in the MSP form no variable appears both complemented and uncomplemented. In the above paragraph y_2 is a unate function of the independent variables while y_1 and y_3 are not.

These are the basic concepts that will be used in the study of ℓ.s. switching functions. In the following sections reduced-function comparability will be related to the values of the coefficients in the separating-function representation, and both will be related to the characteristics of the switching function when represented in minimum-sum-of-products form.

3.2 UNATE FUNCTIONS AND COEFFICIENT SIGNS

This section is concerned with the relationship between the variables that appear in the MSP form of a linearly separable switching function and the signs of the coefficients in the corresponding separating-function representation. For this we use the notation

$$y = F(x_1,x_2,\ldots,x_n) = \langle f(x_1,x_2,\ldots,x_n)\rangle_{u:\ell}$$

where

$$f(x_1,x_2,\ldots,x_n) = a_1x_1 + a_2x_2 + \cdots + a_nx_n$$

and where the MSP form of F is given by

$$y = T_1 \dotplus T_2 \dotplus \cdots \dotplus T_q$$

Suppose that one of the terms in the MSP form of an ℓ.s. function F is

$$T_1 = x_1x_2x_3$$

This certainly implies that

$$a_1 + a_2 + a_3 \geq u$$

Consider now the combination of input values

$$x_1 = 0, \quad x_2 = 1, \quad x_3 = 1$$

For this input combination F must certainly be equal to zero, because otherwise x_2x_3 would be a term of F, contrary to the assumption that $x_1x_2x_3$ is a term in the MSP form. Thus, this input combination implies that

$$0 + a_2 + a_3 \leq \ell$$

Together these two inequalities imply that

$$a_1 > 0$$

Thus we have shown that if x_1 appears in the MSP form of F, then a_1 is positive. Similar reasoning shows that if $\bar{x}_1$ appears in the MSP form, then a_1 is negative. The exact form of the result is set forth in the following theorem.

Theorem 3.2 (*McNaughton*). If F is a linearly separable function, then F is unate. If

$$\sum_{j=1}^{n} a_jx_j$$

is a separating function for F, then, for each j, $a_j > 0$ (or $a_j < 0$) if and only if x_j (or $\bar{x}_j$) appears in the MSP form for F.†

Proof. Assume that a term T_1 is a product of variables, for instance

$$T_1 = x_jx_\alpha \ldots x_\beta\bar{x}_k\bar{x}_\delta \ldots \bar{x}_\gamma$$

and let p_α denote a combination of input values such that each literal in T_1 has unit value and all other terms are zero. Clearly p_α is a point such that $y(p_\alpha) = 1$; hence $f(p_\alpha) \geq u$.

Let p_α' denote the input combination similar to p_α except that x_j has value 0. Then $y(p_\alpha') = 0$, since if $y(p_\alpha)$ and $y(p_\alpha')$ were both unity, then T_1

† We assume that there is no variable in $\sum_{j=1}^{n} a_jx_j$ with sufficiently small coefficient so that F is independent of that variable. This condition is assumed throughout this book whether or not it is explicitly set forth.

would not be a term in the MSP expansion. Thus $f(p_\alpha') \le \ell$. However, since

$$f(p_\alpha) = a_j + f(p_\alpha') \ge u$$

and $u > \ell$, it follows that $a_j > 0$. This establishes that if x_j appears in a term of the MSP form, then a_j is positive.

Now consider the input combination p_α'', which is similar to p_α except that x_k has unit value. As before, $y(p_\alpha'') = 0$ and $f(p_\alpha'') \le \ell$. However

$$f(p_\alpha'') = a_k + f(p_\alpha) \le \ell \qquad \text{where} \quad f(p_\alpha) \ge u > \ell$$

Hence $a_k < 0$. This shows that if $\bar{x}_j$ appears in a term of the MSP form, then a_j is negative. Clearly a weight cannot be simultaneously positive and negative; therefore x_j cannot appear in one term when $\bar{x}_j$ appears in another. This implies that F is unate if it is ℓ.s., and also that the "if" portion of the second part of the theorem is valid. The "only if" part follows immediately by contradiction.▎

Example 3.2. Consider the switching function

$$y = \bar{x}_1 \dotplus x_2\bar{x}_3$$

which can be expressed in terms of a separating function as

$$y = \langle -2x_1 + x_2 - x_3 \rangle_{-1:-2}$$

Note that y is unate and that the weights of x_1 and x_3 are negative since these variables appear complemented in the MSP form of F.

The previous theorem related the form of the variables in the MSP form of an ℓ.s. switching function to the sign of the coefficients in the separating-function representation. We now show how the signs of the coefficients are related to a subset of the reduced switching functions. With this relationship, the signs of the coefficients can be determined independently of whether the Boolean function is in MSP form.

As a simple example of this, consider the ℓ.s. function

$$y = F(x_1,x_2,x_3) = \langle x_1 + x_2 + x_3 \rangle_2$$

From this we can immediately write realizations for the reduced functions $y(x_1 = 0)$ and $y(x_1 = 1)$:

$$y(x_1 = 0) = F(0,x_2,x_3) = \langle 0 + x_2 + x_3 \rangle_2$$

and

$$\begin{aligned} y(x_1 = 1) = F(1,x_2,x_3) &= \langle 1 + x_2 + x_3 \rangle_2 \\ &= \langle 0 + x_2 + x_3 \rangle_1 \end{aligned}$$

where Theorem 2.4B was used to obtain the final realization. Thus $y(x_1 = 0)$ and $y(x_1 = 1)$ are both ℓ.s. and can have the same separating function. Hence they must be comparable (Theorem 3.1).

Furthermore, without even computing the reduced functions, we can use Theorem 3.1 to determine that

$$y(x_1 = 1) > y(x_1 = 0)$$

since the threshold value for $y(x_1 = 1)$ is less than that for $y(x_1 = 0)$. This can easily be verified, since

$$\begin{aligned} y &= x_1(x_2 \dotplus x_3) \dotplus x_2x_3 \\ y(x_1 = 0) &= x_2x_3 \\ y(x_1 = 1) &= x_2 \dotplus x_3 \end{aligned}$$

Looking only at the realizations, the reason that the threshold for $y(x_1 = 1)$ is less than that for $y(x_1 = 0)$ is that the coefficient of x_1 is positive, and hence decreased the threshold value when Theorem 2.4B was applied. If this coefficient had been negative, then the threshold value for $y(x_1 = 1)$ would have been larger than that of $y(x_1 = 0)$ and the functional containment would have been reversed. Thus the direction of the containment between the reduced functions $y(x_1 = 1)$ and $y(x_1 = 0)$ is related to the sign of a_1.

To investigate this in more detail, consider the general ℓ.s. switching function

$$y = F(x_1, x_2, \ldots, x_n) = \left\langle \sum_{j=1}^{n} a_jx_j \right\rangle_{u:\ell}$$

and form the reduced functions $y(x_1 = 0)$ and $y(x_1 = 1)$. It follows immediately that

$$y(x_1 = 0) = F(0, x_2, \ldots, x_n) = \left\langle \sum_{j=2}^{n} a_jx_j \right\rangle_{u:\ell}$$

and

$$y(x_1 = 1) = F(1, x_2, \ldots, x_n) = \left\langle a_1 + \sum_{j=2}^{n} a_jx_j \right\rangle_{u:\ell}$$

It is evident that $y(x_1 = 0)$ and $y(x_1 = 1)$ are ℓ.s. functions of the variables $x_2, x_3, \ldots, x_n$ and that a separating function for each can be obtained from the separating function for F. With the separating function $\sum_{j=2}^{n} a_jx_j$, the map of $y(x_1 = 0)$ is separated by the interval denoted by $u:\ell$ and the map of $y(x_1 = 1)$ is separated by the interval $(u - a_1):(\ell - a_1)$ (Theorem 2.4B). However, the actual gaps of these functions may be somewhat larger than (but must include) these intervals, as will become apparent from the subsequent discussion. The reason for this is that the spaces on which the reduced functions are defined are proper disjoint subcubes of the original n cube.

In terms of the original n cube, the values of

$$\sum_{j=2}^{n} a_j x_j \qquad \text{and} \quad a_1 + \sum_{j=2}^{n} a_j x_j$$

correspond to values of $f(p)$ at points such that $x_1 = 0$ and $x_1 = 1$ respectively. Each, therefore, is defined on (a disjoint) half of the n-cube points. If u and ℓ are determined by values of p such that $x_1 = 0$, then $u:\ell$ is the gap for

$$y(x_1 = 0) = \left\langle \sum_{j=2}^{n} a_j x_j \right\rangle_{u:\ell}$$

Otherwise the gap for

$$y(x_1 = 0) = \left\langle \sum_{j=2}^{n} a_j x_j \right\rangle_{u_0:\ell_0}$$

is $u_0:\ell_0$ where

$$u_0 = \min \left\{ \sum_{j=2}^{n} a_j x_j \right\}$$

such that

$$\sum_{j=2}^{n} a_j x_j \geq u$$

and

$$\ell_0 = \max \left\{ \sum_{j=2}^{n} a_j x_j \right\}$$

such that

$$\sum_{j=2}^{n} a_j x_j \leq \ell$$

In other words, the gap of $y(x_1 = 0)$ is the smallest gap that contains $u:\ell$ and is bounded by points for which $x_1 = 0$.

Similarly, the gap for

$$y(x_1 = 1) = \left\langle a_1 + \sum_{j=2}^{n} a_j x_j \right\rangle_{u:\ell}$$

is $u{:}\ell$ if u and ℓ are determined by points for which $x_1 = 1$, and otherwise

$$y(x_1 = 1) = \left\langle a_1 + \sum_{j=2}^{n} a_j x_j \right\rangle_{u_{1'}:\ell_{1'}}$$

has a gap $u_{1'}{:}\ell_{1'}$ where

$$u_{1'} = \min\left\{a_1 + \sum_{j=2}^{n} a_j x_j\right\}$$

such that

$$a_1 + \sum_{j=2}^{n} a_j x_j \geq u$$

and

$$\ell_1 = \max\left\{a_1 + \sum_{j=2}^{n} a_j x_j\right\}$$

such that

$$a_{1'} + \sum_{j=2}^{n} a_j x_j \leq \ell$$

In other words, the gap of $y(x_1 = 1)$ is the smallest gap that contains $u{:}\ell$ and is bounded by points for which $x_1 = 1$.

In either case, for both reduced functions, the gaps of the given separating functions always contain the gap of the original function when they are not equal. These results are summarized in Theorem 3.3 and illustrated by Example 3.3.

Theorem 3.3. If $y = F(x_1,x_2,\ldots,x_n)$ is an ℓ.s. switching function and if

$$y = \langle f(x_1,\ldots,x_{j-1},x_j,x_{j+1},\ldots,x_n)\rangle_{u:\ell}$$

then for each j the reduced functions $y(x_j = 0)$ and $y(x_j = 1)$ are both ℓ.s. and can have the same separating function $f(x_1,\ldots,x_{j-1},0,x_{j+1},\ldots,x_n)$ where

$$y(x_j = 0) = \langle f(x_1,\ldots,x_{j-1},0,x_{j+1},\ldots,x_n)\rangle_{u_0:\ell_0}$$

$$y(x_j = 1) = \langle f(x_1,\ldots,x_{j-1},0,x_{j+1},\ldots,x_n)\rangle_{u_1:\ell_1}$$

where u_0 = minimum value of $f(x_1,\ldots,x_{j-1},0,x_{j+1},\ldots,x_n) \geq u$
if such a value exists
$= \infty$, otherwise

u_1 = minimum value of $f(x_1,\ldots,x_{j-1},0,x_{j+1},\ldots,x_n) \geq u - a_j$
if such a value exists
$= \infty$, otherwise

ℓ_0 = maximum value of $f(x_1,\ldots,x_{j-1},0,x_{j+1},\ldots,x_n) \leq \ell$
if such a value exists
$= -\infty$, otherwise

ℓ_1 = maximum value of $f(x_1,\ldots,x_{j-1},0,x_{j+1},\ldots,x_n) \leq \ell - a_j$
if such a value exists
$= -\infty$, otherwise.

Proof. The proof for $y(x_j = 0)$ follows immediately from the above discussion. The proof for $y(x_j = 1)$ similarly follows. In this case, however, Theorem 2.4B is used for the transformation

$$\begin{aligned} y(x_j = 1) &= \langle a_j + f(x_1,\ldots,x_{j-1},0,x_{j+1},\ldots,x_n)\rangle_{u_{1'}:\ell_{1'}} \\ &= \langle f(x_1,\ldots,x_{j-1},0,x_{j+1},\ldots,x_n)\rangle_{(u_{1'}-a_j):(\ell_{1'}-a_j)} \end{aligned}$$

where $u_1:\ell_1 = (u_{1'} - a_j):(\ell_{1'} - a_j)$.▌

Example 3.3. Let

$$y = x_1 \dotplus x_2x_3 = \langle 2x_1 + x_2 + x_3\rangle_{2:1}$$

Setting $x_1 = 0$, we get the reduced function

$$y(x_1 = 0) = x_2x_3 = \langle x_2 + x_3\rangle_{u_0:\ell_0}$$

In this case the minimum and maximum values of $x_2 + x_3$ such that $x_2x_3 = 1$ and $x_2x_3 = 0$ are 2 and 1 respectively; hence

$$u_0:\ell_0 = u:\ell = 2:1$$

Setting $x_1 = 1$, we get the reduced function

$$y(x_1 = 1) \equiv 1 = \langle x_2 + x_3\rangle_{u_1:\ell_1}$$

where

$$u_1:\ell_1 = 0:-\infty$$

Note that

$$u_1 = 0 \geq u - a_1 = 0 \quad \text{and} \quad \ell_1 = -\infty \leq \ell - a_1 = -1$$

In this case the gap for $y(x_1 = 1)$ is larger than the gap for y.

The consideration of separating-function representations of reduced functions is simplified considerably if the gap is replaced by the specific threshold value t where

$$u \geq t > \ell$$

The results are given in the following corollary.

Corollary 3.3.1. If

$$y = F(x_1,x_2,\ldots,x_n) = \left\langle \sum_{j=1}^{n} a_jx_j \right\rangle_t$$

then

$$y(x_1 = 0) = \left\langle \sum_{j=2}^{n} a_jx_j \right\rangle_t$$

$$y(x_1 = 1) = \left\langle a_1 + \sum_{j=2}^{n} a_jx_j \right\rangle_t = \left\langle \sum_{j=2}^{n} a_jx_j \right\rangle_{t-a_1}$$

The last identity follows from Theorem 2.4B. The result shows that a T-gate realization for y is converted to a T-gate realization for $y(x_1 = 0)$ by setting $x_1 = 0$. Similarly, it is converted to a T-gate realization for $y(x_1 = 1)$ by setting $x_1 = 0$ and changing the threshold value from t to $t - a_1$. It is evident from the discussion in terms of gaps, however, that if t is selected for maximum allowable parameter fluctuation for the realization of y, the values given for $y(x_1 = 0)$ and $y(x_1 = 1)$ are not necessarily optimal.

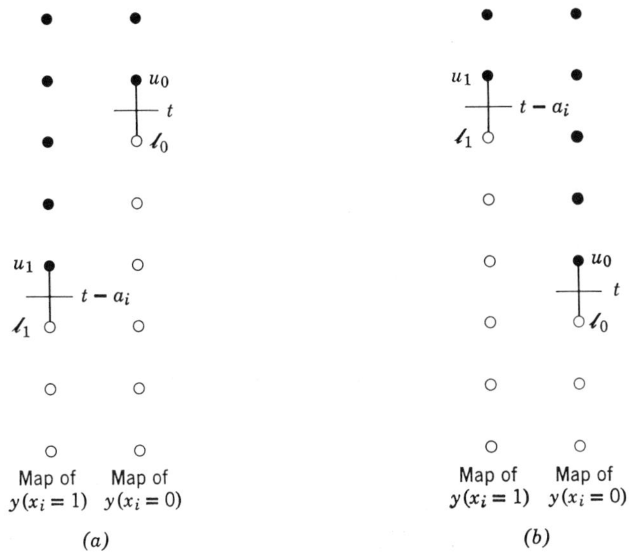

Fig. 3.2. Maps showing the relation between functional containment and coefficient sign. (*a*) Maps when $y(x_i = 1) > y(x_i = 0)$. (*b*) Maps when $y(x_i = 1) < y(x_i = 0)$.

If $y = F(x_1,x_2,\ldots,x_n)$ is an ℓ.s. switching function, then Theorem 3.3 establishes that $y(x_1 = 1)$ and $y(x_1 = 0)$ are ℓ.s. and can have the same separating function, but with different gaps. From Theorem 3.1 it follows that $y(x_1 = 0)$ and $y(x_1 = 1)$ are comparable. These results are combined to establish how the signs of the coefficients of the separating function for F are specified by the comparability of the reduced functions. This is given by Corollary 3.3.2 and depicted in Fig. 3.2.

Corollary 3.3.2. *(Paull and McCluskey).* If $y = F(x_1,x_2,\ldots,x_n)$ is ℓ.s. with separating function $\sum_{j=1}^{n} a_j x_j$, then for each i, $y(x_i = 0)$ and $y(x_i = 1)$ are comparable and $a_i > 0$ (or $a_i < 0$) if and only if

$$y(x_i = 1) > y(x_i = 0) \qquad (\text{or } y(x_i = 1) < y(x_i = 0))$$

Proof. If

$$y(x_i = 1) > y(x_i = 0)$$

then since $y = \langle f(x_1,x_2,\ldots,x_n)\rangle_t$ it follows from Theorem 3.3 and Corollary 3.3.1 that

$$y(x_i = 1) = \langle f(x_i = 0)\rangle_{t-a_i} > y(x_i = 0) = \langle f(x_i = 0)\rangle_t$$

where

$$\ell_1 < t - a_i \leq u_1 \qquad \text{and} \quad \ell_0 < t \leq u_0$$

By Theorem 3.1 $u_1 < u_0$; hence $t - a_i < t$ and $a_i > 0$. Similarly

$$y(x_i = 1) = \langle f(x_i = 0)\rangle_{t-a_i} < y(x_i = 0) = \langle f(x_i = 0)\rangle_t$$

implies that $t < t - a_i$ and $a_i < 0$. The converse follows by the reverse argument.|

Switching functions $y = F(x_1,x_2,\ldots,x_n)$, which have the property that for any variable x_i, $y(x_i = 1)$ and $y(x_i = 0)$ are comparable, are called 1-*monotonic functions* (Winder, 1). We have shown that all ℓ.s. functions are 1-monotonic and that the sign of the coefficients in every separating function can be obtained from the containment. Furthermore, determining this containment does not require that the switching function be in MSP form.

Reference to Example 3.3 shows that

$$y(x_1 = 1) \equiv 1 > y(x_1 = 0) = x_2x_3$$

and that $a_1 > 0$. The signs for a_2 and a_3 could likewise be determined from the corresponding reduced-function comparability.

The following theorem establishes the equivalence of unateness and 1-monotonicity.

Theorem 3.4. (*Paull and McCluskey*). The switching function $y = F(x_1,x_2,\ldots,x_n)$ is 1-monotonic if and only if it is unate.

Proof. Recall that an arbitrary switching function can be expressed in terms of reduced functions as

$$y = F(x_1,x_2,\ldots,x_n) = x_iy(x_i = 1) \dotplus \bar{x}_iy(x_i = 0)$$

where x_i is any one of the variables. If

$$y(x_i = 1) > y(x_i = 0)$$

then

$$y(x_i = 1) = y(x_i = 1) \dotplus y(x_i = 0)$$

in which case the above expression for y becomes

$$\begin{aligned} y &= x_i[y(x_i = 1) \dotplus y(x_i = 0)] \dotplus \bar{x}_i y(x_i = 0) \\ &= x_i y(x_i = 1) \dotplus (x_i \dotplus \bar{x}_i) y(x_i = 0) \\ &= x_i y(x_i = 1) \dotplus y(x_i = 0) \end{aligned}$$

It follows immediately that the final form for y does not contain the variable $\bar{x}_i$. Similarly, if

$$y(x_i = 0) > y(x_i = 1)$$

then

$$y = y(x_i = 1) \dotplus \bar{x}_i y(x_i = 0)$$

and the variable x_i does not appear. Finally, if

$$y(x_i = 1) = y(x_i = 0)$$

then

$$y = (x_i \dotplus \bar{x}_i) y(x_i = 1) = y(x_i = 1)$$

and y is not a function of either x_i or $\bar{x}_i$. It follows immediately, therefore, that F is unate if it is 1-monotonic.

Given that F is unate, then for any variable (for instance, x_1), y can be expressed as

$$y = F(x_1,x_2,\ldots,x_n) = x_1 F_1(x_2,x_3,\ldots,x_n) \dotplus F_0(x_2,x_3,\ldots,x_n)$$

or

$$y = F(x_1,x_2,\ldots,x_n) = F_1(x_2,x_3,\ldots,x_n) \dotplus \bar{x}_1 F_0(x_2,x_3,\ldots,x_n)$$

according as the MSP form for F contains the variable x_1 or $\bar{x}_1$ respectively. From the first form it is evident that

$$\begin{aligned} y(x_1 = 1) &= F_1(x_2,x_3,\ldots,x_n) \dotplus F_0(x_2,x_3,\ldots,x_n) \\ y(x_1 = 0) &= F_0(x_2,x_3,\ldots,x_n) \end{aligned}$$

and that

$$y(x_1 = 1) \geq y(x_1 = 0)$$

From the second expression,

$$\begin{aligned} y(x_1 = 1) &= F_1(x_2,x_3,\ldots,x_n) \\ y(x_1 = 0) &= F_1(x_2,x_3,\ldots,x_n) \dotplus F_0(x_2,x_3,\ldots,x_n) \end{aligned}$$

and

$$y(x_1 = 0) \geq y(x_1 = 1)$$

Therefore, F is 1-monotonic if it is unate.|

The following result was established by the proof of Theorem 3.4.

Corollary 3.4.1. (*Paull and McCluskey*). In the MSP form of a unate function $y = F(x_1,x_2,\ldots,x_n)$, x_i or $\bar{x}_i$ appears according as $y(x_i = 1) > y(x_i = 0)$ or $y(x_i = 0) > y(x_i = 1)$ and conversely.

In summary, the following properties have been established:

(1) If

$$y = F(x_1,x_2,\ldots,x_n) = \langle f(x_1,x_2,\ldots,x_n)\rangle_{u:\ell}$$

is ℓ.s., then for any variable, say x_1, $y(x_1 = 1)$ and $y(x_1 = 0)$ are ℓ.s. and both can have the separating function $f(0,x_2,\ldots,x_n)$.

(2) Unateness and 1-monotonicity are equivalent and are properties of all ℓ.s. functions.

(3) If $y = F(x_1,x_2,\ldots,x_n)$ is ℓ.s., then for each variable x_i,

$$y(x_i = 1) > y(x_i = 0) \quad \text{or} \quad y(x_i = 0) > y(x_i = 1)$$

according as the MSP form for F contains the variable x_i or $\bar{x}_i$ respectively, and conversely.

(4) If

$$y = F(x_1,x_2,\ldots,x_n) = \left\langle \sum_{j=1}^{n} a_j x_j \right\rangle_{u:\ell}$$

is ℓ.s. then for each x_i, $a_i > 0$ or $a_i < 0$ according as the MSP form for F contains the variable x_i or $\bar{x}_i$, and conversely.

3.3 PROPERTIES OF UNATE FUNCTIONS

All ℓ.s. switching functions are unate; therefore a test for determining whether a function is of the ℓ.s. class might first determine whether the function is unate. Although unateness does not imply linear separability, a function is not ℓ.s. if it is not unate. This will be discussed further in later sections.

In considering ℓ.s. switching functions it is convenient to know that the coefficients of the separating function are positive. If the MSP form of the switching function contains only uncomplemented variables, then it is called a *positive unate function* or just a *positive function*. Variables of unate functions are termed positive or negative according as they appear uncomplemented or complemented in the MSP form. The coefficients of the variables $x_1,x_2,\ldots,x_n$ of the separating-function representation of an ℓ.s. function are positive or negative according as the variables in the MSP form are positive or negative respectively. Thus a positive ℓ.s. function has all coefficients positive.

If an arbitrary unate function is not positive in some of its variables, then there is always a positive unate function in its complementing symmetry class (see Section 2.5) that is ℓ.s. if and only if the original function is ℓ.s. Moreover, a simple transformation converts a T-gate realization of the positive ℓ.s. function to a realization of any other member of the class. The positive unate function is called the *representative function* of the class. No loss of generality results by considering only positive ℓ.s. functions.

For an arbitrarily selected unate switching function the representative function is obtained by replacing each complemented variable in the MSP form by the same variable uncomplemented. Of course the two functions are not the same, but they are equivalent for realization purposes. To convert a separating-function representation of the representative function to that for the original switching function, complement each separating-function variable that is complemented in the original function. The resultant is a separating-function representation for the original switching function where some variables are complemented and some are uncomplemented. The following procedure is used to convert this to a separating-function representation in terms of the variables $x_1,x_2,\ldots,x_n$.

Procedure 3.1

(1) Replace each complemented variable $\bar{x}_i$ by $(1 - x_i)$ according to Theorem 2.3. Hence $a_i\bar{x}_i$ becomes $a_i - a_ix_i$.

(2) Alter the gap to remove all constants according to Theorem 2.4.

Example 3.4. The ℓ.s. function

$$y = \bar{x}_1 \dotplus x_2\bar{x}_3$$

has the function

$$y_1 = x_1 \dotplus x_2x_3$$

as the representative function of its complementing symmetry class. A separating-function representation for y_1 is

$$y_1 = \langle 2x_1 + x_2 + x_3 \rangle_{2:1}$$

Therefore, the corresponding representation for y is

$$y = \langle 2\bar{x}_1 + x_2 + \bar{x}_3 \rangle_{2:1}$$

Using Procedure 3.1, the latter representation is expressed in terms of uncomplemented variables as follows:

$$y = \langle 2(1 - x_1) + x_2 + (1 - x_3)\rangle_{2:1} = \langle -2x_1 + x_2 - x_3 + 3\rangle_{2:1}$$

or

$$y = \langle -2x_1 + x_2 - x_3\rangle_{-1:-2}$$

Although all ℓ.s. switching functions are unate, all unate functions are not linearly separable. For all functions of three or fewer variables, however, unateness implies linear separability. This is readily verified by considering all such unate functions. Table 3.1 lists all positive unate switching functions of one, two, and three variables (except for variable permutations) and gives the separating-function representation with maximum relative gap and minimum fan-in when the smallest coefficient is unity. It is interesting to note that all realizations can be obtained by the methods of Chapter 2.

Table 3.1. All one-, two-, and three-variable positive ℓ.s. functions and realizations

Switching Function	Separating-Function Representation
$y = x_1$	$y = \langle x_1\rangle_{1:0}$
$y = x_1 \dotplus x_2$	$y = \langle x_1 + x_2\rangle_{1:0}$
$y = x_1x_2$	$y = \langle x_1 + x_2\rangle_{2:1}$
$y = x_1 \dotplus x_2 \dotplus x_3$	$y = \langle x_1 + x_2 + x_3\rangle_{1:0}$
$y = x_1x_2 \dotplus x_2x_3 \dotplus x_1x_3$	$y = \langle x_1 + x_2 + x_3\rangle_{2:1}$
$y = x_1x_2x_3$	$y = \langle x_1 + x_2 + x_3\rangle_{3:2}$
$y = x_1 \dotplus x_2x_3$	$y = \langle 2x_1 + x_2 + x_3\rangle_{2:1}$
$y = x_1(x_2 \dotplus x_3)$	$y = \langle 2x_1 + x_2 + x_3\rangle_{3:2}$

By considering all possible four-variable, positive unate functions, one finds only three, ignoring variable permutations, that are not ℓ.s. These are the functions

$$y_1 = x_1x_2 \dotplus x_3x_4$$

$$y_1{}^d = (x_1 \dotplus x_2)(x_3 \dotplus x_4) = x_1x_3 \dotplus x_1x_4 \dotplus x_2x_3 \dotplus x_2x_4$$

and

$$y_2 = x_1x_2 \dotplus x_1x_3 \dotplus x_3x_4$$

Note that the dual of y_2 is the same function as y_2 but with the variables permuted.

Occasionally, the knowledge that a particular function is not ℓ.s. is useful in determining that a larger function is likewise not a threshold function. This is illustrated in the following example.

Example 3.5. Consider the switching function

$$y = x_1(x_2x_3 + x_4x_5) + x_2x_3x_4x_5$$

One of its single-variable reduced functions is

$$y(x_1 = 1) = x_2x_3 + x_4x_5$$

which was previously shown to be not ℓ.s. (Example 1.3). Recall, however, that if y is ℓ.s. then this reduced function $y(x_1 = 1)$ must be ℓ.s. It must be concluded, therefore, that the original function is not linearly separable.

If a variable, say x_1, appears uncomplemented in the MSP form of a unate function $y = F(x_1,x_2,\ldots,x_n)$, then y can be written as

$$y = x_1F_1(x_2,x_3,\ldots,x_n) + F_0(x_2,x_3,\ldots,x_n)$$

where $x_1F_1(x_2,x_3,\ldots,x_n)$ consists of those terms of the MSP form which have x_1 as a factor, and $F_0(x_2,x_3,\ldots,x_n)$ consists of the other terms. Clearly the reduced functions are

$$y(x_1 = 1) = F_1(x_2,x_3,\ldots,x_n) + F_0(x_2,x_3,\ldots,x_n)$$
$$y(x_1 = 0) = F_0(x_2,x_3,\ldots,x_n)$$

We now show that the MSP form of a unate function is unique. This, of course, implies that it consists of the sum of all of its prime implicants. Recall that this is not necessarily the case for nonunate functions.

For example,

$$y = \bar{x}_1\bar{x}_3x_4 + x_1\bar{x}_2x_4 + x_1x_3\bar{x}_4 + \bar{x}_1x_2\bar{x}_4$$

and

$$y = \bar{x}_2\bar{x}_3x_4 + x_1\bar{x}_2x_3 + x_2x_3\bar{x}_4 + \bar{x}_1x_2\bar{x}_3$$

are both MSP forms of the same function.

Theorem 3.5. (*Quine*). The MSP form of a unate switching function is unique.

Proof. Assume that $y = F(x_1,x_2,\ldots,x_n)$ is a unate switching function with two MSP forms denoted by

$$y_1 = T_{11} + T_{12} + \cdots + T_{1q}$$
$$y_2 = T_{21} + T_{22} + \cdots + T_{2q}$$

where each T_{ij} is a product of variables corresponding to a prime implicant. No loss of generality results by assuming that the y_1 form has only positive variables. For each i, x_i appears uncomplemented in y_1; hence by Corollary 3.4.1,

$$y_1(x_i = 1) > y_1(x_i = 0)$$

Since $y_1 = y_2$, it is necessary that

$$y_2(x_i = 1) > y_2(x_i = 0)$$

and, according to Corollary 3.4.1, that x_i appear uncomplemented in the y_2 form. Hence all variables in y_2 are uncomplemented.

Now consider any term of y_1, say T_{11}, and for convenience relabel all variables of both forms so that

$$T_{11} = x_1x_2\ldots x_k$$

Furthermore, consider the combination of values p_α such that

$$x_1 = x_2 = \cdots = x_k = 1 \qquad \text{and} \quad x_{k+1} = x_{k+2} = \cdots = x_n = 0$$

Clearly for this p_α, $T_{11} = 1$ and there is at least one term of y_2, say T_{21}, that has unit value. Since T_{21} is a product of uncomplemented variables, either $T_{21} = T_{11}$ for all p or $T_{21} > T_{11}$. In the latter case, however, it would consist of a proper subset of the variables of T_{11}. If, for example, x_1 were a variable not in T_{21}, then $T_{21} = 1$ and $T_{11} = 0$ for the combination of values

$$x_2 = x_3 = \cdots = x_k = 1 \qquad \text{and} \quad x_1 = x_{k+1} = x_{k+2} = \cdots = x_n = 0$$

For this combination, however, $y_1 = 1$; hence there must be some other term of y_1, for instance T_{12}, where

$$T_{12} = x_2x_3\ldots x_k$$

or T_{12} is a product of a subset of the variables $x_2, x_3, \ldots, x_k$. In either case, $T_{12} \geq T_{11}$, and thus T_{11} is an unnecessary term and y_1 is not an MSP form. Since this contradicts the original hypothesis, one must conclude that $T_{11} = T_{21}$, or more generally that to every term in y_1 there is an identical term in y_2 and conversely. This implies that the two forms are identical and that the MSP form for a unate function is unique.|

3.4 REDUCED FUNCTIONS AND COEFFICIENT ORDERING

In the previous section we showed how the comparability of single-variable reduced functions implies the sign of the coefficients in every realization with input variables $x_1, x_2, \ldots, x_n$ when the original switching function is linearly separable. This section is concerned with how reduced-function comparability implies a necessary ordering of the coefficients.

Let

$$y = F(x_1,x_2,\ldots,x_n) = \langle a_1x_1 + a_2x_2 + f'(x_3,x_4,\ldots,x_n)\rangle_t$$

represent a positive ℓ.s. function, and consider the reduced functions

$$y(x_1 = 1,x_2 = 0) = \langle a_1 + f'(x_3,x_4,\ldots,x_n)\rangle_t = \langle f'(x_3,x_4,\ldots,x_n)\rangle_{t-a_1}$$
$$y(x_1 = 0,x_2 = 1) = \langle a_2 + f'(x_3,x_4,\ldots,x_n)\rangle_t = \langle f'(x_3,x_4,\ldots,x_n)\rangle_{t-a_2}$$

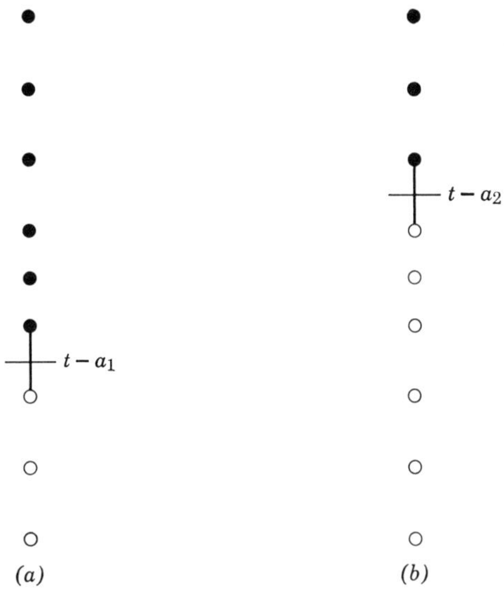

Fig. 3.3. Maps of two-variable reduced functions when $y(x_1 = 1, x_2 = 0) > y(x_1 = 0, x_2 = 1)$. (*a*) Map of $y(x_1 = 1, x_2 = 0) = \langle f'(x_3,x_4,\ldots,x_n)\rangle_{t-a_1}$. (*b*) Map of $y(x_1 = 0, x_2 = 1) = \langle f'(x_3,x_4,\ldots,x_n)\rangle_{t-a_2}$.

It is evident that both of these reduced functions can have the same separating function, but with different thresholds. This is illustrated in Fig. 3.3. According to Theorem 3.1, these reduced functions are comparable, and if

$$y(x_1 = 1,x_2 = 0) > y(x_1 = 0,x_2 = 1)$$

then

$$t - a_1 < t - a_2 \qquad \text{or} \qquad a_1 > a_2$$

Similarly

$$y(x_1 = 1,x_2 = 0) < y(x_1 = 0,x_2 = 1)$$

implies $a_1 < a_2$. Thus, reduced-function comparability implies coefficient ordering, and we have established the following result.

Theorem 3.6. (*Paull and McCluskey*). If $y = F(x_1,x_2,\ldots,x_n)$ is a positive ℓ.s. switching function, then for any x_i and x_j the reduced functions $y(x_i = 1,x_j = 0)$ and $y(x_i = 0,x_j = 1)$ are comparable. Moreover if

$$y(x_i = 1,x_j = 0) > y(x_i = 0,x_j = 1)$$

then $a_i > a_j$ and if

$$y(x_i = 1,x_j = 0) < y(x_i = 0,x_j = 1)$$

then $a_i < a_j$ in every realization for F.

Actually a somewhat stronger statement follows immediately. (For a discussion of the gaps of reduced functions, refer to Theorem 3.3 and the discussion which precedes it.)

Corollary 3.6.1. If

$$y = F(x_1,x_2,\ldots,x_n) = \langle f(x_1,x_2,\ldots,x_n)\rangle_{u:\ell}$$

is an ℓ.s. switching function, then for any pair of variables, say x_1 and x_2, the reduced functions $y(x_1 = 1,x_2 = 0)$ and $y(x_1 = 0,x_2 = 1)$ are ℓ.s. and can have the same separating function. In particular,

$$y(x_1 = 1,x_2 = 0) = \langle f(0,0,x_3,x_4,\ldots,x_n)\rangle_{u_{10}:\ell_{10}}$$

where u_{10} = minimum value of $f(0,0,x_3,x_4,\ldots,x_n) \geq u - a_1$ if such a value exists
= ∞, otherwise

ℓ_{10} = maximum value of $f(0,0,x_3,x_4,\ldots,x_n) \leq \ell - a_1$ if such a value exists
= $-\infty$, otherwise.

and

$$y(x_1 = 0,x_2 = 1) = \langle f(0,0,x_3,x_4,\ldots,x_n)\rangle_{u_{01}:\ell_{01}}$$

where u_{01} = minimum value of $f(0,0,x_3,x_4,\ldots,x_n) \geq u - a_2$ if such a value exists
= ∞, otherwise

ℓ_{01} = maximum value of $f(0,0,x_3,x_4,\ldots,x_n) \leq \ell - a_2$ if such a value exists
= $-\infty$ otherwise

Example 3.6. Consider the positive ℓ.s. switching function

$$y = x_1(x_2 \dotplus x_3 \dotplus x_4) \dotplus x_2x_3x_4$$

and the corresponding reduced functions

$$y(x_1 = 1, x_2 = 0) = x_3 \dotplus x_4$$
$$y(x_1 = 0, x_2 = 1) = x_3 x_4$$

Clearly

$$y(x_1 = 1, x_2 = 0) = x_3 \dotplus x_4 > x_3 x_4 = y(x_1 = 0, x_2 = 1)$$

Hence in every separating function for y, $a_1 > a_2$. It is readily verified that

$$y = \langle 2x_1 + x_2 + x_3 + x_4 \rangle_{3:2}$$

and that

$$y(x_1 = 1, x_2 = 0) = \langle x_3 + x_4 \rangle_{1:0}$$

and

$$y(x_1 = 0, x_2 = 1) = \langle x_3 + x_4 \rangle_{2:1}$$

Example 3.7. Consider the positive unate switching function

$$y = x_1 x_2 \dotplus x_1 x_3 \dotplus x_3 x_4$$

which was listed in Section 3.3 as one of the non-ℓ.s. four-variable functions. To show that it is not ℓ.s., consider the reduced functions

$$y(x_1 = 1, x_3 = 0) = x_2$$

and

$$y(x_1 = 0, x_3 = 1) = x_4$$

Since these functions are not comparable, it follows that the original function y is not ℓ.s.

Example 3.8. Consider the positive unate switching function

$$y = x_1(x_2 \dotplus x_3 x_4 \dotplus x_3 x_5) \dotplus x_3(x_2 \dotplus x_4 x_5)$$

and determine the reduced functions

$$y(x_1 = 1) = x_2 \dotplus x_3 x_4 \dotplus x_3 x_5$$

and

$$y(x_1 = 0) = x_3(x_2 \dotplus x_4 x_5)$$

According to Theorem 3.3, if y is an ℓ.s. function, then both reduced functions are comparable and, in fact, are ℓ.s. It is easily verified that both functions satisfy both conditions. However, Theorem 3.3 also requires that both reduced functions be realizable with the same separating function. Applying Theorem 3.6 to each reduced function yields the following:

(1) $y(x_1 = 1, x_2 = 1, x_3 = 0) = 1 > x_4 + x_5 = y(x_1 = 1, x_2 = 0, x_3 = 1)$

Hence in every separating function for $y(x_1 = 1)$ it is necessary that $a_2 > a_3$.

(2) $y(x_1 = 0, x_2 = 1, x_3 = 0) = 0 < x_4 x_5 = y(x_1 = 0, x_2 = 0, x_3 = 1)$

Hence in every separating function for $y(x_1 = 0)$ it is necessary that $a_2 < a_3$.

Clearly, therefore, $y(x_1 = 1)$ and $y(x_1 = 0)$ can never have the same separating function, and y is not a linearly separable function.

In the function of Example 3.6 the reduced functions

$$y(x_2 = 1, x_3 = 0) = x_1 \quad \text{and} \quad y(x_2 = 0, x_3 = 1) = x_1$$

are found to be equal. Theorem 3.6 does not define an ordering for a_2 and a_3 in this case; however, in the separating function of the example $a_2 = a_3$, and one suspects that this might always be true. Actually this is the case, as shown by the next theorem.

Theorem 3.7. (*Winder*, 1). Let $y = F(x_1, x_2, \ldots, x_n)$ be an ℓ.s. function. If, for any pair of variables x_i and x_j,

$$y(x_i = 1, x_j = 0) = y(x_i = 0, x_j = 1)$$

then F is symmetric in x_i and x_j and conversely. Furthermore, if F is symmetric in m variables, then for every separating-function representation in which the coefficients of these m variables are unequal there is another separating-function representation that is identical to the first with the same relative gap and fan-in except that these m variables have equal coefficients. Finally, if the separating function has m variables with equal coefficients, then the switching function is symmetric in these variables.

Proof. Without loss of generality we let $x_i = x_1$ and $x_j = x_2$. Then the MSP form of the original function can be written as

$$y = x_1 x_2 F_a(x_3, x_4, \ldots, x_n) + x_1 F_b(x_3, x_4, \ldots, x_n) + x_2 F_c(x_3, x_4, \ldots, x_n) + F_d(x_3, x_4, \ldots, x_n)$$

The reduced functions are

$$y(x_1 = 1, x_2 = 0) = F_b(x_3, x_4, \ldots, x_n) + F_d(x_3, x_4, \ldots, x_n)$$
$$y(x_1 = 0, x_2 = 1) = F_c(x_3, x_4, \ldots, x_n) + F_d(x_3, x_4, \ldots, x_n)$$

If F is symmetric in x_1 and x_2, then

$$F_b(x_3,x_4,\ldots,x_n) = F_c(x_3,x_4,\ldots,x_n)$$

and

$$y(x_1 = 1,x_2 = 0) = y(x_1 = 0,x_2 = 1)$$

If

$$y(x_1 = 1,x_2 = 0) = y(x_1 = 0,x_2 = 1)$$

then, since $F_b(x_3,x_4,\ldots,x_n)$ and $F_c(x_3,x_4,\ldots,x_n)$ are unate and in MSP form, the same argument as used in the proof of Theorem 3.5 shows that

$$F_b(x_3,x_4,\ldots,x_n) = F_c(x_3,x_4,\ldots,x_n)$$

and that F is symmetric in x_1 and x_2.

To prove the next part, we need the following auxiliary result. Let the switching function $y_0 = G(x_1,x_2,\ldots,x_n)$ have two different separating-function representations, namely

$$y_0 = \langle f_1(x_1,x_2,\ldots,x_n)\rangle_{u_1:\ell_1}$$

and

$$y_0 = \langle f_2(x_1,x_2,\ldots,x_n)\rangle_{u_2:\ell_2}$$

It is evident that for any positive numbers k_1 and k_2,

$$y_0 = \langle k_1 f_1(x_1,x_2,\ldots,x_n) + k_2 f_2(x_1,x_2,\ldots,x_n)\rangle_{(k_1u_1+k_2u_2):(k_1\ell_1+k_2\ell_2)}$$

is also a separating-function representation of G.

Now assume that F is symmetric in the variables $x_1,x_2,\ldots,x_m$ and that one of the realizations is

$$y = \langle a_1x_1 + a_2x_2 + \cdots + a_mx_m + f'(x_{m+1},x_{m+2},\ldots,x_n)\rangle_{u:\ell}$$

Because of the symmetry of the m variables in F, there are $m - 1$ other realizations that are obtained by cyclically permuting the coefficients. Therefore

$$y = \langle a_2x_1 + a_3x_2 + \cdots + a_1x_m + f'(x_{m+1},x_{m+2},\ldots,x_n)\rangle_{u:\ell}$$
$$y = \langle a_3x_1 + a_4x_2 + \cdots + a_2x_m + f'(x_{m+1},x_{m+2},\ldots,x_n)\rangle_{u:\ell}$$
$$\vdots$$
$$y = \langle a_mx_1 + a_1x_2 + \cdots + a_{m-1}x_m + f'(x_{m+1},x_{m+2},\ldots,x_n)\rangle_{u:\ell}$$

By means of the auxiliary result, a new realization is obtained by adding all of the m separating functions and letting each $k = 1/m$.
Letting

$$\frac{1}{m}\sum_{j=1}^{m} a_j = a$$

we obtain the separating-function representation

$$y = \langle ax_1 + ax_2 + \cdots + ax_m + f'(x_{m+1}, x_{m+2}, \ldots, x_n) \rangle_{u:\ell}$$

The fan-in is the same as in the original realization, since

$$ma = \sum_{j=1}^{m} a_j$$

This part of the proof is therefore completed. The final part of the theorem is evident, since variables in the above representation with equal coefficients can be permuted without changing the value of y; hence the switching function is symmetric in these variables. This completes the proof.|

The results of Theorem 3.7 are actually not quite as strong as those of Theorem 3.6. From the latter, proper containment between reduced functions implies that all realizations of the function must have the indicated coefficient ordering. According to Theorem 3.7, if the reduced functions are equal, the coefficients are not restricted to being equal, nor is a necessary ordering specified. The theorem says only that there exists a realization with the same relative gap and the same fan-in as any other, but in which symmetric variables have equal coefficients. When the choice is ours we select a realization with equal weights for symmetric variables.

3.5 A COEFFICIENT ORDERING PROCEDURE—VARIABLE SIZE

By the repeated use of Theorems 3.6 and 3.7, it is possible to obtain a complete ordering of the coefficients of any separating function of a given ℓ.s. switching function. Checking reduced-function comparability, however, is a tedious and time-consuming activity for functions of more than a few variables. In this section another procedure is specified for determining coefficient ordering. Although this procedure is based upon the previous results, it is computationally much simpler. In fact, it requires only a determination of the number of times the variables appear in groups of terms of the MSP form of the switching function. The procedure is presented first and an example given, and then its validity is established.

Procedure 3.2. Coefficient Ordering (*Winder, 1*). This procedure is applicable to a positive unate ℓ.s. switching function F in MSP form.

(1) Divide the terms of F into subsets $G_1, G_2, \ldots, G_q$, where all terms in each G_i are products of the same number of variables, and for each i the number of variables in the terms of G_i is smaller than the number of variables in the terms of G_{i+1}. Clearly G_1 contains the terms with the fewest variables and G_q the terms with the most.

(2) Consider first the terms in G_1. Divide all variables of F into subsets, called *order-one subsets*, where all variables in each subset occur the same number of times in the terms of G_1. Assign a "greater than" ($>$) ordering to these subsets so that the subset of variables that occur the largest number of times in G_1 is greater than the subset of variables that occur the next-largest number of times in G_1, and so on for each subset. The last subset contains those variables of F that do not appear in the terms of G_1, if such variables exist.

(3) Consider next the terms in G_2. Divide each order-one subset into subsets called *order-two subsets* where all variables in each order-two subset appear the same number of times in the terms of G_2. Order these subsets as follows: maintain the partial ordering implied by the order-one subsets and within this ordering assign an additional greater-than ordering according to the number of times the variables in each subset appear in the terms of G_2 as described in Step (2).

(4) Continue this procedure until each subset contains exactly one variable or until all terms of F have been considered, in which case we have obtained an ordering for the q-order subsets. In either case the coefficients of every realization of F correspond to the ordering of the variables of q-order subsets. If any q-order subset contains more than one variable, then the function is symmetric in these variables, and the corresponding coefficients can be equal.

Example 3.9. Consider the positive ℓ.s. function

$$y = x_1x_2 \dotplus x_1x_3 \dotplus x_1x_4x_5 \dotplus x_1x_4x_6 \dotplus x_1x_5x_6 \dotplus x_2x_3x_4 \dotplus x_2x_3x_5 \dotplus x_2x_4x_5 \dotplus x_3x_4x_5x_6$$

(1) Then

$$\begin{aligned} G_1 &= \{x_1x_2,\ x_1x_3\} \\ G_2 &= \{x_1x_4x_5,\ x_1x_4x_6,\ x_1x_5x_6,\ x_2x_3x_4,\ x_2x_3x_5,\ x_2x_4x_5\} \\ G_3 &= \{x_3x_4x_5x_6\} \end{aligned}$$

(2) The order-one subsets are

$$\{x_1\},\ \{x_2,x_3\},\ \{x_4,x_5,x_6\}$$

which when greater-than ordered become

$$\{x_1\} > \{x_2,x_3\} > \{x_4,x_5,x_6\}$$

since x_1 occurs twice in G_1; x_2, and x_3 each occur once; and x_4, x_5, and x_6 do not appear in G_1.

(3) The order-two subsets are

$$\{x_1\},\ \{x_2\},\ \{x_3\},\ \{x_4,x_5\},\ \{x_6\}$$

The partial order implied by the greater-than ordering of the order-one subsets is

$$\{x_1\} > \{x_2\},\{x_3\} > \{x_4,x_5\},\{x_6\}$$

The final greater-than ordering of the order-two subsets is

$$\{x_1\} > \{x_2\} > \{x_3\} > \{x_4,x_5\} > \{x_6\}$$

where $\{x_2\} > \{x_3\}$ because x_2 appears three times in G_2 and x_3 appears twice, and where $\{x_4,x_5\} > \{x_6\}$ because x_4 and x_5 each appear four times in G_2 and x_6 appears twice.

(4) The order-three subsets are

$$\{x_1\},\ \{x_2\},\ \{x_3\},\ \{x_4,x_5\},\ \{x_6\}$$

The partial order implied by the greater-than ordering of the order-two subsets is

$$\{x_1\} > \{x_2\} > \{x_3\} > \{x_4,x_5\} > \{x_6\}$$

Since the order-three subsets are the same as the order-two subsets, no additional ordering is possible. Moreover, all terms of F have been used; hence the procedure is complete. The coefficient ordering that must be satisfied by all realizations of F is

$$a_1 > a_2 > a_3 > a_4 = a_5 > a_6$$

where coefficients that can be equal are indicated as being equal. This is more fully set forth in Theorem 3.7.

At this point it is necessary to establish the validity of the coefficient ordering procedure. The proof shows that for ℓ.s. functions Procedure 3.2 implies $a_i > a_j$ for any i and j if and only if

$$y(x_i = 1,x_j = 0) > y(x_i = 0,x_j = 1)$$

An exact statement of the theorem that implies this is the following.

Theorem 3.8. Let x_i and x_j be any two variables in the MSP form of a positive ℓ.s. switching function, $y = F(x_1,x_2,\ldots,x_n)$, and let k and K denote the smallest and the largest number of variables, respectively, in the terms of the MSP form of F. If and only if

$$y(x_i = 1,x_j = 0) > y(x_i = 0,x_j = 1)$$

then there exists some real integer λ where $k \leq \lambda \leq K$, so that in the subset of terms having exactly λ variables, x_i appears more times than x_j. Moreover, for every integer m, where $k \leq m < \lambda$, then in those subsets of terms with m variables, x_i and x_j appear the same number of times.

Proof. Since x_i and x_j can be any two variables, no loss of generality results from letting $x_i = x_1$ and $x_j = x_2$ for convenience of notation. Recall that $y = F(x_1,x_2,\ldots,x_n)$ and that it can be factored so that

$$y = x_1x_2F_a(x_3,x_4,\ldots,x_n) \dotplus x_1F_b(x_3,x_4,\ldots,x_n) \dotplus x_2F_c(x_3,x_4,\ldots,x_n) \dotplus F_d(x_3,x_4,\ldots,x_n)$$

Moreover

$$y(x_1 = 1,x_2 = 0) = F_b(x_3,x_4,\ldots,x_n) \dotplus F_d(x_3,x_4,\ldots,x_n) > y(x_1 = 0,x_2 = 1) = F_c(x_3,x_4,\ldots,x_n) \dotplus F_d(x_3,x_4,\ldots,x_n)$$

implies that

$$F_b > F_c$$

Since F is positive ℓ.s. and hence unate, and since it is in MSP form, then by the same reasoning as used in the proof of Theorem 3.5 it follows that to every term of F_c, there corresponds a term of F_b such that the F_b term contains the F_c term. If the containment is proper then the variables of the F_b term are a proper subset of the variables of the F_c term. Otherwise the two terms are identical.

In any step of the ordering procedure the only terms that can yield a different count for the number of times x_1 and x_2 appear are those from x_1F_b and x_2F_c since those from $x_1x_2F_a$ have both variables appearing the same number of times and those from F_d do not contain either x_1 or x_2. Since $F_b > F_c$, it was shown in the preceding paragraph that to each F_c term there corresponds an F_b term where the two are either identical or where the F_b term is a subset of the variables of the F_c term. Among the F_b terms that either have the latter (subset) property or to which there correspond no F_c terms the one with the smallest number of variables defines the number λ. This is immediately evident since there is at least one more F_b term with λ variables than there are F_c terms, and for all m where $k \leq m < \lambda$ there are the same number of terms in each since the terms are identical.

The proof of the converse will not be set forth in detail since it is essentially an inverse argument of the above proof.|

The proof of Theorem 3.8 establishes the validity of the following corollary. With this corollary the validity of Procedure 3.2 is established.

Corollary 3.8.1. Under the conditions of Theorem 3.8, if and only if

$$y(x_i = 1,x_j = 0) = y(x_i = 0,x_j = 1)$$

then in each subset of terms of F with λ variables where $k \leq \lambda \leq K$, the variables x_i and x_j appear the same number of times.

An important hypothesis of Theorem 3.8 is that the switching function is a positive ℓ.s. function. For such functions, the theorem establishes that the coefficient ordering by Procedure 3.2 implies corresponding reduced-function comparability. This is not true, however, if the function is not ℓ.s. Although the procedure assigns a coefficient ordering to non-ℓ.s. functions, the corresponding implied and actual reduced-function comparability may not correspond. Naturally the idea of coefficients is meaningless for non-ℓ.s. functions.

Actually, however, the q-order subset ordering is frequently useful even when the given function is not ℓ.s. In this case it does not imply coefficient ordering, but it does imply an ordering of the variables according to their "frequency of occurrence," which we call *variable size*. As a general terminology we say that the variables are *ordered according to size*, where this is equivalent to coefficient ordering when the function is ℓ.s. In the general terminology we refer to the largest or smallest variable, which for ℓ.s. functions is the variable with the largest or the smallest coefficient, respectively.

Although Procedure 3.2 was defined for positive unate functions, it can readily be generalized so that it is applicable to all switching functions expressed in MSP form. To do this requires only that every appearance of the word "variable" be replaced by the word "literal." In other words, each variable and its complement are treated as the same thing in the application of the procedure to non-ℓ.s. functions.

3.6 THE LARGEST AND THE SMALLEST COEFFICIENTS

The variable with the largest coefficient assumes a particularly important part in the synthesis procedure that will be presented later. This importance is due to a simplification in the form of functions which have been reduced with variables that are ordered according to the size of their corresponding coefficients. Let $y = F(x_1,x_2,\ldots,x_n)$ be a positive ℓ.s. switching function and let x_1 be a variable with corresponding coefficient that is at least as large as all other coefficients. Since F can be factored as

$$y = x_1F_1(x_2,x_3,\ldots,x_n) + F_0(x_2,x_3,\ldots,x_n)$$

it is evident that

$$y(x_1 = 1) = F_1(x_2,x_3,\ldots,x_n) + F_0(x_2,x_3,\ldots,x_n)$$
$$y(x_1 = 0) = F_0(x_2,x_3,\ldots,x_n)$$

We now show that, since x_1 is one of the variables with the largest coefficient, then

$$F_1 \geq F_0$$

and hence that

$$y(x_1 = 1) = F_1(x_2,x_3,\ldots,x_n)$$

Let x_iT be any term of $F_0(x_2,x_3,\ldots,x_n)$. If x_1 is one of the variables with the largest coefficient, then, since $a_1 \geq a_i$, x_1T has unit value for all values of $x_1,x_2,\ldots,x_n$ for which x_iT is unity. Clearly, therefore, T is a term of $F_1(x_2,x_3,\ldots,x_n)$ or is contained in a term of $F_1(x_2,x_3,\ldots,x_n)$. Since this is true for every term of $F_0(x_2,x_3,\ldots,x_n)$ it follows that

$$F_1 \geq F_0$$

which is the desired result.

Actually, a result similar to this is true even if the switching function is not ℓ.s. All that is required is that the function be positive unate and that

$$y(x_1 = 1, x_i = 0) \geq y(x_1 = 0, x_i = 1)$$

for $i = 2,3,\ldots,n$. This is given by Theorem 3.9.

Theorem 3.9 (*Muroga, 1*). Let

$$y = F(x_1,x_2,\ldots,x_n) = x_1F_1(x_2,x_3,\ldots,x_n) \dotplus F_0(x_2,x_3,\ldots,x_n)$$

be a positive unate switching function. If

$$y(x_1 = 1, x_i = 0) \geq y(x_1 = 0, x_i = 1)$$

for $i = 2,3,\ldots,n$; then

$$y(x_1 = 1) = F_1(x_2,x_3,\ldots,x_n)$$

Proof. Let x_iT be any irredundant product term of $F_0(x_2,x_3,\ldots,x_n)$ where x_i is an arbitrarily selected variable of the term. Expand the MSP form of the original function as

$$\begin{aligned} y = {} & x_1x_iF_{A_i}(x_2,\ldots,x_{i-1},x_{i+1},\ldots,x_n) \dotplus x_1F_{B_i}(x_2,\ldots,x_{i-1},x_{i+1},\ldots,x_n) \\ & \dotplus x_iF_{C_i}(x_2,\ldots,x_{i-1},x_{i+1},\ldots,x_n) \dotplus F_{D_i}(x_2,\ldots,x_{i-1},x_{i+1},\ldots,x_n) \end{aligned}$$

If

$$y(x_1 = 1, x_i = 0) \geq y(x_1 = 0, x_i = 1)$$

then

$$F_{B_i} \dotplus F_{D_i} \geq F_{C_i} \dotplus F_{D_i}$$

which implies, since y is unate, that

$$F_{B_i} \geq F_{C_i}$$

Now for each term in F_0, there must be an i such that the function $x_i F_{C_i}$ contains that term. Therefore,†

$$F_0 = \sum_i{}^{\boldsymbol{\cdot}} x_i F_{C_i}$$

Moreover, for each i,

$$F_{B_i} \geq F_{C_i} \quad \text{and} \quad F_1 = x_i F_{A_i} \dotplus F_{B_i}$$

Therefore each term of F_0 is contained in a term of F_1. This implies that

$$F_1 \geq F_0$$

and that

$$y(x_1 = 1) = F_1(x_2,x_3,\ldots,x_n) \dotplus F_0(x_2,x_3,\ldots,x_n) = F_1(x_2,x_3,\ldots,x_n) \quad \blacksquare$$

Example 3.10. The largest coefficient for the positive ℓ.s. switching function of Example 3.9 was shown to be a_1. This function can be expanded in the form

$$\begin{aligned} y &= x_1F_1(x_2,x_3,\ldots,x_6) \dotplus F_0(x_2,x_3,\ldots,x_6) \\ &= x_1(x_2 \dotplus x_3 \dotplus x_4x_5 \dotplus x_4x_6 \dotplus x_5x_6) \dotplus x_2x_3x_4 \dotplus x_2x_3x_5 \\ &\qquad \dotplus x_2x_4x_5 \dotplus x_3x_4x_5x_6 \end{aligned}$$

Since a_1 is the largest coefficient, then $y(x_1 = 1, x_i = 0) \geq y(x_1 = 0, x_i = 1)$ for $i = 2,3,\ldots,6$, and Theorem 3.9 specifies that

$$\begin{aligned} F_1 = x_2 \dotplus x_3 \dotplus x_4x_5 \dotplus x_4x_6 \dotplus x_5x_6 &\geq x_2x_3x_4 \dotplus x_2x_3x_5 \\ &\quad \dotplus x_2x_4x_5 \dotplus x_3x_4x_5x_6 = F_0 \end{aligned}$$

This is easily verified.

The variable with the smallest coefficient is of interest because it provides an upper bound on the gap length. Let

$$y = F(x_1,x_2,\ldots,x_n) = \langle f(x_1,x_2,\ldots,x_n)\rangle_{u:\ell}$$

be an MSP form of a positive ℓ.s. switching function where each of the variables $x_1,x_2,\ldots,x_n$ appears in the MSP expression.

Let x_n be the variable with the smallest coefficient. Since F is a function of the variable x_n, there must be at least one pair of values of the independent variables, p_α, $p_{\alpha'}$ which are identical except that the x_n value is unity for p_α and zero for $p_{\alpha'}$, and where

$$F(p_\alpha) = 1 \quad \text{and} \quad F(p_{\alpha'}) = 0$$

† $\sum^{\boldsymbol{\cdot}}$ denotes Boolean summation.

If this were not the case, then the MSP form of F would not contain the variable x_n. The above implies

$$f(p_\alpha) \geq u \qquad \text{and} \quad f(p_{\alpha'}) \leq \ell$$

where

$$f(p_\alpha) = a_n + f(p_{\alpha'})$$

Therefore

$$g = u - \ell \leq f(p_\alpha) - f(p_{\alpha'}) = a_n + f(p_{\alpha'}) - f(p_{\alpha'}) = a_n$$

Hence the gap length is bounded above by the value of the smallest coefficient. This bound can be used to obtain a bound for the relative gap $(u - \ell)/(u + \ell)$.

Theorem 3.10. If

$$y = F(x_1,x_2,\ldots,x_n) = \langle f(x_1,x_2,\ldots,x_n)\rangle_{u:\ell}$$

is a positive ℓ.s. function in MSP form and if K is the largest number of variables in any term of the MSP form, then in every single-gate realization of F,

$$\frac{u - \ell}{u + \ell} \leq \frac{1}{2K - 1}$$

Proof. When $K = 1$, F is the OR gate, in which case $u:\ell = a_n:0$ where a_n is the smallest coefficient, and the validity is evident.

Assume $K > 1$. Let T_K denote one of the terms of $F(x_1,x_2,\ldots,x_n)$ that contains K variables and let a_n be the smallest coefficient in every realization of F. Now for any combination of values of the independent variables, such that $(K - 1)$ of the variables of T_K are unity and all other variables are zero, the value of $F(x_1,x_2,\ldots,x_n)$ is zero. Since each of the $(K - 1)$ variables with unit values have coefficients greater than or equal to a_n, it follows that

$$(K - 1)a_n \leq \ell \qquad \text{or} \quad a_n \leq \frac{\ell}{K - 1}$$

But $a_n \geq u - \ell$; hence

$$u - \ell \leq \frac{\ell}{K - 1}$$

To complete the proof we must alter the form of this inequality as follows:

For some real number $\epsilon \geq 0$ it follows that

$$u - \ell = \frac{\ell}{K - 1}(1 - \epsilon)$$

Also

$$u + \ell = (u - \ell) + 2\ell = \frac{\ell}{K-1}(1 - \epsilon) + 2\ell = \frac{\ell}{K-1}(2K - 1 - \epsilon)$$

Hence

$$\frac{u-\ell}{u+\ell} = \frac{1-\epsilon}{2K-1-\epsilon}$$

Note, however, that for all $K \geq 1$ and all $\epsilon \geq 0$,

$$\frac{u-\ell}{u+\ell} = \frac{1-\epsilon}{2K-1-\epsilon} \leq \frac{1}{2K-1}$$

since ϵ is a larger percentage of 1 than it is of $2K - 1$. This completes the proof.|

Theorem 3.10 shows that $1/(2K - 1)$ is an upper bound for the relative gap. Given the set of all ℓ.s. functions, it is the *least* upper bound. To show this requires only that an ℓ.s. function be demonstrated for which the relative gap is $1/(2K - 1)$. The $y = M_K(x_1,x_2,\ldots,x_n)$ class of switching functions for all $K \leq n$ are examples. Such functions are ℓ.s. and can be expressed in terms of separating functions as

$$y = \left\langle \sum_{j=1}^{n} x_j \right\rangle_{K:(K-1)}$$

The relative gap in these cases is

$$\frac{u-\ell}{u+\ell} = \frac{K-(K-1)}{K+(K-1)} = \frac{1}{2K-1}$$

which is the same value as the upper bound. This establishes the following corollary.

> ***Corollary 3.10.1.*** The least upper bound for the relative gaps of the realizations of the class of positive ℓ.s. switching functions for which K is the maximum number of variables in any term of the MSP form, is $1/(2K - 1)$.

3.7 2-MONOTONICITY

Let

$$y = F(x_1,x_2,\ldots,x_n) = \langle f(x_1,x_2,\ldots,x_n)\rangle_t$$

be a positive ℓ.s. switching function. There are four reduced functions that can be formed from F by assigning values to any pair of variables, say x_1 and x_2. These are

$$y(x_1 = 1,x_2 = 0),\ y(x_1 = 0,x_2 = 1),\ y(x_1 = 1,x_2 = 1),\ y(x_1 = 0,x_2 = 0)$$

Of these, the first two imply an ordering of the coefficients a_1 and a_2 that must be satisfied by every single-gate realization. This was discussed in previous sections. We now show that many of these results apply to any pair of the four.

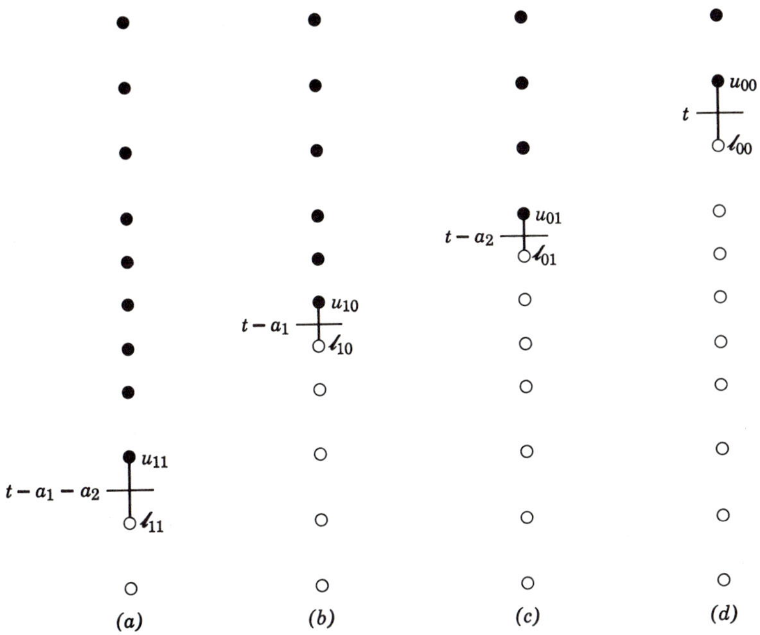

Fig. 3.4. Maps for the four two-variable reduced functions. (*a*) Map of $y(x_1 = 1, x_2 = 1)$. (*b*) Map of $y(x_1 = 1, x_2 = 0)$. (*c*) Map of $y(x_1 = 0, x_2 = 1)$. (*d*) Map of $y(x_1 = 0, x_2 = 0)$.

For a simplification of notation, we assume that x_1 and x_2 are any pair of variables. Also, we let

$$y = \langle f(x_1, x_2, \ldots, x_n)\rangle_t = \langle a_1x_1 + a_2x_2 + f''(x_3, x_4, \ldots, x_n)\rangle_t$$

Note, therefore, that

$$\begin{aligned} y(x_1 = 0, x_2 = 0) &= \langle f''(x_3, x_4, \ldots, x_n)\rangle_t \\ y(x_1 = 1, x_2 = 0) &= \langle a_1 + f''(x_3, x_4, \ldots, x_n)\rangle_t \\ &= \langle f''(x_3, x_4, \ldots, x_n)\rangle_{t-a_1} \qquad (3.2) \\ y(x_1 = 0, x_2 = 1) &= \langle f''(x_3, x_4, \ldots, x_n)\rangle_{t-a_2} \\ y(x_1 = 1, x_2 = 1) &= \langle f''(x_3, x_4, \ldots, x_n)\rangle_{t-a_1-a_2} \end{aligned}$$

Clearly, all four reduced functions are ℓ.s. and can have the same separating function $f''(x_3, x_4, \ldots, x_n)$ with possibly different thresholds and gaps. Maps for these four functions are shown in Fig. 3.4 for the case where

$a_1 > a_2 > 0$. It is evident from Theorem 3.1 that all possible pairs of the four reduced functions are comparable and that corresponding threshold values imply which functions contain which. These results are set forth in the following theorem.

Theorem 3.11. If

$$y = F(x_1,x_2,\ldots,x_n) = \langle a_1x_1 + a_2x_2 + f''(x_3,x_4,\ldots,x_n)\rangle_{u:\ell}$$

is an ℓ.s. switching function, then for any pair of variables, say x_1 and x_2, the reduced functions

$$y(x_1 = 1,x_2 = 1),\ y(x_1 = 1,x_2 = 0),\ y(x_1 = 0,x_2 = 1),\ y(x_1 = 0,x_2 = 0)$$

are ℓ.s. and can have the same separating function. Specifically, the separating-function representations are the following where $\alpha,\beta = 0,1$:

$$y(x_1 = \alpha,x_2 = \beta) = \langle f''(x_3,x_4,\ldots,x_n)\rangle_{u_{\alpha\beta}:\ell_{\alpha\beta}}$$

where $u_{\alpha\beta}$ = minimum value of $f''(x_3,x_4,\ldots,x_n) \geq u - \alpha a_1 - \beta a_2$ if such a value exists

$= \infty$, otherwise

$\ell_{\alpha\beta}$ = maximum value of $f''(x_3,x_4,\ldots,x_n) \leq \ell - \alpha a_1 - \beta a_2$ if such a value exists

$= -\infty$, otherwise

Switching functions are called *2-monotonic* (Winder, 1) if they are 1-monotonic and if for each pair of variables x_i and x_j, where $i \neq j = 1,2,\ldots,n$, the 4 corresponding 2-variable reduced functions are comparable. The following result, therefore, has been established.

Corollary 3.11.1 *(Paul and McCluskey; Muroga et al., 1; Winder, 1).* All ℓ.s. switching functions are 2-monotonic.

Theorem 3.2 and Eqs. 3.2 imply that for all positive ℓ.s. functions

$$t - a_1 - a_2 < t - a_2$$

since $a_1 > 0$; hence by Corollary 3.3.2

$$y(x_1 = 1,x_2 = 1) > y(x_1 = 0,x_2 = 1)$$

Similarly

$$t - a_1 - a_2 < t - a_1$$

because $a_2 > 0$; hence

$$y(x_1 = 1,x_2 = 1) > y(x_1 = 1,x_2 = 0)$$

And finally, for $\alpha \neq \beta = 1,2,$

$$t - a_\beta < t$$

because $a_\beta > 0$; hence

$$y(x_\alpha = 0, x_\beta = 1) > y(x_\alpha = 0, x_\beta = 0)$$

This, together with Theorem 3.6 ,justifies the following summary.

Corollary 3.11.2. If $y = F(x_1, x_2, \ldots, x_n)$ is a positive ℓ.s. switching function, then either

$$(1)\quad y(x_1 = 1, x_2 = 1) > y(x_1 = 1, x_2 = 0) > y(x_1 = 0, x_2 = 1) > y(x_1 = 0, x_2 = 0)$$

$$\text{or } (2)\quad y(x_1 = 1, x_2 = 1) > y(x_1 = 0, x_2 = 1) > y(x_1 = 1, x_2 = 0) > y(x_1 = 0, x_2 = 0)$$

$$\text{or } (3)\quad y(x_1 = 1, x_2 = 1) > y(x_1 = 0, x_2 = 1) = y(x_1 = 1, x_2 = 0) > y(x_1 = 0, x_2 = 0)$$

If (1) holds, then $a_1 > a_2$ and if (2), then $a_1 < a_2$.

Example 3.11. For the ℓ.s. switching function

$$y = x_1x_2 \dotplus x_1(x_3 \dotplus x_4x_5 \dotplus x_4x_6 \dotplus x_5x_6) \dotplus x_2(x_3x_4 \dotplus x_3x_5 \dotplus x_4x_5) \dotplus x_3x_4x_5x_6$$

of Example 3.9 it was determined that $a_1 > a_2$. The four reduced functions obtained by assigning the possible values to x_1 and x_2 are

$$y(x_1 = 1, x_2 = 1) = 1$$
$$y(x_1 = 1, x_2 = 0) = x_3 \dotplus x_4x_5 \dotplus x_4x_6 \dotplus x_5x_6$$
$$y(x_1 = 0, x_2 = 1) = x_3x_4 \dotplus x_3x_5 \dotplus x_4x_5$$
$$y(x_1 = 0, x_2 = 0) = x_3x_4x_5x_6$$

It is readily verified by inspection that

$$y(x_1 = 1, x_2 = 1) > y(x_1 = 1, x_2 = 0) > y(x_1 = 0, x_2 = 1) > y(x_1 = 0, x_2 = 0)$$

as Theorem 3.11 indicates is the case.

Example 3.12. In Section 3.3 we pointed out that the only positive unate functions of four or fewer variables that are not ℓ.s. are

$$y_a = x_1x_2 \dotplus x_3x_4$$
$$y_b = x_1x_3 \dotplus x_2x_3 \dotplus x_1x_4 \dotplus x_2x_4$$
$$y_c = x_1x_2 \dotplus x_3x_4 \dotplus x_1x_3$$

It is readily verified that none of these functions are 2-monotonic. From the function for y_c, for example,

$$y_c(x_1 = 1, x_3 = 0) = x_2$$
$$y_c(x_1 = 0, x_3 = 1) = x_4$$

and these are not comparable.

Note that Example 3.12 establishes a simple characterization of ℓ.s. functions of 4 or fewer variables—namely, that all 2-monotonic functions of 4 or fewer variables are ℓ.s.

3.8 GENERAL MONOTONICITY

By now it should be apparent that the arguments and results of one- and two-variable reduced switching functions can be generalized to any number of variables. Let S_1 and S_2 denote any two disjoint subsets of the variables $x_1, x_2, \ldots, x_n$. For example, let $S_1 = \{x_1, x_2, \ldots, x_i\}$ and $S_2 = x_{i+1}, x_{i+2}, \ldots, x_k$ and let S denote the union of S_1 and S_2. Furthermore, let s_1 and s_2 denote the sums of the coefficients of the variables in S_1 and S_2, respectively. If

$$y = F(x_1, x_2, \ldots, x_n) = \langle f(x_1, x_2, \ldots, x_n) \rangle_t = \Big\langle \sum_{j=1}^{n} a_j x_j \Big\rangle_t$$

is an ℓ.s. switching function, then the reduced function $y(S_1 = 1, S_2 = 0)$ is ℓ.s., and

$$y(S_1 = 1, S_2 = 0) = \langle a_1 + a_2 + \cdots + a_i + a_{k+1}x_{k+1} + a_{k+2}x_{k+2} + \cdots + a_n x_n \rangle_t$$

$$= \Big\langle \sum_{j=k+1}^{n} a_j x_j \Big\rangle_{t-s_1} = \langle f'(x_{k+1}, x_{k+2}, \ldots, x_n) \rangle_{t-s_1}$$

where

$$f'(x_{k+1}, x_{k+2}, \ldots, x_n) = f(0,0,\ldots,0,x_{k+1}, x_{k+2}, \ldots, x_n)$$
$$= \sum_{j=k+1}^{n} a_j x_j$$

When the separating-function representation is in terms of the gap, the results are the following.

Theorem 3.12. Let

$$y = F(x_1,x_2,\ldots,x_n) = \langle f(x_1,x_2,\ldots,x_n)\rangle_{u:\ell}$$

be an ℓ.s. switching function. Then for any subset of the variables, say $S_1 = \{x_1,x_2,\ldots,x_i\}$, and any other disjoint subset, say $S_2 = \{x_{i+1},x_{i+2}, \ldots,x_k\}$, the reduced function $y(S_1 = 1,S_2 = 0)$ is ℓ.s., and

$$y(S_1 = 1,S_2 = 0) = \Big\langle \sum_{j=k+1}^{n} a_j x_j \Big\rangle_{u':\ell'} = \langle f'(x_{k+1},x_{k+2},\ldots,x_n)\rangle_{u':\ell'}$$

where u' = minimum value of $f'(x_{k+1},x_{k+2},\ldots,x_n) \geq u - s_1$
(if no such value exists, then $u' = \infty$)

and ℓ' = maximum value of $f'(x_{k+1},x_{k+2},\ldots,x_n) \leq \ell - s_1$
(if no such value exists, then $\ell' = -\infty$).

Consider any other two disjoint subsets S_3 and S_4 of the variables $x_1,x_2, \ldots,x_n$ such that

$$S_3 \cup S_4 = S_1 \cup S_2 = S$$

Let s_3 and s_4 be the sums of the coefficients of the variables of S_3 and S_4, respectively. As in the above case,

$$y(S_3 = 1,S_4 = 0) \text{ is } \ell\text{.s.}$$

and

$$y(S_3 = 1,S_4 = 0) = \langle f'(x_{k+1},x_{k+2},\ldots,x_n)\rangle_{t-s_3}$$

Thus, $y(S_1 = 1,S_2 = 0)$ and $y(S_3 = 1,S_4 = 0)$ can have the same separating function; hence by Theorem 3.1 they are comparable switching functions. Moreover, if

$$y(S_1 = 1,S_2 = 0) > y(S_3 = 1,S_4 = 0)$$

then the threshold of the former is smaller than that of the latter; that is,

$$t - s_1 < t - s_3 \qquad \text{or} \qquad s_1 > s_3$$

Previously, it was shown that functional containment implies coefficient sign and coefficient ordering. We now see that it implies an ordering for combinations of coefficients. This is summarized by the following theorem.

Theorem 3.13 (*Paull and McCluskey; Muroga et al.*, 1; *Winder*, 1). Let the pairs S_1,S_2 and S_3,S_4 denote disjoint subsets of the variables $x_1,x_2,\ldots,x_n$ where

$$S_1 \cup S_2 = S_3 \cup S_4 = S$$

Let s_j denote the sum of the coefficients of the variables of S_j; $j = 1, 2, 3, 4$. If $y = F(x_1,x_2,\ldots,x_n)$ is an ℓ.s. switching function, then $y(S_1 = 1, S_2 = 0)$ and $y(S_3 = 1, S_4 = 0)$ are comparable ℓ.s. functions and can have the same separating function. Moreover, $y(S_1 = 1, S_2 = 0) > y(S_3 = 1, S_4 = 0)$ or $y(S_1 = 1, S_2 = 0) < y(S_3 = 1, S_4 = 0)$ implies

$$s_1 > s_3 \qquad \text{or} \qquad s_1 < s_3$$

respectively in every realization of $F(x_1,x_2,\ldots,x_n)$.

Example 3.13. Consider the switching function of Example 3.11:

$$y = x_1x_2 \dotplus x_1x_3 \dotplus x_1x_4x_5 \dotplus x_1x_4x_6 \dotplus x_1x_5x_6 \dotplus x_2x_3x_4 \\ \dotplus x_2x_3x_5 \dotplus x_2x_4x_5 \dotplus x_3x_4x_5x_6$$

and determine how $a_1 + a_3$ compares with $a_2 + a_4$ in every realization of y. Let

$$S_1 = \{x_1,x_3\},\ S_2 = \{x_2,x_4\},\ S_3 = \{x_2,x_4\},\ S_4 = \{x_1,x_3\}$$

Then

$$y(S_1 = 1,S_2 = 0) = y(x_1 = x_3 = 1,x_2 = x_4 = 0) \equiv 1$$

and

$$y(S_3 = 1,S_4 = 0) = y(x_2 = x_4 = 1,x_1 = x_3 = 0) = x_5$$

Since

$$y(S_1 = 1,S_2 = 0) > y(S_3 = 1,S_4 = 0)$$

then $s_1 > s_3$ and

$$a_1 + a_3 > a_2 + a_4$$

When the set union S has q members, then there are 2^q pairs of subsets S_i,S_j such that $S_i \cup S_j = S$. The set of all such reduced functions is treated in the following corollary.

Corollary 3.13.1. Let S be any subset of q of the independent variables $x_1,x_2,\ldots,x_n$ and let S_i and S_j be any of the 2^q partitions of S such that $S_i \cup S_j = S$. If

$$y = F(x_1,x_2,\ldots,x_n) = \langle f(x_1,x_2,\ldots,x_n)\rangle_{u:\ell}$$

is an ℓ.s. switching function, then all of the 2^q reduced functions $y(S_i = 1, S_j = 0)$ are ℓ.s., and all can have the same separating function, namely, the separating function $f(x_1,x_2,\ldots,x_n)$ with the variables of S set equal to 0.

As we shall see later, this is a very strong condition. In fact, it is the basis for the test and realization procedures of Chapter 5. A more immediately useful result comes from Corollary 3.13.1 and Theorem 3.1; namely, that all of the 2^q functions $y(S_i = 1,S_j = 0)$ are comparable.

Switching functions are called *n-monotonic* or just *monotonic* (Winder, 1), if for every subset S of the variables $x_1,x_2,\ldots,x_n$ the corresponding set of 2^q reduced functions

$$y(S_i = 1, S_j = 0)$$

where $S_i \cup S_j = S$, are comparable. Monotonicity is an important property of ℓ.s. functions as given by Corollary 3.13.2.

Corollary 3.13.2 *(Paull and McCluskey; Muroga et al., 1; Winder, 1),* All ℓ.s. switching functions are monotonic.

Switching functions such that, for each set S of m or fewer variables, the corresponding reduced functions are comparable are called *m-monotonic.* Stated another way, a switching function is m-monotonic if it is $(m - 1)$-monotonic and if, for every subset S of exactly m variables, the corresponding 2^m reduced functions are comparable. Clearly, ℓ.s. functions are m-monotonic for $m = 1,2,\ldots,n$.

Although monotonicity is a necessary condition for a switching function to be ℓ.s., it is not a sufficient condition. For sufficiently large n, there are monotonic switching functions of n variables that are not ℓ.s. It can be verified by considering all possibilities (Cobham; Winder, 1) that all monotonic switching functions of seven or fewer variables are ℓ.s.; also a monotonic function of nine variables which is not ℓ.s. has been reported (Gabelman, 1). Whether all monotonic functions of eight variables are ℓ.s. remains open, however.

The task of determining whether a given switching function is monotonic is tedious and time-consuming. For a function of n variables, about 2^{n-1} comparisons of reduced functions are required. Fortunately, the principal test and realization procedures presented in Chapter 5 do not require that the given switching function first be tested for monotonicity.

3.9 MONOTONICITIES AND THE SIMULTANEOUS INEQUALITIES

Since ℓ.s. switching functions are monotonic, one can deduce various inequalities among the coefficients of the separating function by means of reduced-function comparability. These inequalities are also implied by the set of simultaneous inequalities that define the T gate which realizes the function (see Eq. 1.1). In this section we study the relationships between the inequalities that result from reduced-function comparability and the original set of simultaneous inequalities.

Consider, as a simple example, switching functions of three variables. The n cube on which these functions are defined consists of the eight points

$p_0, p_1, \ldots, p_7$, and each p_j defines a linear inequality in the coefficients a_1, a_2, and a_3 and the gap boundaries u and ℓ where $u > \ell$.

p	p_0	p_1	p_2	p_3	p_4	p_5	p_6	p_7
x_1,x_2,x_3	000	001	010	011	100	101	110	111

Of these eight inequalities, there are four pairs that together imply the sign of a_1. These pairs are (p_0,p_4), (p_1,p_5), (p_2,p_6), and (p_3,p_7). Note that in each case the pair of points differs only in the x_1-value. Assume that $y = F(p)$ is a particular function where $F(p_0) = 0$, and $F(p_4) = 1$. This implies the two inequalities

$$0 \leq \ell \qquad \text{and} \quad a_1 \geq u$$

or that $a_1 > 0$ when $u > \ell$. Suppose now that $F(p_1) = 1$ and $F(p_5) = 0$. This implies two inequalities

$$a_3 \geq u \quad \text{and} \quad a_1 + a_3 \leq \ell$$

which, together with $u > \ell$, imply that $a_1 < 0$. Since a_1 cannot be both positive and negative, the specified switching function is not ℓ.s. The inequalities are then said to be inconsistent. In particular, the inequalities corresponding to the pair of points (p_0,p_4) are not consistent with those corresponding to the pair (p_1,p_5).

If a set of inequalities is consistent, and if one of the pairs of points (p_0,p_4), (p_1,p_5), (p_2,p_6), or (p_3,p_7) implies that $a_1 > 0$, then none of the other pairs can imply that $a_1 < 0$. This is an internal consistency that is certainly necessary if the entire set of inequalities is consistent. We now show that this is exactly the information implied by the reduced-function comparability:

$$y(x_1 = 1) > y(x_1 = 0)$$

The reduced function $y(x_1 = 1)$ corresponds to the original function, F, when $p = p_4,p_5,p_6,p_7$, while $F(x_1 = 0)$ corresponds to y when $p = p_0,p_1,p_2,$ p_3. The containment relation

$$y(x_1 = 1) > y(x_1 = 0)$$

implies that from the four possible pairs, (p_0,p_4), (p_1,p_5), (p_2,p_6), or (p_3,p_7), there is at least one pair (p_a,p_{a+4}), where

$$F(p_{a+4}) = 1 \qquad \text{and} \quad F(p_a) = 0$$

and for each pair either

$$F(p_{j+4}) = 1 \quad \text{and} \quad F(p_j) = 0$$

or

$$F(p_{j+4}) = F(p_j) = 1$$

or

$$F(p_{j+4}) = F(p_j) = 0$$

A similar interpretation applies to higher-order-reduced-function comparability. For example, consider the reduced functions

$$y(x_1 = 1, x_2 = 0) > y(x_1 = 0, x_2 = 1)$$

The reduced function $y(x_1 = 1, x_2 = 0)$ corresponds to the original function, F, when $p = p_4, p_5$ and $y(x_1 = 0, x_2 = 1)$ corresponds to F when $p = p_2, p_3$. These points can be considered in pairs (p_2, p_4) and (p_3, p_5) so that the coordinate values for each member of a pair are the same except for x_1 and x_2. The inequality

$$y(x_1 = 1, x_2 = 0) > y(x_1 = 0, x_2 = 1)$$

implies that, for one pair (p_a, p_{a+2}),

$$F(p_{a+2}) = 1 \quad \text{and} \quad F(p_a) = 0$$

and for the other pair (p_b, p_{b+2}), either

$$F(p_{b+2}) = 1 \quad \text{and} \quad F(p_b) = 0$$

or

$$F(p_{b+2}) = F(p_b) = 1$$

or

$$F(p_{b+2}) = F(p_b) = 0$$

It should be clear that similar results are implied by all comparable reduced functions. The monotonicity of a function, therefore, implies a certain internal consistency of the simultaneous inequalities that are associated with the points for which the function is defined. For functions of many variables, however, this internal consistency is not sufficient to guarantee that the simultaneous inequalities are consistent.

Thus far we have discussed two sets of inequalities: the so-called simultaneous inequalities that are associated with the points on which the function is defined, and the inequalities due to monotonicity, which are those that result from reduced-function comparability. Since the monotonic inequalities are implied by the simultaneous inequalities and conversely, a knowledge of the former makes some of the latter redundant.

Consider first the 1-monotonic inequalities, which are those implied by the comparability of 1-variable reduced functions and which yield the signs of the coefficients. Recall that coefficient sign can also be determined by whether corresponding variables appear complemented or uncomplemented in the MSP form. Assume that the given switching function is positive ℓ.s., in which case the coefficients of the variables $x_1,x_2,\ldots,x_n$ are all positive. Suppose, for example, that 4 of the simultaneous inequalities are

$$
\begin{array}{lrcl}
(1) & a_2 + a_3 & & \geq u \\
(2) & a_1 + a_2 + a_3 & & \geq u \\
(3) & & a_4 + a_5 & \leq \ell \\
(4) & a_1 + & a_4 + a_5 & \leq \ell
\end{array}
$$

It is evident that if a_1 is known to be positive, then inequality (2) is redundant in view of (1) in the sense that any solution to (1) is a solution to (2) if $a_1 > 0$. Similarly, inequality (3) is redundant in view of (4).

The original set of simultaneous inequalities involving the coefficients and gap boundaries of a function of n variables consists of $(2^n + 1)$ members, and is called the R_0 *set* (see Section 1.9). Every single-T-gate realization of the given switching function is a solution to the R_0 set. If, however, the signs of the coefficients are known, then some of the members of the R_0 set are redundant in view of this information. Delete the redundant members and replace them by the subset $a_j > 0$, $j = 1,2,\ldots,n$. Call this the R_1 *reduced set* or just the R_1 *set.* Any solution to the R_1 set is a solution of the R_0 set, and conversely. In the main, the R_1 set has fewer members than the R_0 set and is, therefore, easier to solve.

Having a reduced set of inequalities considerably simplifies the effort required to obtain a solution. This simplification would be nullified to a large extent, however, if it were necessary to start with the R_0 set and search all of the 2^n members to eliminate those that were redundant in order to obtain the R_1 set. Fortunately this is unnecessary, as was first established by McNaughton. He showed that the R_1 set can be specified directly from the MSP forms of the given switching function and the corresponding complement function.

Consider, for example, a positive linearly separable switching function $y = F(x_1,x_2,x_3,x_4)$ and assume that one of the product terms of the MSP form is

$$T = x_1x_2$$

This term implies that $y = 1$ for the four points (combinations of values)

$$(x_1,x_2,x_3,x_4) = (1,1,0,0),\ (1,1,0,1),\ (1,1,1,0),\text{ and } (1,1,1,1)$$

and, therefore, implies the corresponding inequalities

$$\begin{aligned}
a_1 + a_2 \qquad\qquad &\geq u \\
a_1 + a_2 + \qquad a_4 &\geq u \\
a_1 + a_2 + a_3 \qquad &\geq u \\
a_1 + a_2 + a_3 + a_4 &\geq u
\end{aligned}$$

Since it is known that $a_3 > 0$ and $a_4 > 0$, then the last three inequalities are redundant in view of this information and the first inequality. Hence, given that $a_3 > 0$ and $a_4 > 0$, the only irredundant inequality of the above four is

$$a_1 + a_2 \geq u$$

Since a corresponding argument applies to each product term in the MSP form, we have the following result:

To each term in the MSP form of a positive linearly separable switching function, there corresponds one irredundant inequality, given that $a_j > 0$, $j = 1,2,\ldots,n$. Each inequality consists of the sum of the coefficients that correspond to the variables that appear in the term, and this sum is greater than or equal to the upper-gap boundary. In different words, the irredundant inequalities correspond to those points, one for each term of the MSP form, so that the variables that appear in the term have unit value and all other variable values are zero. These inequalities, together with the set $a_j > 0$ for $j = 1,2,\ldots,n$ constitute the members of the R_1 set that are $\geq u$ or > 0.

To obtain the inequalities of the R_1 set that are $\leq \ell$, consider the MSP form of the complement function $\bar{y} = \bar{F}(x_1,x_2,\ldots,x_n)$. Assume that one of the product terms is

$$\bar{x}_1\bar{x}_3$$

This implies that $\bar{y} = 1$ and that $y = 0$ for the four points

$$(x_1,x_2,x_3,x_4) = (0,0,0,0),\ (0,0,0,1),\ (0,1,0,0), \text{ and } (0,1,0,1)$$

The inequalities that correspond to these points are

$$\begin{aligned}
0 + 0 \quad &\leq \ell \\
a_4 &\leq \ell \\
a_2 \qquad &\leq \ell \\
a_2 + a_4 &\leq \ell
\end{aligned}$$

Since it is known that $a_2 > 0$ and $a_4 > 0$, then the first three inequalities are redundant in view of the fourth, and the only irredundant inequality is

$$a_2 + a_4 \leq \ell$$

Thus to each product term of the MSP form of $\bar{F}$ there corresponds one irredundant inequality, given that the signs of the coefficients are positive. In each case this inequality is the one associated with the point whose variable values are zero if the variable appears in the term and are unity otherwise.

From the definition of the MSP form, none of these inequalities can be redundant given only coefficient sign information; therefore, we have found all of the inequalities of the R_1 set. A formal statement of the result is given in the following theorem.

Theorem 3.14. (*McNaughton*). If $y = F(x_1,x_2,\ldots,x_n)$ is a positive linearly separable switching function then the R_1 set is obtained from the MSP forms for F and $\bar{F}$, as follows:

(1) To each product term of the MSP form of F there is the inequality consisting of the sum of the coefficients that correspond to the variables that appear in the term, and this sum is greater than or equal to u.

(2) To each product term of the MSP form of $\bar{F}$ there is the inequality consisting of the sum of the coefficients that correspond to the variables that do not appear in the term, and this sum is less than or equal to ℓ.

(3) The subset of inequalities

$$u > \ell \qquad \text{and} \quad a_j > 0$$

for $j = 1,2,\ldots,n$.

Example 3.14. Consider the positive linearly separable switching function in MSP form:

$$y = x_1x_2 \dotplus x_1x_3x_4 \dotplus x_1x_3x_5 \dotplus x_2x_3x_4$$

where the MSP form of $\bar{y}$ is

$$y = \bar{x}_1\bar{x}_2 \dotplus \bar{x}_1\bar{x}_3 \dotplus \bar{x}_1\bar{x}_4 \dotplus \bar{x}_2\bar{x}_3 \dotplus \bar{x}_2\bar{x}_4\bar{x}_5$$

The inequalities of the R_1 set that are $\geq u$ correspond to the terms of y and are

$$\begin{array}{lllll} a_1 + a_2 & & & & \geq u \\ a_1 + & & a_3 + a_4 & & \geq u \\ a_1 + & & a_3 + & a_5 & \geq u \\ & a_2 + & a_3 + a_4 & & \geq u \end{array}$$

The inequalities of the R_1 set that are $\leq \ell$ correspond to the terms of $\bar{y}$ and are

$$\begin{aligned} a_3 + a_4 + a_5 &\leq \ell \\ a_2 + a_4 + a_5 &\leq \ell \\ a_2 + a_3 + a_5 &\leq \ell \\ a_1 + a_4 + a_5 &\leq \ell \\ a_1 + a_3 &\leq \ell \end{aligned}$$

These nine inequalities, together with the following six, constitute the R_1 set

$$u > \ell,\ a_1 > 0,\ a_2 > 0,\ a_3 > 0,\ a_4 > 0,\ a_5 > 0$$

In this example the R_1 set consists of fifteen inequalities while the R_0 set has thirty-three.

Higher-order monotonicities yield additional information that makes inequalities in the R_1 set redundant (Winder, 7; Muroga et al., 1). For example, if it were known that, in every realization of a given positive linearly separable function,

$$a_1 > a_2$$

and if two of the inequalities of the R_1 set were

$$\begin{aligned} a_2 + a_3 + a_4 &\geq u \\ a_1 + a_3 + a_4 &\geq u \end{aligned}$$

then every solution to the first inequality where $a_1 > a_2$ is a solution to the second, and the second is redundant in view of the first.

Note that the R_1 set consists of the inequalities obtained from 1-monotonicities (1-variable reduced-function comparability), and the irredundant inequalities of the R_0 set, given the 1-monotonic inequalities. As a general notation let the R_q *reduced set* or simply the R_q *set* consist of the irredundant inequalities that result from knowing that the given switching function is q-monotonic, together with the irredundant inequalities of the R_{q-1} set given the q-monotonic inequalities.

In the main, the number of inequalities in each R_q set is smaller as the value of q increases. Unfortunately for $q > 2$, however, one must partially solve the R_{q-1} set, subject to the q-monotonic inequalities, to get the R_q set. This is more difficult as the value of q increases. Since the computations are fairly time-consuming, they are not particularly useful on other than a selective basis, even though a number of short cuts are available.

A procedure exists for obtaining the R_2 set directly from the MSP form for the given switching function and the complement function. The procedure is similar to the procedure for the R_1 set, and is here omitted.

Example 3.15. For the function of Example 3.14,

$$y = x_1x_2 \dotplus x_1x_3x_4 \dotplus x_1x_3x_5 \dotplus x_2x_3x_4$$

the R_1 set of inequalities was found to be

$$\begin{array}{llr}
(1) & a_1 + a_2 & \geq u \\
(2) & a_1 + \quad + a_3 + a_4 & \geq u \\
(3) & a_1 + \quad a_3 + \quad a_5 & \geq u \\
(4) & a_2 + a_3 + a_4 & \geq u \\
(5) & a_3 + a_4 + a_5 & \leq \ell \\
(6) & a_2 + \quad a_4 + a_5 & \leq \ell \\
(7) & a_2 + a_3 + \quad a_5 & \leq \ell \\
(8) & a_1 + \quad a_4 + a_5 & \leq \ell \\
(9) & a_1 + \quad a_3 & \leq \ell \\
(10) & & u > \ell \\
(11) & & a_1 > 0 \\
(12) & & a_2 > 0 \\
(13) & & a_3 > 0 \\
(14) & & a_4 > 0 \\
(15) & & a_5 > 0
\end{array}$$

By the coefficient ordering procedure, Procedure 3.2, we find that $a_1 > a_2 > a_3 > a_4 > a_5$. Since $a_4 > a_5$, inequality (2) is redundant in view of (3). Similarly, since $a_1 > a_2 > a_3$, then inequalities (5) and (6) are redundant in view of (7). Finally, inequalities (11), (12), (13), and (14) are redundant given (15) and the coefficient ordering. The remaining inequalities are irredundant; hence the R_2 set is

$$\begin{array}{lr}
a_1 + a_2 & \geq u \\
a_1 + \quad a_3 + \quad a_5 & \geq u \\
a_2 + a_3 + a_4 & \geq u \\
a_2 + a_3 + \quad a_5 & \leq \ell \\
a_1 + \quad a_4 + a_5 & \leq \ell \\
a_1 + \quad a_3 & \leq \ell \\
 & u > \ell \\
 & a_5 > 0 \\
 & a_1 > a_2 \\
 & a_2 > a_3 \\
 & a_3 > a_4 \\
 & a_4 > a_5
\end{array}$$

and consists of twelve members.

There is one type of R_2 simplification that is easy to perform and always worthwhile. When it is known that F is symmetric in some subset of the variables, say $x_1,x_2,\ldots,x_q$, then since the coefficients of these variables can always be equal, they can be set equal in all of the inequalities. Thus

$$a_1 = a_2 = a_3 = \cdots = a_q$$

This substitution may make two or more of the inequalities identical, in which case all but one of the identical inequalities can be deleted.

Example 3.16. Consider the function

$$\begin{aligned} F &= x_1 \dotplus x_2x_3 \\ \bar{F} &= \bar{x}_1\bar{x}_2 \dotplus \bar{x}_1\bar{x}_3 \end{aligned}$$

The R_1 set of inequalities is

$$\begin{array}{ll}
(1) & a_1 \qquad\qquad \geq u \\
(2) & \qquad a_2 + a_3 \geq u \\
(3) & \qquad\qquad a_3 \leq \ell \\
(4) & \qquad a_2 \qquad \leq \ell \\
(5) & \qquad\qquad u > \ell \\
(6) & \qquad\qquad a_1 > 0 \\
(7) & \qquad\qquad a_2 > 0 \\
(8) & \qquad\qquad a_3 > 0
\end{array}$$

The coefficient ordering is

$$a_1 > a_2 = a_3$$

Setting $a_2 = a_3$ makes (3) and (4) identical, and thus (4) can be omitted. The irredundant inequalities are then

$$\begin{aligned}
a_1 \qquad &\geq u \\
2a_2 &\geq u \\
a_2 &\leq \ell \\
u &> \ell \\
a_2 &> 0 \\
a_1 &> a_2
\end{aligned}$$

together with the auxiliary relation

$$a_2 = a_3$$

which is not needed in the solution.

3.10 ASUMMABILITY†

Although monotonicity is a fairly strong necessary condition on a switching function in order that it be ℓ.s., it is not a sufficient condition. There is, however, an even stronger class of necessary conditions, where monotonicity is the weakest member of the class and where the class culminates in a condition which is both necessary and sufficient.

Let $y = F(x_1,x_2,\ldots,x_n) = F(p)$ be a switching function defined on the vertices $p_0,p_1,\ldots,p_{2^n-1}$ of the n cube. In this discussion, we consider the vertices or points p_j as vectors and define the sum of such points as the usual (as to components) vector sum. For example, if

$$p_1 = (0,1) \qquad \text{and} \quad p_3 = (1,1)$$

then

$$p_1 + p_3 = (1,2)$$

The switching function F is called *k-summable* (Winder, 1) if and only if there exist a set of k points, $p_a,p_b,\ldots,p_k$, not necessarily distinct, such that $F(p_j) = 1$ for $j = a,b,\ldots,k$ and a set of k points $p_a',p_b',\ldots,p_k'$, again not necessarily distinct, such that $F(p_j') = 0$ for $j = a,b,\ldots,k$ where

$$\sum_{j=a}^{k} p_j = \sum_{j=a}^{k} p_j'$$

As an example, the switching function defined by the table of combinations

p_j	$x_1\ x_2$	$y = F(x_1,x_2)$
p_0	0 0	0
p_1	0 1	1
p_2	1 0	1
p_3	1 1	0

is 2-summable since

$$F(p_1) = F(p_2) = 1, \quad F(p_0) = F(p_3) = 0$$

and

$$p_1 + p_2 = (0,1) + (1,0) = p_0 + p_3 = (0,0) + (1,1) = (1,1)$$

Functions which are not k-summable are called *k-asummable.* We first show that k-asummability is a necessary condition for a function to be ℓ.s.

† This section can be omitted for the first reading.

Theorem 3.15 (*Elgot*). If

$$y = F(p) = \langle f(p) \rangle_{u:\ell}$$

is an ℓ.s. switching function of n variables, then it is k-asummable for all $k \geq 1$.

Proof. The proof depends upon the linearity of the separating function f, in which case

$$f(p_i + p_j) = f(p_i) + f(p_j)$$

for any p_i and p_j.

Assume that the theorem is false. Then there exists some ℓ.s. function

$$y = F(p) = \langle f(p) \rangle_{u:\ell}$$

which for some particular value of k is k-summable. This means that there exist sets of points $p_a, p_b, \ldots, p_k$ and $p_a', p_b', \ldots, p_k'$ such that

$$F(p_j) = 1 \qquad \text{and} \quad F(p_j') = 0$$

for $j = a, b, \ldots, k$ and

$$\sum_{j=a}^{k} p_j = \sum_{j=a}^{k} p_j'$$

Clearly

$$f(p_j) \geq u \qquad \text{and} \quad f(p_j') \leq \ell$$

for each $j = a, b, \ldots, k$; hence

$$\sum_{j=a}^{k} f(p_j) \geq ku > k\ell \geq \sum_{j=a}^{k} f(p_j')$$

However, because f is linear,

$$\sum_{j=a}^{k} f(p_j) = f\left(\sum_{j=a}^{k} p_j\right) = f\left(\sum_{j=a}^{k} p_j'\right) = \sum_{j=a}^{k} f(p_j')$$

which is a contradiction of the previous equation. Therefore, we have established that all ℓ.s. switching functions are k-asummable.|

Theorem 3.15 defines a set of necessary conditions for ℓ.s. functions that become increasingly strong as k is increased. We now show that for $k = 2$ the condition is equivalent to complete monotonicity.

Theorem 3.16 (*Elgot*). The property of 2-asummability is equivalent to complete monotonicity.

Proof. We shall show that a nonmonotonic function is 2-summable, and conversely.

Suppose that $y = F(x_1,x_2,\ldots,x_n)$ is a switching function that is not monotonic. Then there exist subsets of variables S_1, S_2, S_3, and S_4 where

$$S_1 \cup S_2 = S_3 \cup S_4$$

such that $y(S_1 = 1, S_2 = 0)$ and $y(S_3 = 1, S_4 = 0)$ are not comparable. By the same reasoning as used in Section 3.9, there are four points of the original n cube, p_a, p_b, p_c, and p_d such that p_a and p_b together imply that

$$s_1 > s_3$$

while p_c and p_d imply the contradictory inequality

$$s_1 < s_3$$

The points involved are related as follows:

(1) For p_a, the S_1 variables and S_2 variables have unit and zero values, respectively, while for p_b the S_3 and S_4 variables have unit and zero values. Variables not in S_1 or S_2 have the same values for both p_a and p_b. Moreover,

$$F(p_a) = 1 \qquad \text{and} \qquad F(p_b) = 0$$

(2) Similarly, for p_c, the S_1 variables and the S_2 variables have unit and zero values, respectively, while for p_d, the S_3 and S_4 variables have unit and zero values. Variables not in S_1 or S_2 have the same values for both p_c and p_d. Moreover,

$$F(p_c) = 0 \qquad \text{and} \qquad F(p_d) = 1$$

The function F is 2-summable, since

$$p_a + p_d = p_b + p_c$$

Conversely, if the given function is 2-summable, then there exist four points p_a, p_b, p_c, and p_d so that

$$F(p_a) = F(p_d) = 1 \qquad \text{and} \qquad F(p_b) = F(p_c) = 0$$

and

$$p_a + p_d = p_b + p_c$$

By the same reasoning as for the first half of the proof, this implies that $y(S_1 = 1, S_2 = 0)$ and $y(S_3 = 1, S_4 = 0)$ are not comparable where S_1 and S_2 are the subsets of variables which have, respectively, unit and zero values in both p_a and p_c, and S_3 and S_4 are the subsets which have, respectively, unit and zero values in both p_b and p_d. ▮

We have shown that 2-asummability is equivalent to complete monotonicity and that higher order k-asummability is a more general necessary condition. The question naturally arises whether any order of k-asummability constitutes a sufficient, as well as a necessary, condition. Winder, 1, has shown (by counterexample) that there is no fixed value of k for which this is the case. However, k-asummability for all k does constitute a necessary and sufficient condition. A switching function that is k-asummable for all k is said to be *asummable.*

Theorem 3.17 (*Elgot*). A switching function is ℓ.s. if and only if it is asummable.

Unfortunately, the proof of this statement is beyond the scope of this chapter. It follows from certain results in the general theory of linear inequalities, and the proof is therefore deferred until Chapter 9.

The fact that asummability is a necessary and sufficient condition is not particularly useful in formulating a test and realization procedure. Determining asummability is equivalent to solving the inequalities by linear programming. Asummability as a concept, however, is a compact way of specifying the realizability conditions for ℓ.s. functions.

3.11 REDUCED FUNCTIONS IN T-GATE NETWORKS

In the preceding sections we have shown that a single gate realization of an ℓ.s. switching function $y = F(x_1,x_2,\ldots,x_n)$ is easily converted to a realization for any reduced function $y(S_1 = 1,S_2 = 0)$. We now show that a similar result holds when the realization of F requires a network of T gates. Actually, the result should be evident by this time; hence no proof is provided.

Theorem 3.18. Let $y = F(x_1,x_2,\ldots,x_n)$ be a switching function that is realized by a network of T gates where

$$y = \left\langle \sum_{j=1}^{m} a_j y_j \right\rangle_{u:\ell}$$

and where each $y_j = F_j(x_1,x_2,\ldots,x_n)$ is similarly realized, and so on. If S_1 and S_2 are disjoint subsets of the independent variables $x_1,x_2,\ldots,x_n$, then a realization for $y(S_1 = 1,S_2 = 0)$ is

$$y(S_1 = 1,S_2 = 0) = \left\langle \sum_{j=1}^{m} a_j y_j(S_1 = 1,S_2 = 0) \right\rangle_{u:\ell}$$

Theorem 3.18 shows that each gate in the realization of $y = F(x_1,x_2,\ldots,x_n)$ realizes the corresponding reduced function $y_j(S_1 = 1,S_2 = 0)$, and that this result can be iterated to include each gate of the network. Those gates which have some or all of the variables of S_1 and S_2 as inputs will have these inputs assigned fixed values of 1 and 0, respectively. The constant inputs of unit value can then be removed by altering the gap location in the usual way (Theorem 2.4B).

Corollary 3.18.1. A T-gate network realization of $y(S_1 = 1, S_2 = 0)$ can be derived from the realization for $y = F(x_1,x_2,\ldots,x_n)$ by setting the independent input variables of S_1 and S_2 equal to the values 1 and 0, respectively.

Now consider the reduced functions $y(S_1 = 1,S_2 = 0)$ and $y(S_3 = 1, S_4 = 0)$ from the switching function $y = F(x_1,x_2,\ldots,x_n)$, where

$$S_1 \cup S_2 = S_3 \cup S_4$$

Applying Corollary 3.18.1 to both reduced functions gives the following corollary.

Corollary 3.18.2. Let S_1, S_2 and S_3, S_4 each be disjoint pairs of subsets of the variables of the switching function $y = F(x_1,x_2,\ldots,x_n)$ where

$$S_1 \cup S_2 = S_3 \cup S_4 = S$$

Then the reduced functions $y(S_1 = 1, S_2 = 0)$ and $y(S_3 = 1, S_4 = 0)$ have identical realizations except for gaps. By this we mean that each realization will consist of the same number of T gates interconnected in the same way, where corresponding gates have the same inputs and coefficients but may have different gaps. Moreover, if $S = S_1 \cup S_2$ has q members, then all 2^q reduced functions $y(S_i = 1, S_j = 0)$ where $S_i \cup S_j = S$ can have the same realization except for gaps.

Example 3.17. Consider the switching function with realization

$$y = \langle x_1 + \bar{x}_2 + \bar{x}_3 + x_4 + 3\langle \bar{x}_1 + x_2 + x_3 + \bar{x}_4 + 2\langle \bar{x}_1 + \bar{x}_2 + \bar{x}_3 \rangle_{3:2} \rangle_{3:2} \rangle_{4:3}$$

as shown in Fig. 3.5*a*. According to Corollary 3.18.1, the reduced function $y(x_1 = 1,x_2 = 0)$ can be realized by

$$\begin{aligned} y(x_1 = 1,x_2 = 0) &= \langle 1 + 1 + \bar{x}_3 + x_4 + 3\langle 0 + 0 + x_3 + \bar{x}_4 + 2\langle 0 + 1 + \bar{x}_3 \rangle_{3:2} \rangle_{3:2} \rangle_{4:3} \\ &= \langle \bar{x}_3 + x_4 + 3\langle x_3 + \bar{x}_4 + 2\langle \bar{x}_3 \rangle_{\infty:1} \rangle_{3:2} \rangle_{2:1} \end{aligned}$$

as shown in Fig. 3.5*b*. Similarly, the reduced function $y(x_1 = 0,x_2 = 0)$ can be realized by

$$y(x_1 = 0,x_2 = 0) = \langle \bar{x}_3 + x_4 + 3\langle x_3 + \bar{x}_4 + 2\langle \bar{x}_3 \rangle_{1:0} \rangle_{2:1} \rangle_{3:2}$$

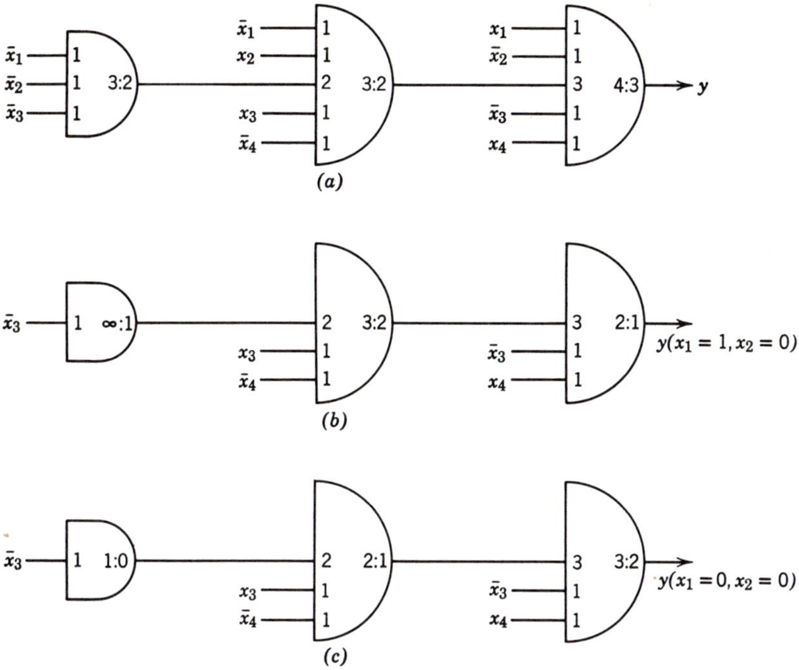

Fig. 3.5. T-gate network realizations of the functions of Example 3.17. (*a*) Realization of $y = F(x_1,x_2,x_2,x_4)$. (*b*) Realization of $y(x_1 = 1,x_2 = 0)$. (*c*) Realization of $y(x_1 = 0,x_2 = 0)$.

as shown in Fig. 3.5*c*. In the case of $y(x_1 = 1,x_2 = 0)$ it is evident that the realization can be simplified to

$$y(x_1 = 1,x_2 = 0) = \langle \bar{x}_3 + x_4 \rangle_{2:1}$$

since

$$\langle \bar{x}_3 \rangle_{\infty:1} \equiv 0$$

and

$$\langle x_3 + \bar{x}_4 + 2\langle \bar{x}_3 \rangle_{\infty:1} \rangle_{3:2} \equiv 0$$

This does not invalidate Corollary 3.18.2, however.

Chapter 4

THE FUNCTION TREE

4.1 INTRODUCTION

The reduced functions of a given switching function yield considerable information concerning its separating-function representations. For linearly separable functions, reduced function comparabilities imply ordering relationships between combinations of separating-function coefficients. For nonlinearly-separable functions, reduced functions define inconsistencies that must be eliminated by the use of additional gates. Actually, the concept of reduced functions has given an embarrassingly large amount of information. What is needed now is a way to organize this information in a form that is useful for defining efficient synthesis procedures.

This chapter introduces the concept of the function tree. This provides a compact way to organize the properties of a switching function that are important for T-gate network realizations and, in fact, provides a structure whereby additional properties become apparent. Moreover, the function tree is the basis for a unified approach to both the single and the multiple T-gate realization procedures.

A function tree is a structure that is associated with an arbitrarily specified switching function $y = F(x_1,x_2,\ldots,x_n)$; it displays a decomposition of F in terms of its reduced functions. The tree is divided into levels $0,1,2,\ldots,n$. Each level includes some of the reduced functions of F. More specifically, for each j, all functions on the j-level are the reduced functions obtained from $(j-1)$-level functions by assigning zero and unit values to x_j. Since permuting the labels of the variables of a switching function produces no fundamental change, the 2^n possible labelings imply 2^n different function trees for a given switching function. Any one of these can be used in the realization procedures defined in the chapters that follow. It will be shown, however, that considerable simplification results from labeling the variables in a particular way.

The description of the function tree is considerably simplified by using a somewhat different notation for the reduced functions

$$y(x_1 = 1), \quad y(x_1 = 1, x_2 = 0), \quad y(x_1 = 0, x_2 = 1, x_3 = 0), \text{ etc.}$$

In each case, there is a value of k such that the reduced function is obtained by assigning values to the variables

$$x_1, x_2, \ldots, x_k$$

An adequate notation for the reduced functions is

$$\begin{aligned} y(x_1 = 1) &= F_1(x_2, x_3, \ldots, x_n) \\ y(x_1 = 1, x_2 = 0) &= F_{10}(x_3, x_4, \ldots, x_n) \\ y(x_1 = 0, x_2 = 1, x_3 = 0) &= F_{010}(x_4, x_5, \ldots, x_n) \end{aligned}$$

or, in general,

$$y(x_1 = \delta_1, x_2 = \delta_2, \ldots, x_k = \delta_k) = F_{\delta_1 \delta_2 \ldots \delta_k}(x_{k+1}, x_{k+2}, \ldots, x_n)$$

where $\delta_i = 0,1$. Using this notation, we describe the procedure for specifying the function tree for an arbitrary switching function.

Procedure 4.1. Function-Tree Formation. The format of the function tree is shown in Fig. 4.1. The formation procedure for the switching function $y = F(x_1, x_2, \ldots, x_n)$ is as follows:

(1) The given switching function F constitutes the 0-level of the tree.

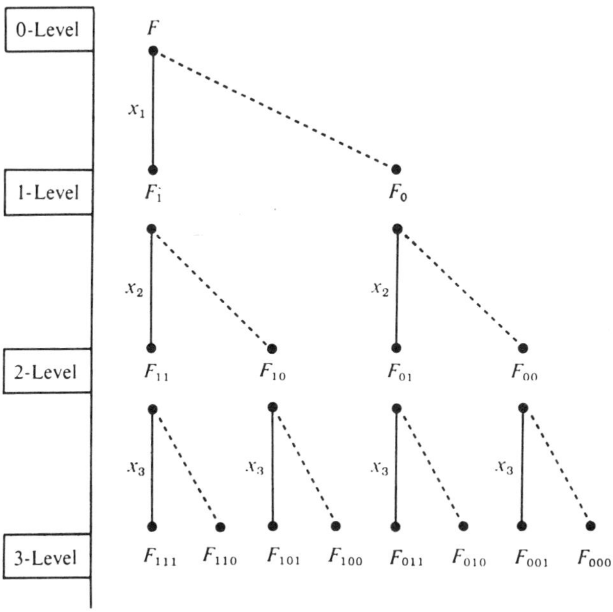

Fig. 4.1. Format of the function tree.

(2) Factor F as

$$F(x_1,x_2,\ldots,x_n) = x_1F_1(x_2,x_3,\ldots,x_n) \dotplus \bar{x}_1F_0(x_2,x_3,\ldots,x_n)$$

The reduced functions F_1 and F_0 constitute the 1-level members of the tree. Between F and F_1 is a solid line labeled x_1. Between F and F_0 is a broken line labeled $\bar{x}_1$. Often the broken-line labeling is not shown on the tree; however, it is implied.

The functions F_1 and F_0 are the reduced functions $y(x_1 = 1)$ and $y(x_1 = 0)$ respectively, and are called the *descendant functions* of F. Sometimes F_1 and F_0 are referred to as the *function pair* descendant from F. Moreover, F is called the *ancestor function* of F_1 and F_0. The procedure continues as follows:

(3) Factor each F_1 and F_0, provided it is not a constant function, as

$$F_1(x_2,x_3,\ldots,x_n) = x_2F_{11}(x_3,x_4,\ldots,x_n) \dotplus \bar{x}_2F_{10}(x_3,x_4,\ldots,x_n)$$
$$F_0(x_2,x_3,\ldots,x_n) = x_2F_{01}(x_3,x_4,\ldots,x_n) \dotplus \bar{x}_2F_{00}(x_3,x_4,\ldots,x_n)$$

The functions F_{11}, F_{10}, F_{01}, F_{00} constitute the 2-level members of the function tree. Between F_1 and F_{11} and between F_0 and F_{01} is a solid line labeled x_2. Between F_1 and F_{10} and between F_0 and F_{00} is a broken line with implied labeling x_2.

Note that pairs F_{11},F_{10} and F_{01},F_{00} are each the descendant functions of F_1 and F_0, respectively. Moreover, observe that each of the functions on the 2-level is a reduced function of a 1-level function and is, therefore, a reduced function of F obtained by assigning values to x_1 and x_2.

The remaining levels of the tree are determined in a similar way. The general step of the procedure given below determines the $(k + 1)$-level given the k-level. (Steps 2 and 3 correspond to $k = 0$ and 1, respectively.)

(4) The set of all possible functions on the k-level of the tree is

$$\{F_{\delta_1\delta_2\ldots\delta_k}(x_{k+1},x_{k+2},\ldots,x_n)\}$$

where $\delta_1,\delta_2,\ldots,\delta_k = 0,1$.
Factor each nonconstant function of this set as

$$F_{\delta_1\delta_2\ldots\delta_k}(\,) = x_{k+1}F_{\delta_1\delta_2\ldots\delta_k 1}(\,) \dotplus \bar{x}_{k+1}F_{\delta_1\delta_2\ldots\delta_k 0}(\,)$$

This yields the set of reduced functions

$$\{F_{\delta_1\delta_2\ldots\delta_k\delta_{k+1}}(x_{k+2},x_{k+3},\ldots,x_n)\}$$

which are the members of the $(k + 1)$-level of the tree. Between each $F_{\delta_1\delta_2\ldots\delta_k}$ and $F_{\delta_1\delta_2\ldots\delta_k 1}$ is a solid line labeled x_{k+1}. Between each $F_{\delta_1\delta_2\ldots\delta_k}$ and $F_{\delta_1\delta_2\ldots\delta_k 0}$ is a broken line, with implied labeling $\bar{x}_{k+1}$.

(5) If $k + 1 < n$, increase the value of k by unity and repeat Step 4. If $k + 1 = n$, the function tree is completed.

Note that all functions on the bottom level (that is, the n-level) of the tree are constant functions.

On each level of the tree, constant functions were explicitly excluded from further reduction; hence, constant functions do not have descendants. This was for convenience rather than necessity, however. If they were not omitted, the procedure would be carried out in the normal way and each constant 0 or 1 k-level function would have descendant functions 0 and 0 or 1 and 1, respectively. These of course have the same properties as the other functions of the tree. Omitting the constant functions from further consideration simplifies the tree to some extent, and if any of them are needed, they can be readily inserted. Function trees to which additional constant functions have been completely or partially supplied are called *complete* or *partially complete* function trees. In the main, the properties of a function tree are independent of whether or not it is complete. No distinction is made between the two kinds of tree, and both are referred to as function trees unless an explicit distinction is required for clarification.

Example 4.1. Figure 4.2 shows the function tree for the switching function

$$y = x_1 \dotplus x_2\bar{x}_3 \dotplus x_2x_4 \dotplus \bar{x}_2x_3\bar{x}_4$$

Although a function tree can be formed for any ordering of the variables, a considerable simplification can be achieved in certain cases by using a particular ordering. If the switching function $y = F(x_1,x_2,\ldots,x_n)$ is positive linearly separable, then the MSP form can be factored as

$$\begin{aligned} y &= F(x_1,x_2,\ldots,x_n) \\ &= x_1F_a(x_2,x_3,\ldots,x_n) \dotplus F_b(x_2,x_3,\ldots,x_n) \end{aligned} \qquad (4.1)$$

Recall that the factoring for the function tree was

$$y = x_1F_1(x_2,x_3,\ldots,x_n) \dotplus \bar{x}_1F_0(x_2,x_3,\ldots,x_n)$$

which in terms of the above function is

$$y = x_1[F_a(\,) \dotplus F_b(\,)] \dotplus \bar{x}_1F_b(\,)$$

Therefore

$$F_1 = y(x_1 = 1) = F_a \dotplus F_b$$

and

$$F_0 = y(x_1 = 0) = F_b$$

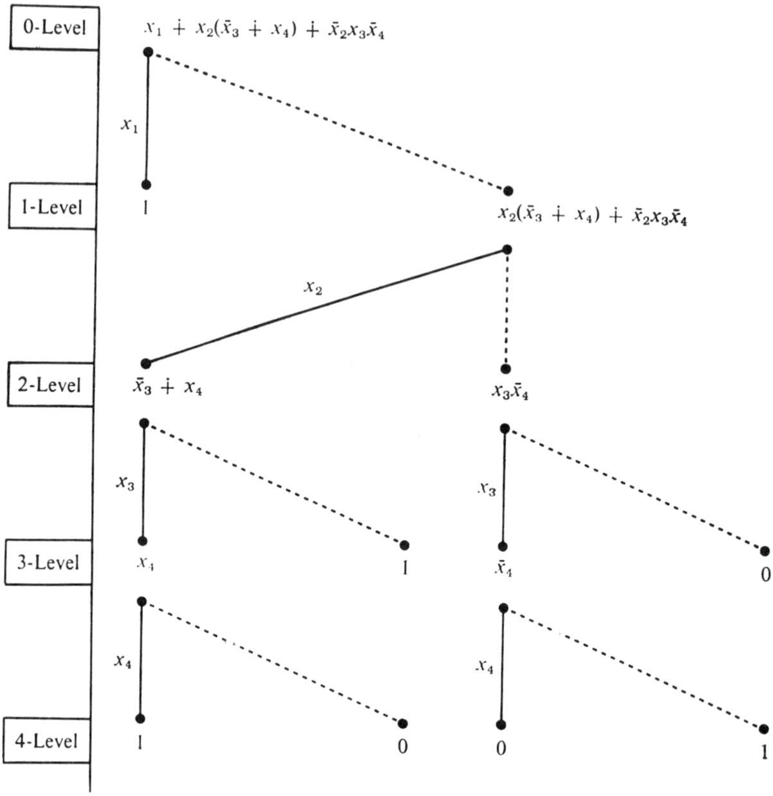

Fig. 4.2. The function tree of Example 4.1.

If x_1 is a variable with the largest coefficient, then by Theorems 3.6 and 3.9,

$$F_a > F_b \tag{4.2}$$

in which case F_1 reduces to

$$y(x_1 = 1) = F_1 = F_a$$

and it is appropriate to express the original function as

$$y = F(x_1,x_2,\ldots,x_n) = x_1F_1(x_2,x_3,\ldots,x_n) + F_0(x_2,x_3,\ldots,x_n)$$

Actually, the function F need not be linearly separable to take advantage of the simplification. The minimal requirement is that the function be positive unate, with the added property for some variable, say x_1, that

$$y(x_1 = 1,x_i = 0) > y(x_1 = 0,x_i = 1) \qquad \text{for} \quad i = 2,3,\ldots,n$$

Given this condition, then by Theorem 3.9,

$$y(x_1 = 1) = F_a(x_2,x_3,\ldots,x_n)$$

When the given function has this property, the formation and utilization of the tree are considerably simplified if the indicated ordering of the variables is used. Whether or not this property is satisfied, experience has shown that, under certain conditions, some orderings of the variables are usually better than others. This experience is summarized in Procedure 4.2, which can be considered as an addition to Procedure 4.1. Function trees constructed using Procedures 4.1 and 4.2 are denoted as *basic function trees.*

Procedure 4.2. Function Tree Variable Selection

(1) In Step 1 of Procedure 4.1, have F in MSP form.

(2) In Step 2 of Procedure 4.1, select x_1 as one of the largest variables as determined by Procedure 3.2. Note that it is unnecessary to order the variables beyond the point of finding the set of largest ones. From this set, determine a variable x_i where the factoring according to Eq. 4.1 is such that Eq. 4.2 is valid. In Step 2 of Procedure 4.1 let this $x_i = x_1$ and let $F_1 = F_a$ and $F_0 = F_b$. If no such x_i exists, then let x_1 be any variable from the set of largest.

(3) Since Step 3 of Procedure 4.1 is included in Step 4 (by setting $k = 2$), we consider only the k-level step.

For each function of the k-level of the tree, determine by the counting procedure (Procedure 3.2) the set of largest variables. The set intersection of these is the subset of largest variables that are common to all k-level functions. If this subset is nonempty, select as x_{k+1} one of the variables for which, in the factoring of each k-level function according to Eq. 4.1, the property of Eq. 4.2 is satisfied the most times. If it is satisfied no times for all variables from the subset, then any one is selected as x_{k+1}. If the subset is empty, then select as x_{k+1} that variable that is common to the most sets of largest variables.

Example 4.2. Figure 4.3 shows the function tree for the linearly separable switching function

$$y = x_1[x_2 \dotplus x_3 \dotplus x_4(x_5 \dotplus x_6) \dotplus x_5x_6] \dotplus x_2[x_3(x_4 \dotplus x_5) \dotplus x_4x_5] \\ \dotplus x_3x_4x_5x_6$$

Since the function is linearly separable, the largest variables of the functions on each level correspond to the variables with the largest coefficients. Also,

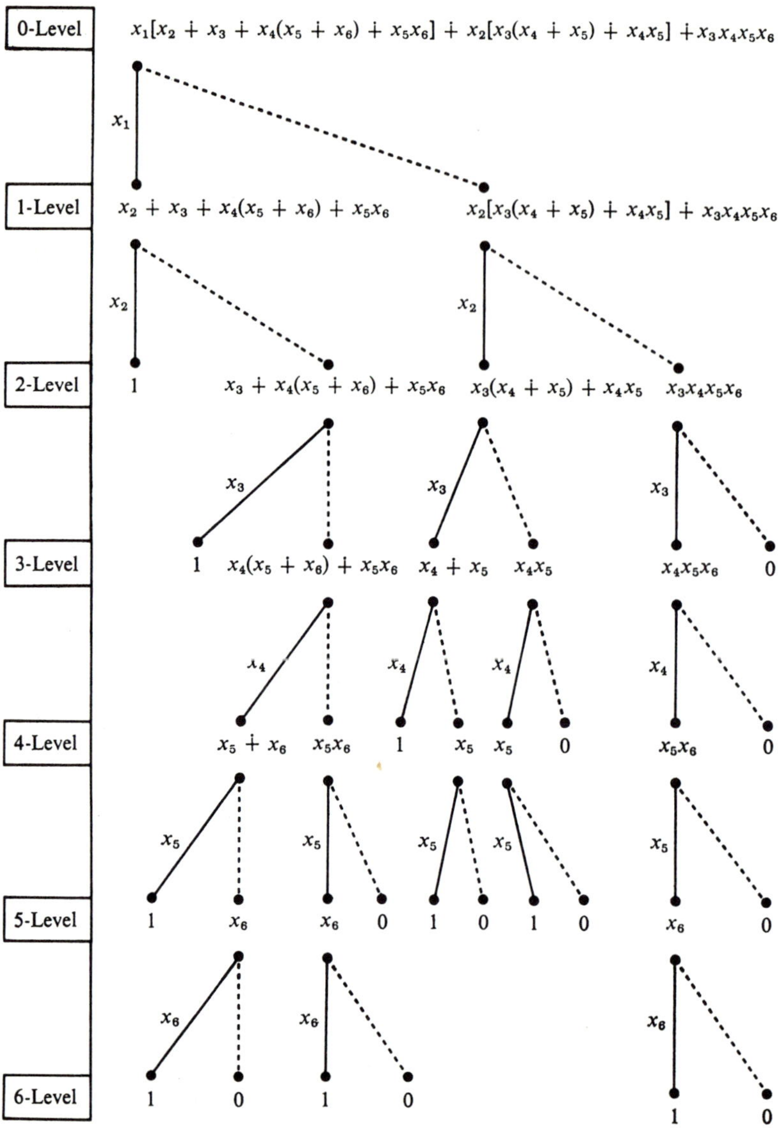

Fig. 4.3. The function tree of Example 4.2.

x_2, for example, is the variable that can have the largest coefficient for all functions on the 2-level, and a similar statement is true for each level.

The basic function tree defined by Procedures 4.1 and 4.2 is used for both single and multigate synthesis. Before using it for this purpose, however, it is necessary to consider some of its properties in detail.

4.2 PROPERTIES OF DESCENDANTS OF LINEARLY SEPARABLE FUNCTIONS

The basic function tree for an ℓ.s. switching function has some useful properties. It is convenient to consider these for positive linearly separable functions. Recall that this constraint does not reduce the generality of the results, since any unate function can be replaced by the positive representative function of its complementing symmetry class.

Consider the arbitrary positive linearly separable switching function

$$y = F(x_1,x_2,\ldots,x_n) = \langle f(x_1,x_2,\ldots,x_n)\rangle_{u:\ell} = \left\langle \sum_{j=1}^{n} a_j x_j \right\rangle_{u:\ell}$$

On the 1-level of the tree are the reduced functions

$$y(x_1 = 1) = F_1 \qquad \text{and} \quad y(x_1 = 0) = F_0$$

According to Theorem 3.3, each of these is linearly separable, and both can have the separating function

$$f_1(x_2,x_3,\ldots,x_n) = f(0,x_2,x_3,\ldots,x_n) = \sum_{j=2}^{n} a_j x_j$$

Moreover, since x_1 is selected according to Procedure 3.2 as a largest variable in the realization of F, it is a variable with a largest coefficient. Therefore, if F is factored as

$$F(x_1,x_2,\ldots,x_n) = x_1 F_a(x_2,x_3,\ldots,x_n) + F_b(x_2,x_3,\ldots,x_n)$$

then

$$F_a > F_b, \; F_1 = F_a, \text{ and } F_0 = F_b$$

Each 2-level function is a 1-level function reduced by assigning values to x_2. The 2-level functions are the 2-variable reduced functions F_{11}, F_{10}, F_{01}, and F_{00}. According to Theorem 3.11, each of these is linearly separable, and all can have the separating function

$$f_2(x_3,x_4,\ldots,x_n) = f(0,0,x_3,x_4,\ldots,x_n) = \sum_{j=3}^{n} a_j x_j$$

Moreover, since x_2 is selected according to Procedure 3.2 as a largest variable in all functions that separate both F_1 and F_0, it is a variable with the largest coefficient in $\sum_{j=2}^{n} a_j x_j$. Therefore, if each 1-level function F_δ is factored as

$$F_\delta(x_2,x_3,\ldots,x_n) = x_2 F_{a_\Delta}(x_3,x_4,\ldots,x_n) \dotplus F_{b_\Delta}(x_3,x_4,\ldots,x_n)$$

where $\delta = 0,1$ and where Δ is the decimal equivalent of δ, then

$$F_{a_\Delta} > F_{b_\Delta}, \quad F_{\delta 1} = F_{a_\Delta} \quad \text{and} \quad F_{\delta 0} = F_{b_\Delta}$$

Continuing this reasoning, each k-level function is a k-variable reduced function of F obtained by assigning values to $x_1,x_2,\ldots,x_k$. This, together with Theorem 3.12 and Procedure 3.2, establishes the following results.

Theorem 4.1. If

$$y = F(x_1,x_2,\ldots,x_n) = \left\langle \sum_{j=1}^{n} a_j x_j \right\rangle_{u:\ell}$$

is a positive linearly separable function, then every function on each k-level of the basic function tree is a linearly separable function, and all functions on each k-level can have the separating function

$$f_k(x_{k+1},x_{k+2},\ldots,x_n) = \sum_{j=k+1}^{n} a_j x_j$$

Moreover, for $j = k + 1, k + 2, \ldots, n$,

$$a_j \le a_k$$

Corollary 4.1.1. Under the conditions of Theorem 4.1, if each $(k - 1)$-level function is factored as

$$F_{\delta_1\delta_2\ldots\delta_{k-1}}(x_k,x_{k+1},\ldots,x_n) = x_k F_{a_\Delta}(x_{k+1}, x_{k+2},\ldots,x_n) \\ \dotplus F_{b_\Delta}(x_{k+1},x_{k+2},\ldots,x_n)$$

where $\delta_i = 0,1$ and where Δ is the decimal equivalent of $\delta_1\delta_2\ldots\delta_{k-1}$ then

$$F_{a_\Delta} > F_{b_\Delta}, \qquad F_{\delta_1\delta_2\ldots\delta_{k-1}1} = F_{a_\Delta}$$

and

$$F_{\delta_1\delta_2\ldots\delta_{k-1}0} = F_{b_\Delta}$$

According to Theorem 3.12, a separating-function representation for each k-level function can be obtained from the separating-function representation for the original switching function F. This is given by Corollary 4.1.2.

Corollary 4.1.2. Given the conditions and notation of Theorem 4.1 and Corollary 4.1.1, each k-level function of the tree can have the separating-function representation

$$F_{\delta_1\delta_2\ldots\delta_k}(x_{k+1},x_{k+2},\ldots,x_n) = \langle f_k(x_{k+1},x_{k+2},\ldots,x_n)\rangle_{u_{\delta_1\delta_2\ldots\delta_k}:\ell_{\delta_1\delta_2\ldots\delta_k}}$$

where

$$u_{\delta_1\delta_2\ldots\delta_k} = \text{minimum value of } f_k(x_{k+1},x_{k+2},\ldots,x_n) \geq u - \sum_{j=1}^{k} \delta_j a_j$$

if such a value exists

$= \infty$, otherwise

$$\ell_{\delta_1\delta_2\ldots\delta_k} = \text{maximum value of } f_k(x_{k+1},x_{k+2},\ldots,x_n) \leq \ell - \sum_{j=1}^{k} \delta_j a_j$$

if such a value exists

$= -\infty$, otherwise

Each function $F_{\delta_1\delta_2\ldots\delta_k}$ of the tree is the k-variable reduced function $y(x_1 = \delta_1,x_2 = \delta_2,\ldots,x_k = \delta_k)$, and there is a unique tree path between it and the 0-level function F. This path has the line between the j- and the $(j + 1)$-levels (where $j = 1,2,\ldots,k - 1$), solid or broken according to whether $\delta_j = 1$ or 0, respectively. Note, therefore, that the path implies the values assigned the variables $x_1,x_2,\ldots,x_k$ to obtain the reduced function and also implies the bounds on the gap in the separating-function representation of the reduced function.

4.3 DESCENDANT PAIR RANGES FOR LINEARLY SEPARABLE FUNCTIONS

Thus far we have considered how the properties of each function of the tree affect the properties of its descendant functions where the latter are reduced functions of the former. Now the reverse problem is considered: how do the properties of various functions of the tree affect the properties of those that are higher in the tree; that is, how are the ancestor functions influenced by their descendants?

If

$$y = F(x_1,x_2,\ldots,x_n) = \langle f(x_1,x_2,\ldots,x_n)\rangle_{u:\ell} = \left\langle \sum_{j=1}^{n} a_j x_j \right\rangle_{u:\ell}$$

is the positive linearly separable function on the 0-level, then on the 1-level are the functions F_1 and F_0, both of which are linearly separable where

$$y = F(x_1,x_2,\ldots,x_n) = x_1F_1(x_2,x_3,\ldots,x_n) \dotplus F_0(x_2,x_3,\ldots,x_n) \qquad (4.3)$$

Moreover, F_1 and F_0 can have the same separating function. The corresponding reverse problem is concerned with the properties that two given functions $F_1(x_2,x_3,\ldots,x_n)$ and $F_0(x_2,x_3,\ldots,x_n)$ must possess in order that the function F, defined by Eq. 4.3, be linearly separable. If F_1 and F_0 are completely unrestricted, then no results are available. If, however, it is required that F_1 and F_0 be comparable, then necessary and sufficient conditions can be set forth. Actually the only case of interest is when $F_1 > F_0$, since, otherwise, $y = F_0(x_2,x_3,\ldots,x_n)$ and no question exists. Given, therefore, that $F_1 > F_0$, then it is immediately evident that F_1 and F_0 are the 1-variable reduced functions of F where

$$y(x_1 = 1) = F_1(x_2,x_3,\ldots,x_n)$$

and

$$y(x_1 = 0) = F_0(x_2,x_3,\ldots,x_n)$$

Thus if F is to be linearly separable, it is necessary by Theorem 3.3 that both F_1 and F_0 be linearly separable. Moreover, by Theorem 4.1, if F can have the separating function

$$f(x_1,x_2,\ldots,x_n) = a_1x_1 + f_1(x_2,x_3,\ldots,x_n)$$

then it is necessary that F_1 and F_0 can both have the separating function

$$f_1(x_2,x_3,\ldots,x_n)$$

but with different gaps.

Now suppose that this is in fact the case: that F_1 and F_0 can both have the same separating function but with different gaps:

$$F_1 = \langle f_1(x_2,x_3,\ldots,x_n)\rangle_{u_1:\ell_1}$$
$$F_0 = \langle f_1(x_2,x_3,\ldots,x_n)\rangle_{u_0:\ell_0}$$

We now show that these are sufficient conditions to guarantee that F is ℓ.s. and that it can have a realization

$$F = \langle a_1x_1 + f_1(x_2,x_3,\ldots,x_n)\rangle_{u:\ell}$$

for some values of a_1, u, and ℓ.

Intuitively, we might try to select a_1, u, and ℓ so that for $x_1 = 0$, the realization for F "looks like" the realization for F_0 [which, remember, is $F(x_1 = 0)$], and for $x_1 = 1$, the realization for F "looks like" the realization for F_1 [which is $F(x_1 = 1)$]. The manner in which this is done is shown in Theorem 4.2. An interesting graphical interpretation of the theorem is given after the proof and the first corollary.

Theorem 4.2 (*Range Theorem*). Let $F_1(x_2,x_3,\ldots,x_n)$ and $F_0(x_2,x_3,\ldots,x_n)$ be arbitrary switching functions such that

$$F_1 > F_0$$

Then the switching function F where

$$y = F(x_1,x_2,\ldots,x_n) = x_1F_1(x_2,x_3,\ldots,x_n) + F_0(x_2,x_3,\ldots,x_n)$$

is linearly separable if and only if F_1 and F_0 are linearly separable and both can have the same separating function. Moreover, if

$$F_1(x_2,x_3,\ldots,x_n) = \langle f_1(x_2,x_3,\ldots,x_n)\rangle_{u_1:\ell_1}$$

and

$$F_0(x_2,x_3,\ldots,x_n) = \langle f_1(x_2,x_3,\ldots,x_n)\rangle_{u_0:\ell_0}$$

then a separating-function representation for F is

$$F(x_1,x_2,\ldots,x_n) = \langle a_1x_1 + f_1(x_2,x_3,\ldots,x_n)\rangle_{u:\ell} = \langle f(x_1,x_2,\ldots,x_n)\rangle_{u:\ell}$$

where the value of a_1 is selected anywhere within the range (interval)

$$(u_0 - \ell_1) > a_1 > (\ell_0 - u_1)$$

and where the resultant gap boundaries are

$$u = \min\{u_0, u_1 + a_1\}$$
$$\ell = \max\{\ell_0, \ell_1 + a_1\}$$

Proof. Let

$$F_1(x_2,x_3,\ldots,x_n) = \langle f_1(x_2,x_3,\ldots,x_n)\rangle_{u_1:\ell_1}$$

and

$$F_0(x_2,x_3,\ldots,x_n) = \langle f_1(x_2,x_3,\ldots,x_n)\rangle_{u_0:\ell_0}$$

Since $F_1 > F_0$, then

$$y(x_1 = 1) = F_1(x_2,x_3,\ldots,x_n), \quad y(x_1 = 0) = F_0(x_2,x_3,\ldots,x_n)$$

and according to Theorem 3.1, $u_1 < u_0$ and $\ell_1 < \ell_0$. We show that F is linearly separable by showing that

$$\langle a_1x_1 + f_1(x_2,x_3,\ldots,x_n)\rangle_{u:\ell}$$

is its separating-function representation.

First observe that $u > \ell$. This follows immediately from the allowable range for a_1.

Divide the points of the n cube on which F is defined into disjoint subsets $\{p^1\}$ and $\{p^0\}$ according to whether the x_1 component has unit or zero value, respectively. For each p^0,

$$F(p^0) = F_0(p^0) = \langle f_1(p^0)\rangle_{u_0:\ell_0}$$

and

$$\langle f(p^0)\rangle_{u:\ell} = \langle f_1(p^0)\rangle_{u:\ell}$$

However, $u \le u_0$ and $\ell \ge \ell_0$; therefore the gap $u:\ell$ is contained within the gap $u_0:\ell_0$ and, since

$$\langle f_1(p^0)\rangle_{u_0:\ell_0}$$

realizes $F(p^0)$, then so does

$$\langle f(p^0)\rangle_{u:\ell}$$

For each p^1,

$$F(p^1) = F_1(p^1) = \langle f_1(p^1)\rangle_{u_1:\ell_1}$$

and

$$\langle f(p^1)\rangle_{u:\ell} = \langle a_1 + f_1(p^1)\rangle_{u:\ell} = \langle f_1(p^1)\rangle_{(u-a_1):(\ell-a_1)}$$

by Theorem 2.4B. However, a_1 is selected so that

$$u \leq u_1 + a_1 \qquad \text{and} \quad \ell \geq \ell_1 + a_1$$

which implies that

$$u_1 \geq u - a_1 \qquad \text{and} \quad \ell_1 \leq \ell - a_1$$

or that the gap $u_1:\ell_1$ contains the gap $(u - a_1):(\ell - a_1)$. Then since

$$\langle f_1(p^1)\rangle_{u_1:\ell_1}$$

realizes $F(p^1)$, so does

$$\langle f_1(p^1)\rangle_{(u-a_1):(\ell-a_1)} = \langle f(p^1)\rangle_{u:\ell}$$

Therefore, $\langle f(p)\rangle_{u:\ell}$ realizes F for both subsets $\{p^0\}$ and $\{p^1\}$ and, hence, is a realization of F for all p of the n cube. This implies that F is linearly separable and establishes the validity of the theorem for all but the "only if" portion of the first part. This, however, is established by Theorem 3.3, since

$$y(x_1 = 1) = F_1(x_2,x_3,\ldots,x_n)$$

and

$$y(x_1 = 0) = F_0(x_2,x_3,\ldots,x_n) \quad |$$

The open interval defined by Theorem 4.2 from which the value of a_1 must be selected is called the *range* of a_1. This is denoted by

$$R(a_1) = U(a_1)*L(a_1)$$

where $U(a_1)$ is the *upper range boundary*

$$U(a_1) = u_0 - \ell_1$$

and $L(a_1)$ is the *lower range boundary*

$$L(a_1) = \ell_0 - u_1$$

In the statement of Theorem 4.2, the function F was formed by introducing the variable x_1, and this same variable was used in the corresponding separating function. It is permissible to use x_1 in the formation of F, but to use $\bar{x}_1$ in specifying its separating function. For this the corresponding coefficient is $-a_1$ or the negative of that of Theorem 4.2.

Corollary 4.2.1. Under the conditions of Theorem 4.2, a realization for F is

$$y = F(x_1,x_2,\ldots,x_n) = \langle -a_1\bar{x}_1 + f_1(x_2,x_3,\ldots,x_n)\rangle_{u':\ell'}$$

where the value of a_1 is selected anywhere within the range

$$R(a_1) = (u_0 - \ell_1)*(\ell_0 - u_1)$$

and where the resultant gap boundaries are

$$u' = \min\{u_1, u_0 - a_1\}$$
$$\ell' = \max\{\ell_1, \ell_0 - a_1\}$$

The range theorem has a convenient graphical interpretation. Since F_1 and F_0 are both linearly separable and have the same separating function and since $F_1 > F_0$, the gap of the map of F_1 is below that of F_0 as shown in Fig. 4.4*a*. For the case where the variable x_1 is used in the separating function for F, the range of a_1 is the set of values by which the gap of F_1 can

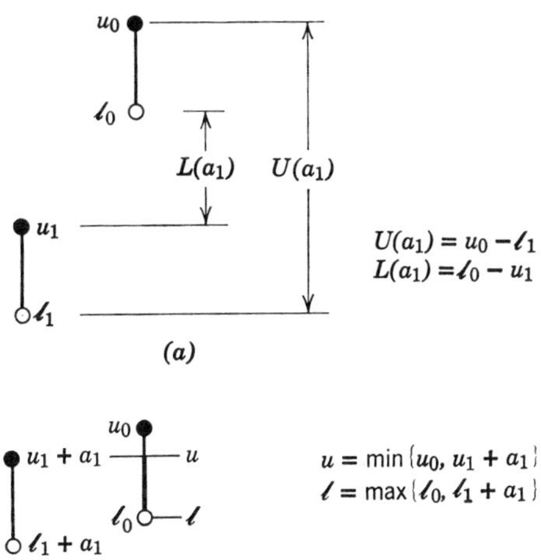

Fig. 4.4. Graphical interpretation of range. (*a*) Graphical interpretation of $U(a_1)$ and $L(a_1)$. (*b*) Relation between gaps of F_1, F_0, and F.

be displaced so that it has a nonvoid intersection with the gap of F_0. By the displaced gap of F_1 is meant the gap $(u_1 + a_1){:}(\ell_1 + a_1)$, and it must have a nonvoid intersection with gap $u_0{:}\ell_0$. It is evident from Fig. 4.4*a* that $U(a_1)$ and $L(a_1)$ represent the bounds on the displacement. Moreover, the gap of F is the intersection of the gap of F_0 and the gap of F_1 displaced by the selected value of a_1 as shown in Fig. 4.4*b*.

Actually, not just the gap but the entire map of F is closely related to the maps of F_1 and F_0. For this we define a displaced map. If a map of F_1 is

$$M_1 = \{p, f_1(p), F_1(p)\}$$

then an a_1-*displacement* of M_1 is the map

$$\{p, (f_1(p) + a_1), F_1(p)\}$$

In other words, each image point of M_1 in the graphical representation is increased by the amount a_1. It is evident from the proof of Theorem 4.2 that the map of F is the union of the map of F_0 with the a_1-displaced map of F_1. As previously indicated, the gap of F is the intersection of the gap of F_0 with the a_1-displaced gap of F.

A similar interpretation is possible when the separating function for F contains the variable $\bar{x}_1$. In this case the map of F is the union of the map of F_1 with the $(-a_1)$-displaced map of F_0 where, as before, the gap is the intersection of the gap of F_1 and the $(-a_1)$-displaced gap of F_0.

Example 4.3. Consider the linearly separable switching functions

$$y_1 = F_1(x_2, x_3) = x_2 \dotplus x_3 = \langle x_2 + x_3 \rangle_{1:0}$$

and

$$y_0 = F_0(x_2, x_3) = x_2 x_3 = \langle x_2 + x_3 \rangle_{2:1}$$

where $F_1 > F_0$. Then

$$y = F(x_1, x_2, x_3) = x_1 F_1(x_2, x_3) \dotplus F_0(x_2, x_3) = x_1(x_2 \dotplus x_3) \dotplus x_2 x_3$$

The maps for these functions are shown in Fig. 4.5*a*. It is clear from these maps that if the map of F_1 is displaced upward by an amount greater

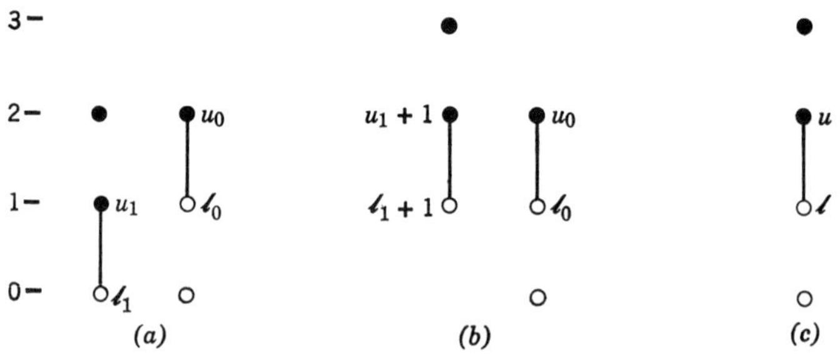

Fig. 4.5. Maps of Example 4.3. (*a*) F_1 map and F_0 map. (*b*) 1-displaced F_1 and F_0 map. (*c*) Map of F.

than 0 but less than 2, then the gap of the displaced F_1 map will have a nonvoid intersection with the gap of the map of F_0. Thus

$$R(a_1) = 2{*}0$$

If we select $a_1 = 1$, then Fig. 4.5*b* shows the map of F_0 and the 1-displaced map of F_1; and Fig. 4.5*c* shows the union of these two maps, which is the map of F. Moreover, the realization of F is

$$y = F(x_1,x_2,x_3) = \langle x_1 + x_2 + x_3 \rangle_{2:1}$$

In Example 4.3, when a_1 was selected as unity, the gap of F turned out to be of unit length, the same length as the gaps of F_1 and F_0. By considering a_1 as a variable within the range $R(a_1)$, it is evident that the a_1-displaced gap of F_1 slides past the gap of F_0 as a_1 takes on values from $L(a_1)$ to $U(a_1)$ and that any value other than $a_1 = 1$ would yield a gap for F of length less than 1. Thus, $a_1 = 1$ is the best choice with regard to sensitivity considerations.

Figure 4.6 shows in the three general situations how the gap length g of F varies as a function of a_1 when

$$R(a_1) = U(a_1){*}L(a_1)$$

When

$$a_1 = L(a_1)$$

the upper boundary of the a_1-displaced F_1 gap equals the lower boundary of the F_0 gap and the gap intersection is void, in which case $g = 0$. As a_1 is increased from $L(a_1)$, g increases linearly with unit slope; when

$$g = \min\{g_1,g_0\}$$

then g no longer increases with increasing a_1 but remains constant until

$$a_1 = U(a_1) - \min\{g_1,g_0\}$$

Increasing a_1 further causes g to decrease linearly (with slope -1) to zero. For

$$a_1 \geq U(a_1), \quad g = 0$$

This is summarized in Corollary 4.2.2.

Corollary 4.2.2. Under the conditions of Theorem 4.2, the value of g as a function of a_1 values within the range

$$R(a_1) = L(a_1){*}U(a_1)$$

is

$$g = \min \begin{Bmatrix} -L(a_1) + a_1 \\ g_1 \\ g_0 \\ U(a_1) - a_1 \end{Bmatrix}$$

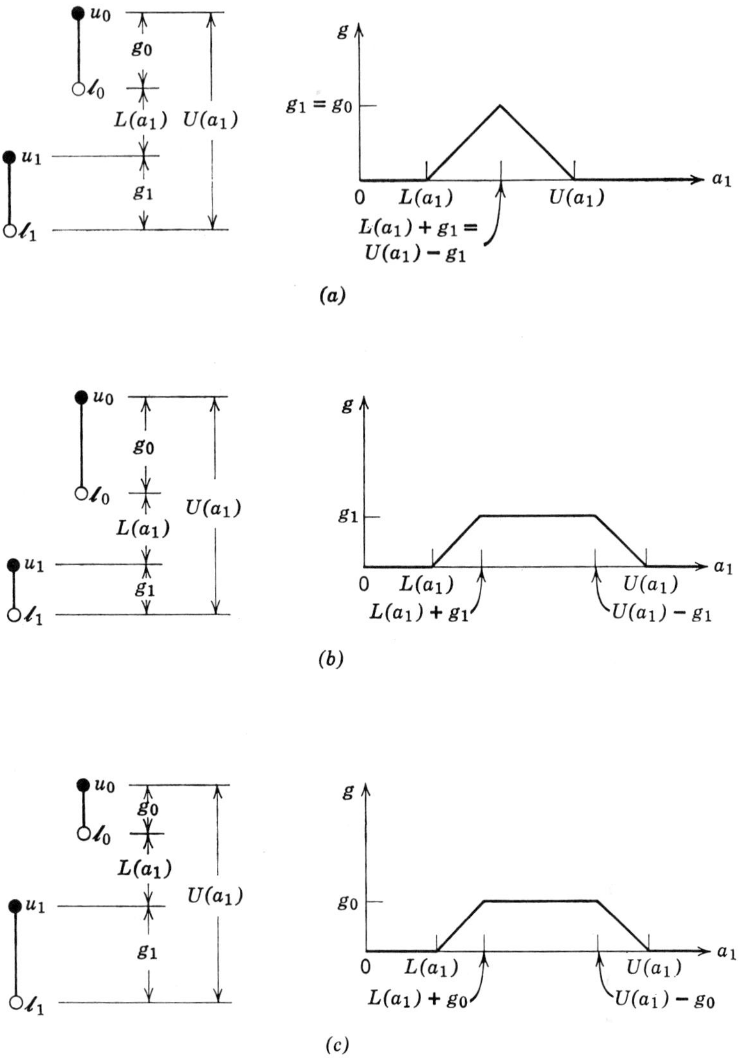

Fig. 4.6. Variation of g as a function of a_1. (*a*) $g_1 = g_0$. (*b*) $g_1 < g_0$. (*c*) $g_1 > g_0$.

An upper bound on the value of g is $\min\{g_1,g_0\}$ and this is attainable only if the length of $R(a_1)$ which is given by

$$| R(a_1) | = U(a_1) - L(a_1)$$

is greater than or equal to twice the $\min\{g_1,g_0\}$. Under these conditions, the smallest value of a_1 for which $g = \min\{g_1,g_0)$ is

$$a_1 = L(a_1) + \min\{g_1,g_0\}$$

and for values of a_1 such that

$$L(a_1) + \min\{g_1,g_0\} \le a_1 \le U(a_1) - \min\{g_1,g_0\}$$

the value of g is constant. However, the maximum relative gap with minimum fan-in is obtained for

$$a_1 = L(a_1) + \min\{g_1,g_0\}$$

and this value is normally used whenever possible.

Example 4.4. Show that, if

$$y = F(x_2,x_3,\ldots,x_n) = \langle f(x_2,x_3,\ldots,x_n)\rangle_{u:\ell}$$

is any positive linearly separable function, then

$$y_a = F_a(x_1,x_2,\ldots,x_n) = x_1 \dotplus F(x_2,x_3,\ldots,x_n)$$

and

$$y_b = F_b(x_1,x_2,\ldots,x_n) = x_1F(x_2,x_3,\ldots,x_n)$$

are both linearly separable; and find a realization for each.

The first two levels of the function trees for F_a and F_b are shown in Fig. 4.7*a*. Since F has realization $\langle f(x_2,x_3,\ldots,x_n)\rangle_{u:\ell}$ then according to Theorem 1.3, the constant functions 1 and 0 are linearly separable and can have separating functions

$$1 = \langle f(x_2,x_3,\ldots,x_n)\rangle_{0:-\infty}$$

and

$$0 = \langle f(x_2,x_3,\ldots,x_n)\rangle_{\infty:\sigma}$$

where

$$\sigma = \sum_{j=2}^{n} a_j$$

The gaps for the functions of the two trees are shown in Fig. 4.7*b*.

For the F_a realization the range for a_1 is

$$R(a_1) = (u + \infty)*(\ell - 0) = \infty*\ell = U(a_1)*L(a_1)$$

and the minimum value of a_1 that yields a maximum value of g_a is

$$a_1 = L(a_1) + g = \ell + (u - \ell) = u$$

The gap $u_a\!:\!\ell_a$ of F_a is the intersection of the gap $u\!:\!\ell$ and the a_1-displaced gap of the constant function 1 which is $u\!:\!-\infty$. Therefore

$$u_a\!:\!\ell_a = u\!:\!\ell$$

and

$$F_a(x_1,x_2,\ldots,x_n) = \langle ux_1 + f(x_2,x_3,\ldots,x_n)\rangle_{u:\ell}$$

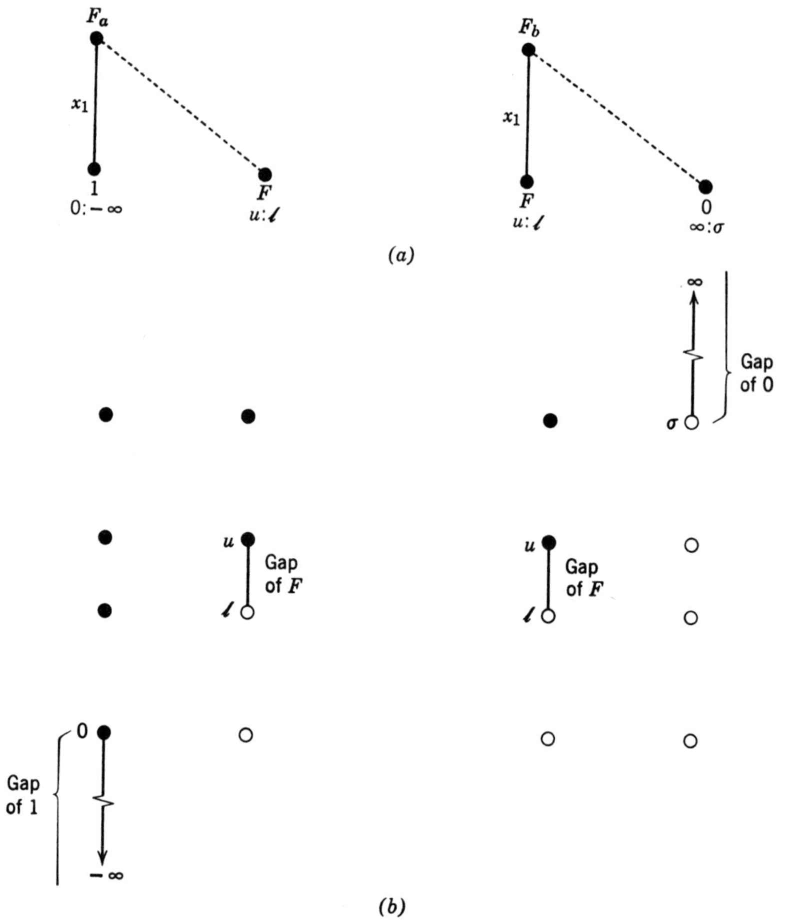

Fig. 4.7. Function tree and maps for Example 4.4. (*a*) Function-tree levels 0 and 1 for F_a and F_b. (*b*) Partial maps for F_a and F_b realizations.

For the F_b realization, the range for a_1 is

$$R(a_1) = (\infty - \ell)*(\sigma - u) = \infty*(\sigma - u) = U(a_1)*L(a_1)$$

and the minimum value of a_1 that yields a maximum value of g_b is

$$a_1 = L(a_1) + g = (\sigma - u) + (u - \ell) = \sigma - \ell$$

The gap $u_b:\ell_b$ of F_b is the intersection of the a_1-displaced gap of F which is $(u + \sigma - \ell):(\ell + \sigma - \ell)$ and the gap $\infty:\sigma$. Therefore

$$u_b:\ell_b = (\sigma + g):\sigma$$

and

$$F_b(x_2,x_3,\ldots,x_n) = \langle(\sigma - \ell)x_1 + f(x_2,x_3,\ldots,x_n)\rangle_{(\sigma+g):\sigma}$$

Example 4.5. Find a single-gate realization for

$$y = F(x_1,x_2,x_3,x_4) = x_1(x_2 \dotplus x_3 \dotplus x_4) \dotplus x_2(x_3 \dotplus x_4)$$

By the counting procedure, Procedure 3.2, x_1 is the largest variable and has the largest coefficient if F is linearly separable.

The corresponding reduced functions are

$$y(x_1 = 0) = x_2(x_3 \dotplus x_4)$$

and

$$y(x_1 = 1) = x_2 \dotplus x_3 \dotplus x_4$$

F is linearly separable if and only if $y(x_1 = 1)$ and $y(x_1 = 0)$ are linearly separable and can have the same separating function. This is in fact the case; for example,

$$y(x_1 = 0) = x_2(x_3 \dotplus x_4) = \langle 2x_2 + x_3 + x_4\rangle_{3:2}$$

and

$$y(x_1 = 1) = x_2 \dotplus x_3 \dotplus x_4 = \langle 2x_2 + x_3 + x_4\rangle_{1:0}$$

By the range theorem

$$R(a_1) = 3{*}1$$

Selecting $a_1 = 2$ gives the realization

$$y = \langle 2x_1 + 2x_2 + x_3 + x_4\rangle_{3:2}$$

With this type of reasoning all linearly separable functions of four or fewer variables and many functions of more variables can be realized almost by inspection. Later an algorithm will be presented for realizing any linearly separable function.

Example 4.6. Show that, if

$$y = F(x_2,x_3,\ldots,x_n) = \langle f(x_2,x_3,\ldots,x_n)\rangle_{u:\ell}$$

is a positive linearly separable function, then the function

$$y_c = F_c(x_1,x_2,\ldots,x_n) = \bar{x}_1F(x_2,x_3,\ldots,x_n) \dotplus x_1F^d(x_2,x_3,\ldots,x_n)$$

is linearly separable, and determine a realization for F_c.

According to Corollary 2.10.1, F^d can have the realization

$$F^d(x_2,x_3,\ldots,x_n) = \langle f(x_2,x_3,\ldots,x_n)\rangle_{(\sigma-\ell):(\sigma-u)}$$

where

$$\sigma = \sum_{j=2}^{n} a_j$$

Since F and F^d can have the same separating function, then either

$$F \geq F^d \qquad \text{or} \qquad F \leq F^d$$

The procedure is similar regardless of which containment exists. It is illustrated for

$$F < F^d$$

Under this condition

$$y_c = F(x_2,x_3,\ldots,x_n) + x_1F^d(x_2,x_3,\ldots,x_n)$$

and

$$y_c(x_1 = 1) = F^d(x_2,x_3,\ldots,x_n)$$
$$y_c(x_1 = 0) = F(x_2,x_3,\ldots,x_n)$$

Since both are linearly separable and can have the same separating function, F_c is linearly separable. The first two levels of the function tree are shown in Fig. 4.8. The range for a_1 is

$$R(a_1) = (u - (\sigma - u))*(\ell - (\sigma - \ell)) = (2u - \sigma)*(2\ell - \sigma)$$

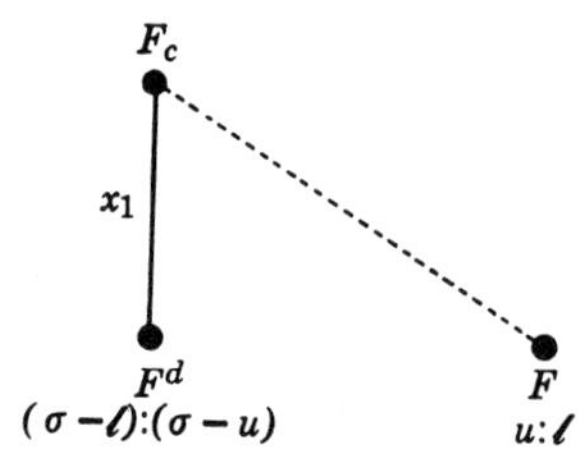

Fig. 4.8. A partial function tree for Example 4.6.

The gap lengths for both F and F^d are

$$g = g^d = u - \ell$$

Hence the value for a_1 is selected as

$$a_1 = (2\ell - \sigma) + (u - \ell) = u + \ell - \sigma$$

The realization for F_c is

$$y_c = F_c(x_1,x_2,\ldots,x_n) = \langle(u + \ell - \sigma)x_1 + f(x_2,x_3,\ldots,x_n)\rangle_{u:\ell}$$

4.4 GENERAL RANGE THEOREMS FOR LINEARLY SEPARABLE FUNCTIONS

In this section the concept of coefficient range is generalized. Let

$$y = F(x_1,x_2,\ldots,x_n) = \langle f(x_1,x_2,\ldots,x_n)\rangle_{u:\ell} = \left\langle \sum_{j=1}^{n} a_j x_j \right\rangle_{u:\ell}$$

be a positive linearly separable switching function with the function tree of Fig. 4.1. Since F is linearly separable, then according to Theorem 4.1 all 2-level functions of the complete tree are linearly separable and can have the same separating function as is the case for the 1-level functions. Moreover, F_{11}, F_{10}, and F_{01}, F_{00} are each descendant pairs of F_1 and F_0, respectively; hence the relations between the maps of the descendant pairs and the corresponding ancestor function is the same as described in Section 4.3. In particular, the map of F_1 is the union of the map of F_{10} with the a_2-displaced map of F_{11}. Similarly, the map of F_0 is the union of the map of F_{00} with the a_2-displaced map of F_{01}. In addition, the map of F is the union of the map of F_0 with the a_1-displaced map of F_1. Combining all of these results defines the map of F as the union of the following maps:

(1) The map of F_{00}
(2) The a_2-displaced map of F_{01}
(3) The a_1-displaced map of F_{10}
(4) The $(a_1 + a_2)$-displaced map of F_{11}

Moreover, the gap of F is the intersection of the gaps of the maps of (1), (2), (3), and (4). Since this intersection must be nonvoid, this places restrictions on the values of a_1 and a_2. The original and displaced gaps of the 2-level functions are illustrated in Fig. 4.9.

One necessary restriction is that each pair of displaced gaps must have a nonvoid intersection. In this section we consider the corresponding constraints on coefficient values. In the next section the complete set of restrictions is studied.

Each pair of 2-level functions implies a constraint on a_1 or a_2 or on a linear combination of a_1 and a_2, because the coefficients must be selected so that the appropriately displaced gaps intersect. Consider the restriction that the gap of F_{00} and the a_2-displaced gap of F_{01} must intersect. This implies a range on a_2, as previously discussed; namely,

$$R(a_2) = (u_{00} - \ell_{01})*(\ell_{00} - u_{01})$$

Similarly, the restriction that the a_2-displaced gap of F_{11} must intersect the gap of F_{10} also implies a range on a_2 given by

$$R(a_2) = (u_{10} - \ell_{11})*(\ell_{10} - u_{11})$$

The fact that the gaps of F_{11} and F_{10} are both displaced by a_1 to form the gap of F does not affect this range for a_2.

In exactly similar fashion, it is clear from Fig. 4.9 that the required intersection of the displaced gaps of F_{11} and F_{01} implies a range on a_1, as does the intersection of the displaced gaps of F_{10} and F_{00}. These ranges are all on individual coefficients.

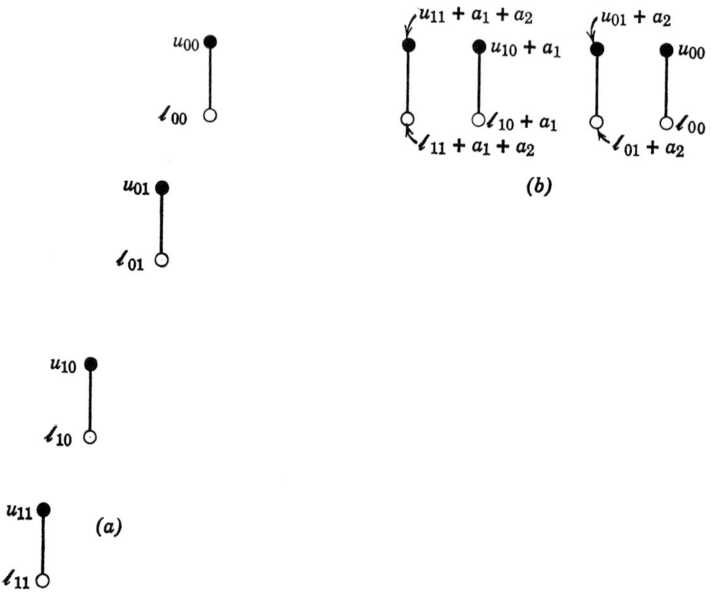

Fig. 4.9. (*a*) Gaps for 2-level functions. (*b*) Displaced 2-level gaps.

In addition, there are ranges on combinations of a_1 and a_2. Because the gap of F_{11} displaced by $a_1 + a_2$ must intersect the gap of F_{00}, there is implied a range on $a_1 + a_2$ where

$$R(a_1 + a_2) = (u_{00} - \ell_{11})*(\ell_{00} - u_{11})$$

Similarly, since the a_1-displaced gap of F_{10} must intersect the a_2-displaced gap of F_{01}, there is a range on $a_1 - a_2$ given by

$$R(a_1 - a_2) = (u_{01} - \ell_{10})*(\ell_{01} - u_{10})$$

Thus each pair of 2-level functions implies a range on some linear combination of a_1 and a_2. In general notation, let $F_{\delta_1\delta_2}$ and $F_{\gamma_1\gamma_2}$ where

$\delta_1,\delta_2,\gamma_1,\gamma_2 = 0,1$ represent any pair of 2-level functions. Then the constraint implied by these functions is

$$R(a_1(\delta_1 - \gamma_1) + a_2(\delta_2 - \gamma_2)) = (u_{\gamma_1\gamma_2} - \ell_{\delta_1\delta_2})*(\ell_{\gamma_1\gamma_2} - u_{\delta_1\delta_2})$$

Note that the above argument can be applied to any pair of functions on any level of the complete function tree. The generalized result is the following:

Theorem 4.3. Let $y = F(x_1,x_2,\ldots,x_n)$ be any positive linearly separable switching function and let

$$F_{\delta_1\delta_2\ldots\delta_k}(x_{k+1},x_{k+2},\ldots,x_n) = \langle f_k(x_{k+1},x_{k+2},\ldots,x_n)\rangle_{u_{\delta_1\delta_2\ldots\delta_k}:\ell_{\delta_1\delta_2\ldots\delta_k}}$$

and

$$F_{\gamma_1\gamma_2\ldots\gamma_k}(x_{k+1},x_{k+2},\ldots,x_n) = \langle f_k(x_{k+1},x_{k+2},\ldots,x_n)\rangle_{u_{\gamma_1\gamma_2\ldots\gamma_k}:\ell_{\gamma_1\gamma_2\ldots\gamma_k}}$$

be any two k-level functions in the complete function tree of F. If a realization for F is of the form:

$$F(x_1,x_2,\ldots,x_n) = \left\langle \sum_{j=1}^{k} a_jx_j + f_k(x_{k+1},x_{k+2},\ldots,x_n)\right\rangle_{u:\ell}$$

then the coefficients $a_1,a_2,\ldots,a_k$ are constrained to the ranges

$$R\left(\sum_{j=1}^{k} a_j(\delta_j - \gamma_j)\right) = (u_{\gamma_1\gamma_2\ldots\gamma_k} - \ell_{\delta_1\delta_2\ldots\delta_k})*(\ell_{\gamma_1\gamma_2\ldots\gamma_k} - u_{\delta_1\delta_2\ldots\delta_k})$$

and the gap $u:\ell$ has the constraint

$$u \le \min\left\{\begin{matrix} u_{\gamma_1\gamma_2\ldots\gamma_k} + \sum_{j=1}^{k}\gamma_ja_j \\ u_{\delta_1\delta_2\ldots\delta_k} + \sum_{j=1}^{k}\delta_ja_j \end{matrix}\right\}$$

$$\ell \ge \max\left\{\begin{matrix} \ell_{\gamma_1\gamma_2\ldots\gamma_k} + \sum_{j=1}^{k}\gamma_ja_j \\ \ell_{\delta_1\delta_2\ldots\delta_k} + \sum_{j=1}^{k}\delta_ja_j \end{matrix}\right\}$$

Example 4.7. Consider the positive ℓ.s. switching function

$$y = F(x_1,x_2,\ldots,x_n) = x_1x_2 \dotplus (x_1 \dotplus x_2)F_2(x_3,x_4,\ldots,x_n)$$

where F_2 is a positive ℓ.s. function with realization

$$y_2 = F_2(x_3,x_4,\ldots,x_n) = \langle f_2(x_3,x_4,\ldots,x_n)\rangle_{u_2:\ell_2} = \left\langle\sum_{j=3}^{n} a_jx_j\right\rangle_{u_2:\ell_2}$$

Let

$$\sigma = \sum_{j=3}^{n} a_j$$

The first three levels of the function tree for F are shown in Fig. 4.10. The four 2-level functions considered by pairs imply 6 ranges:

(1) F_{11} and F_{10} imply $R(a_2) = \infty * \ell_2$
(2) F_{01} and F_{00} imply $R(a_2) = \infty * (\sigma - u_2)$
(3) F_{11} and F_{01} imply $R(a_1) = \infty * \ell_2$
(4) F_{10} and F_{00} imply $R(a_1) = \infty * (\sigma - u_2)$
(5) F_{11} and F_{00} imply $R(a_1 + a_2) = \infty * \sigma$
(6) F_{10} and F_{01} imply $R(a_1 - a_2) = (u_2 - \ell_2) * -(u_2 - \ell_2)$

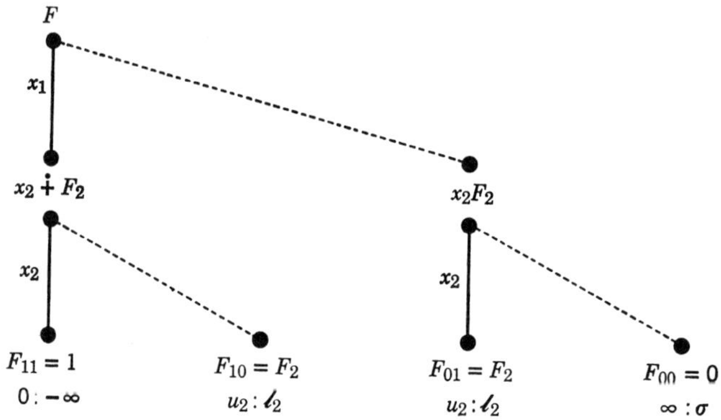

Fig. 4.10. Function tree for Example 4.7.

Since F is symmetric in variables x_1 and x_2, then according to Theorem 3.7, a_1 and a_2 can always be equal. From (6) the range on $a_1 - a_2$ indicates the amounts by which they can differ—essentially by plus or minus the gap length.

4.5 THE FUNCTION TREE AND SIMULTANEOUS INEQUALITIES FOR ℓ.s. FUNCTIONS

The coefficient ranges are closely related to the simultaneous inequalities defined for the points of the n cube. Consider again the four 2-level functions of the complete tree of Fig. 4.1. These functions are related to the 0-level function:

$$F(x_1,x_2,\ldots,x_n) = x_1x_2F_{11}(x_3,x_4,\ldots,x_n) \dotplus x_1F_{10}(x_3,x_4,\ldots,x_n) \dotplus x_2F_{01}(x_3,x_4,\ldots,x_n) \dotplus F_{00}(x_3,x_4,\ldots,x_n)$$

Moreover, it follows from Section 4.4 that the map of F is the union of the maps of all $F_{\delta_1\delta_2}$ displaced by $\delta_1a_1 + \delta_2a_2$ where the gap of F is the intersection of the corresponding displaced gaps. The four displaced gaps are

$$u_{00}:\ell_{00},\ (u_{01}+a_2):(\ell_{01}+a_2),\ (u_{10}+a_1):(\ell_{10}+a_1),$$
$$(u_{11}+a_1+a_2):(\ell_{11}+a_1+a_2)$$

The values of a_1 and a_2 must be selected so that these gaps have a nonvoid intersection in order that F have a nonzero gap.

The four gaps will have a nonvoid intersection if and only if each displaced gap upper boundary is greater than each displaced gap lower boundary. For this, the coefficients a_1 and a_2 must satisfy the following twelve inequalities simultaneously.

$$\begin{aligned}
&u_{00} > \ell_{01} + a_2\\
&u_{00} > \ell_{10} + a_1\\
&u_{00} > \ell_{11} + a_1 + a_2\\
&u_{10} + a_1 > \ell_{00}\\
&u_{10} + a_1 > \ell_{01} + a_2\\
&u_{10} + a_1 > \ell_{11} + a_1 + a_2\\
&u_{01} + a_2 > \ell_{00}\\
&u_{01} + a_2 > \ell_{10} + a_1\\
&u_{01} + a_2 > \ell_{11} + a_1 + a_2\\
&u_{11} + a_1 + a_2 > \ell_{00}\\
&u_{11} + a_1 + a_2 > \ell_{01} + a_2\\
&u_{11} + a_1 + a_2 > \ell_{10} + a_1
\end{aligned}$$

These inequalities can be rewritten as

$$\begin{aligned}
&u_{00} - \ell_{01} > a_2 > \ell_{00} - u_{01}\\
&u_{10} - \ell_{11} > a_2 > \ell_{10} - u_{11}\\
&u_{00} - \ell_{10} > a_1 > \ell_{00} - u_{10}\\
&u_{01} - \ell_{11} > a_1 > \ell_{01} - u_{11}\\
&u_{00} - \ell_{11} > a_1 + a_2 > \ell_{00} - u_{11}\\
&u_{01} - \ell_{10} > a_1 - a_2 > \ell_{01} + u_{10}
\end{aligned}$$

These are exactly the range conditions given by Theorem 4.3. This shows that all range conditions must be simultaneously satisfied to obtain a realization of F and that this is a sufficient as well as a necessary condition.

The same reasoning that was used on the 2-level functions applies to all functions on any other level of the complete function tree. The results are as follows:

Theorem 4.4. Let $y = F(x_1,x_2,\ldots,x_n)$ be a positive ℓ.s. function and let the set of functions on any k-level of the complete tree of F be

$$F_{\delta_1\delta_2\ldots\delta_k} \qquad \text{where} \quad \delta_j = 0,1$$

with corresponding realization

$$F_{\delta_1\delta_2\ldots\delta_k}(x_{k+1},x_{k+2},\ldots,x_n) = \langle f_k(x_{k+1},x_{k+2},\ldots,x_n)\rangle_{u_{\delta_1\delta_2\ldots\delta_k}:\ell_{\delta_1\delta_2\ldots\delta_k}}$$

A necessary and sufficient condition that F have a realization of the form

$$F(x_1,x_2,\ldots,x_n) = \left\langle \sum_{j=1}^{k} a_jx_j + f_k(x_{k+1},x_{k+2},\ldots,x_n)\right\rangle_{u:\ell}$$

is that the values of a_j for $j = 1,2,\ldots,k$ be selected so that all of the displaced gaps

$$\left(u_{\delta_1\delta_2\ldots\delta_k} + \sum_{j=1}^{k} \delta_ja_j\right):\left(\ell_{\delta_1\delta_2\ldots\delta_k} + \sum_{j=1}^{k} \delta_ja_j\right)$$

have a nonvoid intersection. The resultant intersection is the gap of F. The map of F is the union of the $(\sum_{j=1}^{k} \delta_ja_j)$-displaced maps of the k-level functions.

The requirement of Theorem 4.4 (that for any k, all k-level displaced gaps intersect) corresponds to a set of inequalities that must be simultaneously satisfied. These can be expressed in either of two ways:

(1) That each displaced k-level gap upper boundary exceed each displaced k-level gap lower boundary as previously discussed.

(2) That there exist numbers u and ℓ, where

$$u > \ell$$

such that each displaced k-level gap upper boundary equals or exceeds u and each displaced k-level gap lower boundary equals or is exceeded by ℓ. That is,

$$\begin{aligned} u_{\delta_1\delta_2\ldots\delta_k} + \sum_{j=1}^{k} \delta_ja_j &\geq u \\ \ell_{\delta_1\delta_2\ldots\delta_k} + \sum_{j=1}^{k} \delta_ja_j &\leq \ell \end{aligned} \tag{4.4}$$

In the (2) set there are 2^{k+1} inequalities, 2 for each k-level function. If any of the gap boundaries are infinite, however, they can be ignored, since an infinite gap boundary implies no constraint. The (2) set of inequalities is in every way equivalent to the (1) set, and either set can be derived from the other. For our purposes, however, set (2) is the more convenient.

Corollary 4.4.1. Satisfying the inequalities of Eq. 4.4, which are those that must be satisfied to insure that the gaps of Theorem 4.4 have a nonvoid intersection, is equivalent to satisfying simultaneously all paired range requirements of Theorem 4.3.

If the range requirements of Theorem 4.3 or the inequalities implied by Theorem 4.4 are inconsistent and, hence, cannot be satisfied, this does not imply that F is not ℓ.s. The only conclusion that can be drawn is that if F is ℓ.s., a different separating function $f_k(x_{k+1},x_{k+2},\ldots,x_n)$ must be used for the k-level functions.

Example 4.8. Consider again the switching function of Example 4.7:

$$y = F(x_1,x_2,\ldots,x_n) = x_1x_2 \dotplus (x_1 \dotplus x_2)F_2(x_3,x_4,\ldots,x_n)$$

where

$$F_2(x_3,x_4,\ldots,x_n) = \langle f_2(x_3,x_4,\ldots,x_n)\rangle_{u_2:\ell_2}$$

is a positive ℓ.s. function. From Fig. 4.10, the four 2-level displaced gaps are

$$\infty{:}\sigma,\ (u_2 + a_2){:}(\ell_2 + a_2),\ (u_2 + a_1){:}(\ell_2 + a_1),\ (0 + a_1 + a_2){:}-\infty$$

The intersection of the displaced gaps implies the finite inequalities

$$\begin{aligned} u_2 + a_2 &> \sigma \\ u_2 + a_2 &> \ell_2 + a_1 \\[1ex] u_2 + a_1 &> \sigma \\ u_2 + a_1 &> \ell_2 + a_2 \\[1ex] a_1 + a_2 &> \sigma \\ a_1 + a_2 &> \ell_2 + a_2 \\ a_1 + a_2 &> \ell_2 + a_1 \end{aligned}$$

Moreover, these inequalities are satisfied when the coefficients are within the ranges of Example 4.7. Because F is symmetric in x_1 and x_2, the consistency of the inequalities, according to Theorem 3.7, is not affected by letting $a_1 = a_2$. This simplifies the inequalities to the set

$$\begin{aligned} u_2 + a_1 &> \sigma \\ 2a_1 &> \sigma \\ a_1 &> \ell_2 \end{aligned}$$

and these imply that

$$\begin{aligned} a_1 &> \sigma - u_2 \\ a_1 &> \frac{\sigma}{2} \\ a_1 &> \ell_2 \end{aligned}$$

These are the necessary and sufficient conditions on the coefficient $a_1 = a_2$.

Consider the bottom or n-level of the function tree for a positive unate function F and assume that the tree is complete in that all constant functions have been included. There are 2^n n-level functions, each of which is either the 0 or the 1 constant function. Let each n-level function be denoted by

$$F_{\delta_1\delta_2\ldots\delta_n}(\Phi)$$

where Φ denotes the empty subset of the variables $x_1,x_2,\ldots,x_n$ and where $\delta_j = 0$ or 1 according to whether the unique tree path from this function to F has a broken or a solid line between the $(j-1)$- and the j-levels. The n-level functions have the property that they are all ℓ.s. and all can have the same separating function irrespective of whether or not F is ℓ.s. According to Theorem 1.3, all constant functions can be realized by any separating function; therefore, the simplest one that can be selected for the n-level functions is

$$f_n(\Phi) = 0$$

In other words, we select as a separating function for all n-level switching functions the constant function: identically zero. According to Theorem 1.3, the realizations for the n-level functions become

$$F_{\delta_1\delta_2\ldots\delta_n}(\Phi) = 1 = \langle 0\rangle_{0:-\infty} \tag{4.5}$$

and

$$F_{\delta_1\delta_2\ldots\delta_n}(\Phi) = 0 = \langle 0\rangle_{\infty:0} \tag{4.6}$$

Moreover, these realizations for the n-level functions are completely general. According to Theorem 4.1, there are separating functions for each ℓ.s. switching function F such that the separating function for the n-level reduced functions of F is

$$f_n(\Phi) = 0$$

Having selected a general realization for the n-level functions, Theorem 4.4 can now be applied to this level. Because of the generality of the realization, the theorem implies that a necessary and sufficient condition for the 0-level F to be ℓ.s. is that there exist a set of coefficients $a_1,a_2,\ldots,a_n$ such that all of the n-level displaced gaps

$$\left(u_{\delta_1\delta_2\ldots\delta_n} + \sum_{j=1}^{n} \delta_j a_j\right):\left(\ell_{\delta_1\delta_2\ldots\delta_n} + \sum_{j=1}^{n} \delta_j a_j\right)$$

have a nonvoid intersection where that intersection is the gap $u{:}\ell$ of F. Moreover the map of F is the union of the $(\sum_{j=1}^{n} \delta_j a_j)$-displaced maps of the n-level functions.

The condition that these displaced gaps intersect is that the inequalities of Eq. 4.4 be satisfied, and for the n-level these become

$$
\begin{aligned}
u_{\delta_1\delta_2\ldots\delta_n} + \sum_{j=1}^{n} \delta_j a_j \geq u \\
\ell_{\delta_1\delta_2\ldots\delta_n} + \sum_{j=1}^{n} \delta_j a_j \leq \ell
\end{aligned}
\tag{4.7}
$$

Although there are 2^{n+1} inequalities here, half of them are trivial because each gap has one infinite boundary as indicated by Eqs. 4.5 and 4.6. If

$$F_{\delta_1\delta_2\ldots\delta_n}(\Phi) = 1$$

then the gap of this function is $0{:}-\infty$ and the corresponding nontrivial inequality from Eq. 4.7 is

$$\sum_{j=1}^{n} \delta_j a_j \geq u \tag{4.8}$$

Similarily, if

$$F_{\delta_1\delta_2\ldots\delta_n}(\Phi) = 0$$

then its gap is $\infty{:}0$ and the corresponding nontrivial inequality from Eq. 4.7 is

$$\sum_{j=1}^{n} \delta_j a_j \leq \ell \tag{4.9}$$

These are exactly the inequalities of the R_0 set of Eq. 1.24. Recall that $\delta_j = 0$ or 1 according to whether the path from $F_{\delta_1\delta_2\ldots\delta_n}$ to F has a broken or a solid line, respectively, between the $(j-1)$- and the j-levels. Using this, we have the following result.

> ***Corollary 4.4.2.*** Each of the functions on the n-level of a complete function tree of a positive ℓ.s. function F implies one of the inequalities of the R_0 set (Eq. 1.24) that must be satisfied to realize F. Thus, the complete tree contains the necessary and sufficient conditions for realization. The inequality associated with each $F_{\delta_1\delta_2\ldots\delta_n}$ is the sum of the coefficients associated with the solid lines in the path from $F_{\delta_1\delta_2\ldots\delta_n}$ to F, and this sum equals or exceeds u or equals or is exceeded by ℓ, according to whether $F_{\delta_1\delta_2\ldots\delta_n}$ equals 1 or 0, respectively.

4.6 THE INCOMPLETE TREE AND THE MSP FORMS OF F AND $\bar{F}$

In this section the incomplete basic-function tree (in which branches are not continued below constant functions) is shown to also contain sufficient information to realize ℓ.s. functions. Let $y = F(x_1,x_2,\ldots,x_n)$ be a positive

ℓ.s. function in MSP form for which, on each k-level of its function tree, a variable is removed which can have a largest coefficient in the realizations of all k-level functions; that is, the tree is basic. Thus, any particular k-level function is factored as

$$F_{\delta_1\delta_2\ldots\delta_k}(x_{k+1},x_{k+2},\ldots,x_n) = x_{k+1}F_{\delta_1\delta_2\ldots\delta_k 1}(x_{k+2},x_{k+3},\ldots,x_n) \dotplus F_{\delta_1\delta_2\ldots\delta_k 0}(x_{k+2},x_{k+3},\ldots,x_n)$$

where its descendant functions are $F_{\delta_1\delta_2\ldots\delta_k 1}$ and $F_{\delta_1\delta_2\ldots\delta_k 0}$. It is important to note that any particular term of $F_{\delta_1\delta_2\ldots\delta_k}$ contributes a term to either $F_{\delta_1\delta_2\ldots\delta_k 1}$ or $F_{\delta_1\delta_2\ldots\delta_k 0}$ but not to both. The terms of F that do not contain x_{k+1} are the contributions to $F_{\delta_1\delta_2\ldots\delta_k 0}$. The terms of F that have x_{k+1} as a factor are the contributions to $F_{\delta_1\delta_2\ldots\delta_k 1}$ with $x_{k+1} = 1$. Thus each term of F defines a unique path down the tree that terminates in the constant 1 function on some q-level. Since all variables are positive, this term has unit value whenever the x_j variables associated with the solid lines of its path all have unit value. We see, therefore, that each 1 function of the incomplete tree of F corresponds to a unique term of the MSP form of F where the associated term is the product of the variables of the solid lines in the path between the 1 function and F.

A similar property is associated with each 0 function of the incomplete tree of F. In this case each 0 function of the F tree corresponds to a term of the MSP form of $\bar{F}$ where the term is the product of the variables of the broken lines in the path from the 0 function to F. Recall that the variable associated with a broken line is the complement of the corresponding solid-line variable.

To justify this, consider the function tree of $\bar{F}$ where the variables are removed in the same order as for the F tree. On any k-level of the F tree,

$$F_{\delta_1\delta_2\ldots\delta_k}(x_{k+1},x_{k+2},\ldots,x_n) = x_{k+1}F_{\delta_1\delta_2\ldots\delta_k 1}(x_{k+2},x_{k+3},\ldots,x_n) \dotplus F_{\delta_1\delta_2\ldots\delta_k 0}(x_{k+2},x_{k+3},\ldots,x_n)$$

where $F_{\delta_1\delta_2\ldots\delta_k 1}$ and $F_{\delta_1\delta_2\ldots\delta_k 0}$ are the descendant functions and where

$$F_{\delta_1\delta_2\ldots\delta_k 1} > F_{\delta_1\delta_2\ldots\delta_k 0}$$

This implies that

$$\begin{aligned}\bar{F}_{\delta_1\delta_2\ldots\delta_k}(x_{k+1},x_{k+2},\ldots,x_n) &= [\bar{x}_{k+1} \dotplus \bar{F}_{\delta_1\delta_2\ldots\delta_k 1}(x_{k+2},x_{k+3},\ldots,x_n)] \\ &\quad \times \bar{F}_{\delta_1\delta_2\ldots\delta_k 0}(x_{k+2},x_{k+3},\ldots,x_n) \\ &= \bar{x}_{k+1}\bar{F}_{\delta_1\delta_2\ldots\delta_k 0}(x_{k+2},x_{k+3},\ldots,x_n) \\ &\quad \dotplus \bar{F}_{\delta_1\delta_2\ldots\delta_k 1}(x_{k+2},x_{k+3},\ldots,x_n)\end{aligned}$$

since

$$\bar{F}_{\delta_1\delta_2\ldots\delta_k 1} < \bar{F}_{\delta_1\delta_2\ldots\delta_k 0}$$

From this it is evident that each function in the tree of $\bar{F}$ is the complement of the corresponding function in the tree of F. Moreover, since each term of $\bar{F}$ has all variables negative (F has all variables positive), each 1 function of the incomplete tree of $\bar{F}$ corresponds to a term of $\bar{F}$ where this term consists of the product of the variables associated with the broken lines in the path from the 1 function to $\bar{F}$. Since the structure and branch or line labelings of the F and $\bar{F}$ trees are identical and since corresponding functions are complements, the 1 functions of the $\bar{F}$ tree are the 0 functions of the F tree. This establishes the validity of the relationship between the 0 functions of the F tree and the MSP terms or prime implicants of $\bar{F}$. The results are summarized in the following theorem.

Theorem 4.5. Let $y = F(x_1,x_2,\ldots,x_n)$ be a positive ℓ.s. function in MSP form. Then in the basic incomplete tree of F there is a one-to-one correspondence:

(1) between the 1 functions and the terms or prime implicants of F where the corresponding term is the product of the variables of the solid lines in the path between the 1 function and F; and

(2) between the 0 functions and the terms or prime implicants of $\bar{F}$ where the corresponding term is the product of the variables of the broken lines in the path from the 0 function to F.

According to Theorem 4.5, the basic incomplete function tree contains all information about the terms of both F and $\bar{F}$. In fact, the tree furnishes one of the simplest procedures for determining the terms of $\bar{F}$ from the MSP form of F, as is illustrated by Example 4.9. The terms of F and $\bar{F}$ together with coefficient signs define the R_1 set of inequalities, and a solution to these is a necessary and sufficient condition for a single-gate realization of F. It follows, therefore, that the incomplete basic tree for an ℓ.s. function contains enough information to determine ℓ.s. realizations.

Example 4.9. Consider the basic function tree of Fig. 4.3. Clearly the 1 functions of the tree correspond to the terms of F as indicated by Theorem 4.5. The terms of the MSP form of $\bar{F}$ are obtained from the 0 functions of the tree (from left to right) as follows:

$$\begin{aligned}\bar{F}(x_1,x_2,\ldots,x_6) = {} & \bar{x}_2\bar{x}_3\bar{x}_5\bar{x}_6 + \bar{x}_2\bar{x}_3\bar{x}_4\bar{x}_6 + \bar{x}_2\bar{x}_3\bar{x}_4\bar{x}_5 + \bar{x}_1\bar{x}_4\bar{x}_5 \\ & + \bar{x}_1\bar{x}_3\bar{x}_5 + \bar{x}_1\bar{x}_3\bar{x}_4 + \bar{x}_1\bar{x}_2\bar{x}_6 + \bar{x}_1\bar{x}_2\bar{x}_5 + \bar{x}_1\bar{x}_2\bar{x}_4 \\ & + \bar{x}_1\bar{x}_2\bar{x}_3\end{aligned}$$

Regardless of whether or not the tree for a positive-variable function is constructed by removing variables according to their size, the 1 and 0 functions of the incomplete tree imply a set of inequalities, the solution of which is a necessary and sufficient condition for realizability. To each constant function of the tree

$$F_{\delta_1\delta_2\ldots\delta_k}(x_{k+1},x_{k+2},\ldots,x_n) = 1$$

there correspond the inequalities, from Eq. 4.4.,

$$\begin{aligned} u_{\delta_1\delta_2\ldots\delta_k} + \sum_{j=1}^{k} \delta_j a_j &\geq u \\ \ell_{\delta_1\delta_2\ldots\delta_k} + \sum_{j=1}^{k} \delta_j a_j &\leq \ell \end{aligned} \tag{4.10}$$

However, according to Theorem 1.3, since the separating function for all k-level functions has positive coefficients, the gap for a k-level 1 function is $0{:}-\infty$. Therefore only the first of the two inequalities of Eq. 4.10 is nontrivial, and it becomes

$$\sum_{j=1}^{k} \delta_j a_j \geq u \tag{4.11}$$

Note that this is one of the 2^n inequalities of the R_0 set and according to Eq. 4.8 is the one associated with the n-level function of the complete tree $F_{\delta_1\delta_2\ldots\delta_k 00\ldots 0}(\Phi)$. Note, moreover, that the set of inequalities, associated with all 2^{n-k} n-level descendants of $F_{\delta_1\delta_2\ldots\delta_k}$ that appear on the complete tree, is redundant given the inequality of Eq. 4.11 and the fact that all coefficients are positive.

To each constant function of the tree

$$F_{\delta_1\delta_2\ldots\delta_k}(x_{k+1},x_{k+2},\ldots,x_n) = 0$$

there correspond the inequalities from Eq. 4.4 given by Eq. 4.10. Again, according to Theorem 1.3, the gap for a k-level zero function is

$$\infty{:}\left(\sum_{j=k+1}^{n} a_j\right)$$

Therefore only the second of the two inequalities of Eq. 4.10 is nontrivial, and it becomes

$$\sum_{j=k+1}^{n} a_j + \sum_{j=1}^{k} \delta_j a_j \leq \ell \tag{4.12}$$

Again considering the complete tree, Eq. 4.12 is the inequality associated with the n-level function $F_{\delta_1\delta_2\ldots\delta_k 11\ldots 1}(\Phi)$ according to Eq. 4.8. Moreover, the set of inequalities, associated with all 2^{n-k} n-level descendants of

$F_{\delta_1\delta_2\ldots\delta_k}$, is redundant given the inequality of Eq. 4.12 and the fact that all coefficients are positive. We see, therefore, that the 1 and 0 functions of an incomplete tree of a positive ℓ.s. function define for purposes of realizability a necessary and sufficient subset of the R_0 set. The results are summarized as follows.

Corollary 4.5.1. Irrespective of the order in which the variables are removed, the 0 and 1 functions of an incomplete tree for a positive ℓ.s. function specify a necessary and sufficient set of inequalities for realization purposes. To each 1 function $F_{\delta_1\delta_2\ldots\delta_k}$ there corresponds the inequality

$$\sum_{j=1}^{k} \delta_j a_j \geq u$$

To each 0 function $F_{\delta_1\delta_2\ldots\delta_k}$ the corresponding inequality is

$$\sum_{j=k+1}^{n} a_j + \sum_{j=1}^{k} \delta_j a_j \leq \ell$$

4.7 THE TREE FOR ℓ.s. FUNCTIONS WITH SYMMETRIC VARIABLES

It should be increasingly evident that the function tree is useful not only for displaying reduced functions but also for selecting coefficients for the realization. A procedure for this is presented in the next chapter. In the remaining sections of this chapter we consider simplifications that can be made in the tree because of properties of ℓ.s. functions that have been derived previously. The first simplification is the result of variable symmetry.

Theorem 3.7 established that if an ℓ.s. function is symmetric in m variables, then the coefficients of these variables can always be equal. This information can be used to simplify the tree. Let

$$y = F(x_1,x_2,\ldots,x_n)$$

be a positive ℓ.s. function that is symmetric in the m variables $x_1,x_2,\ldots,x_m$ where these variables can have the largest coefficients. Then F can be uniquely factored as

$$F(x_1,x_2,\ldots,x_n) = \sum_{j=0}^{m}\cdot M_j(x_1,x_2,\ldots,x_m)F_j(x_{m+1},x_{m+2},\ldots,x_n)$$

where $M_j(x_1,x_2,\ldots,x_m)$ is the switching function that has unit value if and only if j or more of its variables are unity. It is easily verified that F_j is the reduced function formed from F by assigning unit value to any j of the variables $x_1,x_2,\ldots,x_m$ and zero value to the remaining variables of this set. For example, if $m = 3$, then

$$F_0(x_4,x_5,\ldots,x_n) = y(x_1 = x_2 = x_3 = 0)$$
$$F_1(x_4,x_5,\ldots,x_n) = y(x_1 = 1,x_2 = x_3 = 0) = y(x_1 = x_3 = 0,x_2 = 1) = y(x_1 = x_2 = 0,x_3 = 1)$$
$$F_2(x_4,x_5,\ldots,x_n) = y(x_1 = x_2 = 1,x_3 = 0) = y(x_1 = x_3 = 1,x_2 = 0) = y(x_1 = 0,x_2 = x_3 = 1)$$
$$F_3(x_4,x_5,\ldots,x_n) = y(x_1 = x_2 = x_3 = 1)$$

According to Theorem 3.13, it follows that

$$F_j \geq F_k \qquad \text{if} \quad j > k$$

In the symmetric tree all reduced functions $F_0,F_1,\ldots,F_m$ are on one level rather than on m levels. The following is a procedure for producing the symmetric tree when the function F has symmetric variables. It is very similar to Procedures 4.1 and 4.2.

Procedure 4.3. Symmetric Tree for Positive ℓ.s. Functions with Symmetric Variables

(1) The 0-level function is F.

(2) Determine the variables of F that can have the largest coefficients and denote them as $x_1,x_2,\ldots,x_{m_1}$. Clearly F is symmetric in these variables.

(3) Factor F as

$$F(x_1,x_2,\ldots,x_n) = \sum_{j_1=0}^{m_1}{}^{\bullet}\, M_{j_1}(x_1,x_2,\ldots,x_{m_1})F_{j_1}(x_{m_1+1},x_{m_1+2},\ldots,x_n)$$

The 1-level functions of the tree ordered from left to right are F_{m_1}, $F_{m_1-1},\ldots,F_0$. Between F and each F_{j_1} is a solid line labeled j_1.

Because of the containment properties, if

$$F_i(x_{m_1+1},x_{m_1+2},\ldots,x_n) = 0$$

then

$$F_j(x_{m_1+1},x_{m_1+2},\ldots,x_n) = 0$$

for all $j < i$. Similarly

$$F_i(x_{m_1+1},x_{m_1+2},\ldots,x_n) = 1$$

implies

$$F_j(x_{m_1+1},x_{m_1+2},\ldots,x_n) = 1$$

for all $j > i$. The redundant constant functions are removed from the tree as they were in Procedure 4.1.

(4) Remove from the 1-level of the tree all 1 functions but the rightmost and all 0 functions but the leftmost.

The lower levels of the tree denoted as 2-level, 3-level, etc., are determined in a similar way. The general step of the procedure is as follows:

(5) The functions on the k-level of the tree are $F_{j_1 j_2 \ldots j_k}$ where $j_i = 0,1,\ldots,m_i$. Determine the set of variables in which all nonconstant k-level functions are symmetric and whose coefficients can be the largest in the realizations of all k-level functions. Denote these variables as

$$x_{r_k+1}, x_{r_k+2}, \ldots, x_{r_k+m_{k+1}}$$

where

$$r_k = m_1 + m_2 + \cdots + m_k$$

If no such variables exist, that is, if $m_{k+1} = 0$, then the given function, F, is not ℓ.s. and the symmetric tree cannot be formed.

(6) Factor each nonconstant k-level function as

$$F_{j_1 j_2 \ldots j_k}(x_{r_k+1}, x_{r_k+2}, \ldots, x_n)$$

$$= \sum_{j_{k+1}=0}^{m_{k+1}} M_{j_{k+1}}(x_{r_k+1}, x_{r_k+2}, \ldots, x_{r_k+m_{k+1}}) \times F_{j_1 j_2 \ldots j_{k+1}}(x_{r_{k+1}}, x_{r_{k+1}+1}, \ldots, x_n)$$

For each $j_1 j_2 \ldots j_k$ the ordered set of functions

$$F_{j_1 j_2 \ldots j_k m_{k+1}}, F_{j_1 j_2 \ldots j_k (m_{k+1}-1)}, \ldots, F_{j_1 j_2 \ldots j_k 0}$$

is called the function set descendant from $F_{j_1 j_2 \ldots j_k}$.

(7) Delete from each function set descendant from each $F_{j_1 j_2 \cdots j_k}$ all 1 functions but the rightmost and all 0 functions but the leftmost. The resultant functions are the $(k + 1)$-level descendants of $F_{j_1 j_2 \ldots j_k}$ ordered from left to right in the tree. Between $F_{j_1 j_2 \ldots j_k}$ and $F_{j_1 j_2 \ldots j_k j_{k+1}}$ is a solid line labeled j_{k+1}.

(8) If the $(k + 1)$-level of the tree contains any nonconstant functions, increase the value of k by unity and repeat the procedure starting with Step 5.

Example 4.10 (*Moores Function*). At one time complete monotonicity was believed to be a necessary and sufficient condition for a switching function to be linearly separable. The first counterexample to this conjecture, given by E. F. Moore, is a function of the twelve variables, $a,b,\ldots,\ell$. Partition these variables into subsets

$$A = \{a,b,c\},\ D = \{d,e,f,g\},\ H = \{h,i,j,k,\ell\}$$

In terms of this notation, the function is

$$\begin{aligned} F(A,D,H) = {} & M_3(A)[M_4(D) \dotplus M_3(D)M_2(H) \dotplus M_2(D)M_3(H) \\ & \qquad\qquad\qquad \dotplus M_1(D)M_4(H) \dotplus M_5(H)] \\ & \dotplus M_2(A)[M_4(D)M_2(H) \dotplus M_3(D)M_4(H) \dotplus M_2(D)M_5(H)] \\ & \dotplus M_1(A)[M_4(D)M_4(H) \dotplus M_3(D)M_5(H)] \\ & \dotplus M_4(D)M_5(H) \end{aligned}$$

The function is symmetric in the variables of A, D, and H. Moreover, the ordering procedure (Procedure 3.2) shows that the variables of A are larger than those of D, which in turn are larger than those of H; that is,

$$a_A > a_D > a_H$$

where the variables of A have coefficient a_A, etc. Figure 4.11 shows the symmetric tree for Moore's Function. We shall return to this function in later examples.

The symmetric tree has, in general, the same properties as the other function trees. For convenience some of these are summarized for the functions on any k-level of the tree.

Each function $F_{j_1 j_2 \ldots j_k}$ is the reduced function formed from F by setting equal to unity j_1 of the first subset of m_1 variables, j_2 of the second subset of m_2 variables, and so on up to j_k of the kth subset of m_k variables, and setting the remaining variables of each subset equal to zero. If

$$y = F(x_1,x_2,\ldots,x_n) = \left\langle \sum_{j=1}^{r_k} a_j x_j + f_k(x_{r_k+1},x_{r_k+2},\ldots,x_n) \right\rangle_{u:\ell}$$

where

$$r_k = m_1 + m_2 + \cdots + m_k$$

then all of the k-level functions are ℓ.s. and can have as their separating function $f_k(x_{r_k+1},x_{r_k+2},\ldots,x_n)$. Moreover, the k-level descendant functions of each $F_{j_1 j_2 \ldots j_{k-1}}$ are ordered as

$$F_{j_1 j_2 \ldots j_{k-1} m_k} \geq F_{j_1 j_2 \ldots j_{k-1}(m_k - 1)} \geq \cdots \geq F_{j_1 j_2 \ldots j_{k-1} 0}$$

where the equality holds only between constant functions.

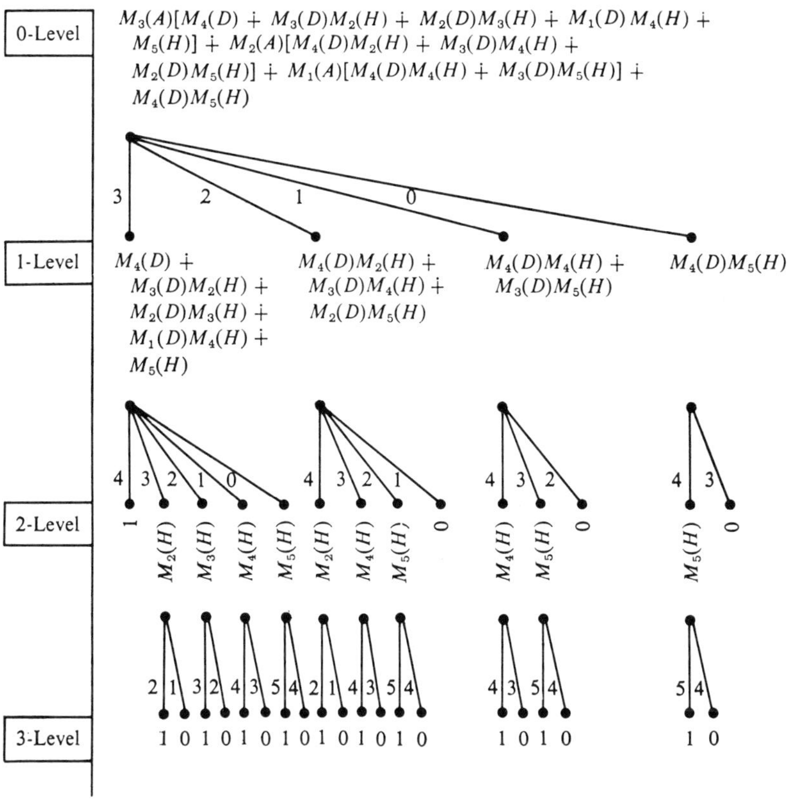

Fig. 4.11 The symmetric tree for Moore's function (Example 4.10).

As in the case of the incomplete tree, the 0 and 1 functions correspond to the terms of the MSP form of F and $\bar{F}$ when the 0-level F is in this form and when the coefficients are removed according to size, as indicated in Procedure 4.3. In this case, however, the correspondence is not as simple as in the previous case. Here, for example, the function

$$F_{j_1 j_2 \ldots j_k}(x_{r_k+1}, x_{r_k+2}, \ldots, x_n) = 1$$

corresponds to the terms of F given by

$$M_{j_1}(x_1, x_2, \ldots, x_{m_1}) M_{j_2}(x_{m_1+1}, x_{m_1+2}, \ldots, x_{r_2}) \ldots M_{j_k}(x_{r_{k-1}+1}, x_{r_{k-1}+2}, \ldots x_{r_{k-1}+m_k})$$

which consists of a total of

$$\binom{m_1}{j_1}\binom{m_2}{j_2}\cdots\binom{m_k}{j_k}$$

product terms of the MSP form of F. A similar relation exists between the 0 functions and the terms of the MSP form of $\bar{F}$.

For purposes of realization, the range theorems are of particular interest. Consider the first two levels of the tree where

$$F(x_1,x_2,\ldots,x_n) = \sum_{j_1=0}^{m_1}\cdot\, M_{j_1}(x_1,x_2,\ldots,x_{m_1})F_{j_1}(x_{m_1+1},x_{m_1+2},\ldots,x_n)$$

Assume that each F_{j_1}, $j_1 = 0,1,\ldots,m_1$, is ℓ.s. and all can have the same separating function; that is,

$$F_{j_1}(x_{m_1+1},x_{m_1+2},\ldots,x_n) = \langle f_1(x_{m_1+1},x_{m_1+2},\ldots,x_n)\rangle_{u_{j_1}:\ell_{j_1}}$$

The symmetric tree was obtained on the assumption that the coefficients of $x_1,x_2,\ldots,x_{m_1}$ would be made equal. The only question that remains concerns the (common) value of these coefficients. By the same reasoning that was used in proving the other range theorems, it follows that F is ℓ.s. and can be realized by the separating function

$$f(x_1,x_2,\ldots,x_n) = a_1\sum_{j=1}^{m_1} x_j + f_1(x_{m_1+1},x_{m_1+2},\ldots,x_n)$$

if and only if a_1 is selected so that, when each gap $u_{j_1}:\ell_{j_1}$ is displaced by j_1a_1, all displaced gaps have a nonvoid intersection. This intersection is the gap of F. In essence, the 1-level functions of the symmetric tree represent the irredundant m_1-level functions of the basic tree previously considered under the assumption that

$$a_1 = a_2 = \cdots = a_{m_1}$$

The j_1 labels of the symmetric tree correspond to the number of solid lines in the path of the basic tree between F and the corresponding m_1-level functions. The following range theorem, therefore, is the result of applying Theorem 4.3 to the first m_1 levels of the basic tree under the assumption that

$$a_1 = a_2 = \cdots = a_{m_1}$$

Theorem 4.6. If for $j = 0,1,\ldots,m$, the switching functions

$$F_j(x_{m+1},x_{m+2},\ldots,x_n) = \langle f_1(x_{m+1},x_{m+2},\ldots,x_n)\rangle_{u_j:\ell_j}$$

are each ℓ.s. and all have the same separating function, then

$$F(x_1,x_2,\ldots,x_n) = \sum_{j=0}^{m}\cdot\, M_j(x_1,x_2,\ldots,x_m)F_j(x_{m+1},x_{m+2},\ldots,x_n)$$

is ℓ.s. with realization

$$F(x_1,x_2,\ldots,x_n) = \left\langle a\sum_{j=1}^{m} x_j + f_1(x_{m+1},x_{m+2},\ldots,x_n)\right\rangle_{u:\ell}$$

if and only if a is selected in the range

$$\overline{R(a)} = \overline{U(a)} * \overline{L(a)}$$

where

$$\overline{U(a)} = \min_{j>k} \left\{ \frac{u_k - \ell_j}{j - k} \right\}$$

$$\overline{L(a)} = \max_{j>k} \left\{ \frac{\ell_k - u_j}{j - k} \right\}$$

where $j,k = 0,1,\ldots,m$. The gap of F is

$$u = \min_j \{u_j + ja\}$$

$$\ell = \max_j \{\ell_j + ja\}$$

Example 4.11. Consider the positive ℓ.s. function

$$y = M_2(x_1,x_2,x_3) \dotplus M_1(x_1,x_2,x_3)M_1(x_4,x_5,x_6) \dotplus M_3(x_4,x_5,x_6)$$

Since the function is symmetric in x_1,x_2,x_3 and x_4,x_5,x_6, it is permissible to let

$$a_1 = a_2 = a_3 = a_A \quad \text{and} \quad a_4 = a_5 = a_6 = a_B$$

According to Procedure 3.2, it follows that

$$a_A > a_B$$

The first two levels of the symmetric function tree are shown in Fig. 4.12. All 1-level functions are separated by

$$f_1(x_3,x_4,x_5) = x_3 + x_4 + x_5$$

with the corresponding gaps indicated on the map.

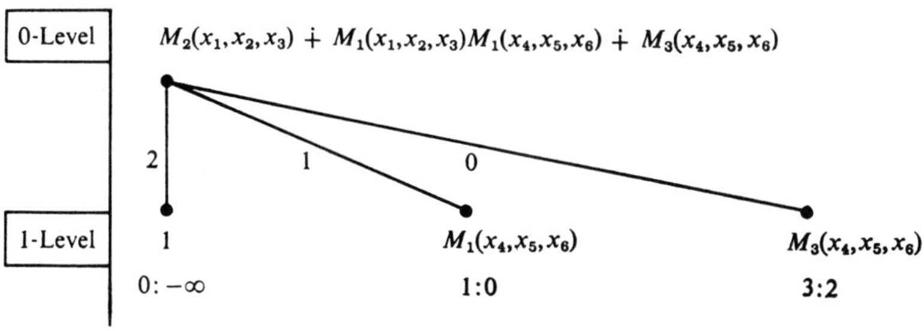

Fig. 4.12. Symmetric tree for Example 4.11.

According to Theorem 4.6,

$$\overline{U(a_A)} = \min\left\{\frac{1-(-\infty)}{2-1}, \frac{3-0}{1-0}, \frac{3-(-\infty)}{2-0}\right\} = 3$$

$$\overline{L(a_A)} = \max\left\{\frac{0-0}{2-1}, \frac{2-1}{1-0}, \frac{2-0}{2-0}\right\} = 1$$

hence $\overline{R(a_A)} = 3*1$. We select $a_A = 2$, in which case the realization for y has gap $u{:}\ell$ where

$$u = \min\{0 + 2\cdot 2, 1 + 2\cdot 1, 3 + 0\} = 3$$

$$\ell = \max\{-\infty + 2\cdot 2, 0 + 2\cdot 1, 2 + 0\} = 2$$

and the realization is

$$y = \langle 2x_1 + 2x_2 + 2x_3 + x_4 + x_5 + x_6\rangle_{3:2}$$

Although† a can be selected anywhere within its range $\overline{R(a)}$, it is desirable for sensitivity considerations to select the smallest value which maximizes the gap length of F. Determining this value is somewhat more difficult for the symmetric tree than for the trees previously considered because we must deal with more than two displaced gaps which are being displaced at different rates as a is varied. As before, if g_s is the smallest gap length of the 2-level functions, then it is an upper bound on the gap length g of F. The actual value may be somewhat less, however.

As an example, Fig. 4.13*b* shows the gaps of the four 2-level functions of Fig. 4.13*a* which must be displaced by amounts $3a$, $2a$, a, and 0 so that their intersection is nonvoid. Note that each pair of adjacent gaps defines a range

$$R_k(a) = U_k(a)*L_k(a) = (u_k - \ell_{k+1})*(\ell_k - u_{k+1}), \quad k = 0,1,2$$

while pairs of gaps one-removed define ranges

$$R_k(2a) = U(2a)*L(2a) = (u_k - \ell_{k+2})*(\ell_k - u_{k+2}), \quad k = 0,1$$

and pairs of gaps two-removed define the range

$$R_k(3a) = U(3a)*L(3a) = (u_k - \ell_{k+3})*(\ell_k - u_{k+3}), \quad k = 0$$

Therefore, in general, each pair of gaps defines the range

$$R_k(\alpha a) = U(\alpha a)*L(\alpha a) = (u_k - \ell_{k+\alpha})*(\ell_k - u_{k+\alpha})$$

† The remainder of this section can be omitted on a first reading.

where $\alpha = 1,2,\ldots,m$, and $k = 0,1,\ldots,(m-\alpha)$. Moreover, each range $R_k(\alpha a)$ on αa implies a range on a denoted as $R_k^{\alpha}(a)$ where

$$R_k^{\alpha}(a) = \frac{R_k(\alpha a)}{\alpha}$$

Figure 4.13*c* shows the range implied by each pair of gaps and shows how the intersection of each pair of gaps varies as a function of a.

In order that the intersection of all gaps be nonvoid, all pairs of gaps must intersect simultaneously; hence a must be selected within all pair-wise ranges. Clearly, therefore,

$$\overline{R(a)} = \overline{U(a)} * \overline{L(a)}$$

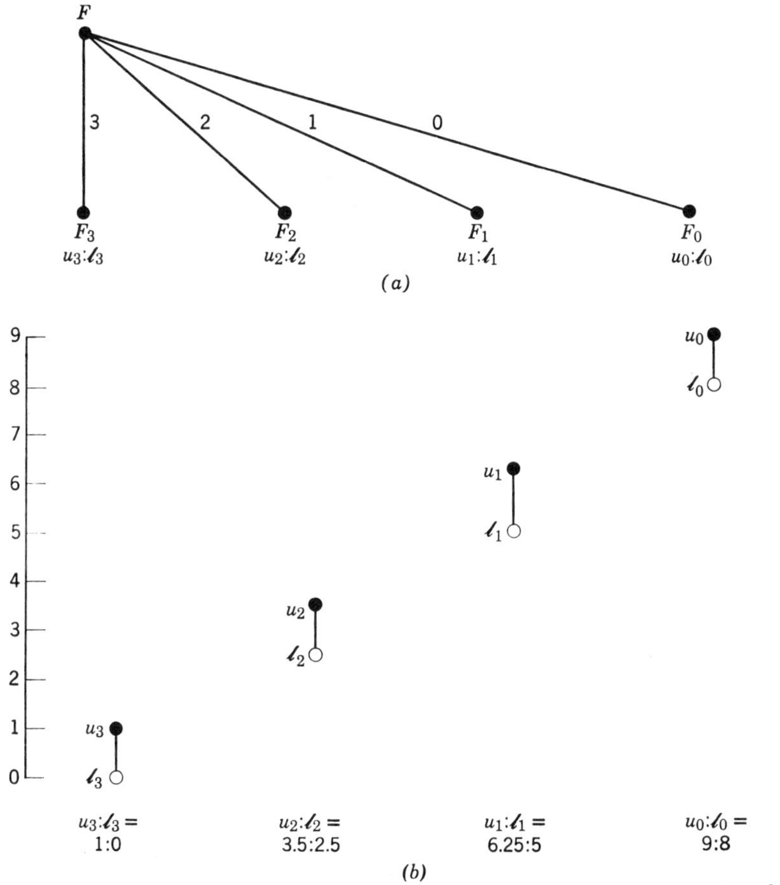

(continued)

Fig. 4.13. The gap length of F as a function of a. (*a*) 0- and 1-levels of symmetric tree. (*b*) Gaps of the 1-level functions of the tree of (*a*). (*c*) Component gap variations as a function of a. (*d*) Variation of g as a function of a.

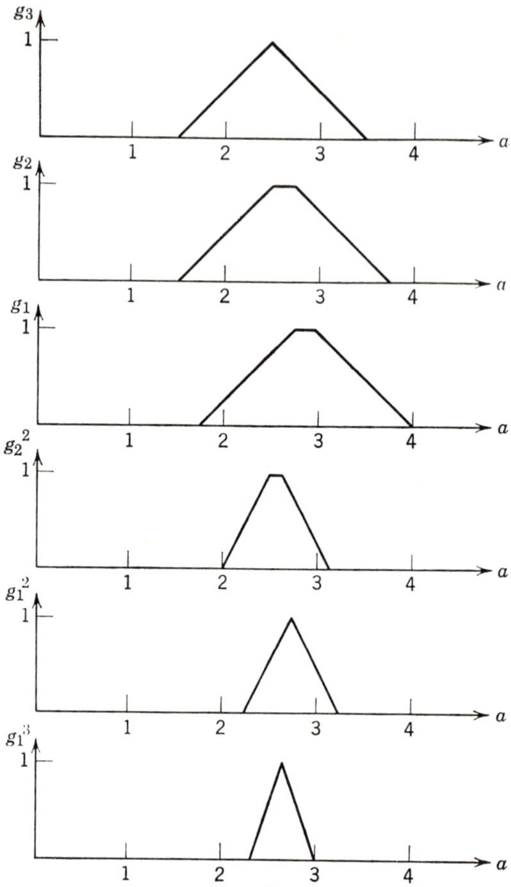

$$R_3(a) = (u_2 - \ell_3)*(\ell_2 - u_3)$$
$$= 3.5*1.5$$

$$R_2(u) = (u_1 - \ell_2)*(\ell_1 - u_2)$$
$$= 3.75*1.5$$

$$R_1(a) = (u_0 - \ell_1)*(\ell_0 - u_1)$$
$$= 4*1.75$$

$$R_2(2a) = (u_1 - \ell_3)*(\ell_1 - u_3)$$
$$= 6.25*4$$
$$R_2{}^2(a) = \tfrac{1}{2}R_2(2a) = 3.125*2$$

$$R_1(2a) = (u_0 - \ell_2)*(\ell_0 - u_2)$$
$$= 6.5*4.5$$
$$R_1{}^2(a) = \tfrac{1}{2}R_1(2a) = 3.25*2.25$$

$$R_1(3a) = (u_0 - \ell_3)*(\ell_0 - u_3)$$
$$= 9*7$$
$$R_1{}^3(a) = \tfrac{1}{3}R_1(2a) = 3*2.33$$

(c)

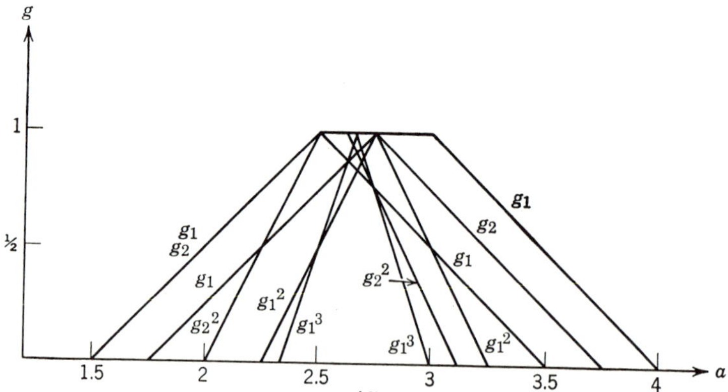

(d)

where

$$\overline{U(a)} = \min \{U_3(a), U_2(a), U_1(a), U_2^{\,2}(a), U_1^{\,2}(a), U_1^{\,3}(a)\} = U_1^{\,3}(a) = 3$$
$$\overline{L(a)} = \max \{L_3(a), L_2(a), L_1(a), L_2^{\,2}(a), L_1^{\,2}(a), L_1^{\,3}(a)\} = L_1^{\,3}(a) = 2.33$$

The gap length g of F is the intersection of the 1-level gaps and is given by

$$g = \min \{g_3, g_2, g_1, g_2^{\,2}, g_1^{\,2}, g_1^{\,3}\}$$

Figure 4.13*d* shows how g varies as a function of a by superimposing all curves of Fig. 4.13*c*. It shows that the component gap that determines the value of g depends upon the particular value of a as follows:

$$\begin{array}{ll} 2.33 < a < 2.5 & g = g_1^{\,3} \\ 2.5 \;\; < a < 2.67 & g = g_1^{\,2} \\ 2.67 < a < 2.75 & g = g_1 \\ 2.75 < a < 3 & g = g_1^{\,3} \end{array}$$

Moreover, maximum g occurs for $a = 2.67$, for which $g \approx 0.84$.

In general, each pair of 1-level gaps defines a component gap of g which is a linear function of a. If the pair of gaps defines the range $R_k(\alpha a)$, then the component gap $g_k^{\,\alpha}$ is a linear function of a with slope α. Since the rate of variation of $g_k^{\,\alpha}$ with a depends upon α, it is convenient to group the $R_k(\alpha a)$ ranges according to α for purposes of determining the value of a that yields maximum g for F. Clearly, a sketch similar to that of Fig. 4.13*d* is often helpful in this regard. An algorithm for determining this value of a when a sketch is inadequate is given in Procedure 4.4.

Procedure 4.4

(1) For each α determine the set

$$\mathscr{U}_1 = \{U^\alpha(a)\} \quad \text{and the set} \quad \mathscr{L}_1 = \{L^\alpha(a)\}$$

where

$$U^\alpha(a) = \min_j \left\{\frac{u_j - \ell_{j+\alpha}}{\alpha}\right\} \qquad j = 0,1,2,\ldots,(m_1 - \alpha)$$

$$L^\alpha(a) = \max_j \left\{\frac{\ell_j - u_{j+\alpha}}{\alpha}\right\} \qquad j = 0,1,2,\ldots,(m_1 - \alpha)$$

(2) Determine $\overline{U_1(a)}$ and $\overline{L_1(a)}$ where

$$\overline{U_1(a)} = \min_\alpha \{U^\alpha(a)\}$$

$$\overline{L_1(a)} = \max_\alpha \{L^\alpha(a)\}$$

(3) Determine the minimum value of α such that $U^\alpha(a) = \overline{U_1(a)}$ and denote it by $\alpha_U{}^1$.

Similarly, determine the minimum value of α such that $L^\alpha(a) = \overline{L_1(a)}$ and denote it by $\alpha_L{}^1$. Set $k = 1$ and continue.

(4) Let g be the smallest 1-level gap length and let

$$\Delta_k = \min\left\{g, [\overline{U_k(a)} - \overline{L_k(a)}]\frac{\alpha_U{}^k\alpha_L{}^k}{\alpha_U{}^k + \alpha_L{}^k}\right\}$$

(5) If $\mathscr{U}_k$ and $\mathscr{L}_k$ each consist of single members, then go to Step 8. Otherwise go to Step 6.

(6) Let each $U^\alpha(a)$ from set $\mathscr{U}_k$ be a member of set $\mathscr{U}_{k+1}$ if

$$\alpha\left[U^\alpha(a) - \overline{U_k(a)} + \frac{\Delta_k}{\alpha_U{}^k}\right] \leq \Delta_k$$

Let each $L^\alpha(a)$ from set $\mathscr{L}_k$ be a member of set $\mathscr{L}_{k+1}$ if

$$\alpha\left[\overline{L_k(a)} - L^\alpha(a) + \frac{\Delta_k}{\alpha_L{}^k}\right] \leq \Delta_k$$

(7) From $\mathscr{U}_{k+1}$ determine the member for which

$$\alpha\left[U^\alpha(a) - \overline{U_k(a)} + \frac{\Delta_k}{\alpha_U{}^k}\right]$$

is minimum and denote this as $\overline{U_{k+1}(a)}$. Denote the corresponding value of α as $\alpha_U{}^{k+1}$. From $\mathscr{L}_{k+1}$ determine the member for which

$$\alpha\left[\overline{L_k(a)} - L^\alpha(a) + \frac{\Delta_k}{\alpha_L{}^k}\right]$$

is minimum and denote this as $\overline{L_{k+1}(a)}$. Denote the corresponding value of α as $\alpha_L{}^{k+1}$. Increase k by unity and go to Step 4.

(8) The value of a for which g of F is maximum is

$$a = \overline{L_k(a)} + \frac{\Delta_k}{\alpha_L{}^k}$$

The corresponding value of g is

$$g = \Delta_k$$

Steps 1 and 2 have been explained in the previous section.

In Step 3 a first estimate is made that the gap length is determined by the lower range boundary due to $\alpha_L{}^1$ and the upper range boundary due to $\alpha_U{}^1$.

Step 4 then considers the kth estimate of those range boundaries that determine the gap boundaries. If $\alpha_L{}^k$ and $\alpha_U{}^k$ were in fact the determining range boundaries, the gap length would be, as a function of a,

$$g = \min \begin{Bmatrix} g_j \\ \alpha_L{}^k(a - \overline{L_k(a)}) \\ \alpha_U{}^k(\overline{U_k(a)} - a) \end{Bmatrix}$$

Assuming that the gap length is determined by the last two expressions, the length is maximum when a is picked so that the two expressions are equal.

$$g = \alpha_L{}^k(a - \overline{L_k(a)}) = \alpha_U{}^k(\overline{U_k(a)} - a)$$

Eliminating a yields

$$g = [\overline{U_k(a)} - \overline{L_k(a)}]\frac{\alpha_U{}^k\alpha_L{}^k}{\alpha_U{}^k + \alpha_L{}^k}$$

Thus the gap length is given by Δ_k of Step 4. Step 5 states that if there are no other range boundaries which could imply a smaller gap length the best value of a has been found, as described in Step 8.

Step 6 considers each of the other range boundaries which could cause the gap length to be smaller and determines if, for the particular value of a selected, they do in fact imply a smaller gap length. Those that do imply a smaller gap length are then used in the next estimate in Step 7.

4.8 REDUCED TREES FOR LINEARLY SEPARABLE FUNCTIONS AND REDUCED SETS OF INEQUALITIES

We have shown that any of the function trees that have been discussed thus far contain enough information to determine single-gate realizability. This was established by showing, among other things, that the tree implies a necessary and sufficient set of inequalities. In the last chapter we showed how the R_0 set of inequalities can be reduced to the R_1 set because of coefficient sign information. In this chapter we showed how the basic tree that corresponds to the R_1 set is obtained directly from the MSP form of the given switching function F. Moreover we showed that this tree can be obtained by removing from the complete tree all descendants of constant functions.

At this point, we wish to make additional tree simplifications that correspond to the reduction of the R_1 set of inequalities to the R_2 set, the R_3 set, etc., when coefficient-sign and higher-order-monotonicity information is available. Actually, it is usually impractical to use monotonicity of order higher than two (coefficient ordering) to reduce the inequalities; therefore, we cover this case in some detail. Similar reasoning, however, applies to the use of all order monotonicities.

Consider the incomplete basic function tree formed from the MSP form of the positive linearly separable switching function $y = F(x_1,x_2,\ldots,x_n)$ where the variables are removed according to their size. Recall that each constant function of the tree implies an inequality of the R_1 set. Consider now a pair of 1 functions, $F_{\delta_1\delta_2\ldots\delta_k}$ and $F_{\gamma_1\gamma_2\ldots\gamma_\ell}$. We wish to know when the inequality

$$\sum_{i=1}^{k} \delta_i a_i \geq u$$

associated with the first, implies that the inequality

$$\sum_{i=1}^{\ell} \gamma_i a_i \geq u$$

associated with the second, is redundant if the coefficient ordering information

$$a_1 \geq a_2 \geq \cdots \geq a_n$$

is known. For example, the 1 functions F_{01101} and $F_{0101010}$ imply the inequalities

$$\begin{aligned} a_2 + a_3 + \quad\quad\quad a_5 \quad &\geq u \\ a_2 + \quad\quad a_4 + \quad a_6 &\geq u \end{aligned}$$

respectively. The first inequality is implied by the second and the ordering inequalities

$$a_3 > a_4 \qquad \text{and} \quad a_5 > a_6$$

From the example it follows that the inequality implied by the 1 function $F_{\gamma_1\gamma_2\ldots\gamma_\ell}$ is redundant in view of the coefficient ordering relations and the inequality associated with $F_{\delta_1\delta_2\ldots\delta_k}$ if

(1) The number of unit values is the same among the sets of numbers $\delta_1,\delta_2,\ldots,\delta_k$ and $\gamma_1,\gamma_2,\ldots,\gamma_\ell$ and

(2) There is a one-to-one correspondence between the unit values of the two sets so that $\delta_i = 1$ implies $\gamma_j = 1$ where $i \geq j$.

Similar reasoning yields the redundancy of the inequalities implied by the 0 functions of the tree. According to Corollary 4.5.1 the 0 function $F_{\delta_1\delta_2\ldots\delta_k}$ implies

$$\sum_{j=k+1}^{n} a_j + \sum_{j=1}^{k} \delta_j a_j \leq t$$

The inequality implied by any other 0 function $F_{\gamma_1\gamma_2\ldots\gamma_\ell}$ is redundant in view of the first and coefficient-ordering information if

(1) The number of zero values is the same among the sets of numbers $\delta_1,\delta_2,\ldots,\delta_k$ and $\gamma_1,\gamma_2,\ldots,\gamma_\ell$; and

(2) There is a one-to-one correspondence between the zero values of the two sets so that $\delta_i = 0$ implies $\gamma_j = 0$ where $i \geq j$.

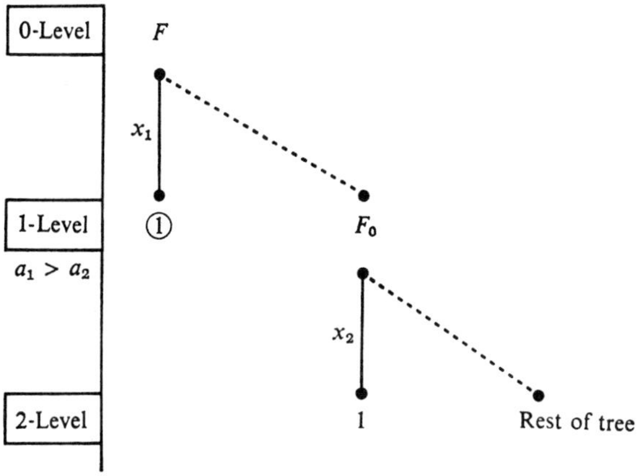

Fig. 4.14. A portion of a function tree that has redundant functions.

For example, the inequality implied by the 0 function F_{001} is redundant in view of the inequality implied by F_{0101} and the inequality $a_2 > a_3$.

If each k-level of the tree $k = 1,2,\ldots,n$ is such that

$$a_1 \geq a_2 \geq \cdots \geq a_n$$

then it is usually a simple procedure to determine the redundant constant functions. In fact, the redundant function and the function that implies the redundancy are often adjacent to each other in the tree. Fig. 4.14 shows a portion of a function tree with two 1 functions where the encircled 1 is redundant in view of the other 1 function and the inequality

$$a_1 > a_2$$

Since this is a frequently occurring condition, it should be checked first in a search for redundancies.

When a redundant constant function is found in the tree, it is encircled, and the coefficient-ordering inequality that implies the redundancy is written on the left of the tree under the name of the level. When all redundant constant functions have been found, the *reduced tree* consists of the remaining constant functions (those not encircled) and their unique tree paths to F, together with the coefficient-ordering inequalities that were entered at the side of the tree. It is evident that the reduced tree contains sufficient information for determining realizability.

Example 4.12. Consider the function tree of Fig. 4.3. An ordering of the coefficients shows that a_4 can equal a_5. We can simplify the tree to that shown in Figure 4.15 where the 4- and 5-levels have been changed because of the stipulation that $a_4 = a_5$. The redundant 1 functions of Fig. 4.15 due to coefficient ordering are given in Table 4.1.

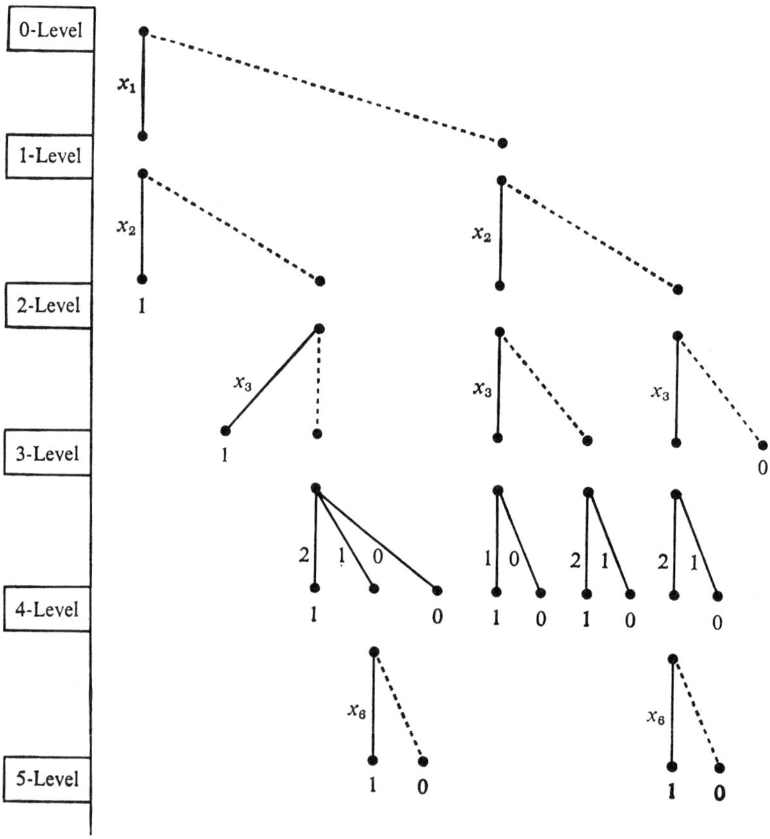

Fig. 4.15. The simplified tree of Fig. 4.3 when $a_4 = a_5$.

Table 4.1.

Redundant Function	Function Implying Redundancy	Coefficient Inequality
F_{11}	F_{101}	$a_2 > a_3$
F_{1002}†	F_{10011}	$a_4 > a_6$
F_{0111}	F_{0102}	$a_3 > a_4$

† Note that the 2 in F_{1002} counts as two 1's because $a_4 = a_5$.

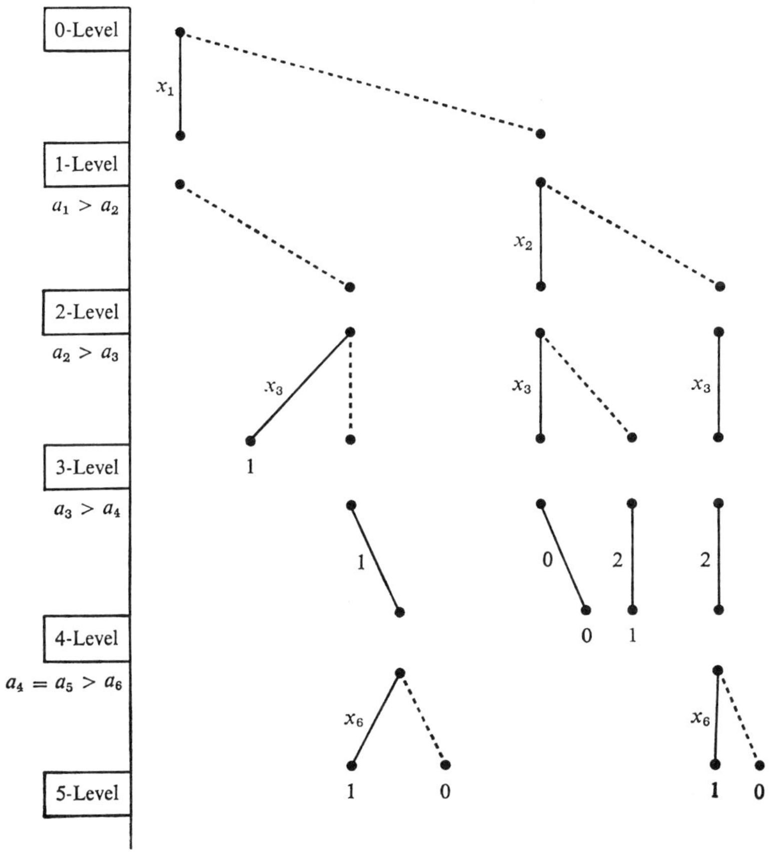

Fig. 4.16. The reduced tree of Example 4.12.

Similarly, the redundant 0 functions are given in Table 4.2.

Table 4.2.

Redundant Function	Function Implying Redundancy	Coefficient Inequaltiy
F_{000}	F_{00120}	$a_3 > a_6$
F_{0011}†	F_{00120}	$a_4 > a_6$
F_{0101}	F_{0110}	$a_3 > a_4$
F_{1000}	F_{10010}	$a_4 > a_6$

† Note that in counting the number of 0's, F_{0011} counts as having three 0's since one of the variables x_4 or x_5 has zero value.

Deleting the redundant functions from the tree leaves the reduced tree of Fig. 4.16.

Figure 4.17 shows the reduced tree of the symmetric tree (Fig. 4.11) for Moore's function.

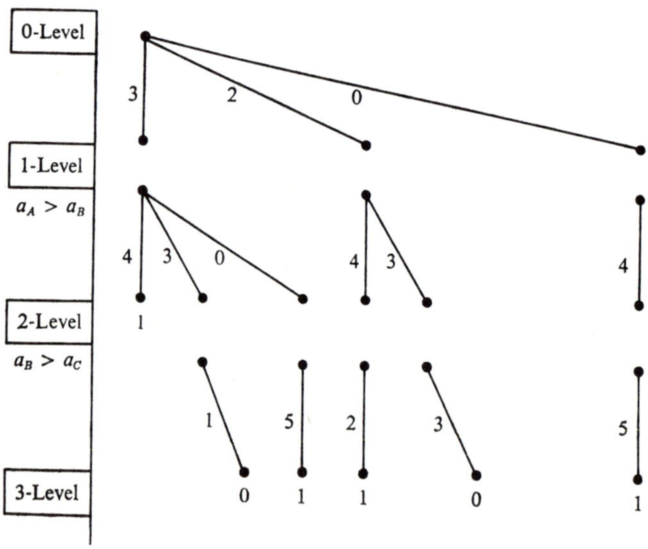

Fig. 4.17. The reduced tree of Fig. 4.11.

4.9 RANGE THEOREMS AND FUNCTION TREES FOR ARBITRARY SWITCHING FUNCTIONS

Whether or not the given switching function F is linearly separable or even unate, a function tree can be found by means of Procedure 4.1. The functions of the 1 level are F_1 and F_0 where

$$F(x_1,x_2,\ldots,x_n) = x_1F_1(x_2,x_3,\ldots,x_n) + \bar{x}_1F_0(x_2,x_3,\ldots,x_n)$$

Moreover, the variables can be removed according to size as indicated in Procedure 4.2.

To realize such a function requires a network of T gates; therefore, the constant functions of the tree do not correspond in a simple way to a set of inequalities whose solution is a realization. Nevertheless, range theorems analogous to Theorem 4.2 can be derived for the functions of the general tree. This is given by Theorem 4.7.

Theorem 4.7 (*General Range Theorem*). Consider the arbitrary switching functions

$$F_1(x_2,x_3,\ldots,x_n) = \langle f_1(y_1',y_2',\ldots,y_m')\rangle_{u_1:\ell_1} = \Big\langle \sum_{j=1}^{m} a_j'y_j'\Big\rangle_{u_1:\ell_1}$$

and

$$F_0(x_2,x_3,\ldots,x_n) = \langle f_0(y_1'',y_2'',\ldots,y_p'')\rangle_{u_0:\ell_0} = \Big\langle \sum_{j=1}^{p} a_j''y_j''\Big\rangle_{u_0:\ell_0}$$

which have the indicated realizations in terms of y_j' and y_j'', where

$$y_j' = F_j'(x_2,x_3,\ldots,x_n) \qquad j = 1,2,\ldots,m$$
$$y_j'' = F_j''(x_2,x_3,\ldots,x_n) \qquad j = 1,2,\ldots,p$$

Then

$$y = F(x_1,x_2,\ldots,x_n) = x_1F_1(x_2,x_3,\ldots,x_n) \dotplus \bar{x}_1F_0(x_2,x_3,\ldots,x_n)$$

can have the realization

$$y = \langle a_1x_1 + x_1f_1(y_1',y_2',\ldots,y_m') + \bar{x}_1f_0(y_1'',y_2'',\ldots,y_p'')\rangle_{u:\ell}$$

or

$$y = \langle -a_1\bar{x}_1 + x_1f_1(y_1',y_2',\ldots,y_m') + \bar{x}_1f_0(y_1'',y_2'',\ldots,y_p'')\rangle_{u':\ell'}$$

where a_1 can have any value in the range

$$R(a_1) = (u_0 - \ell_1)*(\ell_0 - u_1)$$

where

$$u = \min\{u_0,\ u_1 + a_1\}$$
$$\ell = \max\{\ell_0,\ \ell_1 + a_1\}$$
$$u' = \min\{u_1,\ u_0 - a_1\}$$
$$\ell' = \max\{\ell_1,\ \ell_0 - a_1\}$$

and where

$$x_1f_1(y_1',y_2',\ldots,y_m') \quad \text{and} \quad \bar{x}_1f_0(y_1'',y_2'',\ldots,y_p'')$$

denote respectively

$$\sum_{j=1}^{m} a_j'[x_1y_j'] \quad \text{and} \quad \sum_{j=1}^{p} a_j''[\bar{x}_1y_j'']$$

Proof. The proof is virtually identical to that of Theorem 4.2. It involves partitioning the points of the n cube according to whether $x_1 = 0$ or 1 and observing that the corresponding values of F are

$$F(0,x_2,x_3,\ldots,x_n) = F_0(x_2,x_3,\ldots,x_n) = \langle f_0(y_1'',y_2'',\ldots,y_p'')\rangle_{u:\ell}$$

and

$$F(1,x_2,x_3,\ldots,x_n) = F_1(x_2,x_3,\ldots,x_n) = \langle a_1 + f_1(y_1',y_2',\ldots,y_m')\rangle_{u:\ell}$$

respectively. The final arguments are the same as for the proof of Theorem 4.2 and will not be given.|

The application of this theorem to function synthesis is given in succeeding chapters; however, a number of comments are appropriate here. Note that the terms of $x_1f_1(y_1',y_2',\ldots,y_m')$ and $\bar{x}_1f_0(y_1'',y_2'',\ldots,y_p'')$ can be combined according to the results of Chapter 2 to simplify the network. If, for example, $m = p$ and there is a one-to-one correspondence between the nonzero coefficients of f_1 and f_0 so that corresponding coefficients are equal (that is, if the variables can be relabeled so that $a_j' = a_j''$) then the realization of F can be expressed as

$$F(x_1,x_2,\ldots,x_n) = \langle a_1x_1 + \sum_{j=1}^{m} a_j'(x_1y_j' + \bar{x}_1y_j'')\rangle_{u:\ell}$$

By Corollary 2.5.1, this can be written as

$$F(x_1,x_2,\ldots,x_n) = \langle a_1x_1 + \sum_{j=1}^{m} a_j'[x_1y_j' \dotplus \bar{x}_1y_j'']\rangle_{u:\ell}$$

since x_1y_j' and $\bar{x}_1y_j''$ are disjoint.

The realization is now in the form of a single gate that realizes F where this gate has $m + 1$ inputs, m of which are Boolean functions

$$[x_1y_j' \dotplus \bar{x}_1y_j'']$$

To obtain a realization with inputs $x_1,x_2,\ldots,x_n$, each of these Boolean function inputs must be realized by T-gate networks. In this case, they already are in the form to which the range theorem can again be applied, assuming that realizations for each y_j' and y_j'' are known. This is illustrated in the following example.

Example 4.13. Assume that there is a known multigate realization of the switching function

$$y_1 = F_1(x_2,x_3,\ldots,x_n) = \left\langle \sum_{j=2}^{m} a_j y_j \right\rangle_{u_1:\ell_1}$$

where each

$$y_j = \left\langle \sum_{i=1}^{m} a_{ij} y_{ij} \right\rangle_{u_j:\ell_j}$$

and so on, and that a realization of the function

$$y = F(x_1,x_2,\ldots,x_n) = \bar{x}_1 F_1(x_2,x_3,\ldots,x_n) \dotplus x_1 F_1^d(x_2,x_3,\ldots,x_n)$$

is desired. According to Theorem 2.11, F_1^d can be realized by

$$F_1^d(x_2,x_3,\ldots,x_n) = \left\langle \sum_{j=2}^{m} a_j y_j^d \right\rangle_{(\sigma-\ell_1):(\sigma-u_1)}$$

where

$$\sigma = \sum_{j=2}^{m} a_j$$

From Theorem 4.7,

$$R(a_1) = (2u_1 - \sigma)*(2\ell_1 - \sigma)$$

and we select

$$a_1 = 2\ell_1 - \sigma + (u_1 - \ell_1) = u_1 + \ell_1 - \sigma$$

to maximize the gap. This gives the realization

$$F(x_1,x_2,\ldots,x_n) = \langle (u_1 + \ell_1 - \sigma)x_1 + \sum_{j=2}^{m} a_j(\bar{x}_1 y_j + x_1 y_j^d) \rangle_{u:\ell}$$

which according to Corollary 2.5.1 can be expressed as

$$F(x_1,x_2,\ldots,x_n) = \langle (u_1 + \ell_1 - \sigma)x_1 + \sum_{j=2}^{m} a_j[\bar{x}_1 y_j \dotplus x_1 y_j^d] \rangle_{u:\ell}$$

Note that the value of a_1 is such that when $x_1 = 0$, the gap $u{:}\ell$ is such that F is realized, and when $x_1 = 1$ the effective gap is displaced symmetrically about the center of the map by the amount $u_1 + \ell_1 - \sigma$ so as to realize F_1^d. Each Boolean function

$$[\bar{x}_1 y_j \dotplus x_1 y_j^d]$$

is in the form permitting Theorem 4.7 to be applied again with the same result. The coefficient a_1 is added to the second level of gates so as to displace their gaps symmetrically about the centers of the corresponding maps, and so on.

The general form of the resultant network that realizes

$$F(x_1,x_2,\ldots,x_n) = \bar{x}_1F_1(x_2,x_3,\ldots,x_n) + x_1F_1{}^d(x_2,x_3,\ldots,x_n)$$

should be evident. The realization of F is exactly the realization of F_1 with the variable x_1 added to each gate where the corresponding coefficient is

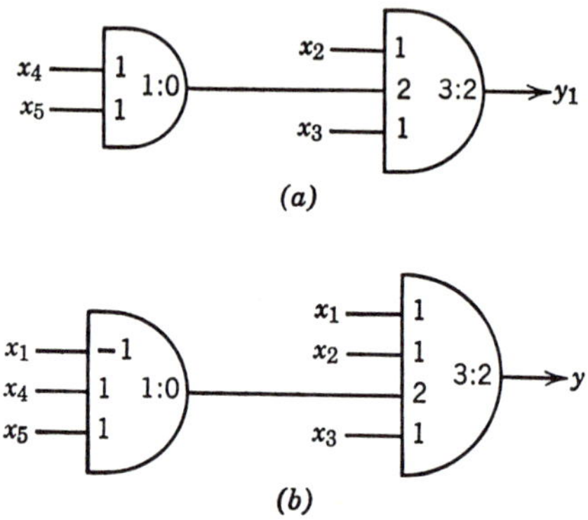

Fig. 4.18. Realization of Example 4.13. (*a*) Realization of $y_1 = (x_2 + x_3)(x_4 + x_5)$. (*b*) Realization of $y = \bar{x}_1y_1 + x_1y_1{}^d$.

such that the effect of $x_1 = 1$ is to displace the gap symmetrically about the center of the map. Figure 4.18*a* shows the network for

$$y_1 = F_1(x_2,x_3,x_4,x_5) = (x_2 + x_3)(x_4 + x_5)$$

and Fig. 4.18*b* shows the corresponding realization for

$$\begin{aligned} y &= \bar{x}_1F_1(x_2,x_3,x_4,x_5) + x_1F_1{}^d(x_2,x_3,x_4,x_5) \\ &= \bar{x}_1(x_2 + x_3)(x_4 + x_5) + x_1(x_2x_3 + x_4x_5) \end{aligned}$$

This is an example of a situation that occurs frequently in our synthesis procedures: both F_1 and F_0 have the same realizations except for gaps. That is, F_1 and F_0 are realized by the same number of gates, interconnected in the same way, with corresponding gates having the same coefficients but different gaps. Consider another example of this situation.

Example 4.14. Let

$$F_1(x_2,x_3,x_4,x_5) = x_2x_3 + x_4x_5 = \langle 2x_2 + x_3 + 3y_2\rangle_{3:2}$$

where

$$y_2 = x_4x_5 = \langle 2x_4 + x_5 \rangle_{3:2}$$

and

$$F_0(x_2,x_3,x_4,x_5) = x_2x_4 = \langle 2x_2 + x_3 + 3y_3 \rangle_{5:4}$$

where

$$y_3 = x_4 = \langle 2x_4 + x_5 \rangle_{2:1}$$

Note that both functions have the same realizations except for gaps. To find the realization for

$$\begin{aligned} y &= x_1F_1(x_2,x_3,x_4,x_5) \dotplus \bar{x}_1F_0(x_2,x_3,x_4,x_5) \\ &= x_1x_2x_3 \dotplus x_1x_4x_5 \dotplus \bar{x}_1x_2x_4 \end{aligned}$$

apply the range theorem to the final gates for F_1 and F_0, which gives

$$y = \langle 2x_1 + 2x_2 + x_3 + 3[x_1y_2 \dotplus \bar{x}_1y_3] \rangle_{5:4}$$

Applying the range theorem to the Boolean function in the brackets gives

$$[x_1y_2 \dotplus \bar{x}_1y_3] = \langle \bar{x}_1 + 2x_4 + x_5 \rangle_{3:2}$$

in which case

$$y = \langle 2x_1 + 2x_2 + x_3 + 3\langle \bar{x}_1 + 2x_4 + x_5 \rangle_{3:2} \rangle_{5:4}$$

Note in the above example that the range theorem is applied to each gate and that this could have been done directly. This is a general property of realizations of this type as indicated in the next corollary.

Corollary 4.7.1. If $F_1(x_2,x_3,\ldots,x_n)$ and $F_0(x_2,x_3,\ldots,x_n)$ are switching functions with the same realization except for gaps, then a realization for

$$F(x_1,x_2,\ldots,x_n) = x_1F_1(x_2,x_3,\ldots,x_n) \dotplus \bar{x}_1F_0(x_2,x_3,\ldots,x_n)$$

is obtained by applying Theorem 4.7 to each gate in the realization of F_1.

This range theorem and the equivalent ones for single-gate realizations are the basis for the realization procedures developed in the next chapter.

4.10 THE FUNCTION TREE FOR FUNCTIONS WITH DON'T-CARE POINTS

Many practical synthesis problems involve switching functions which are incompletely specified (*incomplete* functions) or, what is the same thing, which include don't-care conditions. Such functions are characterized by the property that, for certain input combinations, the value of the function is immaterial and hence is not defined. More specifically, the input combinations (points) of an incomplete switching function $\mathscr{F}(x_1,x_2,\ldots,x_n)$ are divided into three sets: (1) those points $\{p^1\}$ for which the value of $\mathscr{F}$ is 1, (2) those points $\{p^0\}$ for which the value of $\mathscr{F}$ is 0, and (3) those points $\{p^\phi\}$ for which the value of $\mathscr{F}$ is unspecified—the don't-care points.

A completely specified (*complete*) function $F(x_1,x_2,\ldots,x_n)$ is said to *satisfy* $\mathscr{F}$ if

$$\mathscr{F} = 1 \text{ implies } F = 1 \text{ and } \mathscr{F} = 0 \text{ implies } F = 0$$

Each function with don't-care conditions is satisfied by a number of complete functions.

The don't-care synthesis problem can be stated thus: given an incomplete function $\mathscr{F}$, synthesize a complete function F which satisfies $\mathscr{F}$. Moreover, it is usually appropriate to use that function F which requires the fewest gates or satisfies other engineering requirements.

Consider first the single-gate realization problem. Using the simultaneous inequality approach of Section 1.9, it follows that an incomplete function $\mathscr{F}(x_1,x_2,\ldots,x_n)$ can be realized with a single T gate if and only if there exists a set of real numbers $(a_1,a_2,\ldots,a_n,u,\ell)$ so that for each p^0 and p^1

$$\sum_{i=1}^{n} a_i x_i^{p^1} \geq u \tag{4.13}$$

$$\sum_{i=1}^{n} a_i x_i^{p^0} \leq \ell$$

$$u > \ell$$

where the coordinates of each p^1 are $(x_1^{p^1},x_2^{p^1},\ldots,x_n^{p^1})$ and of each p^0 are $(x_1^{p^0},x_2^{p^0},\ldots,x_n^{p^0})$.

These inequalities are analogous to those of Eq. 1.24 except that there are fewer than $2^n + 1$ of them. Since don't-care points place no restrictions on the value of the function, the inequalities corresponding to these points can be omitted. Thus Eq. 4.13 specifies a set of inequalities which have been reduced because of don't-care conditions.

These reductions are in many ways similar to reductions due to coefficient sign, ordering, etc. Therefore one might suspect that there would be a reduced tree corresponding to the don't-care reductions just as there is a reduced tree corresponding to the reductions discussed earlier. This in fact is the case; the don't-care tree is quite similar to the usual tree except that branches corresponding to don't-care points are omitted, since they place no constraint on the realization.

This same type of reasoning applies to the multigate as well as the single-gate realization, and so the subsequent discussion applies equally to both situations.

To describe the tree formation process, it is convenient to introduce some additional notation. Of all the functions which satisfy $\mathscr{F}$, define:

(1) The *minimum function* F_m such that for all p^1, p^0 and p^ϕ

$$F_m(p^1) = 1 \text{ and } F_m(p^0) = F_m(p^\phi) = 0$$

(2) The *maximum function* F_M such that for all p^1, p^0 and p^ϕ

$$F_M(p^0) = 0 \text{ and } F_M(p^1) = F_M(p^\phi) = 1$$

A convenient way to specify an incomplete function is to specify F_m and F_M.

The tree formation procedure is then to construct two trees simultaneously, one for F_m and one for F_M. The tree for $\mathscr{F}$ consists of those branches of the F_m tree which terminate on 1 and those branches of the F_M tree which terminate on 0. More specifically, the procedure is described below and illustrated by Example 4.15.

Procedure 4.5. The Don't-Care Tree

(1) The 0-levels of the F_m and F_M trees consist of F_m and F_M, respectively.

(2) Let x_1 denote the most frequent variable† (Procedure 4.2) of the set $\{F_m, F_M\}$. Factor

$$F_m = x_1 F_{m1} + \bar{x}_1 F_{m0}$$
$$F_M = x_1 F_{M1} + \bar{x}_1 F_{M0}$$

The 1-levels of the F_m and F_M trees are then F_{m1}, F_{m0} and F_{M1}, F_{M0} respectively. Solid and dotted lines are added as in Procedure 4.1.

(3) To form the $(k + 1)$-level of each tree $(k + 1 \leq n)$, let x_{k+1} be the most frequent variable in the set of all k-level functions of both the

† Although any variable can be selected for the factoring, the complexity is usually least if the most frequent variable is used.

For the case of single-gate realizations, it is sometimes possible to determine at least a partial ordering of the coefficients or partial information as to their sign. For example, it is evident that if

$$F_M(x_1 = 1, x_2 = 0) > F_m(x_1 = 0, x_2 = 1)$$

but $F_M(x_1 = 0, x_2 = 1)$ and $F_m(x_1 = 1, x_2 = 0)$ are not comparable, then in any ℓ.s. function F which satisfies $\mathscr{F}$,

$$F(x_1 = 1, x_2 = 0) > F(x_1 = 0, x_2 = 1)$$

Similarly, if neither of these pairs of reduced functions are comparable, then there is no monotonic (and hence no ℓ.s.) F which satisfies $\mathscr{F}$. This additional information is not needed for the subsequent synthesis procedure and can usually be ignored.

F_m and the F_M trees. The $(k + 1)$-level of each tree consists of the descendant function pairs which result from factoring the nonconstant k-level functions with respect to x_{k+1}.

The validity of the following properties of the F_m and F_M trees are evident from their definition.

(a) Each function on the F_m tree which is identically 1 defines a subset of the points of $\mathscr{F}$ (specified by the product of the literals on the path from that 1 to F_m) for which the value of $\mathscr{F}$ is unity.

(b) Each function on the F_M tree which is identically 0 defines a subset of the points of $\mathscr{F}$ (specified by the product of the literals on the path from that 0 to F_M) for which the value of $\mathscr{F}$ is zero.

(c) The 1's on the F_m tree and the 0's on the F_M tree define all of the specified points of $\mathscr{F}$.

The functions at corresponding locations on the F_m and F_M trees are correspondingly reduced functions of F_m and F_M. Since $F_M \geq F_m$, then to each F_m-tree location with function 1 there corresponds an F_M-tree location with function 1. Similarly, to each F_M-tree location with function 0 there corresponds an F_m-tree location with function 0.

The function tree for $\mathscr{F}$ consists of that portion of the F_m tree which terminates in the functions identically 1 and that portion of the F_M tree which terminates in the functions identically 0. More precisely,

(4) The $\mathscr{F}$-function tree or $\mathscr{F}$ tree is composed of that part of the F_m tree between the 0-level and the functions identically 1 and that part of the F_M tree between the 0-level and the functions identically 0. The only functions indicated on the $\mathscr{F}$ tree are the functions identically 1 from the F_m tree and the functions identically 0 from the F_M tree.

It is evident that the constant functions of the $\mathscr{F}$ tree define the points for which $\mathscr{F}$ is specified and that those constant functions which define don't-care points have been deleted.

Example 4.15. Consider the incomplete function $\mathscr{F}$ defined by

$$F_m = x_1[x_2(x_3 + \bar{x}_4) + \bar{x}_2\bar{x}_3x_5(x_4\bar{x}_6 + \bar{x}_4x_6)] + x_2x_4(x_3\bar{x}_5 + \bar{x}_1\bar{x}_3x_5x_6)$$

$$F_M = x_1(x_2 + x_3 + x_5 + \bar{x}_4\bar{x}_6) + \bar{x}_1x_4(\bar{x}_3 + \bar{x}_5) + x_2(x_3 + x_4 + \bar{x}_5) + x_3\bar{x}_5 + x_4x_6 + \bar{x}_3x_4x_5$$

The F_m and F_M trees are shown in Fig. 4.19 and the $\mathscr{F}$ tree in Fig. 4.20.

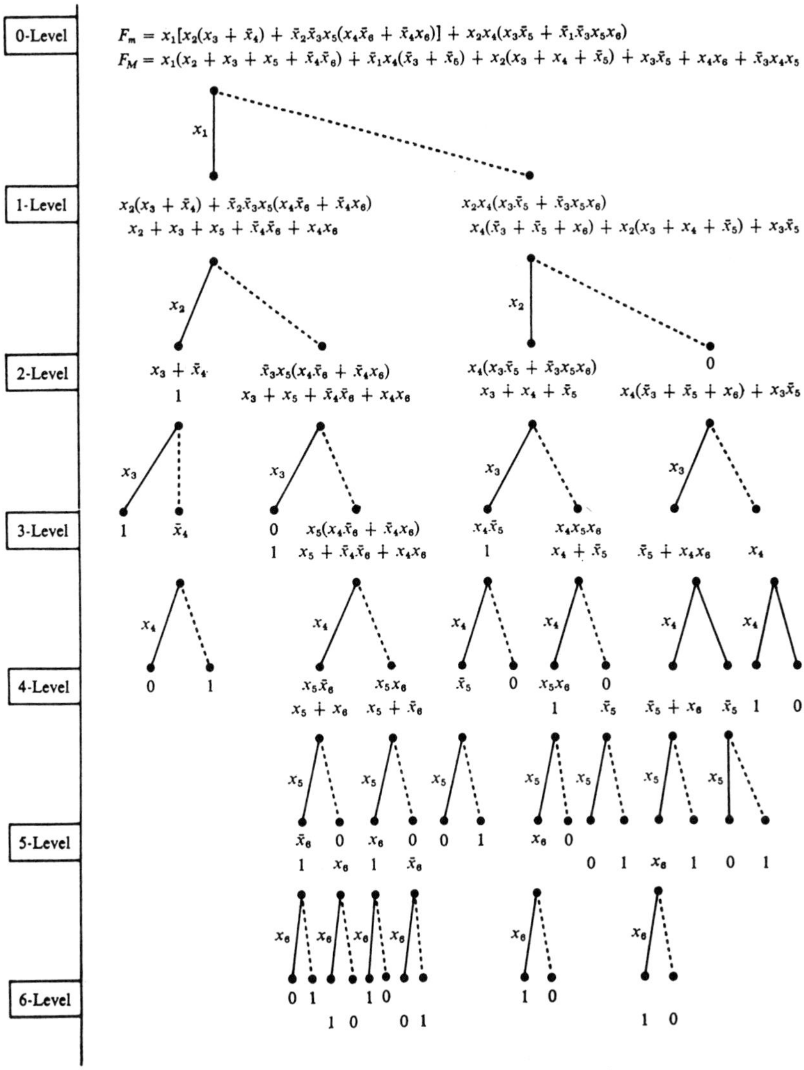

Fig. 4.19. F_m and F_M trees of Example 4.15.

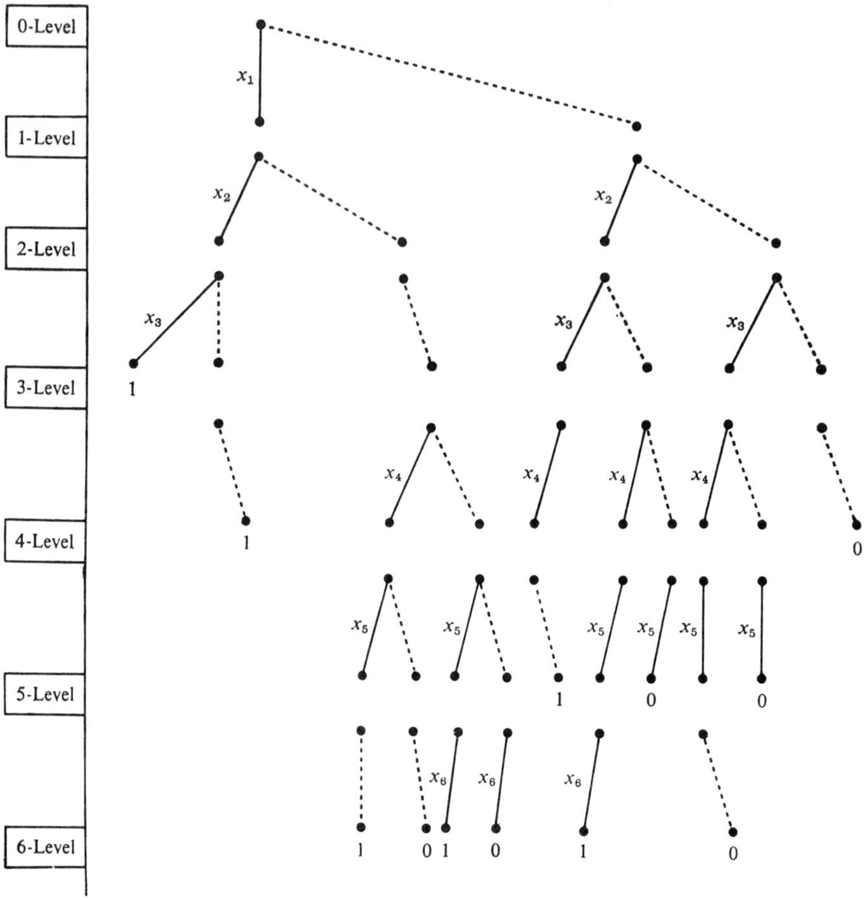

Fig. 4.20. $\mathscr{F}$ tree for Example 4.15.

The range theorems for don't-care trees are identical to those for the usual trees. If only one function in a function pair is defined, that function pair places no constraint on the pertinent coefficient. If both functions are defined, the range theorems (Theorems 4.2, 4.3, 4.4, and 4.7) apply in their usual form.

Chapter 5

REALIZATION PROCEDURES USING THE FUNCTION TREE

5.1 INTRODUCTION

This chapter is concerned with single and multigate realization procedures that utilize the function trees. They actually constitute a single procedure that can be divided into two parts. The first part is an algorithm that determines for a linearly separable switching function a set of coefficients and a gap that represents a single-gate realization. The second part is a continuation of the first that is needed for determining realizations for functions that are known to be not linearly separable.

The procedure begins by optimistically assuming that the arbitrarily specified switching function is ℓ.s. and can be realized by the first part of the procedure. If the function is not ℓ.s., this will be indicated at some later stage, at which point the second part is used to complete the realization. The second part allows control of both relative gap and fan-in for both ℓ.s. and non-ℓ.s. functions. That is, one can specify in advance a minimum permissible value for relative gap and/or a maximum permissible value for normalized fan-in, and all gates in the realization will satisfy these requirements. Specifically, therefore, our objective is the following:

To determine a network of T gates that will realize an arbitrarily specified switching function and will satisfy, when desired, specified relative gap and fan-in constraints.

The procedure begins with the function in MSP form. If the function is unate, it is converted to the positive representative function of its complementing symmetry class, and this is used thereafter. Otherwise the original function is used. In either case, the first step of the procedure is to form the function tree using Procedure 4.1, with whatever simplifications are possible from Procedure 4.2. If the given function is unate and if according to

Procedure 4.2 there is a variable on each level of the tree that can have the largest coefficient for all functions on that level, then one begins to think that the given function may be ℓ.s. Under this assumption, therefore, all additional tree simplifications due to variable symmetry (Procedure 4.3) and all tree reductions due to coefficient ordering (Section 4.8) are determined. The resultant reduced tree is used in the single-gate realization procedure described in Sections 5.2 to 5.5.

If the given switching function is not unate or if application of Procedure 4.2 shows that on some level of the tree there is no variable that can have the largest coefficient in all functions on that level, then a single-gate realization is impossible. The tree that results from Procedures 4.1 and 4.2 is used in the realization procedure described in Sections 5.6 and 5.7 to obtain a multigate realization.

5.2 SINGLE-GATE PROCEDURE: GENERAL DESCRIPTION

The single-gate realization procedure can be applied to any function tree with or without don't-cares: the tree of Procedures 4.1 and 4.2, the tree simplified by variable symmetry (Procedure 4.3), or any tree reduced because of known coefficient ordering. In the description of the procedure, it is assumed that the given function is positive unate, but even this is unnecessary. With obvious modifications, the same procedure can be used to realize any ℓ.s. function.

The general idea is to start at the bottom level of the tree and proceed to the top. We determine whether all functions on each level can have the same separating function where this separating function is determined from that for all functions on the previous level. That is, assuming that all functions on the k-level are realized with the same separating function, it is used, together with the range theorem (Theorem 4.2), to determine whether all $(k - 1)$-level functions can be realized with the same separating function. When the 0-level is attained, the resultant is a realization of the given function F.

To illustrate what is involved, consider the first two levels of the function tree of Fig. 5.1. Assume that we have previously shown that all four functions on the 2-level are ℓ.s. and are realized with the separating function $f_2(x_3,x_4,\ldots,x_n)$ and the gaps shown on the tree. The procedure continues as follows:

It must be shown that both $F_1(x_2,x_3,\ldots,x_n)$ and $F_0(x_2,x_3,\ldots,x_n)$ are ℓ.s. and can have the same separating function $f_1(x_2,x_3,\ldots,x_n)$. Since

$$F_1(x_2,x_3,\ldots,x_n) = x_2F_{11}(x_3,x_4,\ldots,x_n) + F_{10}(x_3,x_4,\ldots,x_n)$$

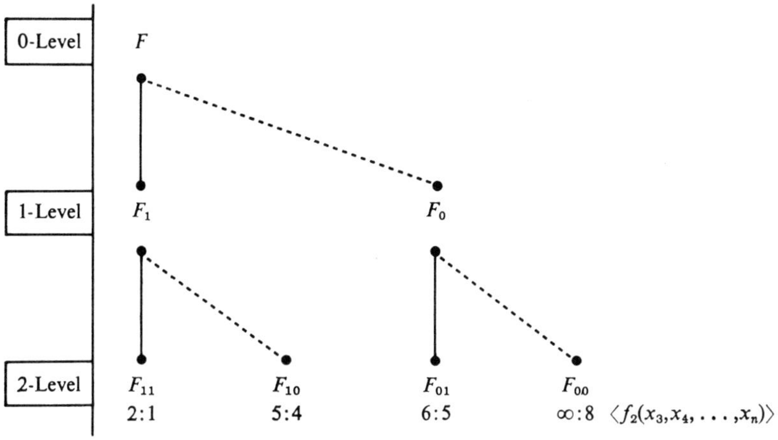

Fig. 5.1. Two levels of a function tree.

and

$$F_0(x_2,x_3,\ldots,x_n) = x_2F_{01}(x_3,x_4,\ldots,x_n) \dotplus F_{00}(x_3,x_4,\ldots,x_n)$$

then, according to Theorem 4.2, F_1 and F_0 are each ℓ.s. with separating functions

$$a_2{}^1x_2 + f_2(x_3,x_4,\ldots,x_n) \qquad \text{and} \quad a_2{}^0x_2 + f_2(x_3,x_4,\ldots,x_n)$$

respectively. Clearly these will be the same if $a_2{}^1 = a_2{}^0$. Again, according to Theorem 4.2, the ranges for $a_2{}^1$ and $a_2{}^0$ are

$$R(a_2{}^1) = 4{*}2 \qquad \text{and} \quad R(a_1{}^0) = \infty{*}2$$

Since these ranges have a common intersection, there are some values such that $a_2{}^1 = a_2{}^0 = a_2$, in which case F_1 and F_0 can both have the separating function $f_1(x_2,x_3,\ldots,x_n)$ where

$$f_1(x_2,x_3,\ldots,x_n) = a_2x_2 + f_2(x_3,x_4,\ldots,x_n)$$

We define the *common upper range boundary*, denoted by $\overline{U(a_2)}$, as the smallest upper bound on a_2. In this case,

$$\overline{U(a_2)} = \min\{4,\infty\} = 4$$

Similarly, we define the *common lower range boundary*, denoted by $\overline{L(a_2)}$, as the largest lower bound on a_2. In this example,

$$\overline{L(a_2)} = \max\{2,2\} = 2$$

We define the *common range* for a_2, denoted by $\overline{R(a_2)}$ as

$$\overline{R(a_2)} = \overline{U(a_2)}*\overline{L(a_2)}$$

by which we mean the open interval bounded above by $\overline{U(a_2)}$ and below by $\overline{L(a_2)}$. The common range is void if

$$\overline{U(a_2)} \leq \overline{L(a_2)}$$

In the present example,

$$\overline{R(a_2)} = 4*2$$

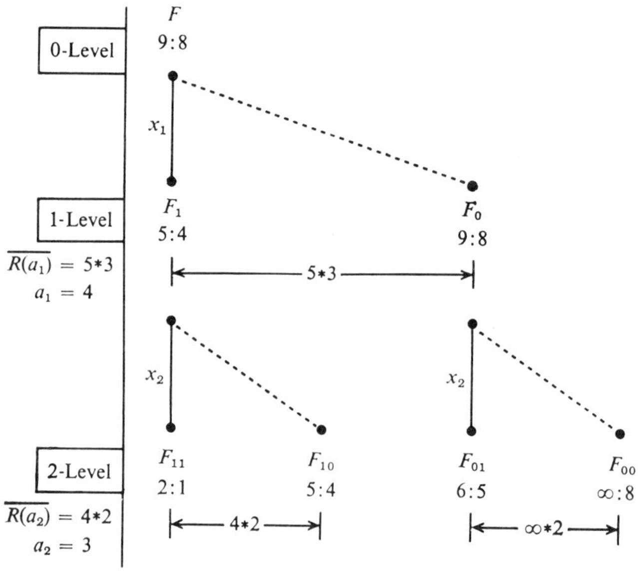

Fig. 5.2. The realization of the tree of Fig. 5.1.

We know that any value of a_2 within the common range can be used to realize both F_1 and F_0; therefore, we select $a_2 = 3$ for both realizations. Figure 5.2 shows how range and gap information is included on the tree.

Having selected $a_2 = 3$, we must determine the gaps of F_1 and F_0 when they have separating function

$$f_1(x_2,x_3,\ldots,x_n) = 3x_2 + f_2(x_3,x_4,\ldots,x_n)$$

According to Theorem 4.2, the gaps of F_1 and F_0 respectively are

$$u_1:\ell_1 = 5:4 \qquad \text{and} \quad u_0:\ell_0 = 9:8$$

Now Theorem 4.2 can be applied to the 1-level functions F_1 and F_0 to yield

$$\overline{R(a_1)} = 5*3$$

Selecting $a_1 = 4$ gives the realization for F of

$$\begin{aligned} F(x_1,x_2,\ldots,x_n) &= \langle 4x_1 + f_1(x_2,x_3,\ldots,x_n)\rangle_{9:8} \\ &= \langle 4x_1 + 3x_2 + f_2(x_3,x_4,\ldots,x_n)\rangle_{9:8} \end{aligned}$$

The basic idea for proceeding up the tree is as follows. On some k-level, it is known that all functions are ℓ.s. and can have the same separating function. For each function pair that descends from a common $(k - 1)$-level function, a range on a_k is determined according to Theorem 4.2 so as to determine a separating function for the $(k - 1)$-level functions. (For the tree simplified because of symmetric variables, each set of k-level functions that descends from a common $(k - 1)$-level function is used to determine a range on a_k according to Theorem 4.6). In order that all $(k - 1)$-level functions have the same separating function, a common range

$$\overline{R(a_k)} = \overline{U(a_k)}*\overline{L(a_k)}$$

is determined where $\overline{U(a_k)}$ is the smallest of the upper range boundaries for a_k and where $\overline{L(a_k)}$ is the largest of the lower range boundaries. Any a_k value from this common range, if it is nonvoid, will yield a separating function for the $(k - 1)$-level functions with corresponding gaps determined according to Theorem 4.2 (Theorem 4.6 for symmetric trees). Thus all $(k - 1)$-level functions of the tree are ℓ.s. and have the same separating function, and the process can be repeated.

This is the basic idea of the procedure. Before it is set forth in a step-by-step form, a number of details must be considered.

5.3 SINGLE-GATE PROCEDURE: PRELIMINARY CONSIDERATIONS

Starting

The general step of the procedure is based upon the assumption that on some level of the tree all functions are ℓ.s. and have the same separating function. Moreover, this separating function must be known so that it can be used in the manner of Theorems 4.2 and 4.6 to determine a separating function for all $(k - 1)$-level functions. We have yet to discuss how the procedure starts.

The procedure can begin on any level of the tree for which all functions are ℓ.s. and where a realization for each is known, so that all realizations have the same separating function. For all function trees, the required information is at least available for the n-level as discussed in Section 4.5. Here all switching functions are constants, and a general separating function for all is $f_n(\Phi) = 0$. According to Eqs. 4.5 and 4.6, the n-level realizations are, therefore,

$$1 = \langle 0 \rangle_{0:-\infty} \quad \text{and} \quad 0 = \langle 0 \rangle_{\infty:0}$$

Hence the procedure can always begin on the level of the tree for which all reduced functions are constants.

For trees which have not been simplified because of symmetry or don't-care's, the procedure can always begin on the $(n - 1)$-level. The functions on this level are either constant functions or the Boolean function x_n. Since for any ℓ.s. function one separating-function coefficient can be selected arbitrarily without loss of generality, we select $a_n = 1$. As a result, all $(n - 1)$-level functions can have as a separating function

$$f_{n-1}(x_n) = x_n$$

and the realizations with the aid of Theorem 1.3 are

$$x_n = \langle x_n \rangle_{1:0}, \quad 1 = \langle x_n \rangle_{0:-\infty}, \quad 0 = \langle x_n \rangle_{\infty:1}$$

Symmetric trees have q levels where $q < n$. If the tree has not been reduced because of coefficient-ordering information and if it does not have don't-care points, then the procedure can always begin on the $(q - 1)$-level. On this level the functions are either constants or are the symmetric functions

$$M_k(x_{r_{q-1}+1}, x_{r_{q-1}+2}, \ldots, x_n)$$

where $1 \leq k \leq m_q$ and where

$$r_{q-1} = m_1 + m_2 + \cdots + m_{q-1}$$

It follows from Theorem 3.7 and the fact that one coefficient can be selected arbitrarily that no loss of generality results from making

$$a_{r_{q-1}+1} = a_{r_{q-1}+2} = \cdots = a_n = 1$$

in the separating function for the $(q - 1)$-level functions. The corresponding realizations are the following

$$M_k(x_{r_{q-1}+1}, x_{r_{q-1}+2}, \ldots, x_n) = \langle x_{r_{q-1}+1} + x_{r_{q-1}+2} + \cdots + x_n \rangle_{k:(k-1)}$$

$$1 = \langle x_{r_{q-1}+1} + x_{r_{q-1}+2} + \cdots + x_n \rangle_{0:-\infty}$$

$$0 = \langle x_{r_{q-1}+1} + x_{r_{q-1}+2} + \cdots + x_n \rangle_{\infty:m_q}$$

With these the procedure can begin on the $(q - 1)$-level.

Constant Functions

In all trees, branches terminate on 1 and 0 functions at various levels. These functions place no restrictions upon the lower-level coefficient selections. But when the procedure reaches the level on which a constant function appears, then a realization for that function must be specified in order to proceed, and that realization must have the same separating function as all switching functions on that level. According to Theorem 1.3, this is always possible. If $f_k(x_{k+1},x_{k+2},\ldots,x_n)$ is the separating function of all nonconstant k-level switching functions, then k-level constant functions can have realizations

$$1 = \langle f_k(x_{k+1},x_{k+2},\ldots,x_n)\rangle_{0:-\infty}$$
$$0 = \langle f_k(x_{k+1},x_{k+2},\ldots,x_n)\rangle_{\infty:\sigma_k}$$

where

$$\sigma_k = \sum_{j=k+1}^{n} a_j$$

when all coefficients of $f_k(x_{k+1},x_{k+2},\ldots,x_n)$ are nonnegative. This defines the appropriate gaps that are needed to continue the realization.

Example 5.1. Show that if $F_2(x_3,x_4,\ldots,x_n)$ is a positive ℓ.s. function and if $F_2 > F_2{}^d$, then

$$y = F(x_1,x_2,\ldots,x_n) = x_1x_2 \dotplus x_1F_2(x_3,x_4,\ldots,x_n) \dotplus x_2F_2{}^d(x_3,x_4,\ldots,x_n)$$

is ℓ.s., and determine a realization using the realization for F_2.

Figure 5.3 shows the first three levels of the tree for F. Since F_2 is ℓ.s., there exist coefficients and gaps such that

$$F_2(x_3,x_4,\ldots,x_n) = \Big\langle \sum_{j=3}^{n} a_jx_j \Big\rangle_{u_2:\ell_2}$$

and the gaps of the 2-level functions are as shown where

$$\sigma_2 = \sum_{j=2}^{n} a_j$$

The range for each descendant pair of 2-level functions is obtained and the common range for a_2 is determined. The value of a_2 is selected as

$$a_2 = \overline{L(a_2)} + (u_2 - \ell_2) = u_2$$

in order to maximize the gap lengths on the 1-level. Gaps for F_1 and F_0 are obtained according to Theorem 4.2 and are indicated below the corresponding 1-level functions. Then the range for a_1 is determined and

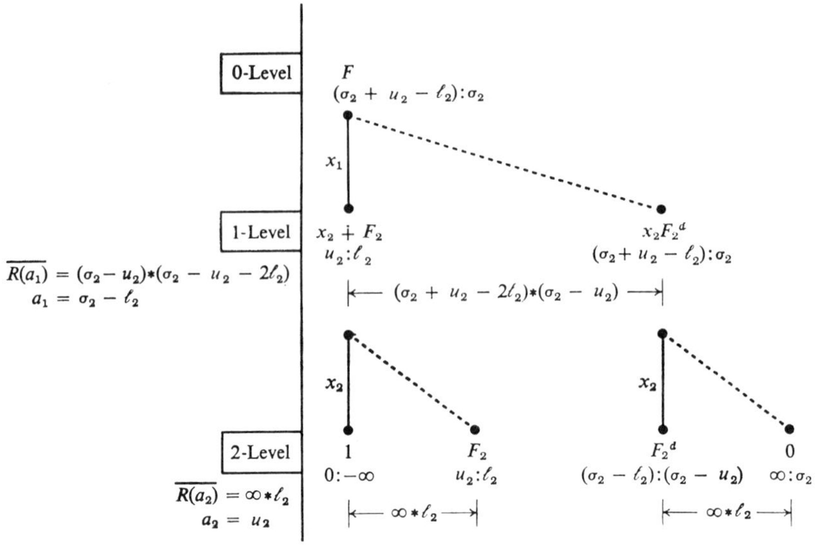

Fig. 5.3. The function tree for Example 5.1.

the value of $(\sigma_2 - \ell_2)$ is selected. This is the minimum value of a_1 so that the 0-level gap is as large as the smallest 1-level gap.

The realization for F is

$$F(x_1,x_2,\ldots,x_n) = \left\langle(\sigma_2 - \ell_2)x_1 + u_2x_2 + \sum_{j=3}^{n} a_jx_j\right\rangle_{(\sigma_2+u_2-\ell_2):\sigma_2}$$

Coefficient Selection and Displaced Gaps

In Example 5.1 the value of a_2 is selected as

$$a_2 = \overline{L(a_2)} + (u_2 - \ell_2) = u_2$$

and this is the smallest value of a_2 which maximizes the gaps (gap lengths) of both 1-level functions. According to Section 1.7, this is clearly the best choice with regard to sensitivity considerations. In a more general case, however, the choice may not be so evident.

In Section 3.6 we showed that the gap length of any realization is bounded above by the value of the smallest coefficient provided the corresponding variable appears in the MSP form of the switching function. Therefore, $g \leq 1$ when $a_n = 1$ and for each ℓ.s. function there is a maximum possible value of g when the smallest coefficient has unit value.

In general we wish to select the smallest possible value for each coefficient so that the gap of F (the 0-level function) is as large as possible. These values can be very difficult to determine, however, and so we select the coefficients by the following simple procedure. In some cases it does not yield the optimum realization in the above sense, but its computational simplicity makes it attractive. This is particularly true for nonsymmetric trees; hence they are considered first.

On a function tree the gaps of the k-level functions are not necessarily of the same length. According to Section 4.4, the gap $u:\ell$ of F is the intersection of all k-level displaced gaps; therefore, an upper bound on the gap length g of F is the smallest gap length of the k-level functions, and we denote this as $\bar{g}_k$ and call it the *minimum gap*. Our policy is to select the smallest value for a_k such that $\bar{g}_{k-1}$ is as large as possible where

$$k = n - 1, n - 2, \ldots, 1$$

It frequently occurs that

$$\bar{g}_{k-1} < \bar{g}_k$$

for some particular value of k. This comes about as shown in Fig. 5.4. Figure 5.4a shows the two k-level function pairs that determine the common range for a_k, and their corresponding ranges. The common range is

$$\overline{U(a_k)} * \overline{L(a_k)} = 2*1$$

In Fig. 5.4b are shown the gaps of the maps of these k-level functions after they have been displaced by the amount corresponding to $a_k = 1\frac{1}{2}$. Figure 5.4c shows the variation of the gap lengths g_1 and g_2 of F_1 and F_2, respectively, as a function of a_k where the variation of each is as described in Section 4.3. Since, for each value of a_k,

$$\bar{g}_{k-1} = \min \{g_1, g_2\}$$

the variation of $\bar{g}_{k-1}$ as a function of a_k shown in Fig. 5.4d is immediately evident. Note that $\bar{g}_{k-1} < \bar{g}_k$ if the common range for a_k has length less than $2\,\bar{g}_k$. Under these conditions the maximum value of $\bar{g}_{k-1}$ occurs for a_k in the center of the common range, and this corresponds to equal intersections between (1) the gap of F_b and the a_k-displaced gap of F_a and (2) the gap of F_d and the a_k-displaced gap of F_c. The results are summarized for nonsymmetric trees by the following theorem. The validity follows from the above discussion.

Theorem 5.1. *Coefficient-Selection Procedure* (*Nonsymmetric Tree*). If $\bar{g}_k$ is the smallest k-level gap length and

$$\overline{R(a_k)} = \overline{U(a_k)} * \overline{L(a_k)}$$

is the a_k-common range, then the smallest value of a_k that maximizes $\bar{g}_{k-1}$ is

$$a_k = \min \{\overline{L(a_k)} + \bar{g}_k,\ \ \tfrac{1}{2}[\overline{L(a_k)} + \overline{U(a_k)}]\}$$

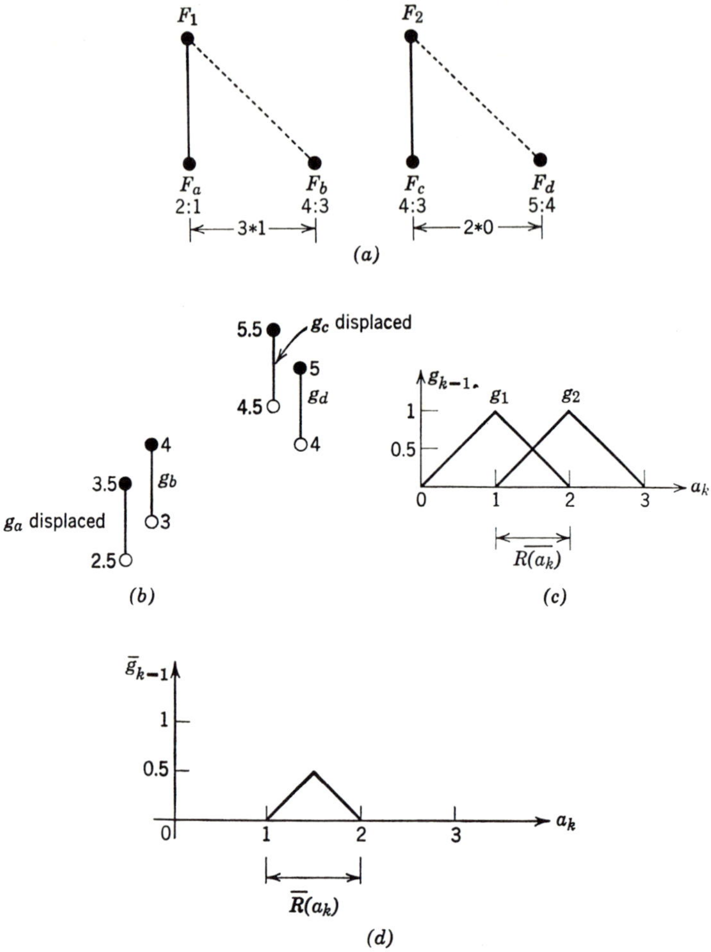

Fig. 5.4. The minimum gap g_{k-1} as a function of a_k. (*a*) Two pairs of k-level functions. (*b*) Displaced gaps. (*c*) Gap lengths as a function of a_k. (*d*) $\bar{g}_{k-1}$ as a function of a_k.

For the symmetric tree, the coefficient-selection procedure of Section 4.7 (Procedure 4.4) should be used where the common range boundaries $\overline{U_j(a)}$ and $\overline{L_j(a)}$ are defined from all of the functions of the level rather than from just one function set. This will be clarified by Example 5.3.

Note again that this selection procedure as well as that implied by Theorem 5.1 does not necessarily yield a maximum value for the gap of F. It only insures that the smallest gap one level up is maximized. This may or may not yield a maximum g. The problem of maximally reliable realizations is discussed again in Section 8.5 where a different and more sophisticated coefficient selection procedure is presented.

After the coefficient is selected, all gaps on the level are displaced in the usual way according to the labels of the branches between the gaps and the functions on the next higher level. This is the case for all trees including those that have been reduced because of variable symmetry, coefficient ordering, and don't-care's, in which only one member of a function pair or of a function set can occur without the others. Recall that broken lines imply a zero label which corresponds to a zero displacement of their gaps.

5.4 SINGLE-GATE PROCEDURE: STEP-BY-STEP DESCRIPTION

The single-gate realization procedure begins with one of the function trees. We assume that the tree has q levels where $q = n$ or $q < n$ according to whether or not the tree is symmetric. The actual description begins on the $(q - 1)$-level. If the tree is reduced because of don't-care's or because of coefficient ordering, it may be more convenient to begin on the q-level. (See Section 5.3 (p. 175) for determining the q-level gaps.) Although the description assumes that the function tree has not been reduced because of don't-care's or coefficient ordering, the changes that are required when this is not the case are minor.

Procedure 5.1

(1) Select the smallest coefficient

$$a_n = 1$$

If the function is symmetric in the variables $x_n, x_{n-1}, \ldots, x_{n-m_q+1}$ and the tree has been simplified accordingly, then let

$$a_n = a_{n-1} = \cdots = a_{n-m_q+1} = 1$$

(2) Below each $(q - 1)$-level function, list the corresponding gap where, for nonsymmetric trees,

$$1 = \langle x_n \rangle_{0:-\infty}$$

$$0 = \langle x_n \rangle_{\infty:1}$$

$$x_n = \langle x_n \rangle_{1:0}$$

For symmetric trees,

$$1 = \left\langle \sum_{j=n-m_q+1}^{n} x_j \right\rangle_{0:-\infty}$$

$$0 = \left\langle \sum_{j=n-m_q+1}^{n} x_j \right\rangle_{\infty:m_q}$$

$$M_i(x_{n-m_q+1}, x_{n-m_q+2}, \ldots, x_n) = \left\langle \sum_{j=n-m_q+1}^{n} x_j \right\rangle_{i:(i-1)}$$

Proceed to the general step (Step 3), and initially set $k = q - 1$.

(3) Each k-level function has separating function

$$f_k(x_{k+1}, x_{k+2}, \ldots, x_n) = \sum_{j=k+1}^{n} a_j x_j$$

with the appropriate gap listed below the function. For every k-level function pair (or function set) which contains more than one member, compute a range on the coefficient to be added to the separating function. Denote this coefficient by a_k.

(4) From the set of k-level ranges and any coefficient-ordering inequalities that were used in the tree reduction and are indicated on that level, let $\overline{U(a_k)}$ denote the smallest upper bound and $\overline{L(a_k)}$ the largest lower bound. If

$$\overline{U(a_k)} \leq \overline{L(a_k)}$$

then the common range is void and this procedure cannot be continued. This is discussed in the subsequent Section 5.5. If

$$\overline{U(a_k)} > \overline{L(a_k)}$$

then the common range

$$\overline{R(a_k)} = \overline{U(a_k)} * \overline{L(a_k)}$$

defines the allowable values for a_k. Select the value of a_k (according to the method of Section 5.3, p. 178) so that the smallest gap on the $(k - 1)$-level is as large as possible. The separating function for all $(k - 1)$-level functions is

$$f_{k-1}(x_k, x_{k+1}, \ldots, x_n) = a_k x_k + f_k(x_{k+1}, x_{k+2}, \ldots, x_n)$$

$$= \sum_{j=k}^{n} a_j x_j$$

(5) Below each $(k - 1)$-level function, list the gap† with which it is realized by the separating function of Step 4. These are determined as follows:

(a) Constant 1 functions have gap $0:-\infty$.
(b) Constant 0 functions have gap $\infty:\sigma_{k-1}$ where

$$\sigma_{k-1} = \sum_{j=k}^{n} a_j$$

(c) Gaps for nonconstant functions are determined from the gaps of their k-level descendant pair (or descendant set) by Theorem 4.2 (Theorem 4.6 for a descendant set).

(6) If $k - 1 = 0$, we are at the 0-level of the tree and the function is realized. Otherwise reduce k by unity and repeat the procedure starting with Step 3.

Example 5.2. Figure 5.5 shows the unreduced tree of Example 4.2 and its realization by means of the tree procedure. Note that the gap $\bar{g}_2$ was reduced from $\bar{g}_3$ in going from the 3- to the 2-level.

Figure 5.6 shows, for the function of Example 4.2, the reduced tree from Example 4.12 and the realization by means of the tree procedure. On the 4-level there are no ranges except those implied by coefficient ordering. Note how the gaps are displaced on this level to determine the next level gap. There are two ranges on the 3-level, and the common range is such that the minimum gap of the 2-level is reduced from its 3-level value. The realization is the same as that obtained with the unreduced tree of Fig. 5.6; namely,

$$y = \langle 4\tfrac{1}{2}x_1 + 3x_2 + 2\tfrac{1}{2}x_3 + 2x_4 + 2x_5 + x_6 \rangle_{7:6\frac{1}{2}}$$

However, the amount of computation is considerably less.

† If the tree has been reduced or includes don't-care points, then the gap listed below a particular function may not be the actual gap for the specified separating function. All gap boundaries that determine the common range boundaries are correct, however. At the top of the tree, the gap for the function F is, of course, correct for the particular separating function that has been determined.

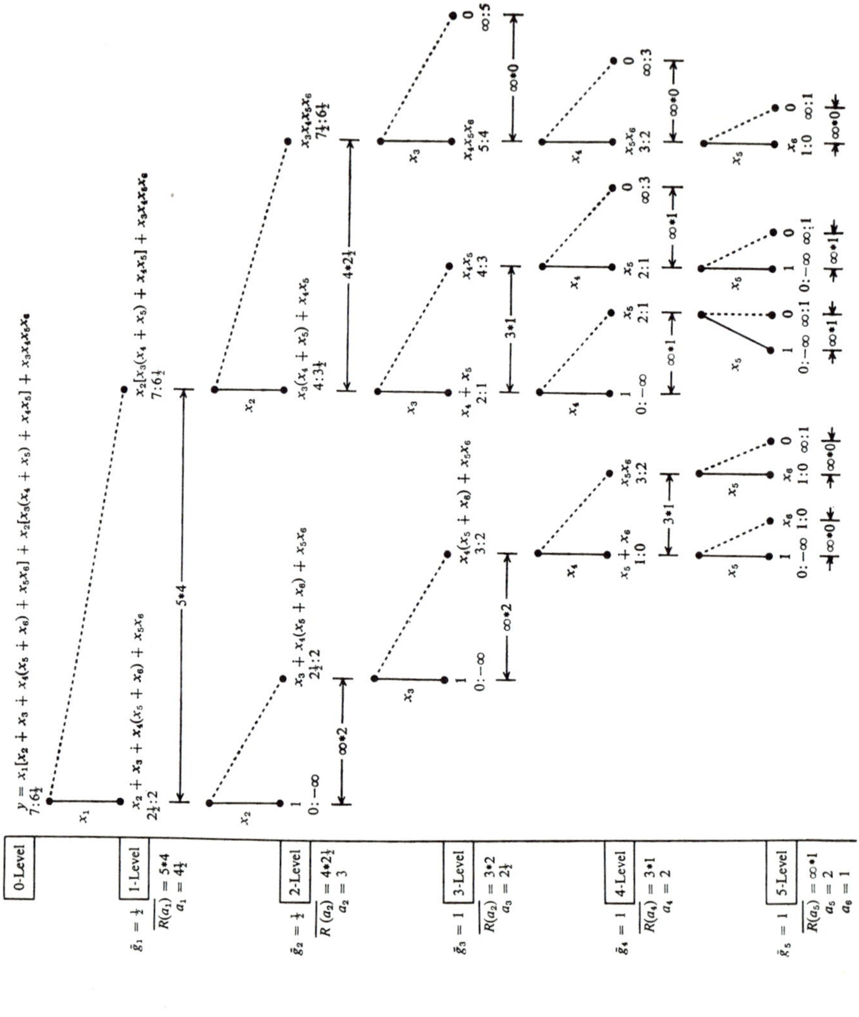
0-Level
1-Level
2-Level
3-Level
4-Level
5-Level

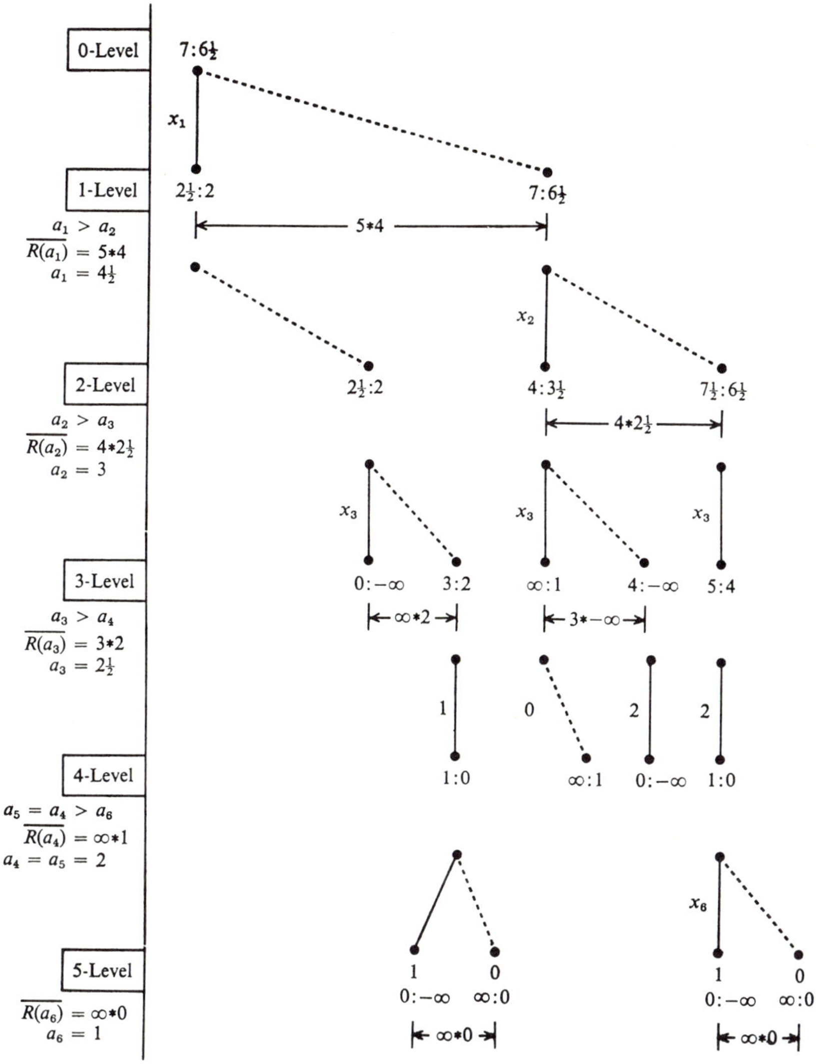

Fig. 5.6. Realization of Example 5.2 by means of the reduced tree.

5.5 SINGLE-GATE PROCEDURE: VOID COMMON RANGE

If on any level of a tree the common range is void, then Procedure 5.1 will fail to give a realization. This will of course occur for non-ℓ.s. functions; however, it can also occur for sufficiently complex ℓ.s. functions. This is illustrated in the following two examples.

Example 5.3. Figure 5.7 shows the attempt to apply Procedure 5.1 to

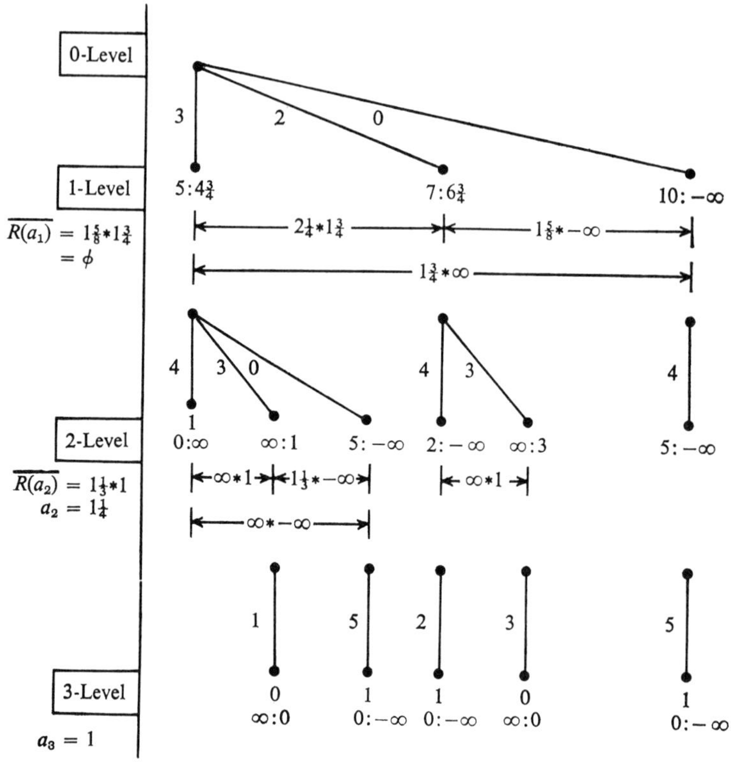

Fig. 5.7. Tree-realization procedure for Example 5.3.

the reduced symmetric tree of Moore's function given in Fig. 4.17. On the 1-level, two of the ranges are

$$R(a_1) = 1\tfrac{5}{8}*-\infty \qquad \text{and} \quad R(a_1) = 2\tfrac{1}{4}*1\tfrac{3}{4}$$

Hence the common range is void. It is shown in Example 8.2 that Moore's function is in fact not ℓ.s.

Example 5.4 (*Winder 9,*). Consider the switching function

$$y = x_1[x_2(x_3 \dotplus x_4 \dotplus x_5 \dotplus x_6) \dotplus x_3(x_4 \dotplus x_5) \dotplus x_4x_5x_6]$$
$$\dotplus x_2[x_3(x_4 \dotplus x_5x_6) \dotplus x_4x_5x_6] \dotplus x_3x_4x_5x_6$$

Figure 5.8*a* shows the tree for this function and Fig. 5.8*b* shows this tree reduced and the result of applying Procedure 5.1 in an attempt to obtain

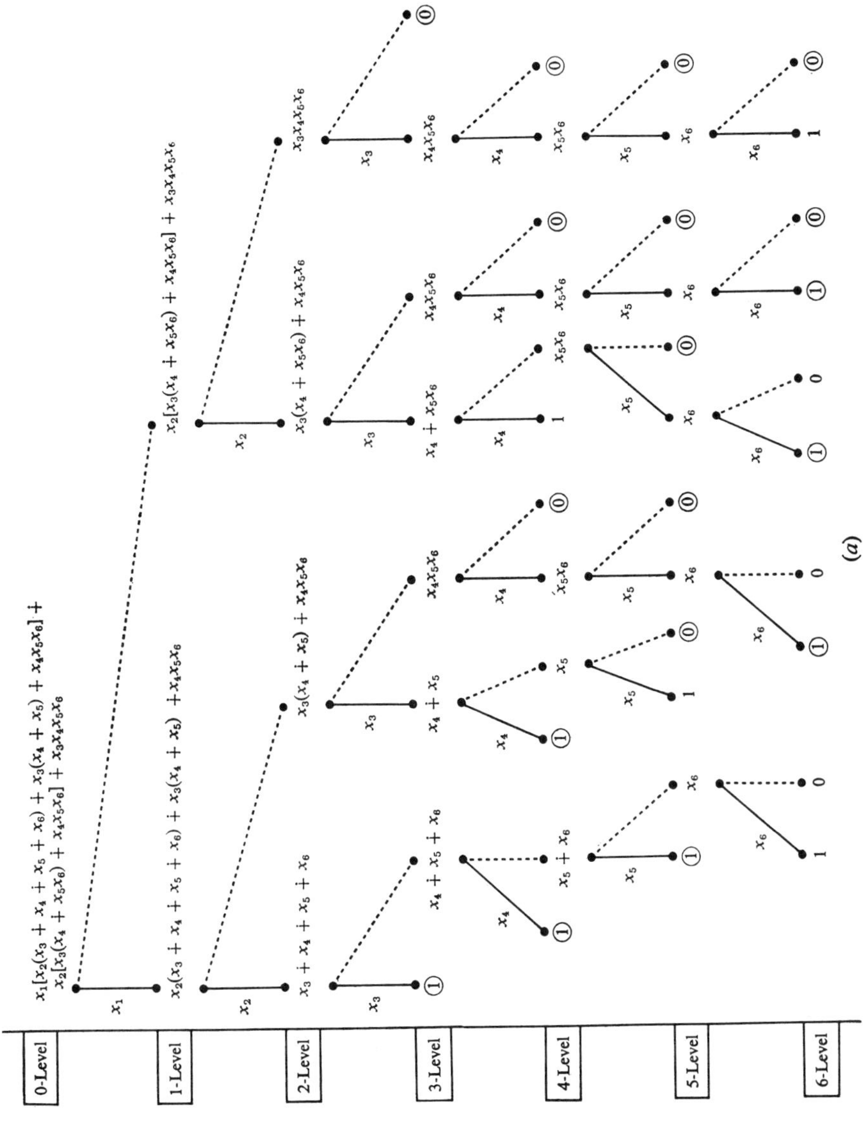

Fig. 5.8. Tree for Example 5.4. (*a*) Unreduced tree; (*b*) tree realization procedure applied to the reduced tree of (*a*).

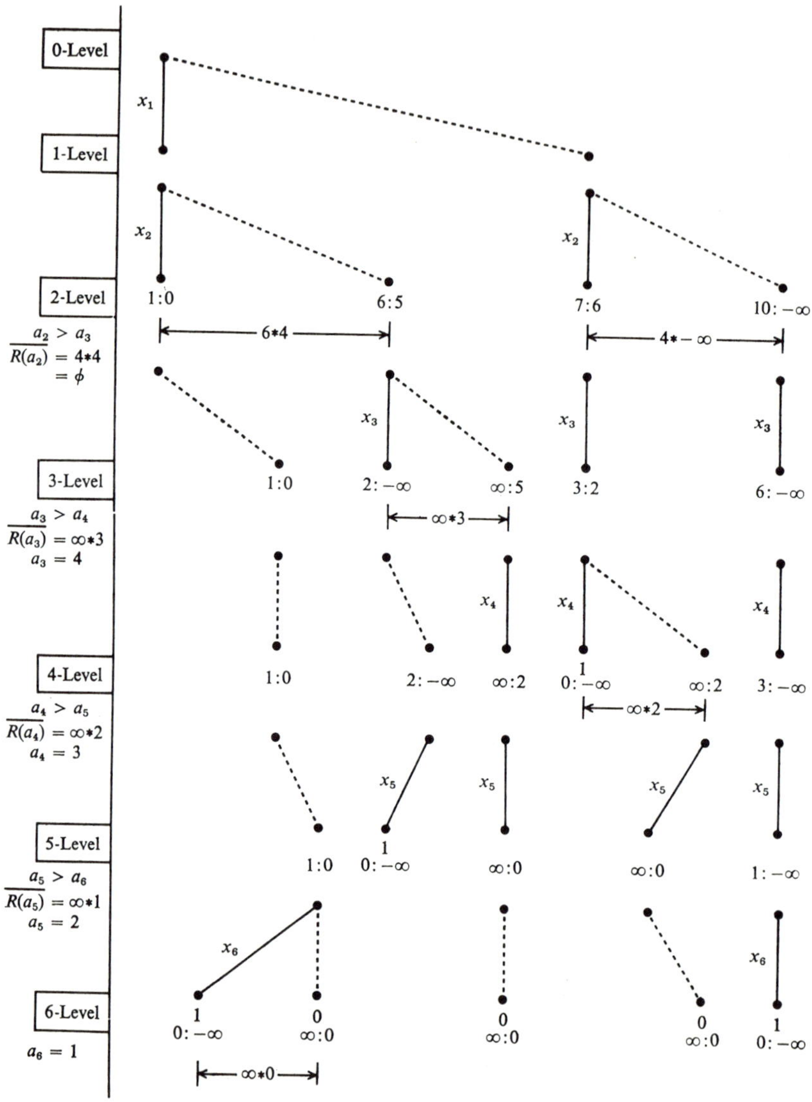
0-Level
1-Level
2-Level
$a_2 > a_3$
$\overline{R(a_2)} = 4*4 = \phi$
3-Level
$a_3 > a_4$
$\overline{R(a_3)} = \infty*3$
$a_3 = 4$
4-Level
$a_4 > a_5$
$\overline{R(a_4)} = \infty*2$
$a_4 = 3$
5-Level
$a_5 > a_6$
$\overline{R(a_5)} = \infty*1$
$a_5 = 2$
6-Level
$a_6 = 1$
x_1
x_2
1:0
6:5
7:6
10: −∞
6*4
4* − ∞
x_3
1:0
2: −∞
∞:5
3:2
6: −∞
∞*3
x_4
1:0
2: −∞
∞:2
0: −∞
∞:2
3: −∞
∞*2
x_5
1:0
0: −∞
∞:0
∞:0
1: −∞
x_6
0: −∞
∞:0
∞:0
∞:0
0: −∞
∞*0

Fig. 5.8b.

a realization. A realization is not obtained since $\overline{R(a_2)} = 4{*}4$ is void. Nevertheless the function is ℓ.s. and has the realization

$$y = \langle 8x_1 + 7x_2 + 6x_3 + 5x_4 + 4x_3 + 3x_2 \rangle_{18:17}$$

Example 5.4 shows that the tree procedure can fail to give a realization even though the given function is ℓ.s. The reason for this failure follows from the discussion of Chapter 4. On any k-level of the tree, the remaining coefficients must be selected so that all k-level gaps, appropriately displaced, have a nonvoid intersection. This is a necessary and sufficient condition for realization. Determining the ranges of values for the coefficients that satisfy this condition requires a simultaneous solution of inequalities. The result would be the "actual range" on each remaining coefficient such that any choice within that range will guarantee a realization if one exists. Simultaneous solutions, however, require extensive computation, and our objective is eventually to accomplish the same results with much less effort.

The range that the tree procedure determines does not involve simultaneous inequalities. It is only a range that insures that all functions on the next higher level will have the same separating function. It does not guarantee that these will then define a nonvoid range for the next coefficient if one exists or that this range will be such that the next coefficient has a nonvoid range and so on. Only the "actual range" insures this. We know, however, that the tree procedure range must contain the actual range.

Now we are at the root of the difficulty. If on any k-level of the tree the selected value of the coefficient a_k within the common range is outside of the "actual range," then the procedure will fail on some higher level even though the given function is ℓ.s. A different choice for a_k such that it is within both the common and actual ranges would lead to a realization, but we have no way of determining this with the tree procedure as described in Procedure 5.1. In Example 5.4 a realization would have been obtained if the choice of either $a_6 = 3$ or $a_6 = 1$ and $a_5 = \frac{4}{3}$ had been made.

Hence the tree procedure by itself does not constitute a test for ℓ.s. functions. If the procedure fails, we do not know that the given function is not ℓ.s. However, a relatively simple continuation of the procedure does constitute a test. Additional computations can be performed on the tree to determine whether there exist different choices for any of the coefficients which will allow a realization. The final result of this procedure is either a realization or a proof that no such realization exists and that the function is not ℓ.s. These additional computations which constitute a test for ℓ.s. functions are considered in Chapter 8.

Experience has shown that when the tree procedure fails to give a realization for an ℓ.s. function, the sensitivity of the realization to component and signal fluctuations is probably too high for a practical realization. For instance, the function of Example 5.4, which is the simplest known ℓ.s. function for which the procedure fails, has a gap of 18 : 17. This is a relative gap of .0286, a value that would be difficult to implement economically.

The severe sensitivity requirement is a direct consequence of the reason for the failure of Procedure 5.1. This procedure fails because there are some very restrictive requirements on the values of the coefficients that are not immediately apparent on the tree. But these kinds of severe restrictions are almost certain to imply correspondingly tight tolerances on the components that implement the coefficients and on the signals that represent the logic variables. Thus, if the tree procedure fails to give a realization for an ℓ.s. function, it is highly probable that a single-gate realization would be impractical.

It is for this reason that the test for ℓ.s. functions is postponed until Chapter 8. The remainder of this chapter is concerned with a different continuation of the tree procedure, one that yields a multigate realization in which the sensitivity parameters of relative gap and fan-in can be controlled.

5.6 MULTIGATE PROCEDURE: PRELIMINARY CONSIDERATIONS

General Description

The multigate realization procedure allows the sensitivity of the resulting network to be controlled by specifying lower bounds for the relative gap and upper bounds for the fan-in of all gates. Actually, this procedure is a continuation of the single-gate procedure. It assumes that a function tree has been formed and that a single-gate realization has been attempted by means of Procedure 5.1. This procedure has failed because, on some level of the tree, any choice of coefficient within that range violates the sensitivity constraints or because the range is void. Actually, it may be known from the beginning that the given function is not unate and hence is not ℓ.s. Nevertheless a tree can be formed (but it cannot be reduced), and Procedure 5.1 can be applied. It may fail on the $(n - 1)$-level because the range on that level is void, or it may proceed up the tree to some higher level before a void range is encountered. In either case, our objective is to add additional gates and thereby obtain a realization of the given switching function for which the fan-in and relative gap of each gate do not exceed the specified bounds.

The multigate procedure is performed on an unreduced tree with or without don't-care points. Assume that the single-gate procedure has proceeded up the tree to some k-level and that at this level all functions have been realized by the same separating function. Suppose however, that on this level the common range for a_k is void, or any choice within the range violates the sensitivity constraints. We propose to *add an additional gate* to the realization in order to eliminate the difficulty. If the difficulty is a void common range, then the two-gate realization will have nonvoid common ranges for both gates. If the difficulty is sensitivity, then the two-gate realization will satisfy the given requirements. The additional gate must be added, however, so that all k-level functions have the same separating-function representation except for gaps. Then, according to Corollary 4.7.1, the range theorem can be applied to corresponding gates of the function-pair realizations to proceed up the tree. If, on some higher level, additional range or sensitivity difficulties are encountered, additional gates are added as before to eliminate the problem. When the top of the tree is reached, a multigate realization of the given function has been obtained.

Adding Gates to Remove Void Ranges

The crux of the multigate procedure is the method of gate addition in order to eliminate range or sensitivity difficulties. We consider first the elimination of range difficulties.

Example 5.5. Figure 5.9 shows the first three levels of a function tree for which all 2-level functions have been realized by the separating function $f_2(x_3,x_4,\ldots,x_n)$. From the indicated gaps the two ranges for a_2 are determined to be

$$R(a_2) = 2{*}0 \qquad \text{and} \quad R(a_2) = 4{*}2$$

with the result that the common range is void.

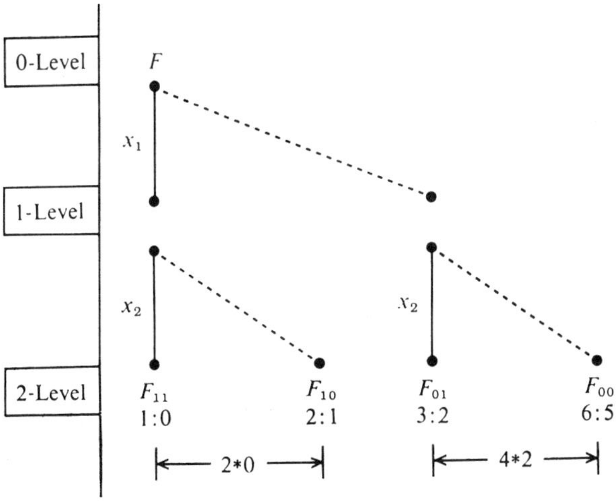

Fig. 5.9. A tree with void common range: Example 5.5.

It is readily observed, however, that if the gap of F_{10} were 4:3 instead of 2:1, then both ranges would be 4*2 and the common range would be nonvoid. According to Theorem 2.4B, this change can be effected by changing the realization of F_{10} from

$$\langle f_2(x_3,x_4,\ldots,x_n)\rangle_{2:1}$$

to

$$\langle 2 + f_2(x_3,x_4,\ldots,x_n)\rangle_{4:3} = \langle 2[1] + f_2(x_3,x_4,\ldots,x_n)\rangle_{4:3}$$

This change corresponds to adding an additional input with coefficient 2 to the realization where the value of the input variable is always unity.

To preserve the property that all 2-level functions have the same realization except for gaps, it is necessary to add an input with coefficient 2 to the three other 2-level realizations. In order that this not alter their gaps, the input variables to these must have zero value. Thus

$$\begin{aligned}
F_{11}(x_3,x_4,\ldots,x_n) &= \langle f_2(x_3,x_4,\ldots,x_n)\rangle_{1:0} = \langle 2[0] + f_2(x_3,x_4,\ldots,x_n)\rangle_{1:0}\\
F_{10}(x_3,x_4,\ldots,x_n) &= \langle f_2(x_3,x_4,\ldots,x_n)\rangle_{2:1} = \langle 2[1] + f_2(x_3,x_4,\ldots,x_n)\rangle_{4:3}\\
F_{01}(x_3,x_4,\ldots,x_n) &= \langle f_2(x_3,x_4,\ldots,x_n)\rangle_{3:2} = \langle 2[0] + f_2(x_3,x_4,\ldots,x_n)\rangle_{3:2}\\
F_{00}(x_3,x_4,\ldots,x_n) &= \langle f_2(x_3,x_4,\ldots,x_n)\rangle_{6:5} = \langle 2[0] + f_2(x_3,x_4,\ldots,x_n)\rangle_{6:5}
\end{aligned}$$

Now the common range for a_k is 4*2 and, hence, is nonvoid. Note, however, that the separating-function representations are still not identical except for gaps. Some have the input variable with 0 value and others have the input value 1. Since these are both constant functions, they can be realized by any separating function f_2' the simplest of which is

$$f_2'(\Phi) = 0$$

According to Theorem 1.3, the realizations are

$$[1] = \langle 0\rangle_{0:-\infty} \qquad \text{and} \quad [0] = \langle 0\rangle_{\infty:0}$$

Therefore,

$$\begin{aligned}
F_{11}(x_3,x_4,\ldots,x_n) &= \langle 2\langle 0\rangle_{\infty:0} + f_2(x_3,x_4,\ldots,x_n)\rangle_{1:0}\\
F_{10}(x_3,x_4,\ldots,x_n) &= \langle 2\langle 0\rangle_{0:-\infty} + f_2(x_3,x_4,\ldots,x_n)\rangle_{4:3}\\
F_{01}(x_3,x_4,\ldots,x_n) &= \langle 2\langle 0\rangle_{\infty:0} + f_2(x_3,x_4,\ldots,x_n)\rangle_{3:2}\\
F_{00}(x_3,x_4,\ldots,x_n) &= \langle 2\langle 0\rangle_{\infty:0} + f_2(x_3,x_4,\ldots,x_n)\rangle_{6:5}
\end{aligned}$$

and all 2-level functions have the same separating function except for gaps. The inner gate may appear somewhat peculiar since it has no inputs; however, inputs will be acquired as the procedure progresses up the tree.

Figure 5.10 shows how the new gap information is listed on the tree. The gaps for the outer and inner gates are written one above the other.

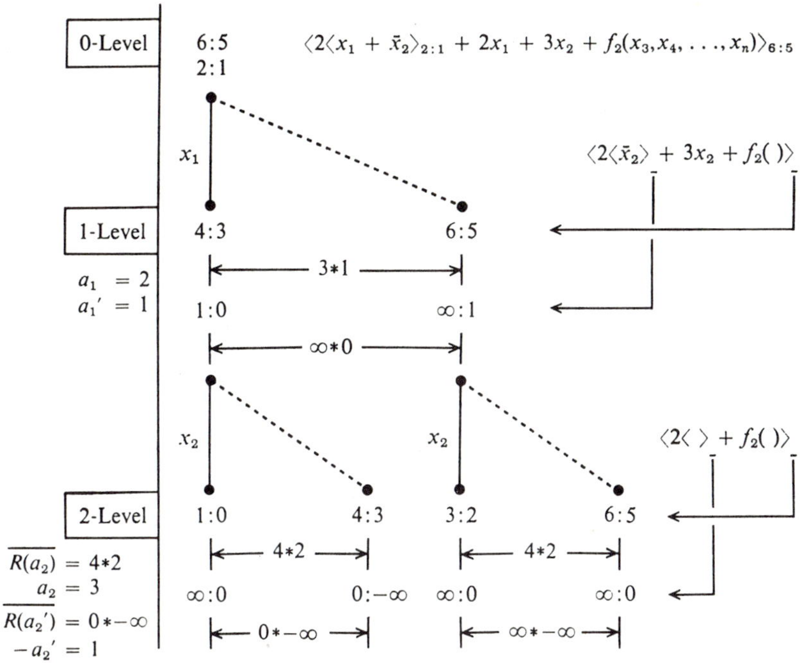

Fig. 5.10. A two-gate realization of Example 5.5.

The arrows on the realization at the right indicate which gaps are associated with which gate.

For each function pair on the 2-level, the set of ranges for each gate is determined in the usual way. The common range for each set is computed and listed on the left of the tree together with the selected values for the coefficients. Observe that the inner gate has a common range in the negative region only. It follows from Corollary 4.2.1, however, that $-a_2' = 1$ can be used as the coefficient of $\bar{x}_2$. The gaps of the 1-level functions are now determined, where gaps for each gate on the 1-level are determined from the corresponding pair of 2-level gaps in the usual way. The procedure is repeated on the 1-level as shown in Fig. 5.10, and this yields a 2-gate realization for the 0-level function without additional difficulty.

This illustrates how a gate can be added to a realization to eliminate a void common range. The procedure is straightforward, except that at the beginning there is an element of choice. We eliminated the void common range by displacing the gap of F_{10} upward two units. (See Fig. 5.9.) The same thing could have been accomplished by displacing the gap of F_{01} upward by two. This realization is shown in Fig. 5.11 and leads to a slightly different network. Another alternative would have been to displace the gaps of F_{11} and F_{00} downward and a 2-gate realization would again be obtained.

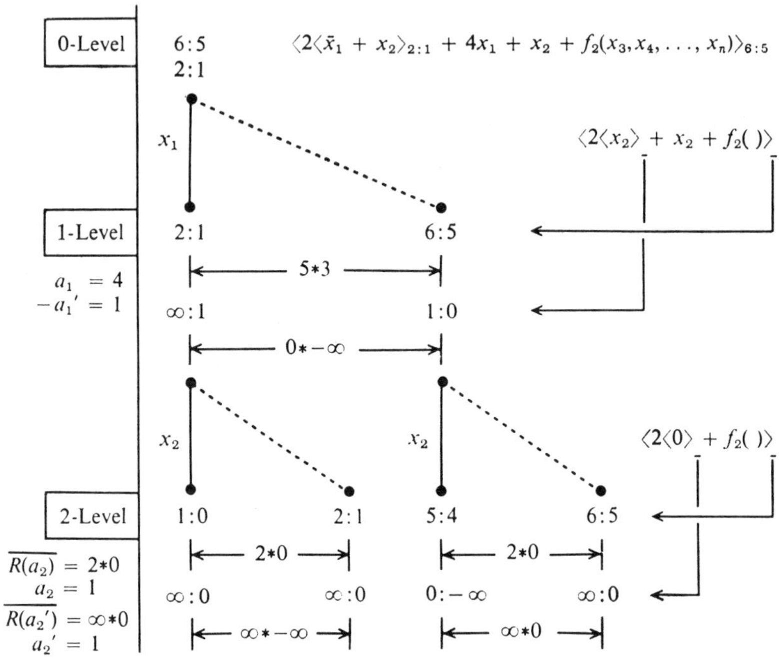

Fig. 5.11. Another two-gate realization for Example 5.5.

The realizations of Figs. 5.10 and 5.11 both have the same fan-in and relative gap; thus they are equivalent with regard to sensitivity. In more complex cases there may be more choices available, and the resultant networks may differ considerably. Regardless of the selection, a realization can be obtained; however, experience, intuition, and a little trial and error computation often yield a simpler realization than would otherwise be obtained.

Adding Gates to Increase Relative Gaps

Relative gap has only been defined for definite realizations (all coefficients of the same sign). However, since every function can be realized with all coefficients positive (by using complemented variables where necessary), this is not a restriction. Assume that a lower bound has been set for the relative gap of the realization. As the realization proceeds up the tree, we do not know what the final relative gap will be until all coefficients are selected. However, on each k-level of the tree, an upper bound on the relative gap can be determined as follows:

Let ℓ_M^k be the largest lower boundary of all k-level gaps and, as before, let $\bar{g}_k$ denote the length of the smallest k-level gap. Since the gap $u{:}\ell$ of F is the intersection of all k-level gaps suitably displaced upward, it is evident that

$$\ell \geq \ell_M^k \qquad \text{and} \quad u - \ell \leq \bar{g}_k$$

Hence,

$$\text{relative gap} = \frac{u - \ell}{u + \ell} = \frac{u - \ell}{2\ell + (u - \ell)} \leq \frac{\bar{g}_k}{2\ell_M^k + \bar{g}_k} \tag{5.1}$$

On any level, therefore, an upper bound on the final relative gap can be determined. If in the course of the procedure there is some level for which this bound is less than the specified allowable value, then an additional gate must be added to increase the bound which the level imposes. This procedure is repeated on each level where the bound on the relative gap becomes too small. When the 1-level is reached, the bound is actually attainable. This follows from the fact that the final gap is the intersection of the suitably displaced 1-level gaps.

We now consider how to alter the relative gap bound. Suppose that on some k-level of the tree the bound on the relative gap is within the specified value. We wish to consider how this bound can decrease in going to the next higher level and how this decrease can be prevented if the new bound will be less than the specification.

1. If the k-level range on $\overline{R(a_k)}$ is less than $2\bar{g}_k$, then the maximum possible value for $\bar{g}_{k-1}$ is $\frac{1}{2}\bar{g}_k$, and the bound on the relative gap is decreased by a factor of 2. This reduction can be prevented by adding a gate so that the length of $\overline{R(a_k)}$ is at least $2\bar{g}_k$. For example, if the 4 gaps that determine $\overline{R(a_k)}$ are shown in Fig. 5.12a, then the length of $\overline{R(a_k)}$ can be increased to 2 by adding a gate so as to displace either the gap of F_a or the gap of F_d upward by unit distance. Displacing the gap of F_d upward may decrease the relative gap bound on the next level if its gap (5:4) is the highest one on the k-level. Therefore, it is probably better to displace upward the gap of F_a. Figure 5.12b shows the 2-gate realizations for this and the resultant ranges. Naturally, it is necessary to add the new gate to the realizations of all k-level functions.

2. A second situation that can occur is that

$$\ell_M^{k-1} > \ell_M^k$$

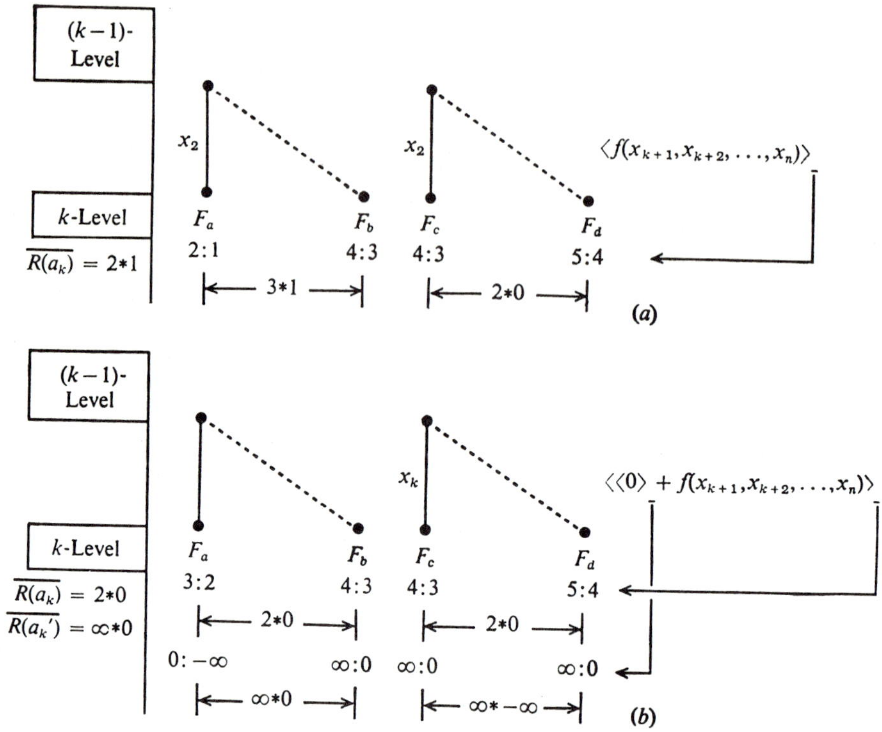

Fig. 5.12. Two function pairs demonstrating gate addition to eliminate gap reduction. (*a*) Gaps on the k-level that cause $\bar{g}_{k-1} = \frac{1}{2}\bar{g}_k$. (*b*) Gaps after adding gate.

for any choice of a_k such that

$$\bar{g}_{k-1} = \bar{g}_k$$

This is illustrated in Fig. 5.13a, where four k-level gaps are shown and where it is assumed that gap 6:4 has the largest lower bound; that is,

$$\ell_M^k = 4$$

The range $\overline{R(a_k)}$ is such that

$$a_k = 2$$

is the only value for which

$$\bar{g}_{k-1} = \bar{g}_k = 1$$

but for this value of a_k,

$$\ell_M^{k-1} = 5$$

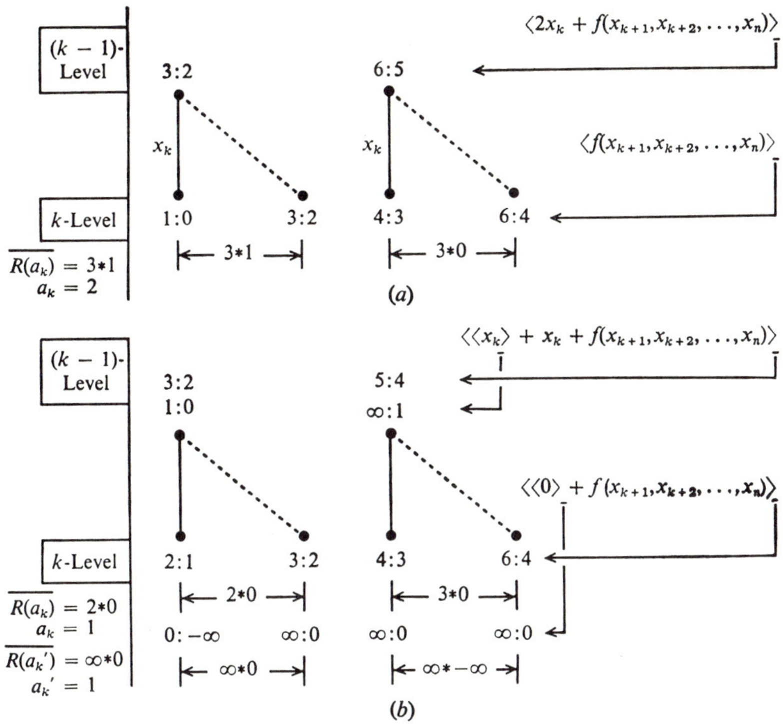

Fig. 5.13. Two function pairs demonstrating gate addition so that $\ell_M^{k-1} = \ell_M^k$.
(*a*) Gaps such that $\ell_M^{k-1} > \ell_M^k$.
(*b*) Adding a gate so that $\ell_M^{k-1} = \ell_M^k$.

and the bound on the relative gap is reduced. This reduction can be eliminated by displacing the 1:0 gap upward by unit amount so that $a_k = 1$ is an allowable choice such that

$$\bar{g}_{k-1} = \bar{g}_k = 1$$

In this case

$$\ell_M^{k-1} = \ell_M^k = 4$$

A realization of the k-level functions that accomplishes this is shown in Fig. 5.13*b*.

3. A third situation wherein the bound on relative gap is reduced comes about because a 0 function exists on the $(k - 1)$-level. After the value for a_k is selected, the gap for this function is $\infty : \sigma_k$ where σ_k is the maximum value of the separating function. Clearly

$$\ell_M^{k-1} = \sigma_k$$

Therefore the value of ℓ_M^{k-1} must be reduced by adding a gate if the $(k-1)$-level bound on relative gap is too small. The situation is illustrated by Fig. 5.14*a*, where the 0 function is shown on the 2-level of the tree. The realizations for all 2-level functions are obtained with the indicated separating

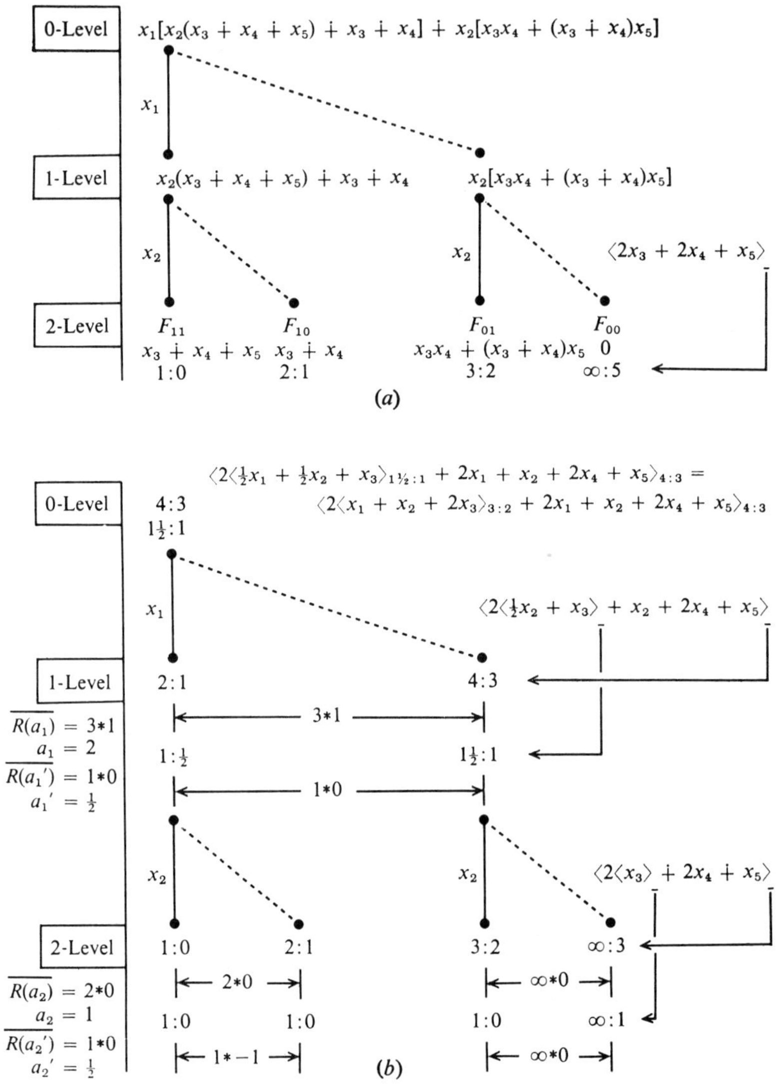

Fig. 5.14. Gate addition to lower 0-function gap. (*a*) A tree for which the 0-function gap is too high. (*b*) Adding a gate to lower the 0-function gap.

function and the gaps that are listed where the 0 function has gap $\infty:5$. Suppose, however, that $\ell_M^2 = 5$ is too large and, in fact, $\ell_M^2 = 3$ is the largest permissible value for $\bar{g}_2 = 1$. We must, therefore, add a gate in such a way that the gap for the 0 function is $\infty:3$. This means that a gate must be added so as to reduce the maximum value of the separating function, σ_2, since the 0 function has gap $\infty:\sigma_2$. This is done by replacing one of the input variables with coefficient 2, say x_3, by a gate. Thus, in the realization of F_{11}, F_{10}, and F_0,

$$x_3 = \langle x_3 \rangle_{1:0}$$

This is a trivial change for these functions; however, in F_{00} the gap can be changed so that $\langle x_3 \rangle_{u:\ell}$ realizes the 0 function; that is,

$$0 = \langle x_3 \rangle_{\infty:1}$$

and the separating function for F_{00} is

$$2\langle x_3 \rangle_{\infty:1} + 2x_4 + x_5$$

where now the maximum value is 3. The realization for F_{00} has gap $\infty:3$ and is

$$F_{00}(x_3,x_4,x_5) = 0 = \langle 2\langle x_3 \rangle_{\infty:1} + 2x_4 + x_5 \rangle_{\infty:3}$$

This together with the final realization is shown in Fig. 5.14*b*. Note that the inner gap had to be reduced in going from the 2- to the 1-level but that this did not violate the sensitivity constraints.

In a more general case it might have been necessary to replace more than one input variable with a gate in order to achieve the desired relative gap. This might, in turn, produce other difficulties such as void common ranges or require gap length reduction beyond an acceptable amount. Each of these, however, can be corrected by adding additional gates. Thus, it is always possible to add gates to control the relative gap.

Adding a Gate When the Normalized Fan-In is Too Large

The normalized fan-in is the ratio of fan-in, σ, to gap length, g. We assume that a maximum fan-in and a minimum gap length have been specified. The preceding section shows how to control gap length. Here we consider controlling fan-in.

The basic idea of adding a gate in order to control fan-in is quite similar to that for controlling relative gap. When the procedure has reached some level of the tree, we can again not tell exactly what the fan-in of each gate in the final realization will be, but we can get a bound, this time a lower bound, on the fan-in that can be achieved after selecting the remaining weights. If on any level this lower bound exceeds the specified fan-in, the specifications can not be met, and so we add a gate in order to decrease the lower bound. On the 1-level the bound can always be achieved, and thus we can meet the specifications.

We now derive the lower bound. Let ℓ_M^k be the largest lower gap boundary and u_m^k be the smallest upper gap boundary of all k-level gaps. The sum of the additional coefficients that will be added on higher levels must certainly be at least large enough to displace the lowest gap until it intersects the highest gap. Furthermore, if there is a constraint that the final gap length cannot be less than some value $\bar{g}$, the sum of the additional coefficients must be at least large enough to displace u_m^k to $\ell_M^k + \bar{g}$. Thus if σ_k' is the sum of the coefficients yet to be chosen

$$\sigma_k' = \sum_{i=1}^{k} a_i$$

then σ_k' is bounded by

$$\sigma_k' \geq \ell_M^k + \bar{g} - u_m^k$$

Again, in general, this is only a bound, since it only involves two gaps, the highest and the lowest. The other gaps may complicate the situation and cause the bound to be exceeded. However, on the 1-level there are only two functions, and the bound can always be achieved.

The total bound on fan-in is obtained by adding the sum of the coefficients already chosen to the bound on σ_k', the sum of the coefficients yet to be chosen.

Now suppose that on the k-level the lower bound on fan-in is within specifications. We wish to study how this bound could be increased in going to the next higher level and how we can add a gate to prevent this increase if it causes the bound to exceed the specifications.

1. Suppose that on the next higher level [the $(k - 1)$-level] there is a 0 function. After we select the weight and assign gaps on the $(k - 1)$-level, the gap of the 0 function will be $\infty : \sigma_k$ where σ_k is the sum of the weights already selected. This new gap will be the highest gap on the $(k - 1)$-level and will be higher than the gap that was previously highest on the k-level. If this new highest gap causes the bound to increase so that it exceeds the specifications, a gate must be added to cause the bound to decrease somewhat.

Consider again the function shown in Fig. 5.14a. On the 2-level, the sum of the coefficients previously chosen is

$$2 + 2 + 1 = 5$$

Neglecting for a moment the 0 function on the 2-level, and assuming that the gap length must be 1, the remaining weights must be at least large enough to displace the 1:0 gap until it intersects the 3:2 gap; that is, the sum of the remaining weights must be at least 2. Thus the bound on fan-in is

$$5 + 2 = 7$$

This is the bound that would have been computed on the next lower level.

If we now consider the 0 function, its gap is $\infty:5$. The sum of the remaining weights must be large enough to displace the 1:0 gap until it intersects the $\infty:5$ gap; that is, this sum must be at least 5. The bound on fan-in is thus

$$5 + 5 = 10$$

Assume that this is too large. We must then add a gate in such a way that the height of the gap of the 0 function is decreased. This was done (for a different purpose) in Fig. 5.14*b*. After we add the gate, the gap of the 0 function is $\infty:3$ and the bound on fan-in is

$$5 + 3 = 8$$

If this bound is still too high, another input must be converted into a gate in order to make the gap of the 0 function even lower. This may cause the common range to be void, in which case still another gate must be added to make the common range nonvoid. It may appear that this new gate might increase the fan-in even more, but we shall see later (in 3) that this is not the case.

2. Suppose that on the next higher level there is a 1 function. Its gap will be $0:-\infty$ and it will be the lowest gap and may be lower than the previously lowest gap. If this new low gap causes the bound on fan-in to become too high, a gate must be added so as to cause the bound to decrease.

The method for adding the gate is the dual of the method of 1. Consider the function of Fig. 5.15*a*. The sum of the weights previously chosen is again

$$2 + 2 + 1 = 5$$

Neglecting the 1 function, we see that the sum of the weights yet to be chosen must be at least 2. Thus the bound on fan-in (and this is the bound that would be computed on the next lower level) is

$$5 + 2 = 7$$

If we now consider the 1 function, its gap is $0:-\infty$ and the new bound on fan-in is

$$5 + 5 = 10$$

If this is too high, we must add a gate in order to displace the gap of the 1 function upward.

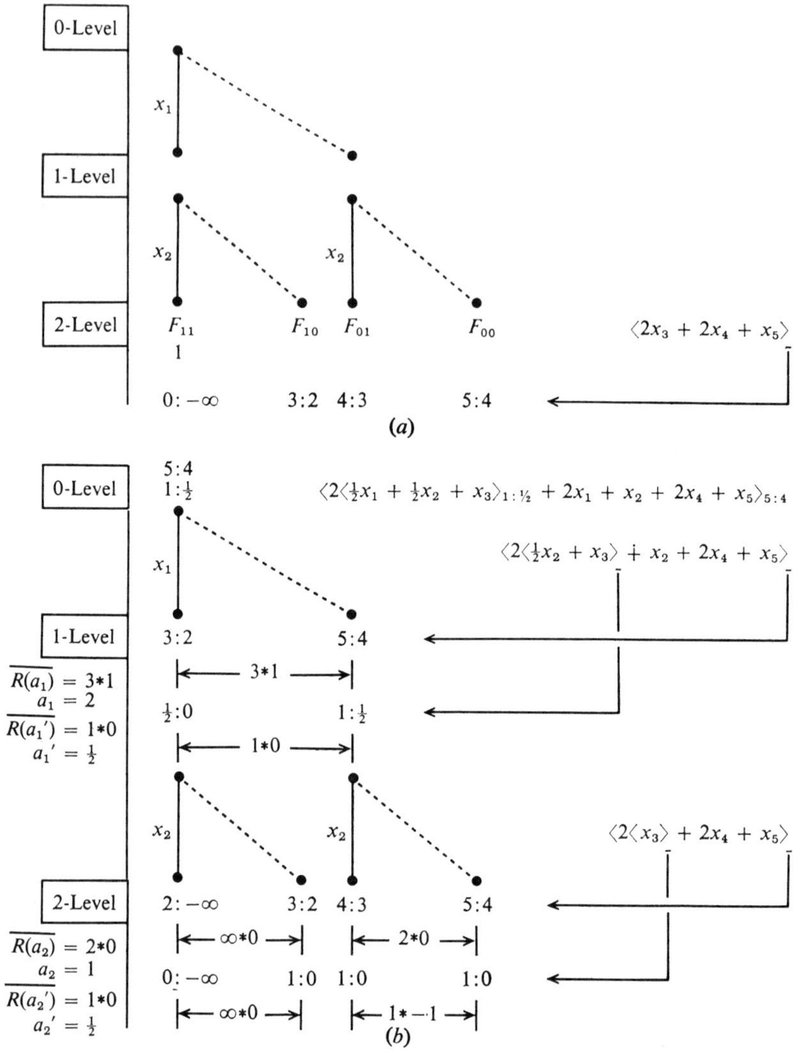

Fig. 5.15. Gate addition to raise a 1-function gap. (*a*) A tree for which the gap of a 1-function is too low. (*b*) Adding a gate to raise the 1-function gap.

We do this by changing one of the inputs, in this case x_3, to a gate. Thus in the separating functions for the three nonconstant functions

$$x_3 = \langle x_3 \rangle_{1:0}$$

In the constant function F_{11}, we let this new gate realize the function 1; that is, we make its gap $0\!:\!-\infty$

$$1 = \langle x_3 \rangle_{0:-\infty}$$

Thus the realization for F_{11} is

$$F_{11} = 1 = \langle 2\langle x_3 \rangle_{0:-\infty} + x_4 + x_5 \rangle_{2:-\infty}$$

Observe that since one of the inputs always contributes a weight of 2, the gap of the final gate has been increased to $2\!:\!-\infty$. The new bound on fan-in is now

$$5 + 3 = 8$$

The 2-gate realizations for the four 2-level functions and the rest of the realization are shown in Fig. 5.15*b*. Note the duality between this realization and that of Fig. 5.14.

If the resulting bound is still too high, another input can be converted into a gate. In any event, gates can always be added to displace the lowest gap as high as necessary.

3. Suppose that on the k-level it is necessary to add a gate either to make the common range nonvoid or to avoid gap reduction. This new gate may increase the bound on fan-in. It will certainly increase that part of the bound due to the sum of the weights previously selected. If it does not decrease by the same amount that part of the bound due to the highest and lowest k-level gaps, then the bound will increase. We shall show that the new gate can always be added in such a way that the lowest gap will be displaced upward by the value of the weight of the new gate, and the highest gap will remain unchanged. Hence the bound on fan-in will remain the same.

Consider first the situation in which both the highest and the lowest gaps are included in the four gaps which require the new gate to be added. The only two cases of this situation are exemplified in Fig. 5.16*a* and *b*. In Fig. 5.16*a* the higher-valued range is between the lower gaps, while in Fig. 5.16*b* the higher-valued range is between the higher gaps.

In Fig. 5.16*a* the range can be made nonvoid by displacing upward by 2 the gap of either F_a or F_d. We should displace the gap of F_a, thereby decreasing the distance between the highest and lowest gap by the same amount that we increase the sum of the weights previously chosen.

In Fig. 5.16*b* the range can be made non-void by displacing upward by 2 the gap of either F_f or F_g. Neither of these yields the desired result of raising upward the gap of F_e (which we assume is the lowest gap). We must therefore add the gate somewhat differently.

What we really want to do is change the relative distances between the

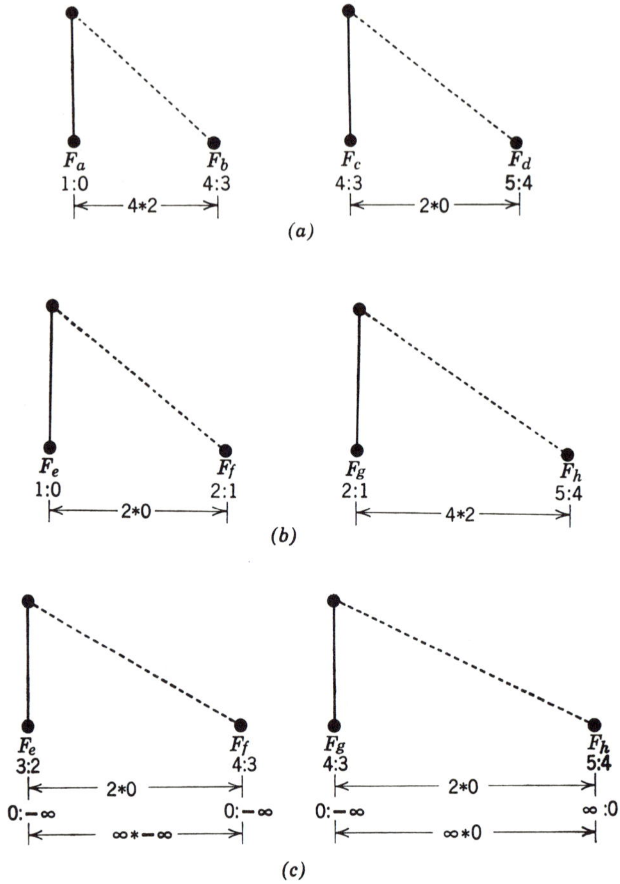

Fig. 5.16. Adding a gate in order to raise the lowest gap. (*a*) The range of the lower gaps on the left is higher than the range on the right. (*b*) The range of the higher gaps on the right is higher than the range on the left. (*c*) Adding a gate in order to raise the lowest gap while making the range nonvoid.

two gaps in each function pair; the absolute height of the gaps is not important. We therefore displace the gaps of both F_e and F_f by 2, which does not change their relative distance and hence their range, and at the same time we displace the gap of F_g upward by 2 to change its range. We accomplish all of this by adding a single gate with a weight of 2 as an additional input to the final gate. The new gate realizes the function 1 in F_e, F_f and F_g and the function 0 in F_h. This is shown in Fig. 5.16*c*. Thus we have again added the gate in such a way as not to increase the bound on fan-in.

Now suppose that the highest or lowest gap is not included in the four functions that require the additional gate. We can still add the gate so as to raise up the lowest gap while keeping the highest gap unchanged. We do this by choosing the gaps of the new gate so that (1) it realizes the function 0 in the realization of the function with the highest gap and also in the other function in its function pair and (2) it realizes the function 1 in the realization of the function with the lowest gap and also in the other function in its function pair (in a manner similar to Fig. 5.16*c*). Thus both function pairs still contribute the same range, but the bound on fan-in remains the same.

5.7 STARTING THE PROCEDURE

The only part of the procedure that remains to be explained is how to begin. We must somehow find, for all functions on some level of the tree, realizations which are the same except for gaps. Then the various range theorems and gate-addition results can be applied to proceed up the tree.

The procedure can always be started on the bottom level (the n-level), where the only possible functions are 1 and 0. Since these functions can have realizations

$$1 = \langle 0 \rangle_{0:-\infty}$$
$$0 = \langle 0 \rangle_{\infty:0}$$

which are the same except for gaps, the procedure can be started on this level.

A difficulty arises, however, when the n-level common range is void for these single-gate realizations, and an additional gate must be added in order to proceed to the $(n - 1)$-level. This situation occurs whenever one n-level function pair is 1,0 and another is 0,1. An additional gate can always be added as described in Section 5.6 (p. 191) so as to make the common range nonvoid, and the procedure can then be continued up the tree. In this regard there is usually considerable freedom. Although a realization will be obtained no matter how the freedom is exercised, the simplicity of the realization is markedly affected by the initial choices, and there is often an embarrassingly large number of possibilities.

The n-level functions 1 and 0 can have the two-gate realizations

$$\begin{aligned} 1 &= \langle 0 + K\langle 0 \rangle_{\infty:0} \rangle_{0:-\infty} \\ &= \langle 0 + K\langle 0 \rangle_{0:-\infty} \rangle_{K:-\infty} \\ 0 &= \langle 0 + K\langle 0 \rangle_{\infty:0} \rangle_{\infty:0} \\ &= \langle 0 + K\langle 0 \rangle_{0:-\infty} \rangle_{\infty:K} \end{aligned}$$

where K is any real number which can be normalized to 1 if desired without loss of generality. For the first realization in each case, the inner gate realizes the Boolean function 0, and for the second realization it realizes the Boolean function 1. Thus for each n-level constant function there are 2 possible realizations. If there are q n-level functions, there are 2^q possible choices for their realizations. Of course, some of these are removed as possibilities by requiring that the n-level common ranges be nonvoid, but there is still a large number of possible selections each of which has a nonvoid n-level common range. This difficulty will be described in the next example.

Example 5.6. Consider the Boolean function for one stage of a binary adder,

$$S = \bar{k}(x\bar{y} + \bar{x}y) + k(xy + \bar{x}\bar{y})$$

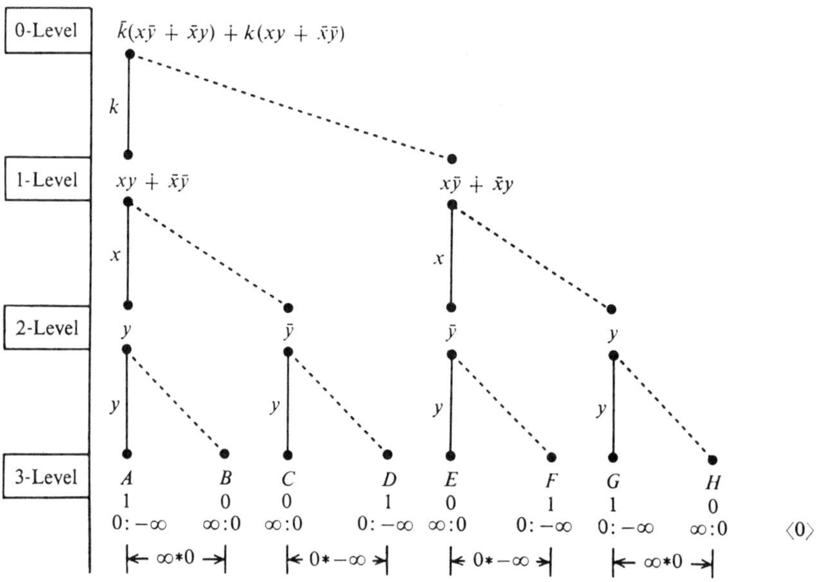

Fig. 5.17. The function tree for a binary adder with a tentative single-gate realization.

where x and y are the two digits to be added and k is the carry from the preceding stage. A tree for this function, together with the gaps of a tentative 1-gate realization for the 3-level constant functions is shown in Fig. 5.17.

The common range is void since two of the ranges include only positive values and two only negative values. Hence an additional gate must be added on this level. There is immediately a choice; the new gate can be used to make all the ranges positive or all negative (that is, either the gaps of functions D and F, or those of functions A and G, can be raised).

Suppose that we decide to make the ranges positive. Then we must add the gate to functions C, D, E, and F as shown on the 3-level of Fig. 5.18 so as to raise the gaps of D and F while leaving the gaps of C and E unchanged. Below the gaps are shown the constant functions realized by the inner gate. Note also that with no loss of generality the coefficient of the inner gate has been selected as unity.

Now consider the function pairs A,B and G,H. We want to keep their ranges the same, $\infty * 0$, but this can be done in two ways: either by leaving their gaps unchanged or by raising them both equally. In other words, the inner gates can both realize the Boolean function 0 or both realize the Boolean function 1 as shown inside the parentheses on the 3-level of Fig. 5.18. There are four choices† in all (two for each function pair), each of which allows the 3-level common range to be nonvoid.

However, some of the choices may lead to void common ranges on higher levels, whereas other choices may cause no difficulties. In this very simple case, we can carry all choices up the tree at least one more level. This is possible because the 3-level ranges for all choices are the same: $\infty * 0$ for the outer gate and $\infty * -\infty$ for the inner gate.

We therefore choose $a_y = \frac{1}{2}$ for the outer gate and $-a_y{}' = 1$ for the inner gate and carry both choices for each function pair up to the next level inside the parentheses, as shown in Fig. 5.18. Each choice leads to different 2-level gaps and ranges.

Now we see that not all choices lead to nonvoid 2-level common ranges. In fact, there are only two valid choices: the top choice for function pair A,B together with the bottom choice for G,H or the bottom choice for A,B together with the top choice for G,H—in other words, the 0,0 choice for A,B and the 1,1 choice for G,H or the 1,1 choice for A,B and the 0,0 choice for G,H.

It turns out that either of these choices leads to a two-gate realization. We decide, as shown by the checks, to try the top realization for A,B and the bottom realization for G,H. The remainder of the procedure causes no difficulty, as shown in Fig. 5.18, and yields the realization given in the

† There is actually another choice that needs to be considered in the general case. The inner gates of A and B (or of G and H) could realize the Boolean functions 1 and 0, respectively. This would make the common range of the outer gate nonvoid, but the common range on the inner gate would be void, because of the 1,0 and 0,1 function pairs. Hence at least 3 gates would be needed. In this example we neglect this possibility.

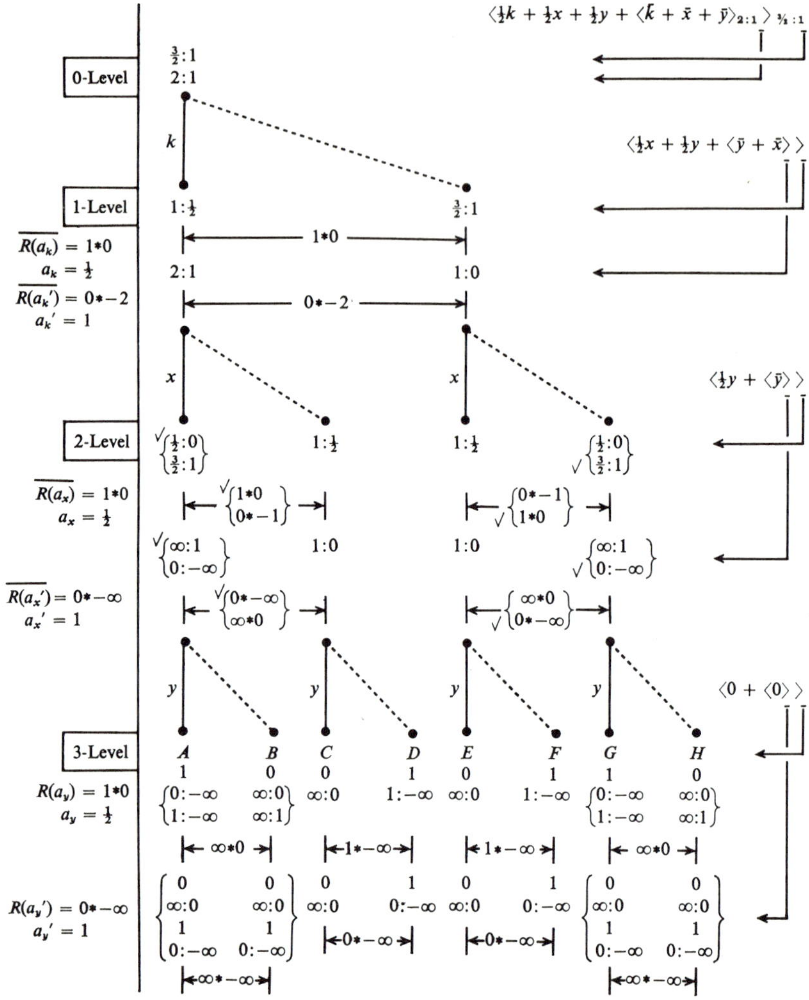

Fig. 5.18. A two-gate realization of the binary adder.

figure. Observe that the inner gate realizes the complement of the carry function needed for the next stage of the adder.

In this simple example the allowable freedom was easily disposed of. In a more complicated example it may not be practical to carry all choices up the tree and keep track of which combinations of choices are allowable. However, it is frequently possible to eliminate some of the choices on the bottom level of the tree, by doing some additional computations on that level.

Figure 5.19 repeats the bottom level of the tree of Fig. 5.18. Again we decide to raise the gaps of D and F. This results in the indicated values for the functions realized by the inner gate, where the remaining choices are displayed in the brackets.

These initial choices were made by observing the function pairs A and B, C and D, E and F, and G and H and requiring their individual ranges to have a nonvoid intersection. This was particularly easy since the ranges were either $\infty * 0$ for a pair 1,0 or $0 * -\infty$ for a pair 0,1 (or $\infty * -\infty$ for 0,0 and 1,1). Thus the only combinations that cause difficulties are a 1,0 and a 0,1.

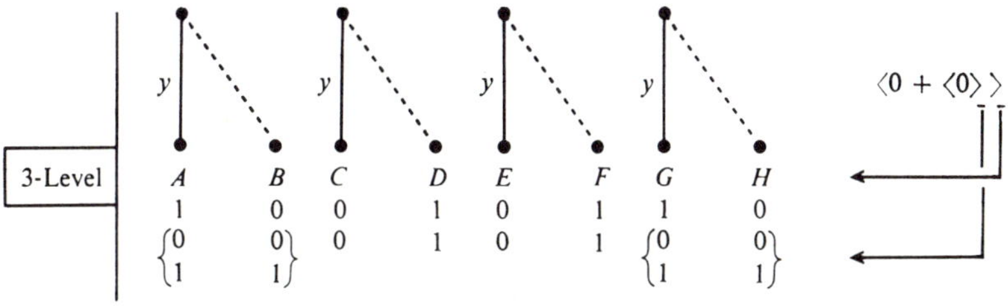

Fig. 5.19. The bottom level of the tree for the binary adder.

This same type of reasoning can be used to investigate other pairs of 3-level functions that imply general ranges on other coefficients or combinations of coefficients as described in Section 4.4. Consider, for example, the functions A and C. As seen by the labels on the tree branches, this pair of functions implies a range on a_x. Similarly, the functions E and G also imply a generalized range on a_x. Observe that these two ranges have a void intersection since A,C is 1,0 and E,G is 0,1. Thus we know that unless this difficulty is removed, the 2-level common range on a_x will be void. However, we can use some of the choices we have on the 3-level to eliminate this inconsistency.

Since the gaps of C and E have already been fixed, the difficulty can be eliminated in only two ways: either the gap of A can be raised, leaving the gap of G unchanged, or the gap of G can be raised, leaving the gap of A unchanged. In other words, either A,B can be selected as 1,1 and G,H as 0,0, or A,B can be selected as 0,0 and G,H as 1,1. Observe that this is exactly the result that was obtained by carrying all choices up to the 2-level. The present line of reasoning, however, is in many respects simpler and more straightforward, and in this simple example is sufficient to dispose of all of the choices.

In a more general case, checking the general ranges in both the inner and the outer gates can often be used to advantage in making choices on the n-level and, in fact, on any other level at which a gate must be added. Sometimes this kind of reasoning can be used to show that two gates are not sufficient and that at least three are necessary. This occurs when all of the range inconsistencies cannot be eliminated with the available choices.

In any fairly complex function tree, there are a great many general ranges that can be investigated. The designer can check as many or as few as he desires. Each inconsistency that is found and corrected on the bottom level usually means one fewer gate in the final realization. Thus these computations are time well spent.

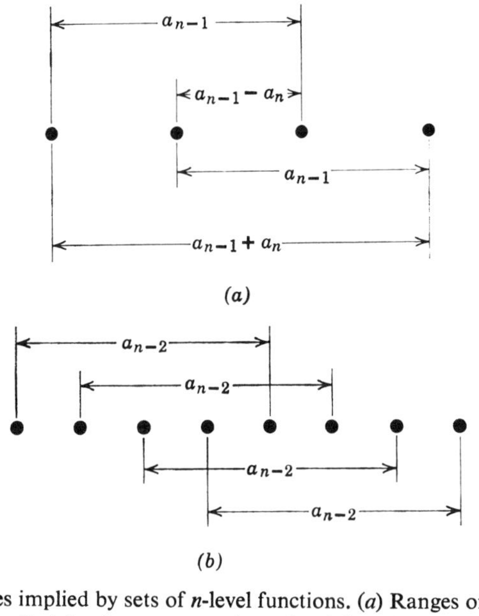

Fig. 5.20. Ranges implied by sets of n-level functions. (a) Ranges on a_{n-1}, $a_n + a_{n-1}$, and $a_{n-1} - a_n$ implied by four consecutive n-level functions (two function pairs). (b) Ranges on a_{n-2} implied by eight consecutive n-level functions (four function pairs).

Furthermore, because of the regularity of the structure of the tree, the necessary computations can be easily systematized either for computer or for hand calculations.

For example, in an n-level tree, the ranges on a_n are obtained by dividing the n-level functions into sets of two (function pairs). Similarly the general implied ranges on a_{n-1}, $a_{n-1} + a_n$, and $a_{n-1} - a_n$ can be obtained by dividing the n-level functions into sets of four (two function pairs) as shown in Fig. 5.20a. Thus by looking at each such set of four functions, it is easy to determine whether corresponding ranges are consistent. Again, the only difficulty occurs when one pair that implies a particular range is 1,0 and another pair that implies the same range is 0,1.

In a similar way, the sets of eight consecutive functions (four function pairs) imply ranges on sums and differences of a_{n-2}, a_{n-1}, and a_n. For example, Fig. 5.20*b* shows just the ranges on a_{n-2}.

Thus it is relatively straightforward to investigate the consistency of as many (or as few) of these implied ranges as desired. It is somewhat more complex in a hand computation to keep track of the choices that eliminate the various difficulties. Thus for hand computation the designer can usually check only a relatively few ranges.

Example 5.7. Consider the function

$$F = x_1x_2x_3\bar{x}_4 + \bar{x}_2\bar{x}_3x_4 + \bar{x}_1x_2x_3x_4 + \bar{x}_1\bar{x}_3\bar{x}_4 + \bar{x}_1\bar{x}_2\bar{x}_4 + \bar{x}_1\bar{x}_2\bar{x}_3$$

for which a tree is given in Fig. 5.21. On the 4-level there is an inconsistency; three of the function pairs are 0,1 and two are 1,0. We therefore elect to add

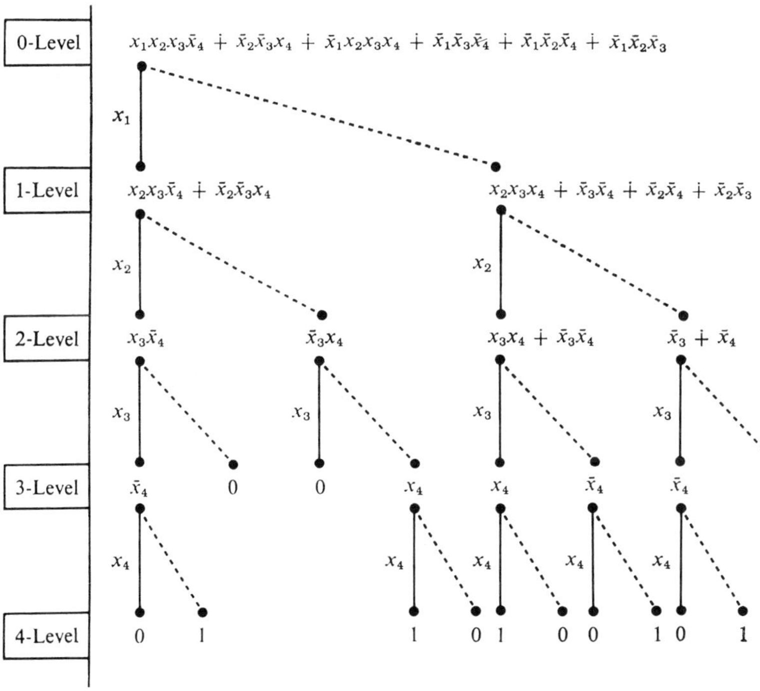

Fig. 5.21. The tree for the function of Example 5.7.

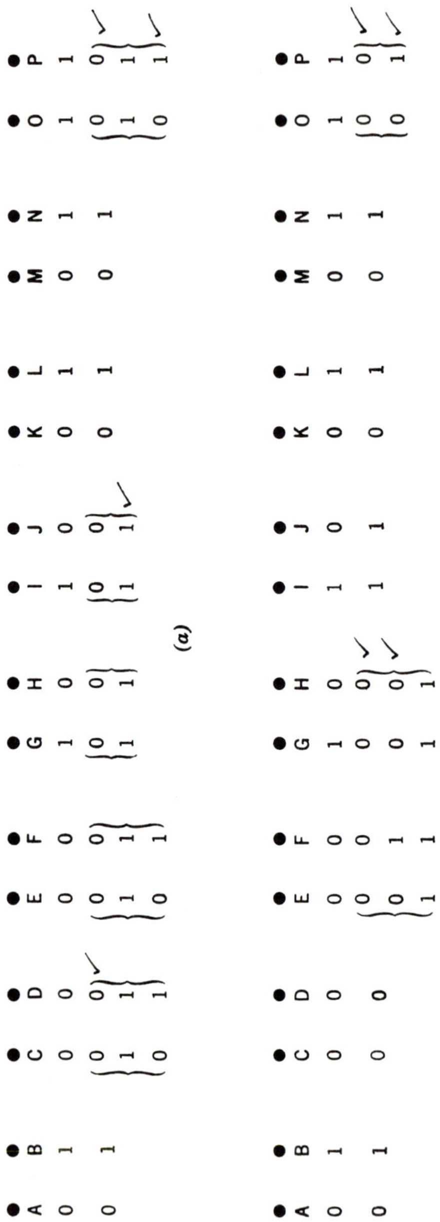
(a)
(b)

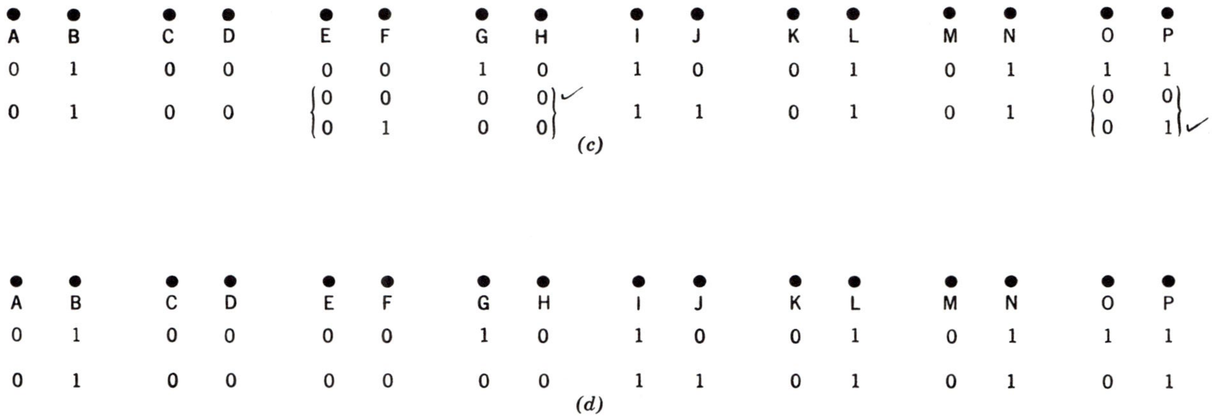

Fig. 5.22. The bottom level of the tree of Fig. 5.21. (*a*) The initial choices. (*b*) The second choices. (*c*) The third choice. (*d*) The final choice.

a gate which will raise the rightmost function in the 0,1 pairs, thus making the common range $1*-\infty$. This fixes some of the values of the functions realized by the inner gate and leaves some choices for the remaining values.

To determine how to use these choices, we investigate the consistency of some of the generalized ranges. However as Fig. 5.21 is drawn, many of the generalized ranges are not evident because the tree is incomplete—descendants of constant functions are omitted. Figure 5.22*a* shows the bottom row of the complete tree, with the various choices for the functions realized by the inner gate.

Note that there are three choices in the brackets of the constant functions; the only choice that is omitted is 1,0, since this would cause an inconsistency with the 0,1 pair of (say) A,B. In the 1,0 function pairs (such as G,H) there are only two choices. The 1,0 is omitted for the same reason as described above, and the 0,1 is omitted because we want to eliminate the range inconsistency with the 0,1 pairs in the outer gate.

Consider now the outer-gate general ranges implied by the functions two apart in each group of four (that is, ranges on a_3). In particular, observe the functions

$B\ D$	$E\ G$	$I\ K$	$J\ L$	MO
1 0	0 1	1 0	0 1	0 1

which imply an inconsistent range. This inconsistency can be eliminated in a number of ways. We elect to (1) raise the gap of B, and leave that of D; (2) raise the gap of I and leave that of K; (3) keep the relative spacing of the gaps of E and G; (4) keep the relative spacing of the gaps of J and L; (5) keep the relative spacing of the gaps of M and O. These choices are indicated with checks on Fig. 5.22*a* and lead to the situation shown in Fig. 5.22*b*.

Consider now the same general ranges (on a_3) in the inner gate. It is readily verified that for all choices within the brackets these ranges are consistent; that is, they are all either 0,0, 1,1, or 1,0. Thus, considering this range gives no new information.

Consider next the outer-gate implied ranges on a_2 (these are implied by pairs of functions four apart which are both in a set of eight so that one function is in each subset of four). In particular, observe

$B\ F$	$C\ G$	$I\ M$	$J\ N$	$K\ O$
1 0	0 1	1 0	0 1	0 1

Again there is a range inconsistency. This time there is no choice as to how we eliminate the inconsistency. Since the inner-gate functions of functions I, M, J, and N have been fixed, we must select the remaining choices in order to make all ranges negative. Thus we must select one of the 0 choices for G and one of the 0 choices for O. The situation is now as shown in Fig. 5.22*c*.

Consider now the same general range (on a_2) in the inner gate. Again these ranges are consistent and give no additional information.

Now we consider the general range on a_1 in the outer gate. (This is determined by functions eight apart.) In particular, observe

AI	BJ	DL	FN	HP
0 1	1 0	0 1	0 1	0 1

These ranges are inconsistent, but the inconsistency is partially eliminated by the previous choices. To complete the elimination, we must select the inner-gate function of F to be 0 and that of P to be 1.

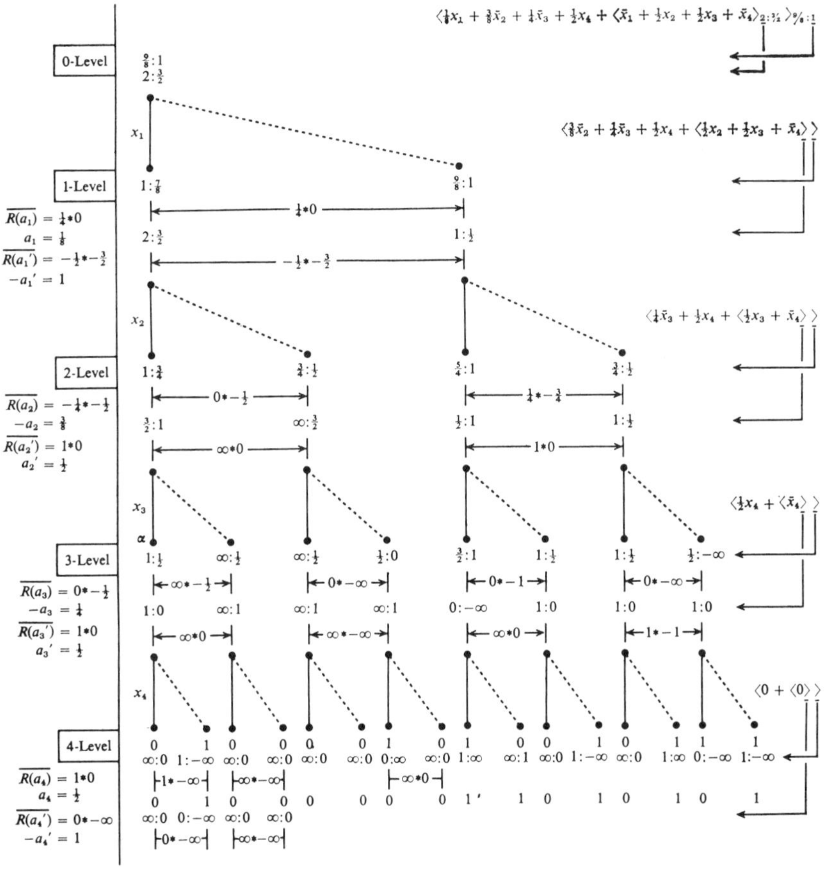

Fig. 5.23. The realization of the function of Example 5.7.

The final situation is as shown in Fig. 5.22*d*. All choices have been made. We could now check to see if there are any other uncorrected inconsistencies (such as, for example, in the range on $a_4 - a_3$, etc.), in which case we would know that the proposed realization will not work. But it is easier at this point actually to attempt the realization. This is done in Fig. 5.23. No further difficulties are encountered and a realization is obtained.

It should be evident that these manipulations on the bottom level of the tree are in many ways analogous to the map methods used in AND- and

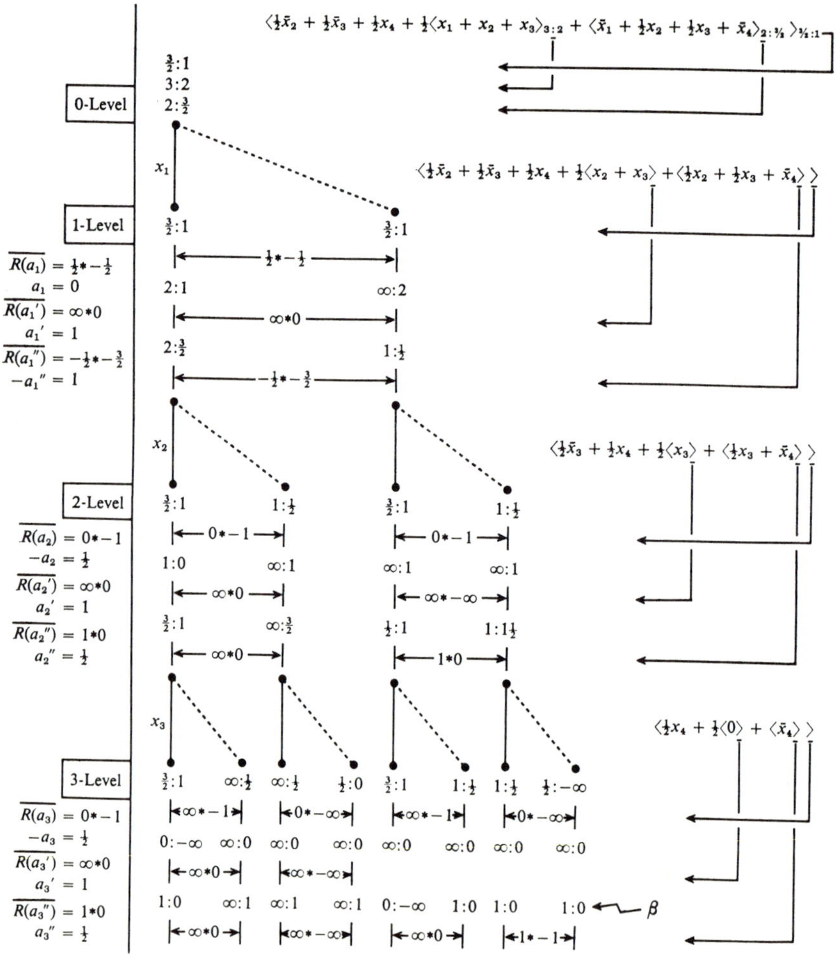

Fig. 5.24. Another realization of the function of Example 5.7.

OR-gate logic. In both cases computations are performed on arrays of truth table points arranged in a convenient manner. The tree arrangement is particularly suitable for T-gate realizations. In addition, as described in Section 5.6, the designer is not limited to use only the bottom level of the tree.

For example, consider again the function of Example 5.7 as realized in Fig. 5.23. The outer gate has a gap of $\frac{9}{8}$:1, which is equivalent to 9:8. Suppose that this relative gap is too small for sensitivity considerations, and that another gate must be added to increase the relative gap. On the

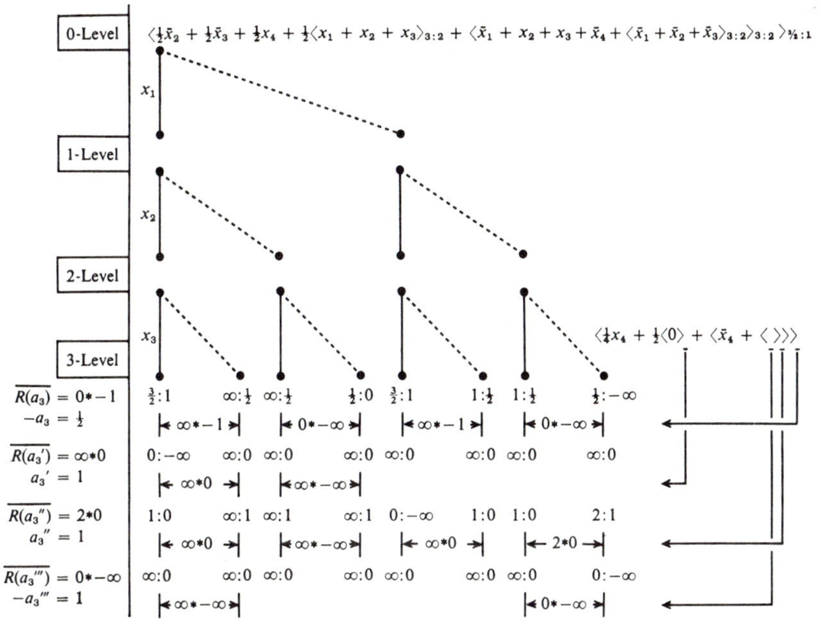

Fig. 5.25. A realization with gaps 3:2 for the function of Example 5.7.

3-level, one outer-gate gap reduction can be eliminated if the gap of the leftmost function, denoted by α in Fig. 5.23, is raised by $\frac{1}{2}$. This can be accomplished by adding a gate with a weight of $\frac{1}{2}$ which realizes the function 1 in the leftmost function, and zero elsewhere. This is done in Fig. 5.24 and results in the realization given in that figure.

Now the outer gate has gap $\frac{3}{2}$:1 or 3:2, but the second inner gate has a gap 2:$\frac{3}{2}$ or 4:3. If this 4:3 gap is too high, it can be lowered by adding still another gate. For example, the gap reduction in that gate on the 3-level can be eliminated if the gap of the rightmost function, denoted by β in

Fig. 5.24, is raised by 1, keeping the other gaps the same. This has been done in Fig. 5.25 by adding a fourth gate, resulting in the three-layer realization given in the figure. The details of the realization have been omitted.

This example points out a number of situations that may arise in this type of synthesis. However, one important type of situation has thus far not come up, and this is described in the next example.

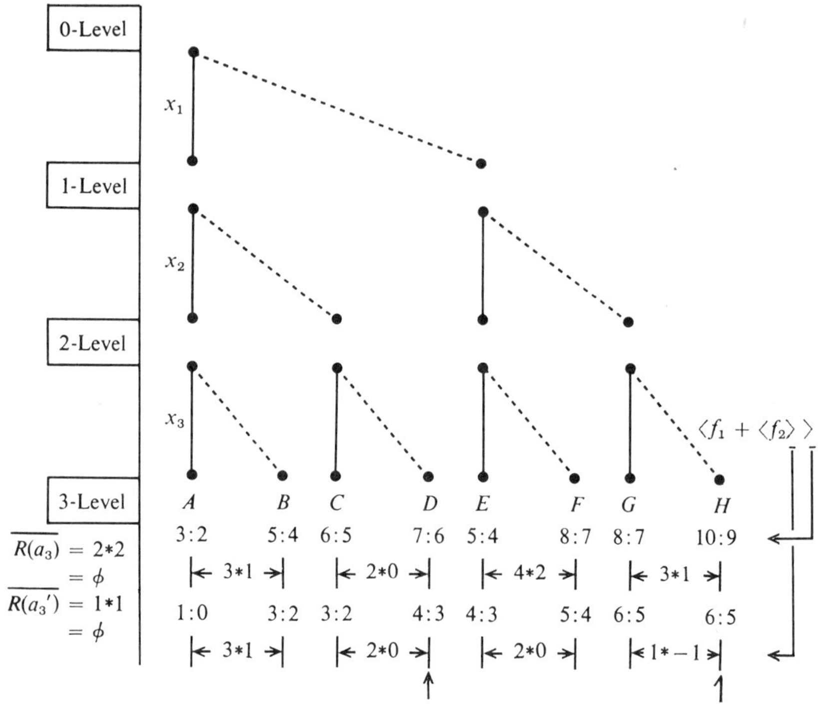

Fig. 5.26. The function tree for Example 5.8. The 3-level common range is void for both gates.

Example 5.8. Figure 5.26 shows the 3-level of a tree on which a 2-gate realization is being attempted, the gaps having arisen because of computations on lower levels. On the 3-level, the range for both gates is void. The range on the outer gate can be made consistent by raising the gap of D by 1, keeping that of C unchanged. The inner gate range can be made nonvoid by raising the gap of H by 1, keeping that of G unchanged. Thus it might be thought that two additional gates are needed. However, it is readily verified that both ranges can be made nonvoid if the gaps of D and H in both gates are raised by 1, the other gaps remaining the same. Thus both inconsistencies can be eliminated by adding only one additional gate, which feeds both original gates. This is shown in Fig. 5.27. Note that, in the ⟨ ⟩ notation, the third gate must be written twice, with interconnecting arrows to indicate that it refers to the same gate. The topology of the resulting network is also indicated in the figure. The remainder of the realization is omitted.

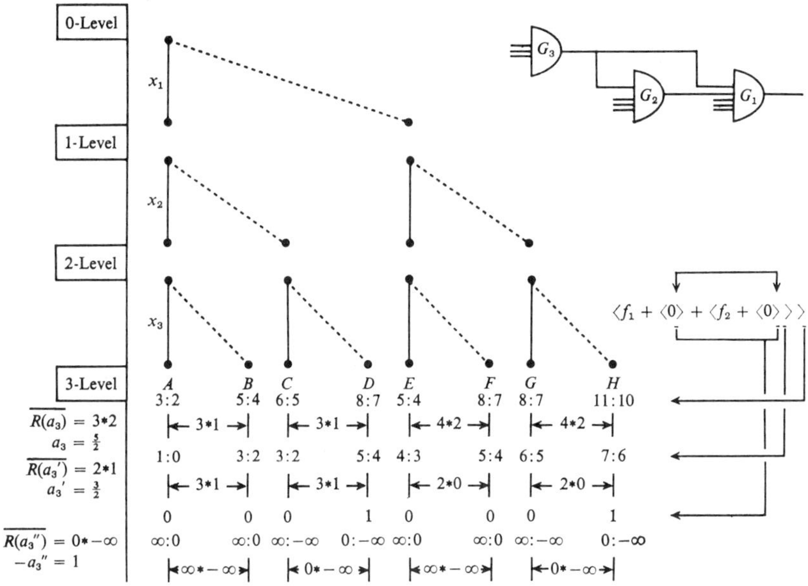

Fig. 5.27. A three-gate realization for the function of Example 5.8.

In summary, the procedures described in this section are algorithmic in that they always yield a realization, but there is considerable room for intuition and cleverness in reducing the number of required gates.

Chapter 6

THREE-INPUT MAJORITY GATES AND SYMMETRIC FUNCTIONS

6.1 INTRODUCTION

The simplest nontrivial T gate is the *three-input majority gate* (M_3 gate) which realizes the function

$$y = M(y_1,y_2,y_3) = y_1y_2 \dotplus y_2y_3 \dotplus y_1y_3 \tag{6.1}$$

and has the separating-function representation

$$y = \langle y_1 + y_2 + y_3 \rangle_{2:1} \tag{6.2}$$

If the number of inputs were decreased to 2, the M_3 gate would become an AND gate; if its gap were decreased to 1:0, it would become an OR gate. In this discussion these are classed as trivial T gates. Therefore the M_3 gate is the nontrivial T gate with the smallest fan-in and largest relative gap. As such it is the most insensitive to component and signal fluctuations.

Because of their simplicity and consequent ease of construction, M_3 gates are given special consideration in threshold logic, and special techniques of analysis and synthesis have been developed for networks using such gates. Naturally the general procedures considered previously are applicable with the appropriate constraints on gap and fan-in, but these turn out to be unnecessarily cumbersome. The special procedures that have been developed are almost invariably simpler, and their use makes the synthesis of networks from M_3 gates computationally easier than when the more general techniques are used.

This simplicity may be costly, however. For example, if the sensitivity constraints are such that an M_3 gate can be constructed, then it is likely that a three-input OR gate can also be built† since its realization is

$$y = \langle y_1 + y_2 + y_3 \rangle_{1:0}$$

If, however, we limit ourselves to M_3 gates, then we give up the possibility of using three-input OR gates when they are needed. This implies that more gates will probably be required since two M_3 gates are needed to realize the three-input OR function

$$y = x_1 \dotplus x_2 \dotplus x_3$$

Therefore in complex functions, any time that a three-input OR gate can be efficiently used in the realization, the use of M_3 gates necessitates using an extra gate.

This increase is probably not substantial when averaged over a large number of functions, and for many applications the inherent simplification in the synthesis procedures and in the fact that only one kind of gate need be constructed outweighs the small disadvantage of a few extra gates. At any rate, before this trade-off can be evaluated in any particular situation, one must have a detailed understanding of the special procedures for M_3 gates.

The first part of this chapter is concerned with synthesis techniques for M_3 gates. Although many of the results apply equally well to majority gates with more inputs (such as a 5-input majority gate for which the gap is 3:2), the necessary extensions or rewordings are not included. There appears to be little justification for restricting networks to only majority gates when more than three inputs are allowed. For example, if sensitivity restrictions are such that a gap of 3:2 and a fan-in of 5 are achievable, then there are a number of T gates other than the 5-input majority gates that can be constructed, and it is almost certainly not a wise choice to use only majority gates under these circumstances.

The second part of this chapter is concerned with the realization of symmetric functions (that is, functions that are symmetric in all variables). It is shown that a canonic form for the realization of such functions by means of M_3 gates can be defined.

6.2 SIMPLE REALIZATIONS

There are a number of elementary results that often yield realizations in

† This may not be true in some technologies. For example, parametrons are fundamentally majority devices. If the fan-in of such a device is restricted to three including biasing, then only the M_3 gate can be built.

simple situations by inspection. For example, the two input OR and AND functions can be realized† as

$$y = x_1 \dotplus x_2 = M(x_1,x_2,1)$$
$$y = x_1x_2 = M(x_1,x_2,0)$$

Note the constant inputs 1 and 0. These are certainly allowable and are often necessary to obtain a realization. Actually the above results are special cases of the following general theorem.

Theorem 6.1. If $y = F(x_2,x_3,\ldots,x_n)$ is realized by a network of M_3 gates so that

$$y = F(x_2,x_3,\ldots,x_n) = M(y_1,y_2,y_3)$$

where y_1, y_2, and y_3 are Boolean functions that are realized by other M_3 gates and so on, then

A. $y_a = x_1 \dotplus F(x_2,x_3,\ldots,x_n) = M(1,x_1,M(y_1,y_2,y_3))$
B. $y_b = x_1F(x_2,x_3,\ldots,x_n) = M(0,x_1,M(y_1,y_2,y_3))$
C. $y_c = x_0x_1 \dotplus (x_0 \dotplus x_1)F(x_2,x_3,\ldots,x_n) = M(x_0,x_1,M(y_1,y_2,y_3))$

Proof. The validity follows immediately by substituting y for $M(y_1,y_2,y_3)$ and then expressing the resultant in terms of the Boolean function that it realizes.▮

With the use of this theorem, a number of simple realizations can be obtained by inspection, as in the following example.

Example 6.1.

(1) $x_1 \dotplus x_2x_3 = M(1,x_1,M(x_2,x_3,0))$

(2) $x_1(x_2 \dotplus x_3) = M(0,x_1,M(x_2,x_3,1))$

(3) $x_1 \dotplus x_2 \dotplus x_3 = M(1,x_1,M(x_2,x_3,1))$

(4) $x_1 \dotplus x_2 \dotplus \cdots \dotplus x_n = M(1,x_1,M(1,x_2,$
$M(\ldots, M(1,x_{n-2},M(1,x_{n-1},x_n))\ldots)$

(5) $x_1x_2x_3 = M(0,x_1,M(x_2,x_3,0))$

(6) $x_1x_2\ldots x_n = M(0,x_1,M(0,x_2,$
$M(\ldots,M(0,x_{n-2},M(0,x_{n-1},x_n))\ldots)$

(7) $x_1 \dotplus M(x_2,x_3,x_4) = M(1,x_1,M(x_2,x_3,x_4))$

(8) $x_1M(x_2,x_3,[x_4x_5]) = M(0,x_1,M(x_2,x_3,M(0,x_4,x_5)))$

† We use the notation $M(\,)$ as the separating-function representation of the switching function that is realized by an M_3 gate.

Since M_3 gates are special T gates, the results of Chapter 2 on the realization of complement and dual functions are applicable. In fact, these results can be simplified somewhat when applied to networks composed of M_3 gates, since the function realized by an M_3 gate is self-dual.

Theorem 6.2. If $y = F(x_1,x_2,\ldots,x_n)$ is realized by a network of M_3 gates, then

A. $\bar{y} = \bar{F}(x_1,x_2,\ldots,x_n)$ can be realized by the network that realizes F when each independent variable (including constant inputs) is replaced by its complement.

B. $y^d = F^d(x_1,x_2,\ldots,x_n)$ can be realized by the network that realizes F if each constant input is replaced by its complement. (Recall that the dual of a constant function is defined as its complement.)

As an example of this theorem, observe that the first two functions of Example 6.1 are duals of each other and that their realizations are related as defined by the theorem.

The following corollary is a direct consequence of Part B of Theorem 6.2.

Corollary 6.2.1. A function can be realized by a network of majority gates without constant inputs if and only if the function is self-dual.

An example of this is, of course, the majority function itself.

Corollary 6.2.1 shows that constant inputs are a necessary part of any synthesis procedure that involves only majority gates. They are required, in fact, in the realization of all non-self-dual functions.

Example 6.2. Assume that a function $y_1 = F_1(x_2,x_3,\ldots,x_n)$ has been realized by a network of M_3 gates. Then the function

$$y = \bar{x}_1F_1(x_2,x_3,\ldots,x_n) + x_1F_1{}^d(x_2,x_3,\ldots,x_n)$$

can be realized by the network for F_1 with each constant input of 0 replaced by x_1 and each constant input of 1 replaced by $\bar{x}_1$. This is easily verified by replacing x_1 by 1 and 0 in the expression for F and then using Theorem 6.2.

6.3 SIMPLIFICATION METHODS

Before we develop the general M_3 gate synthesis procedures, we present some simplification theorems. These are results that can be used to obtain equivalent M_3 gate separating-function representations of the same switching function. They are similar to the corresponding Boolean algebra identities that yield equivalent AND-OR-gate realizations of a given switching function. The validity of the theorems is verified by replacing the appropriate M_3-gate separating-function representation by its Boolean-function representation.

In the following theorem, the first five identities define simplifications that are possible when both a variable and its complement appear in the realization. The sixth identity shows how variables can be "factored" from the realization.

Theorem 6.3. (*Cohn and Lindaman*).

A. $M(y_1,\bar{y}_1,y_2) = y_2$
B. $M(y_1,y_2,M(\bar{y}_1,y_2,y_3)) = y_2$
C. $M(y_1,y_2,M(\bar{y}_1,\bar{y}_2,y_3)) = M(y_1,y_2,y_3)$
D. $M(y_1,y_2,M(\bar{y}_1,y_3,y_4)) = M(y_1,y_2,M(y_2,y_3,y_4))$
E. $M(y_1,M(\bar{y}_1,y_2,y_3),M(y_1,y_2,y_3)) = M(y_1,y_2,y_3)$
F. $M(M(y_1,y_2,y_3),M(y_1,y_2,y_4),y_5) = M(y_1,y_2,M(y_3,y_4,y_5))$

Other identities are readily derived, and in fact the literature contains a number of additional ones.

Example 6.3. The above results can be used to directly simplify the separating-function representation of a switching function without first converting it to the Boolean representation. Consider for example

$$y = (x_1 \dotplus x_2)(x_1 \dotplus x_3)$$

which can be simplified to

$$y = x_1 \dotplus x_2x_3$$

and which is realized in Example 6.1 as

$$y = M(1,x_1,M(x_2,x_3,0))$$

If, however, the original Boolean representation were obtained by the application of Theorem 6.1 as

$$y = M(0,M(1,x_1,x_2),M(1,x_1,x_3))$$

then Theorem 6.3F yields directly the simpler realization

$$y = M(1,x_1,M(x_2,x_3,0))$$

by letting

$$y_1 = 1,\ y_2 = x_1,\ y_3 = x_2,\ y_4 = x_3, \text{ and } y_5 = 0$$

6.4 REALIZATION PROCEDURES: SINGLE PASS

In this and the following sections we present two realization procedures† for M_3-gate networks, both of which are most conveniently performed on

† In Chapter 7 a canonic form procedure is presented after the required background in dual and self-dual functions has been developed. Miiller and Winder give another procedure based on map methods.

treelike structures. We denote these as the *one-pass* and *two-pass* procedures. The two-pass procedure is similar to the tree procedure of Chapter 5; the tree is constructed from top to bottom on the first pass and then the realization is obtained in the second pass going from bottom to top. In the one-pass procedure, on the other hand, the realization is obtained in one pass as the tree is constructed from top to bottom. The one-pass procedure has the advantage of being quick and easy. But the two-pass procedure usually gives better realizations since more of the structure of the function can be utilized. We first present the one-pass procedure in its simplest form as proposed by Miyata. This is followed by some modifications due to Lindaman and to Akers which can be used to obtain simpler realizations at the expense of added computations.

In the simple one-pass procedure, the function tree is formed in a manner similar to that of Chapter 4, but each node of the tree corresponds to an M_3 gate. Thus on the 0-level there is the original function F that can be factored as

$$F(x_1,x_2,\ldots,x_n) = x_1F_1(x_2,x_3,\ldots,x_n) \dotplus \bar{x}_1F_0(x_2,x_3,\ldots,x_n)$$

This expression can be realized with the M_3 gate

$$F(x_1,x_2,\ldots,x_n) = M(1,x_1F_1(x_2,x_3,\ldots,x_n),\bar{x}_1F_0(x_2,x_3,\ldots,x_n))$$

Actually this is just the OR function of

$$x_1F_1(x_2,x_3,\ldots,x_n) \quad \text{and} \quad \bar{x}_1F_0(x_2,x_3,\ldots,x_n)$$

The 1-level functions are $x_1F_1(x_2,x_3,\ldots,x_n)$, 1, and $\bar{x}_1F_0(x_2,x_3,\ldots,x_n)$ as shown in Fig. 6.1. The 0-level node represents the M_3 gate that realizes

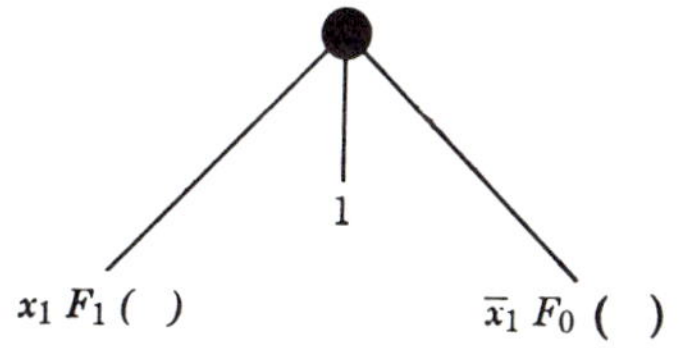

Fig. 6.1. The first two levels of the one-pass tree.

the 0-level function F where the 1-level functions are the inputs to the 0-level M_3 gate.

After the first two levels the tree is formed somewhat differently. The functions F_1 and F_0 are factored as

$$F_1(x_2,x_3,\ldots,x_n) = x_2F_{11}(x_3,x_4,\ldots,x_n) \dotplus \bar{x}_2F_{10}(x_3,x_4,\ldots,x_n)$$
$$F_0(x_2,x_3,\ldots,x_n) = x_2F_{01}(x_3,x_4,\ldots,x_n) \dotplus \bar{x}_2F_{00}(x_3,x_4,\ldots,x_n)$$

and expressed as

$$x_1F_1(x_2,x_3,\ldots,x_n) = x_1(x_2F_{11}(x_3,x_4,\ldots,x_n) + \bar{x}_2F_{10}(x_3,x_4,\ldots,x_n))$$
$$\bar{x}_1F_0(x_2,x_3,\ldots,x_n) = \bar{x}_1(x_2F_{01}(x_3,x_4,\ldots,x_n) + \bar{x}_2F_{00}(x_3,x_4,\ldots,x_n))$$

These expressions can be realized by M_3 gates as follows:

$$x_1F_1(x_2,x_3,\ldots,x_n) = M(x_1,x_2F_{11}(x_3,x_4,\ldots,x_n),\bar{x}_2F_{10}(x_3,x_4,\ldots,x_n))$$
$$\bar{x}_1F_0(x_2,x_3,\ldots,x_n) = M(\bar{x}_1,x_2F_{01}(x_3,x_4,\ldots,x_n),\bar{x}_2F_{00}(x_3,x_4,\ldots,x_n))$$

This is entered on the tree as shown in Fig. 6.2. Again each 1-level node represents an M_3 gate which realizes its 1-level label where the inputs are the 2-level functions that have a branch to the 1-level node.

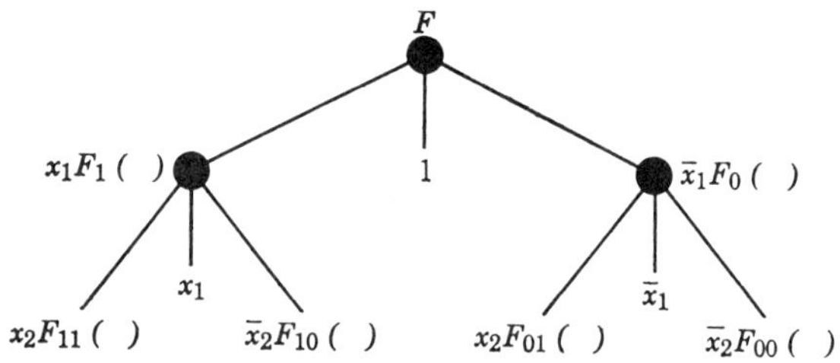

Fig. 6.2. The first three levels of the one-pass tree.

The procedure is continued in the same way. Each path down the tree stops when the function associated with its bottom end is a constant function or a single variable. The resulting tree is a picture of the network of M_3 gates that realizes the given function. The inputs to the network are the variables and constant functions on which the paths down the tree terminate. The nodes of the tree correspond to the M_3 gates and the branches to the interconnecting lines. The output of each gate is the label of the corresponding node. The output of the top gate is the output of the network.

This procedure is simple and straightforward because there is virtually no freedom except in the order in which the variables are selected. As might be expected, therefore, the realizations that result often have more gates than are actually needed. This shortcoming is demonstrated in Example 6.4, where the three-input majority function is realized by the procedure.

Example 6.4. Figure 6.3 shows the one-pass tree for realizing the majority function

$$y = x_1x_2 + x_1x_3 + x_2x_3$$

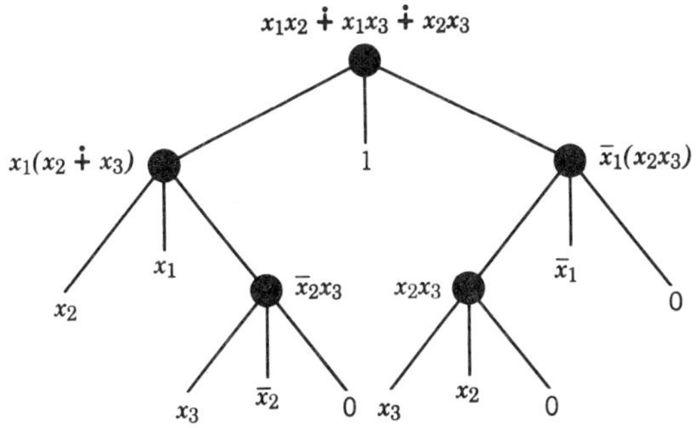

Fig. 6.3. One-pass tree for three-variable majority function.

The resultant realization is

$$y = M(1,M(x_1,x_2,M(\bar{x}_2,x_3,0)),M(M(x_3,x_2,0),\bar{x}_1,0))$$

Five gates are used whereas only one is required. The simplification theorems can be used to eliminate some of these gates but, as we shall see, the simple realization can be obtained in a more straightforward way.

We now consider three theorems that can be used on any level of the tree to obtain a simpler realization.

Theorem 6.4 (*Lindaman, 2*). To each Boolean representation of $F(x_1,x_2,\ldots,x_n)$ of the form

$$\begin{aligned} F(x_1,x_2,\ldots,x_n) &= x_1F_1(x_2,x_3,\ldots,x_n) \dotplus \bar{x}_1F_0(x_2,x_3,\ldots,x_n) \\ &= x_1[F_a(\ ) \dotplus F_b(\ )] \dotplus \bar{x}_1[F_c(\ )F_d(\ )] \end{aligned}$$

there corresponds the M_3-gate representation

$$F(x_1,x_2,\ldots,x_n) = M(x_1,[x_1F_a(\ ) \dotplus \bar{x}_1F_c(\ )], [x_1F_b(\ ) \dotplus \bar{x}_1F_d(\ )])$$

Proof. The validity follows by expressing the separating-function representation by its Boolean representation and simplifying.|

Observe that F_1 has been split into the sum of two functions and F_0 into the product of two functions and that the realization is in terms of these four new functions. If it is possible to select these functions so that

$$F_a = F_c \qquad \text{and} \qquad F_b = F_d$$

then the result simplifies to the following corollary.

Corollary 6.4.1. Under the hypothesis of Theorem 6.4, if

$$F_a = F_c \quad \text{and} \quad F_b = F_d$$

then

$$F(x_1,x_2,\ldots,x_n) = M(x_1,F_a(\ ),F_b(\ ))$$

Example 6.5. Consider again the majority function

$$y = x_1x_2 \dotplus x_1x_3 \dotplus x_2x_3 = x_1(x_2 \dotplus x_3) \dotplus x_2x_3$$

Applying Corollary 6.4.1 where

$$F_a(\ ) = F_c(\ ) = x_2$$

$$F_b(\ ) = F_d(\ ) = x_3$$

gives

$$y = M(x_1,x_2,x_3)$$

Hence Corollary 6.4.1 yields a realization on the first level of the tree.

In this example Theorem 6.4 causes a drastic simplification. In more complicated situations it may not be clear how to split F_1 and F_0 to obtain any simplification at all.

A somewhat similar theorem depends upon F being expanded about two variables.

Theorem 6.5 (*Cohn and Lindaman*). To each Boolean representation of $F(x_1,x_2,\ldots,x_n)$ of the form

$$F(x_1,x_2,\ldots,x_n) = x_1x_2F_{11}(x_3,x_4,\ldots,x_n) \dotplus x_1\bar{x}_2F_{10}(x_3,x_4,\ldots,x_n) \dotplus \bar{x}_1x_2F_{01}(x_3,x_4,\ldots,x_n) \dotplus \bar{x}_1\bar{x}_2F_{00}(x_3,x_4,\ldots,x_n)$$

there corresponds the separating-function representation

$$F(x_1,x_2,\ldots,x_n) = M(x_1,M(\bar{x}_1,x_2,[x_2F_{11}(\ ) \dotplus \bar{x}_2F_{00}(\ )]), M(\bar{x}_1,\bar{x}_2,[x_2F_{01}(\ ) \dotplus \bar{x}_2F_{10}(\ )]))$$

Proof. Again the proof follows immediately by replacing each separating-function representation by its Boolean-function representation and simplifying.|

The theorem is simplified considerably when

$$F_{11} = F_{00} \quad \text{and} \quad F_{01} = F_{10}$$

Corollary 6.5.1. Under the conditions of Theorem 6.5, if

$$F_{11} = F_{00} \quad \text{and} \quad F_{01} = F_{10}$$

then

$$F(x_1,x_2,\ldots,x_n) = M(x_1,M(\bar{x}_1,x_2,F_{11}(\ )),M(\bar{x}_1,\bar{x}_2,F_{01}(\ )))$$

Figure 6.4 shows the one-pass tree when Corollary 6.5.1 is used. Note that the theorem yields a network with two gates on the third level, unlike the original simple procedure that required four gates. This saving is propagated downward through the tree. Each lower level will have only half as many gates as are needed with the simple procedure; thus a substantial saving has been realized.

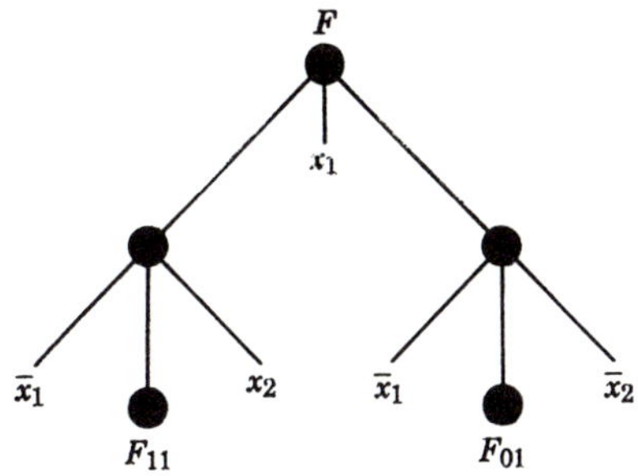

Fig. 6.4. A simplified tree due to the application of Corollary 6.5.1.

The use of this theorem is straightforward in simple situations. In more complicated cases, however, it may not be clear how it can be applied in an efficient way.

The following theorem is particularly useful for unate functions. Its validity is immediately evident.

Theorem 6.6 (*Akers, 2*). If $F(x_1,x_2,\ldots,x_n)$ is unate in the variable x_1, then it can be written as

$$\begin{aligned} F(x_1,x_2,\ldots,x_n) &= x_1F_a(x_2,x_3,\ldots,x_n) + F_b(x_2,x_3,\ldots,x_n) \\ &= M(x_1,[F_a(\,) + F_b(\,)],F_b(\,)) \end{aligned}$$

Observe that both functions of Theorem 6.6 that remain to be realized, namely $F_a(\,) + F_b(\,)$ and $F_b(\,)$, are not functions of x_1. Thus one variable has been eliminated. For completely unate functions the procedure can be repeated until all variables have been removed and a realization obtained.

Example 6.6 (*Akers, 2*). Figure 6.5 shows the realization of the function

$$y = x_1x_2 \dotplus x_1x_3 \dotplus x_3x_4$$

using Theorem 6.6 as the tree formation rule. On the top level of the tree x_1 was removed, and from the two functions on the next level x_2 and x_3 were removed. The resultant realization has three gates.

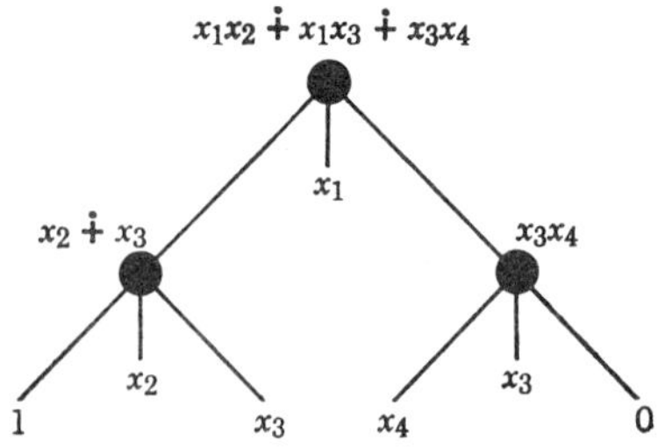

Fig. 6.5. The realization of a unate function—Example 6.6.

In summary, the single-pass realization techniques are quick and simple but frequently lead to inefficient realizations. The simplifying theorems, which are really alternative tree formation rules, are sometimes helpful. In complicated cases, however, it is not at all apparent which theorem to use, or when to use it, for the most efficient realization relative to the number of gates that are used.

The two-pass procedure circumvents many of these disadvantages. One reason for this is that freedom exists whereby the form of the realization can be controlled. Moreover, it is unnecessary to make choices until the entire structure of the function, as depicted by the tree, is known.

6.5 REALIZATION PROCEDURE: TWO-PASS

Basic Theorems

The two-pass realization procedure is performed on the function tree defined by Procedure 4.1. In fact, it is the multigate procedure of Chapter 5 especially adapted to M_3-gate realizations. The procedure is based upon two theorems concerning M_3-gate networks.

Recall that if

$$y = F(x_1,x_2,\ldots,x_n)$$

is realized by an M_3-gate network, then in terms of the final gate

$$y = F(x_1,x_2,\ldots,x_n) = M(y_1,y_2,y_3)$$

where y_1, y_2, and y_3 are switching functions defined by the remainder of the network. The following theorem shows how the network is altered so that it realizes a reduced function of F.

Theorem 6.7. If

$$y = F(x_1,x_2,\ldots,x_n) = M(y_1,y_2,y_3)$$

is realized by an M_3-gate network, and if F is factored so that

$$y = x_1F_1(x_2,x_3,\ldots,x_n) + \bar{x}_1F_0(x_2,x_3,\ldots,x_n) = x_1y(x_1 = 1) \dotplus \bar{x}_1y(x_1 = 0)$$

then

$$y(x_1 = 1) = F_1(x_2,x_3,\ldots,x_n) = M(y_1(x_1 = 1),y_2(x_1 = 1),y_3(x_1 = 1))$$

and

$$y(x_1 = 0) = F_0(x_2,x_3,\ldots,x_n) = M(y_1(x_1 = 0),y_2(x_1 = 0),y_3(x_1 = 0))$$

The validity of this theorem is evident. Moreover, it is clear that it can be iteratively applied to all M_3 gates of the network. Those gates which have x_1 or $\bar{x}_1$ as inputs will have constant inputs for the reduced functions. These constant inputs cannot be used to alter the gap in the manner of Theorem 2.4, however, since we are confined to M_3 gates only.

We now define two M_3-gate networks as being *identical except for constant inputs* if they are identical except that corresponding constant input functions may be equal or may be complements of each other. With this definition the following corollary is implied by Theorem 6.7.

Corollary 6.7.1. All functions on any k-level of the function tree can have the same M_3-gate network realizations except for constant inputs.

Theorem 6.7 describes how the reduced functions of the function tree are realized by the M_3-gate network that realizes the original function on the 0-level. We now consider a range theorem that describes how to proceed up the tree to obtain a realization of an ancestor function.

Theorem 6.8 (*Range Theorem*). If

$$F_a(x_2,x_3,\ldots,x_n) = M(1,y_2,y_3)$$

and

$$F_b(x_2,x_3,\ldots,x_n) = M(0,y_2,y_3)$$

then

A. $$F_1(x_1,x_2,\ldots,x_n) = x_1F_a(x_2,x_3,\ldots,x_n) + \bar{x}_1F_b(x_2,x_3,\ldots,x_n) = M(x_1,y_2,y_3)$$

B. $$F_2(x_1,x_2,\ldots,x_n) = \bar{x}_1F_a(x_2,x_3,\ldots,x_n) + x_1F_b(x_2,x_3,\ldots,x_n) = M(\bar{x}_1,y_2,y_3)$$

This is a special case of the more general range theorems of Chapter 4. Here the difference in the constant inputs plays the part of the difference in gaps in the more general case.

Actually Theorem 6.8 can be generalized as given by the following corollary.

Corollary 6.8.1. If

$$F_a(x_2,x_3,\ldots,x_n) = M(y_1,y_2,y_3)$$

and

$$F_b(x_2,x_3,\ldots,x_n) = M(y_4,y_5,y_6)$$

then

$$\begin{aligned} F(x_1,x_2,\ldots,x_n) &= x_1F_a(x_2,x_3,\ldots,x_n) + \bar{x}_1F_b(x_2,x_3,\ldots,x_n) \\ &= M([x_1y_1 + \bar{x}_1y_4],[x_1y_2 + \bar{x}_1y_5],[x_1y_3 + \bar{x}_1y_6]) \end{aligned}$$

Theorem 6.8 and Corollary 6.8.1 can be iteratively applied to all gates that realize the y_j functions to give the following corollary.

Corollary 6.8.2 If $F_a(x_2,x_3,\ldots,x_n)$ and $F_b(x_2,x_3,\ldots,x_n)$ are realized by the same M_3-gate network except for constant function inputs, then

$$F(x_1,x_2,\ldots,x_n) = x_1F_a(x_2,x_3,\ldots,x_n) + \bar{x}_1F_b(x_2,x_3,\ldots,x_n)$$

can be realized by the same network, except that

(1) All corresponding inputs that are the constant 1 function in the realization for F_a and the 0 function in the realization for F_b are x_1 in the realization for F, and

(2) All corresponding inputs that are the 0 function in the realization for F_a and the 1 function in the realization for F_b are $\bar{x}_1$ in the realization for F.

These results are the basis of the two-pass realization procedure. First the function tree is formed. Then, assuming that for all functions on some level of the tree a realization can be found which is the same except for constant inputs, the range theorem (Theorem 6.8) is applied to each pair of descendant functions. This yields realizations for all functions on the next level up the tree. If on any level there is difficulty in obtaining realizations that are identical except for constant inputs, gates are added to remove the difficulty and the process is repeated. Thus the general idea of the procedure is quite similar to that of the multigate procedure of Chapter 5. Before we can describe it in detail, however, we must first discuss how to start the procedure and how to add gates.

Starting the Procedure

The procedure starts on the $(n-1)$-level of the tree where the only Boolean functions are 0, 1, x_n, and $\bar{x}_n$. If x_n and $\bar{x}_n$ are not both present, then all functions on the $(n-1)$-level can be realized with a single gate and all realizations can be the same except for constant inputs. Specifically, if only x_n, 0, and 1 are present, then

$$\begin{aligned} x_n &= M(1,x_n,0) \\ 1 &= M(1,x_n,1) \\ 0 &= M(0,x_n,0) \end{aligned}$$

If x_n and $\bar{x}_n$ both appear, then it is necessary to use three gates in order that all functions have the same realization except for constant inputs. The realization is

$$y = M(C_1,M(C_2,x_n,C_3),M(C_4,\bar{x}_n,C_5))$$

where $C_i = 0,1$. By selecting the constant inputs properly, the network can be made to realize any of the functions x_n, $\bar{x}_n$, 0, or 1. For example,

$$\begin{aligned} y = x_n &= M(1,M(1,x_n,0),M(0,\bar{x}_n,0)) \\ y = \bar{x}_n &= M(1,M(0,x_n,0),M(1,\bar{x}_n,0)) \\ y = 1 &= M(1,M(1,x_n,1),M(1,\bar{x}_n,1)) \\ y = 0 &= M(0,M(0,x_n,0),M(0,\bar{x}_n,0)) \end{aligned}$$

Table 6.1 lists all of the choices for the values of the constants, except for C_2,C_3 and C_4,C_5 permutations, that yield particular realizations. The choices should be utilized so that the fewest number of gates need be added on the next level.

Table 6.1. Three-Gate Realizations for $(n - 1)$-Level Functions

	$y = M(C_1,M(C_2,x_n,C_3),M(C_4,\bar{x}_n,C_5))$				
y	C_1	C_2	C_3	C_4	C_5
x_n	1	1	0	0	0
	0	1	0	1	1
$\bar{x}_n$	1	0	0	1	0
	0	1	1	1	0
1	1	1	1	1	1
	1	1	1	1	0
	1	1	1	0	0
	1	1	0	1	1
	1	0	0	1	1
	0	1	1	1	1
0	0	0	0	0	0
	0	0	0	1	0
	0	0	0	1	1
	0	1	0	0	0
	0	1	1	0	0
	1	0	0	0	0

Adding Gates for Constant Functions

Just as in the multigate procedure of Chapter 5, gates must be added whenever necessary to make all realizations on any level of the tree the same except for constant inputs. Assume that all functions on the k-level of the tree have the same realization except for constant inputs. We are interested in knowing (1) why the procedure may yield $(k - 1)$-level functions that do not satisfy this condition and (2) how to add gates to eliminate the difficulty.

Suppose that on the $(k - 1)$-level of the tree there are some 1 functions or 0 functions. At this point these must be realized by the same M_3-gate networks, except for constant inputs, as the other functions on the level. This produces the difficulty that the realization of a 1 function requires that two of the inputs to the final gate be 1 functions, and the realization of a 0 function requires that two of the inputs be 0 functions. That is,

$$1 = M(1,1,y_1)$$
$$0 = M(0,0,y_1)$$

for any y_1. The nonconstant functions on the level will not, in general, have two constant function inputs; therefore additional gates must be added so that all realizations on that level are the same except for constant inputs.

For example, if the realizations are of the form

$$y = M(C_1,y_1,y_2)$$

where C_1 is the constant 0 or 1, this cannot be made to realize a constant function by selecting a value for C_1. A gate can be added, however, to eliminate the difficulty. In particular, the input of the gate with variable y_2 is replaced by the gate $M(C_2,y_2,C_3)$, in which case the realization for y becomes

$$y = M(C_1,y_1,M(C_2,y_2,C_3))$$

In all cases for which y is not a constant function, $M(C_2,y_2,C_3)$ must realize the function y_2. This is obtained by selecting

$$C_2 = \bar{C}_3 = 1 \qquad \text{or} \qquad C_2 = \bar{C}_3 = 0$$

in which case

$$y = M(C_1,y_1,M(1,y_2,0)) = M(C_1,y_1,y_2)$$

When y is the 0 function or the 1 function, then $M(C_2,y_2,C_3)$ must realize the 0 function or the 1 function, respectively. That is,

$$y = 1 = M(1,y_1,M(1,y_2,1))$$
$$y = 0 = M(0,y_1,M(0,y_2,0))$$

By this means it is always possible to add gates so that the constant functions on any level of the tree have the same realizations, except for constant inputs, as all other functions on the level. At most, two gates are required since the maximum number of inputs that need be changed to constants is two.

Adding Gates to Yield the Same Realizations

Although all functions on some k-level of the tree have the same realization except for constant inputs, the application of the range theorem may yield $(k - 1)$-level realizations that are not the same. In one realization a particular input may be the variable x_k, for example, and in another the corresponding input may be a constant function or $\bar{x}_k$. This situation is equivalent to a void common range in the general T-gate realization procedure.

Consider, for example, the two function pairs on the k-level of the tree of Fig. 6.6a. All have the same realization except for constant inputs. When

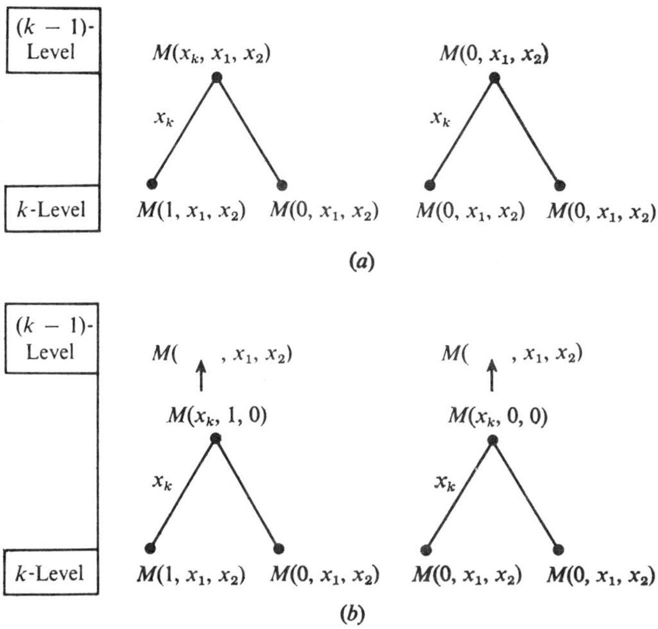

(a) A situation that yields different $(k - 1)$-level realizations. (b) Eliminating the difficulty of (a) by adding a gate.

Fig. 6.6. The equivalent of a void common range, Case 1.

the range theorem is applied, the $(k - 1)$-level realizations shown in Fig. 6.6a are obtained, and these are not the same realization except for constant inputs.

To eliminate this difficulty the input with variable x_k is replaced by a gate as shown in Fig. 6.6b. Note the way in which this is indicated. In this procedure the $(k - 1)$-level realizations of Fig. 6.6a would first be obtained. These would then be replaced by the $(k - 1)$-level realizations of Fig. 6.6b.

A related situation is shown in Fig. 6.7*a*. In this case the $(k - 1)$-level realizations are such that both x_k and $\bar{x}_k$ must be realized by additional gates in order for all $(k - 1)$-level realizations to be the same except for constant inputs. This requires the addition of three more gates, as shown in Fig. 6.7*b* and as discussed under "Starting the Procedure," p. 232. As a result, we try to avoid this situation whenever possible by appropriate selections from among the choices available to the designer, as described under "Starting the Procedure."

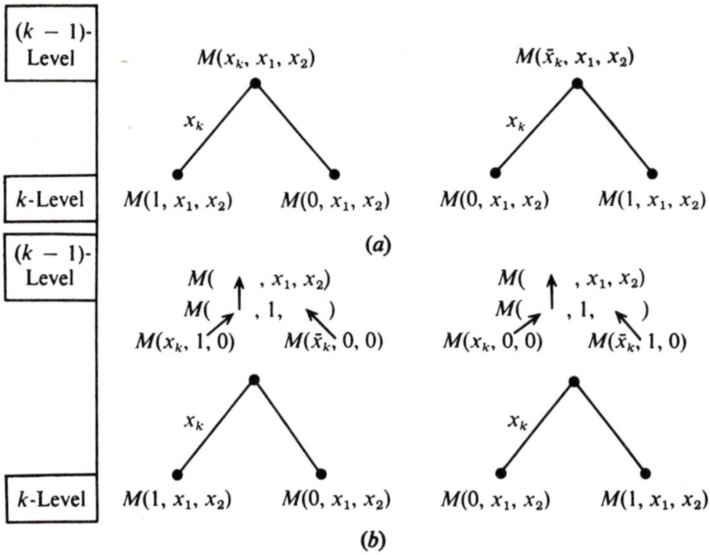

(*a*) Another situation in which $(k - 1)$-level realizations differ
(*b*) Eliminating the difficulty of (*a*) by adding three gates.
Fig. 6.7. The equivalent of a void common range, Case 2.

This covers all situations in which gates must be added. Because of the logical simplicity of the M_3 gates, it is usually necessary to add several gates on each level. However, this is far superior to the one-pass procedure, which, without the simplification theorems, requires an additional gate for each node of the tree.

6.6 TWO-PASS REALIZATION PROCEDURE: STEP-BY-STEP DESCRIPTION

The M_3-gate realization procedure is now described in step-by-step form and followed by a detailed example.

Procedure 6.2 (Two-Pass Realization Procedure)

(1) Form the function tree as described by Procedure 4.1.

(2) Select M_3-gate realizations, which are the same except for constant inputs, for all $(n - 1)$-level functions of the tree. See Table 6.1. To continue, let $k = n - 1$.

(3) For each k-level function pair, use the range theorem (Theorem 6.8) to obtain a realization for the $(k - 1)$-level ancestor function.

(4) Using the methods of Section 6.5 (pp. 233–236), add gates if necessary so that all $(k - 1)$-level functions have the same realization except for constant inputs. Assign realizations to $(k - 1)$-level constant functions, again adding gates if necessary so that all $(k - 1)$-level realizations are the same except for constant inputs.

(5) If $k - 1 = 0$, the realization is completed. Otherwise reduce k by unit amount and return to Step 3.

Example 6.7 (The Five-Variable Majority Function). This example is concerned with the realization of the five-variable majority function by means of M_3 gates. The Boolean representation is

$$\begin{aligned} y = {} & x_1[x_2(x_3 \dotplus x_4 \dotplus x_5) \dotplus x_3(x_4 \dotplus x_5) \dotplus x_4x_5] \\ & \dotplus x_2[x_3(x_4 \dotplus x_5) \dotplus x_4x_5] \dotplus x_3x_4x_5 \end{aligned}$$

The tree for this function is shown in Fig. 6.8 and redrawn in Fig. 6.9 with all functions omitted except those on the $(n - 1)$-level and all constant functions. Each 4-level function can be realized with a single M_3 gate, and all realizations are the same except for constant inputs. These are entered on the 4-level of the tree of Fig. 6.9. The range theorem is then applied and yields the 3-level realizations of Fig. 6.9.

A 0 function and a 1 function appear on the 3-level of the Fig. 6.9 tree. In order to assign realizations to these, we must add a gate as shown in Fig. 6.10. Observe that this new gate realizes the function x_4 for all but the constant functions. The range theorem is then applied to the 3-level functions and yields the 2-level realizations of Fig. 6.10.

These 2-level realizations are not the same except for constant inputs. In the output gate of these realizations there is an input x_3 for two of them, but not for the other two. Also, in the second gate of each realization, there is an x_3 input for only two of the four. Each of these gates requires that its x_3 input be replaced by a gate. This requires two additional gates.

There is some choice as to how the gate is added to the second of the two gates of the realization. For F_{10} and F_{01} of Fig. 6.10 the second gate has two constant inputs. Either of these can be realized by the additional gate. We choose to realize the 0 input of F_{10} and the 1 input of F_{01}, since then additional gates are not required on the next level. (An additional gate is also not required if the 1 input of F_{10} and the 0 input of F_{01} are realized by the added gate.)

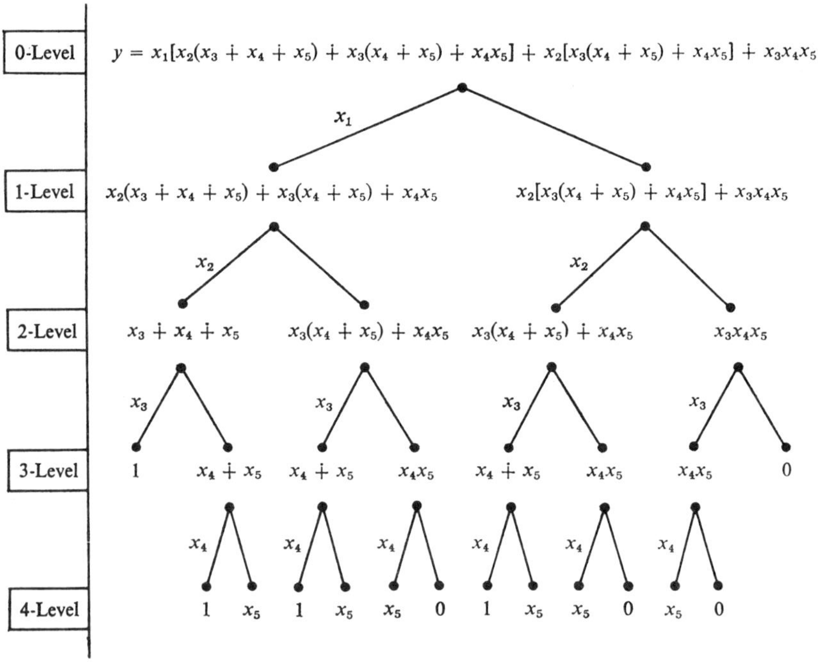

Fig. 6.8. The function tree for the five variable majority function.

The two additional gates that are required on the 2-level are shown in Fig. 6.11. The remainder of the procedure is also shown in this figure. No further difficulties are encountered and the following four-gate realization is obtained:

$$y = M(x_5,M(x_1,x_2,x_3),M(x_1,x_4,M(x_2,x_3,\bar{x}_1)))$$

It has been shown (Amarel, Cooke, and Winder) that this is the minimum number of M_3 gates with which the function can be realized.

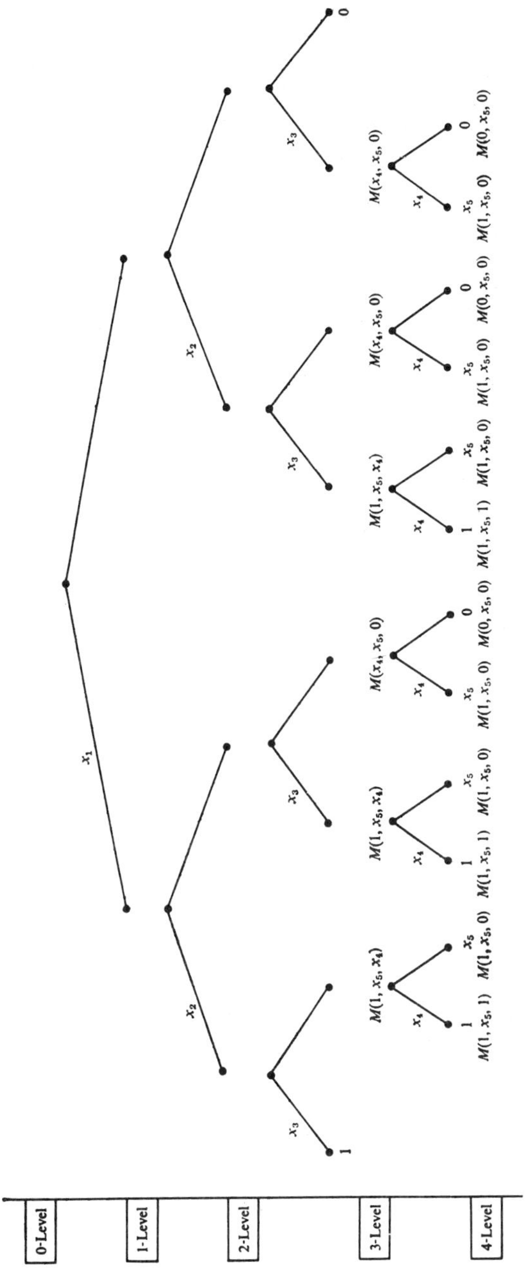

Fig. 6.9. The first step of the procedure.

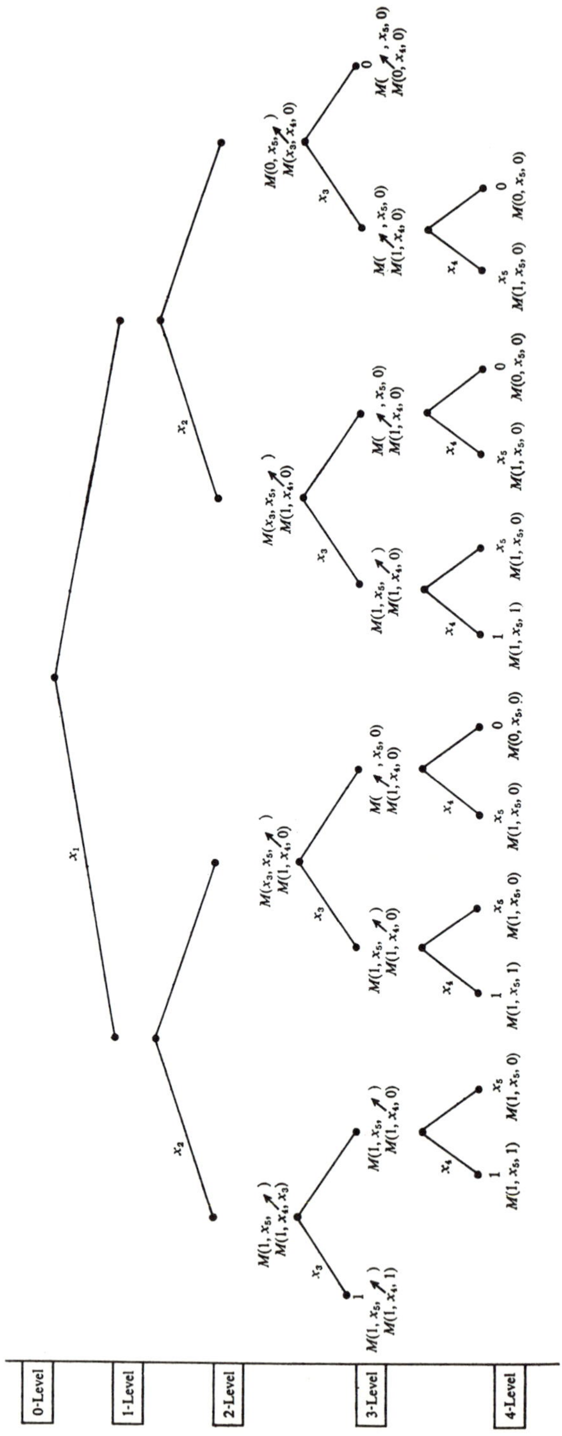

Fig. 6.10. Adding a gate on the 3-level.

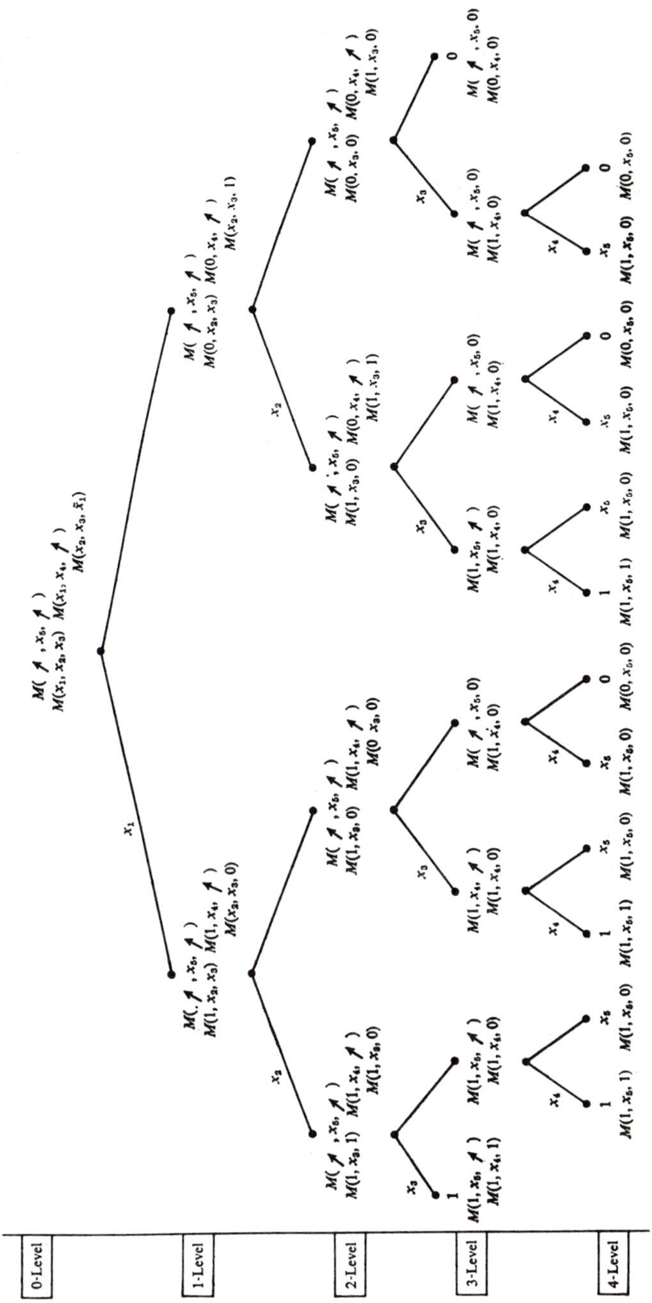

Fig. 6.11. Adding two gates on the 2-level and completing the procedure.

6.7 M$_3$-GATE CANONIC REALIZATIONS OF SYMMETRIC FUNCTIONS

Symmetric switching functions are an important class that deserves special consideration with regard to synthesis techniques. A switching function is called symmetric if it is symmetric in all of its literals. As a result there are positive integers $\alpha_1,\alpha_2,\ldots,\alpha_k$ such that the function has unit value whenever exactly α_j, $j = 1,2,\ldots,k$, of the variables $x_1,x_2,\ldots,x_n$ have unit value. In addition to its Boolean representation such a function is represented by

$$S_{\alpha_1,\alpha_2,\ldots,\alpha_k}(x_1,x_2,\ldots,x_n)$$

An example is the switching function

$$y = S_{1,3}(x_1,x_2,x_3) = x_1\bar{x}_2\bar{x}_3 + \bar{x}_1x_2\bar{x}_3 + \bar{x}_1\bar{x}_2x_3 + x_1x_2x_3$$

T gates in general, as well as M$_3$ gates, are efficient in the realization of symmetric functions as compared with AND-OR-gate realizations. This section is concerned with M$_3$-gate realizations of such functions. The remaining sections of the chapter deal with general T-gate realizations.

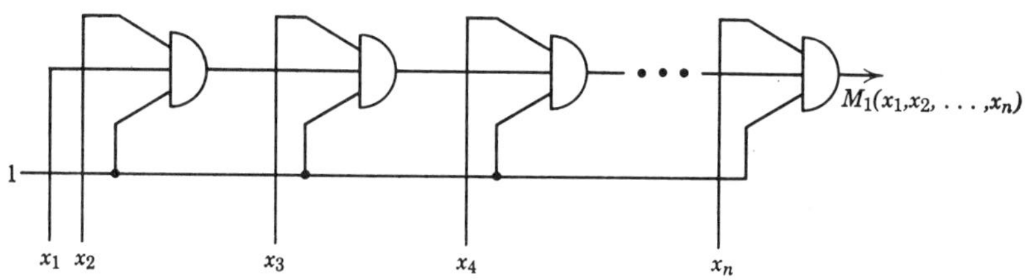

Fig. 6.12. M$_3$-gate realization of $M_1(x_1,x_2,\ldots,x_n)$.

Canonic M$_3$-gate network realizations of symmetric functions (Amarel, Cooke, and Winder) are similar to the conventional contact network realizations (Caldwell). Consider the network of Fig. 6.12. Here the half-circles represent M$_3$ gates where each has one input that is the constant 1 function, and the other two are variables. In all but the leftmost gate, one input is an input variable to the network, and the other is the output of the preceding gate. Observe that if any one of the input variables has unit value, the output of the corresponding gate is unity because of the constant 1-function input. Moreover all gates to the right of this one also have unit output. As a result the network realizes the switching function†

$$y = M_1(x_1,x_2,\ldots,x_n)$$

† Here the function is represented by the notation of Chapter 1 where $M_k(x_1,x_2,\ldots,x_n)$ denotes the function that has unit value when k or more inputs have unit value.

Consider now the M_3-gate network of Fig. 6.13, which consists of two layers of gates. The bottom layer is the same as the network of Fig. 6.12 and its output is $M_1(x_1,x_2,\ldots,x_n)$. The top layer is similar to the bottom layer (neglecting for a moment the leftmost gate) except that the third input of each gate is connected to the output of the preceding bottom-layer gate instead of to the constant 1 function. As a result, if any input variable has value 1, then the output of the top-layer gate for which it is an input will be 1 if and only if a variable to its left also has value 1. This is true because, if the variable to the left has value 1, then it causes its bottom-layer gate and all bottom-layer gates to its right to have output 1. Thus two

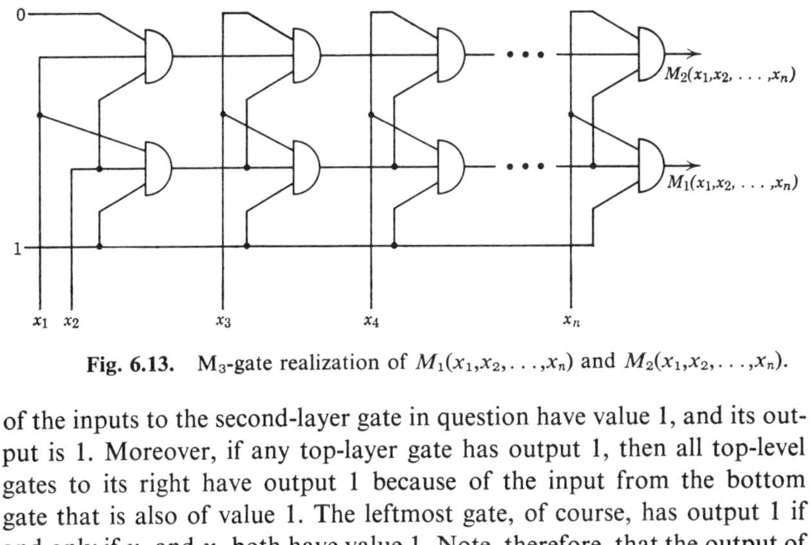

Fig. 6.13. M_3-gate realization of $M_1(x_1,x_2,\ldots,x_n)$ and $M_2(x_1,x_2,\ldots,x_n)$.

of the inputs to the second-layer gate in question have value 1, and its output is 1. Moreover, if any top-layer gate has output 1, then all top-level gates to its right have output 1 because of the input from the bottom gate that is also of value 1. The leftmost gate, of course, has output 1 if and only if x_1 and x_2 both have value 1. Note, therefore, that the output of the top layer is the function $M_2(x_1,x_2,\ldots,x_n)$.

Additional layers can be added in the same way so that the network realizes simultaneously the functions

$$M_1(x_1,x_2,\ldots,x_n),\ M_2(x_1,x_2,\ldots,x_n),\ \ldots,\ M_n(x_1,x_2,\ldots,x_n)$$

Fig. 6.14 shows the complete network for $n = 7$. In general, the number of gates required for the complete network is

$$1 + 2 + 3 + \cdots + (n-1) + (n-1) = \frac{n(n+1)}{2} - 1$$

Thus far we have shown how to realize all of the $M_k(x_1,x_2,\ldots,x_n)$ functions for $k = 1,2,\ldots,n$. We now show how to realize the symmetric functions that are not of the $M_k(x_1,x_2,\ldots,x_n)$ class.

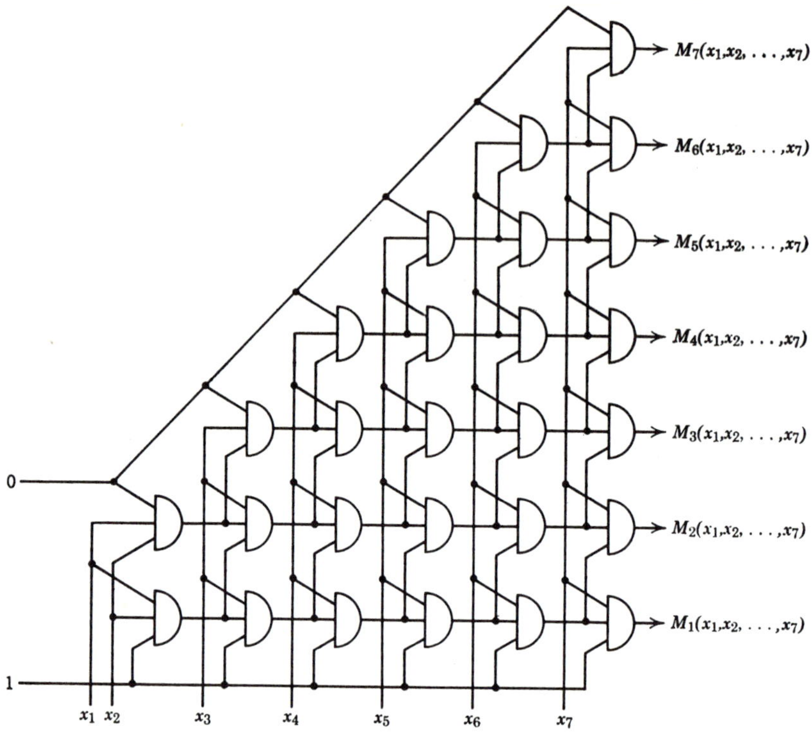

Fig. 6.14. Canonic M_3-gate networks for $n = 7$.

The procedure is to express the given symmetric function

$$S_{\alpha_1,\alpha_2,\ldots,\alpha_k}(x_1,x_2,\ldots,x_n)$$

as a Boolean function of the $M_k(x_1,x_2,\ldots,x_n)$ functions. For example,

$$\begin{aligned} S_{1,3,5,7}(x_1,x_2,\ldots,x_7) = {} & M_1(x_1,x_2,\ldots,x_7)\overline{M}_2(x_1,x_2,\ldots,x_7) \\ & \dotplus M_3(x_1,x_2,\ldots,x_7)\overline{M}_4(x_1,x_2,\ldots,x_7) \\ & \dotplus M_5(x_1,x_2,\ldots,x_7)\overline{M}_6(x_1,x_2,\ldots,x_7) \\ & \dotplus M_7(x_1,x_2,\ldots,x_7) \end{aligned}$$

Thus $S_{1,3,5,7}(x_1,x_2,\ldots,x_7)$ has value 1 if one or more AND NOT two or more of the variables have value 1, OR if three or more AND NOT four or more have value 1, OR if, etc. This type of function can be realized with the M_3-gate network of Fig. 6.15. The $*$ on the input leads denotes a complementation.

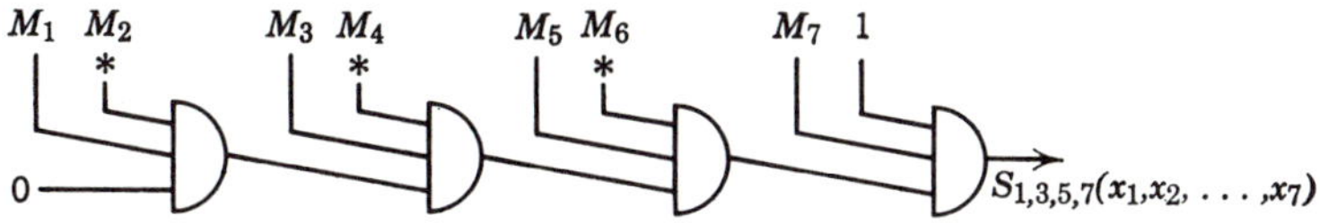

Fig. 6.15. M_3-gate realization of $S_{1,3,5,7}(x_1,x_2,\ldots,x_7)$.

The operation of this network is strongly dependent upon the functional containment relations

$$M_1 > M_2 > \cdots > M_7$$
$$\overline{M}_1 < \overline{M}_2 < \cdots < \overline{M}_7$$

Thus any input such that

$$M_1(x_1,x_2,\ldots,x_7) = \overline{M}_2(x_1,x_2,\ldots,x_7) = 1$$

implies

$$\overline{M}_4(x_1,x_2,\ldots,x_7) = \overline{M}_6(x_1,x_2,\ldots,x_7) = 1$$

Hence all gates in the network have output 1 for this input.

Connecting the network of Fig. 6.15 to the output of the network of Fig. 6.14 gives a realization of the odd parity function of seven variables $S_{1,3,5,7}(x_1,x_2,\ldots,x_7)$ that requries 31 M_3 gates. The even parity function $S_{0,2,4,6}(x_1,x_2,\ldots,x_7)$ can be obtained from the odd parity realization by the rules for complementing realizations, since

$$S_{0,2,4,6}(x_1,x_2,\ldots,x_7) = \overline{S}_{1,3,5,7}(x_1,x_2,\ldots,x_7)$$

Any other symmetric function can be similarly realized and in general will require fewer gates. For example, $S_{1,4}(x_1,x_2,\ldots,x_7)$ can be expressed as

$$S_{1,4}(x_1,x_2,\ldots,x_7) = M_1(x_1,x_2,\ldots,x_7)\overline{M}_2(x_1,x_2,\ldots,x_7)$$
$$\dotplus M_4(x_1,x_2,\ldots,x_7)\overline{M}_5(x_1,x_2,\ldots,x_7)$$

Hence only two M_3 gates are required to realize the output network. Moreover, only M_1, M_2, M_4, and M_5 are needed as inputs; therefore, only a part of the canonic network need be realized. The complete realization requires 25 M_3 gates and is shown in Fig. 6.16.

For an arbitrary symmetric function of n variables, a maximum of $(n + 1)/2$ M_3 gates is required† in the output network. An upper bound on the total number of M_3 gates that are needed to realize any n-variable symmetric function is

$$\frac{n(n+1)}{2} - 1 + \frac{n+1}{2} = \frac{(n+1)^2}{2} - 1$$

† Frequently a saving in gates can be achieved by realizing the complement function and then using the usual rules for complementing realizations.

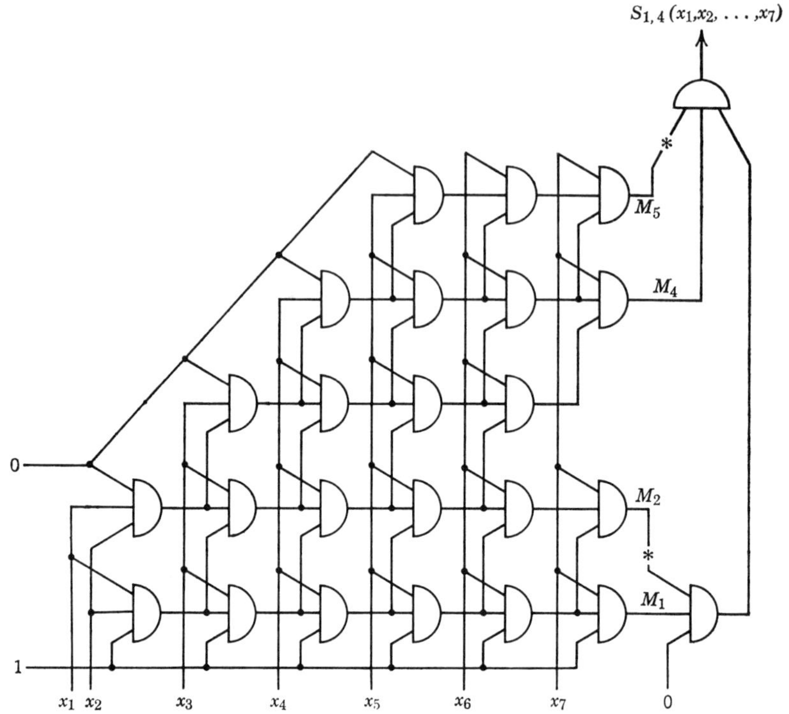

Fig. 6.16. Canonic M_3-gate network realization of $S_{1,4}(x_1,x_2,\ldots,x_7)$.

6.8 CANONIC T-GATE NETWORKS FOR SYMMETRIC FUNCTION REALIZATIONS

If general T gates rather than M_3 gates are used to realize symmetric functions, then fewer gates are required. This section considers two canonic T-gate realizations.

The first of these (Muroga, 10) is a direct realization of the expansion of the symmetric function in terms of $M_k(\,)$ functions. For example,

$$S_{1,4}(x_1,x_2,\ldots,x_7) = M_1(x_1,x_2,\ldots,x_7)\overline{M}_2(x_1,x_2,\ldots,x_7) + M_4(x_1,x_2,\ldots,x_7)\overline{M}_5(x_1,x_2,\ldots,x_7)$$

Both product terms are easily realized as shown in Fig. 6.17*a*. The notation $\{x\}_7$ denotes the 7 inputs, each with coefficient 1 and input variable x_j, $j = 1,2,\ldots,7$. In the realization of

$$M_1(x_1,x_2,\ldots,x_7)\overline{M}_2(x_1,x_2,\ldots,x_7)$$

the 1:0 gate will have output 1 if any one of the x_j variables has value 1,

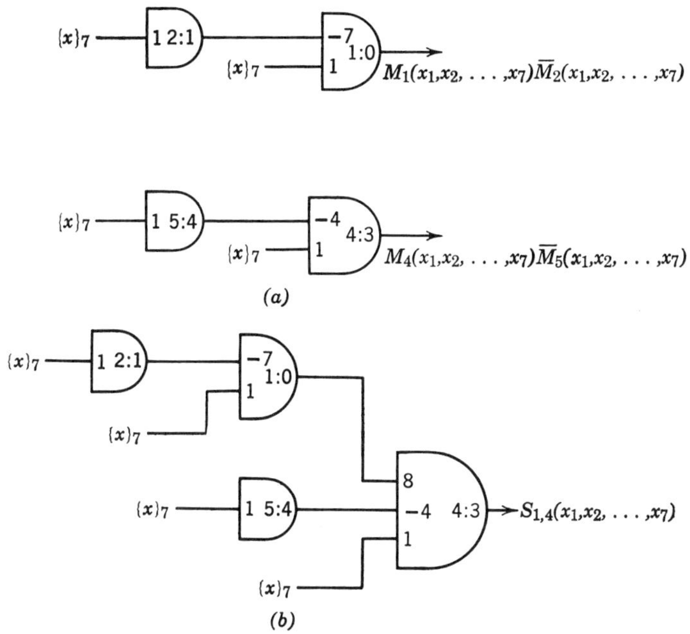

Fig. 6.17. First canonic T-gate realization of $S_{1,4}(x_1,x_2,\ldots,x_7)$. (*a*) Realization of the two product terms. (*b*) The T-gate realization.

provided two or more are not 1. In this latter case the 2:1 gate will have output 1, and the resultant input of -7 to the 1:0 gate is sufficient to keep its output 0 even if all seven x's have value 1. The realization of the other product term is similar.

The realizations of the two product terms are combined as shown in Fig. 6.17*b*. When the output of the 1:0 gate has value 1, then its coefficient of 8 units is sufficient to cause an output of 1 for the 4:3 gate regardless of the values of its other inputs. Negative coefficients are used in the realizations for the sake of simplicity. The usual theorems can be used to convert these to positive coefficients if desired.

An alternate canonic form for this realization is shown in Fig. 6.18. Again there is one gate for each $M_k(\,)$ function but this time they are arranged in only two levels. If one and only one of the x's has value 1, then the 1:0 gate is the only second-level gate with output 1. Since its coefficient in the final gate has value 4, the output of the 4:3 gate is 1. However, if exactly two x's have value 1, then the 2:1 gate has output 1, and its 4:3 gate coefficient of -4 cancels the value 4 from the 1:0 gate

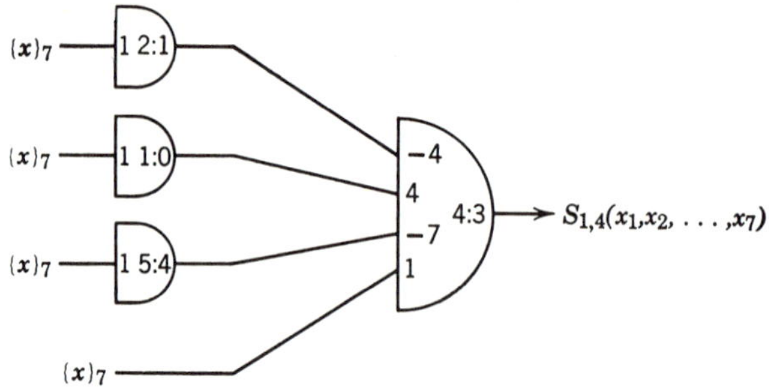

Fig. 6.18. An alternate canonic realization of $S_{1,4}(x_1,x_2,\ldots,x_7)$.

so that the 4:3 gate has output 0. The remainder of the function is realized as in the previous case.

Although the two canonic realization forms considered thus far were given only by example, the general procedure should be apparent without further explanation. Both forms require a number of gates equal to the number of $M_k(\ )$ functions that are required to express the given function. We now consider another form (Minnick, 1) that requires only about half as many gates. More precisely, it requires a number of gates one greater than the number of complemented $M_k(\ )$ functions that are required in the expression for the given function. For example,

$$S_{1,4}(x_1,x_2,\ldots,x_n) = M_1(x_1,x_2,\ldots,x_7)\overline{M}_2(x_1,x_2,\ldots,x_7) + M_4(x_1,x_2,\ldots,x_7)\overline{M}_5(x_1,x_2,\ldots,x_7)$$

has two $\overline{M}_k(\ )$ functions in its expression ($\overline{M}_2(\ )$ and $\overline{M}_5(\ )$) and hence three gates are required for the realization.

The canonic realization for $S_{1,4}(x_1,x_2,\ldots,x_7)$ is shown in Fig. 6.19. The second-level gates realize the $M_k(\ )$ functions that are complemented in the expression for the $S(\ )$ function. The first-level gate has its upper gap boundary equal to α_1 where α_1 is the smallest α_j in the $S_{\alpha_1,\alpha_2,\ldots,\alpha_k}(\ )$ function that is being realized

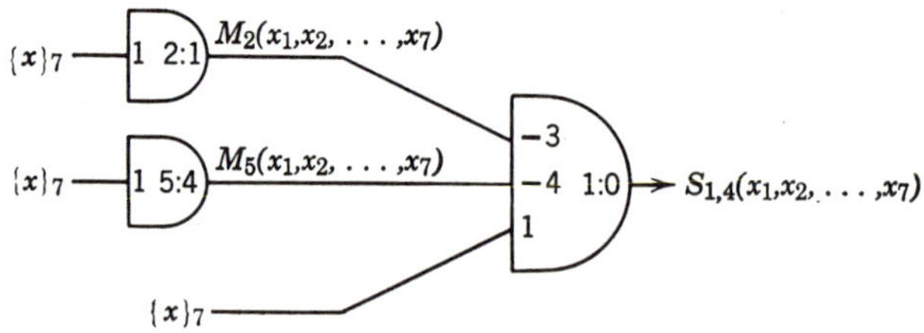

Fig. 6.19. The second canonic realization of $S_{1,4}(x_1,x_2,\ldots,x_7)$.

If exactly one of the x's has value 1, then both second-level gates have output 0, and the first-level gate has output 1. However, as soon as at least two x's have value 1, the 2:1 gate has output 1. Since it drives a coefficient of -3, the first-level gate has an effective gap of 4:3. As a result the first-level gate has output 0 until four of the x's have value 1, in which case its output becomes 1 again. Thus when exactly one or four of the x's have value 1, the first-level or output gate has output 1, but when two or three of the x's have value 1, it has output 0. When five or more x's have value 1, the 5:4 gate as well as the 2:1 gate has output 1. Since collectively the sum of the coefficients which they drive is -7, then the first level gate has output 0.

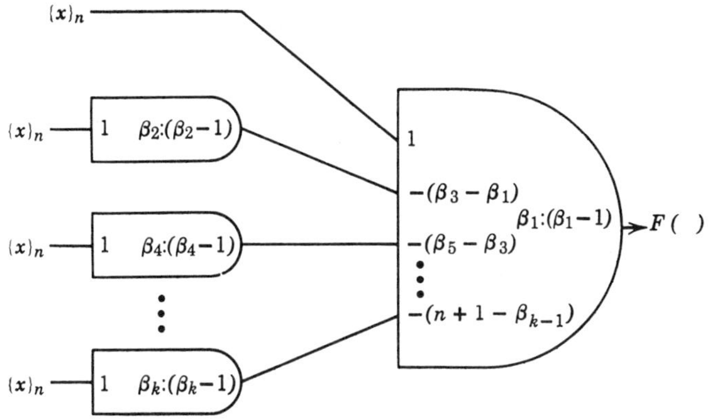

Fig. 6.20. Another general canonic realization.

The general form of the canonic realization is obtained as follows. Express the given function as

$$F(x_1,x_2,\ldots,x_n) = M_{\beta_1}(\ )\overline{M}_{\beta_2}(\ ) + M_{\beta_3}(\ )\overline{M}_{\beta_4}(\ ) + \cdots + M_{\beta_{k-1}}(\ )\overline{M}_{\beta_k}(\ )$$

where

$$\beta_1 < \beta_2 < \cdots < \beta_k$$

Observe that the β subscripts do not have the same meaning as the α subscripts defined earlier. The subscript β_1 indicates the smallest number of unit-valued inputs for which $F = 1$; β_2 indicates the smallest number of unit-valued inputs, greater than β_1, for which $F = 0$; β_3 indicates the smallest number of unit-valued inputs greater than β_2 for which $F = 1$; etc. (If $\beta_k = n + 1$, that is, if $\overline{M}_{\beta_k}(\ )$ is the constant 1 function, it can be omitted.)

The canonic realization for this function is shown in Fig. 6.20. The final gate of the network has gap $\beta_1:(\beta_1 - 1)$. Each of the variables $x_1,x_2, \ldots,x_n$ is an input to the final gate, and all have coefficient 1. In addition, there is an input corresponding to each $\overline{M}_{\beta_{2j}}(\,)$ in the above expression, and each is driven with a gate that realizes $M_{\beta_{2j}}(\,)$ with corresponding coefficient value $-(\beta_{2j+1} - \beta_{2j-1})$. (The coefficient for M_{β_k} is computed assuming $\beta_{k+1} = n + 1$.)

Observe that this canonic form has exactly two logic levels. This is quite important for those applications in which high-speed operation is desired.

This canonic form gives an upper bound on the number of T gates that are required to realize symmetric functions. If k is the number of $\overline{M}_k(\,)$ functions that are needed to express the symmetric function, then the number of T gates that are needed for realization is given by the expression†

$$r \leq k + 1$$

A somewhat looser bound is obtained by observing that

$$k \leq \frac{n}{2}$$

When n is an even number, the function for which k is the largest is the odd parity function, for example

$$S_{1,3}(x_1,x_2,x_3,x_4) = M_1(x_1,x_2,x_3,x_4)\overline{M}_2(x_1,x_2,x_3,x_4) \dotplus M_3(x_1,x_2,x_3,x_4)\overline{M}_4(x_1,x_2,x_3,x_4)$$

and hence

$$k = \frac{n}{2} = 2$$

When n is an odd number, the function for which k is the largest is the even parity function, for example

$$S_{0,2}(x_1,x_2,x_3) = M_0(x_1,x_2,x_3)\overline{M}_1(x_1,x_2,x_3) \dotplus M_2(x_1,x_2,x_3)\overline{M}_3(x_1,x_2,x_3)$$

In this case one can realize the complement function (which is the odd parity function)

$$S_{1,3}(x_1,x_2,x_3) = M_1(x_1,x_2,x_3)\overline{M}_2(x_1,x_2,x_3) \dotplus M_3(x_1,x_2,x_3)$$

for which

$$k = \frac{n}{2} - 1 = 1$$

† Again, a saving can frequently be achieved by realizing the complement function and then converting the network by the usual rules for complementing realizations.

and then use the usual rules for complementing a realization to obtain the original function.

This shows that

$$r \leq \left[\frac{n}{2}\right] + 1$$

where $[n/2]$ denotes the largest integer less than or equal to $n/2$.

6.9 CANONIC T-GATE REALIZATIONS FOR THE PARITY FUNCTIONS

Many symmetric functions can be realized with significantly fewer gates than the upper bound of Section 6.8. A particularly important class in this regard is the parity functions for which a canonic realization (due to Kautz) requires a number of gates

$$r = 1 + \lceil \log_2 n \rceil$$

which can be shown to be minimal.

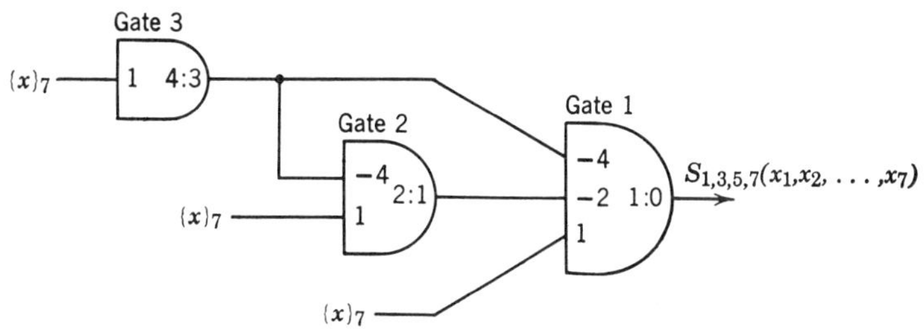

Fig. 6.21. The canonic realization for the odd 7-variable parity function.

Figure 6.21 shows the canonic realization for the 7-variable parity function $S_{1,3,5,7}(x_1,x_2,\ldots,x_7)$ which can be written as

$$S_{1,3,5,7}(\,) = M_1(\,)\overline{M}_2(\,) + M_3(\,)\overline{M}_4(\,) + M_5(\,)\overline{M}_6(\,) + M_7(\,)$$

When exactly one of the x's has value 1, Gates 2 and 3 have output 0 and Gate 1 has output 1. When exactly two of the x's have value 1, Gate 2 has output 1, which in effect sets the threshold of Gate 1 to 3. Thus for exactly two x's with value 1, Gate 1 has output 0, and for exactly three x's with value 1, the output is 1. So far this is just like the previous canonic form. When exactly four x's have value 1, Gate 3 has output 1, and this output does double duty. First it causes Gate 2 to have output 0 and in effect it causes the threshold of Gate 1 to be 5. Hence, for four x's with value 1, Gate 1 has output 0, and for five x's with value 1 it has output 1.

The reduction in the number of gates with this form comes from the double duty of Gate 2. When Gate 3 has output 1 it, in effect, sets the threshold of Gate 2 to 6. Thus when six x's have value 1, Gate 2 has output 1, and this together with Gate 3, in effect, sets the threshold of Gate 1 to 7. Therefore Gate 2 performs the functions that would have been performed by a 2:1 and a 6:5 gate in the previous canonic form.

To study this realization further, Fig. 6.22*a* (Sheng, 2) shows the sum

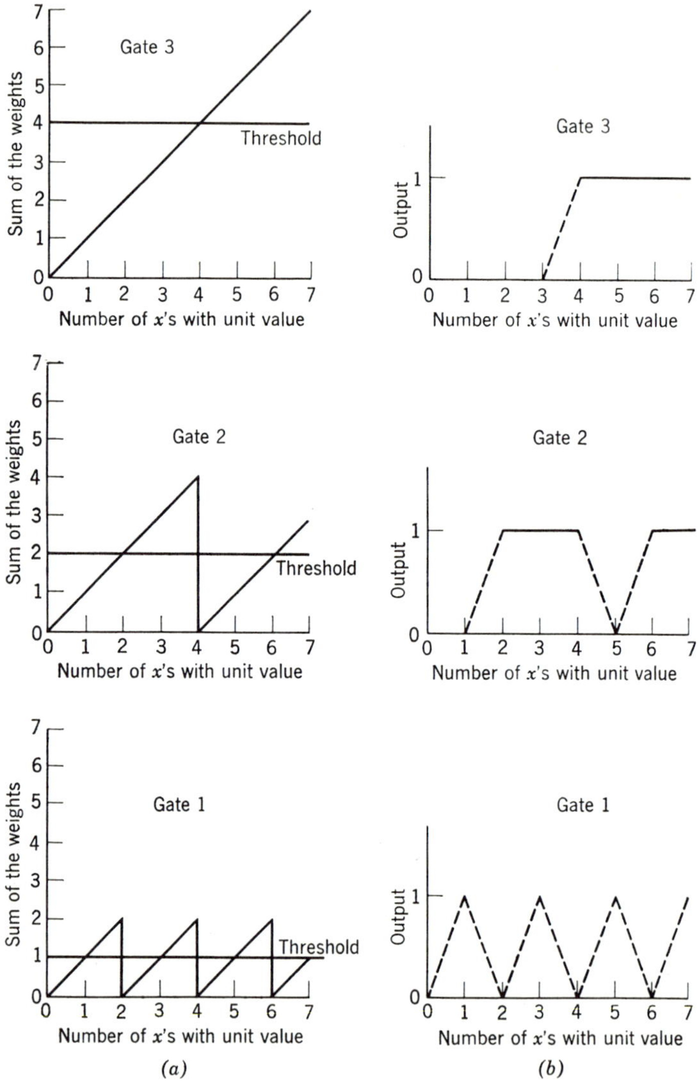

Fig. 6.22. A further study of the gates of Fig. 6.21. (*a*) Sum of the weights. (*b*) Gate output functions.

of the weights of these three gates as a function of the number of x's with value 1, and Fig. 6.22*b* (Kautz) shows their outputs as a similar function of the number of unit valued x's. The output of Gate 3 changes from 0 to 1 when the number of x's with value 1 changes from three to four. Gate 2 behaves similarly with threshold 2 as long as the number of x's with value 1 is less than four. However, when Gate 3 has output 1, the sum of the weights of Gate 2 abruptly decreases by 4, thus increasing the apparent threshold for x inputs to $2 + 4 = 6$. Thus Gate 2 has two positive transitions as the number of x's with value 1 increases, where a positive transition means the gate changes from output 0 to output 1. In Gate 1 the number of positive transitions is double that of Gate 2. The effective threshold of Gate 1 changes from 1 to 3 to 5 and finally to 7 as Gates 2 and 3 have output values shown in Fig. 6.22.

For parity functions of more variables, it is clear that this process can be continued to achieve a doubling of the number of positive transitions of each gate that is added, provided the number of x's is sufficient. In general the output of the rth gate will have 2^{r-1} positive transitions and $2^{r-1} - 1$ negative transitions or a total of $2^r - 1$ transitions of both types providing the coefficients and threshold are selected correctly.

The parity function requires exactly n transitions; therefore the number of gates required is bounded by

$$2^r - 1 \geq n$$

or

$$r \geq \log_2 (n + 1)$$

Since r must be an integer, then

$$\begin{aligned} r &= \lfloor \log_2 (n + 1) \rfloor \\ &= 1 + \lceil \log_2 n \rceil \end{aligned}$$

where $\lceil K \rceil$ and $\lfloor K \rfloor$ denote, respectively, the largest integer equal to or less than K and the smallest integer equal to or greater than K.

It is readily verified that the form of the realization of the parity function of n variables is as given by Fig. 6.23. The kth gate from the output gate has gap $2^k : (2^k - 1)$ and its output drives an input for each of the gates $k - 1, k - 2, \ldots, 1$ and each of these inputs has coefficient -2^k. It is interesting to observe that if the outputs of the gates are ordered from top to bottom as shown in Fig. 6.23, they form the base-two representation of the number of x's that have value 1.

This canonic network realizes the parity function with the minimum number of gates for a feedforward (nonfeedback) network. Irrespective of the arrangement of a feedforward network of r gates, the $r - 1$ gates

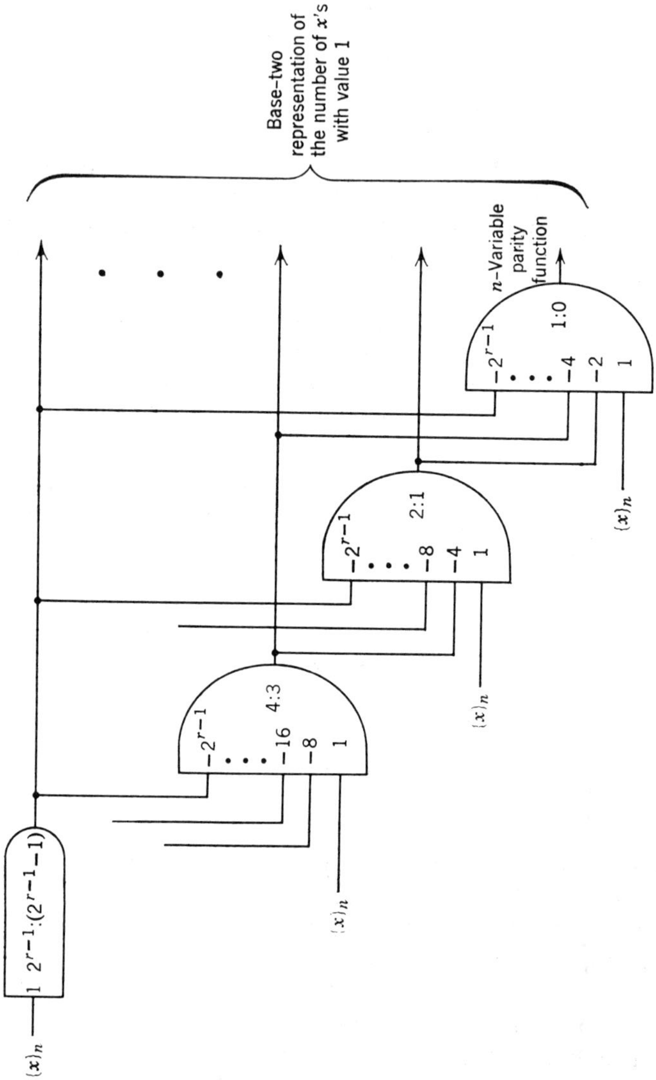

Fig. 6.23. Canonic realization of the n-variable odd parity function.

that drive the final gate can have at most 2^{r-1} different output combinations that can change the effective threshold of the final gate. Each of these effective threshold changes can produce no more than one positive transition. Thus a network of r gates can have at most 2^{r-1} positive transitions. But this is just the number in the network of Fig. 6.23; hence no feed-forward network can realize the parity function with fewer T gates.

6.10 ANOTHER BOUND ON THE NUMBER OF REQUIRED T GATES

The reasoning of the previous section can be carried one step further. Since a network with r gates can realize at most 2^{r-1} transitions in one direction and $2^{r-1} - 1$ transitions in the other direction, or $2^r - 1$ transitions in all, a bound on r to realize a given symmetric switching function of n variables that has τ transitions is

$$2^{r-1} \geq \tau$$

or

$$r \geq 1 + \lceil \log_2 \tau \rceil$$

Recall the bound of Section 6.8 that was based upon the canonic realization of Fig. 6.20. In particular it depended upon the number of $\overline{M}_k(\ )$ functions in the M_k-function expression for the given symmetric function. Clearly the gates that correspond to the $\overline{M}_k(\ )$ functions in that formulation produce what are equivalent to the negative transitions here. By realizing either the given function or its complement, the number of negative transitions can always be made less than or equal to $\lceil \tau/2 \rceil$. Thus the bound on the number of gates r to realize a symmetric function with τ transitions is

$$1 + \lceil \log_2 \tau \rceil \leq r \leq 1 + \lceil \tau/2 \rceil$$

We have just shown that the lower bound is $1 + \lceil \log_2 \tau \rceil$ and we showed previously that the second canonic form of Section 6.8 always can yield a realization with $1 + \lceil \tau/2 \rceil$ gates.

For $\tau \leq 5$, the upper and lower bounds are equal; hence the second canonic realization of Section 6.8 is minimal. For $6 \leq \tau \leq 9$ the two bounds differ by one, and for $10 \leq \tau \leqslant 11$ they differ by two. Thus, for problems where the number of transitions is less than 12, the second canonic realization of Section 6.8 is quite efficient.

In addition, as noted earlier, the canonic realization of Section 6.8 has only two logic levels, which results in high-speed operation. In contrast to this, the networks of Section 6.9 may have as many as $1 + \lceil \log_2 n \rceil$ levels which will cause slower operation.†

6.11 SUMMARY: SYNTHESIS OF SYMMETRIC FUNCTIONS

Most of the discussion of symmetric function synthesis has been in terms of canonic networks, that is, certain fixed configurations with which any function can be realized. In certain cases the number of gates required is minimal. However, we have not considered the sensitivity of the realizations to signal and parameter fluctuations. With the exception of the M_3-gate synthesis, virtually all of the canonic forms require large fan-ins and have small relative gaps when all of the coefficients are made positive. (Even if the coefficients are allowed to remain negative, it is shown in Appendix A that the allowable fluctuations are small.)

If sensitivity is an important factor, as is often the case, then the M_3-gate synthesis is the most useful realization procedure presented in the chapter even though it requires on the order of $n^2/2$ rather than $n/2$ or $\log_2 n$ gates. Another attribute of M_3-gate canonic realizations is their extensive regularity, which in turn makes them particularly useful for mass fabrication techniques. Other M_3-gate realizations are given in the next chapter.

Naturally, one is not limited to canonic structures. It would be possible to derive special synthesis techniques for symmetric functions in which sensitivity is controlled. At present no such results exist, although the works of Kautz and of Sheng (which include more results than discussed here) appear the most promising in this regard.

† Speed is not always dependent only on the total number of logic levels. In those implementations (such as parametrons) in which the logic signals are clocked from level to level by an auxiliary clock pulse, the speed is determined by the depth of the right-most level of the network to which independent variable inputs are applied (assuming the network is drawn with the input levels on the left). This depth determines the number of clock pulses for which the values of the input variables must be maintained in order that the network output will be correct at the (later) clock time when the effect of the leftmost input reaches the output gate. For example if the network of Fig. 6.17*b* were implemented with clocked logic, the x signals would have to be maintained for three clock pulses, while the logic signals propagated through the three levels. However if there had been no x inputs to the rightmost gate, then the x inputs would have to be maintained for only two clock pulses and then the correct signals would be available to the rightmost gate. During the third clock pulse the rightmost gate would give the correct network output and at the same time the x signals could be changed in preparation for the next use of that network.

However even in the case of clocked logic, the networks of Section 6.8 are considerably faster than those of Section 6.9, since the latter require independent variable inputs to all levels.

Chapter 7

DUALITY AND RELATED T-GATE REALIZATIONS

7.1 INTRODUCTION

Previous discussions and examples have pointed out the close relationship between the T-gate realizations of a switching function and of its dual. In this chapter we consider in detail the properties of dual functions and their application to the realization of self-dual functions. This in turn provides other realization procedures for arbitrary switching functions by means of general T gates or M_3 gates.

The basic definitions and properties of dual and self-dual functions are set forth in Section 2.6. Let $F(x_1,x_2,\ldots,x_n)$ be an arbitrary switching function and $F^d(x_1,x_2,\ldots,x_n)$ be the corresponding dual function. Then F and F^d are *dual comparable* if

$$F \geq F^d$$

or

$$F^d \geq F$$

The "larger" of a dual comparable pair is called the *major function* and the "smaller" the *minor function* (Muroga, 2). According to Section 2.7, a dual comparable pair is self-dual when

$$F = F^d$$

We frequently denote a self-dual function by the Greek letter Φ.

Example 7.1.

(1) Let

$$y = F(x_1,x_2) = x_1x_2 \dotplus \bar{x}_1\bar{x}_2$$

Then

$$y^d = F^d(x_1,x_2) = x_1\bar{x}_2 + \bar{x}_1x_2$$

For the combination of values $x_1 = x_2 = 1$,

$$y = 1 \quad \text{and} \quad y^d = 0$$

while for $x_1 = 1$ and $x_2 = 0$,

$$y = 0 \quad \text{and} \quad y^d = 1$$

Hence F and F^d are not dual comparable.

(2) Let

$$y = F(x_1,x_2,x_3) = x_1 \dotplus x_2x_3$$

Then

$$y^d = F^d(x_1,x_2,x_3) = x_1(x_2 \dotplus x_3)$$

Clearly $y^d = 1$ implies $y = 1$; hence F and F^d are dual comparable. Moreover $F > F^d$; therefore F is the major function and F^d is the minor.

(3) Let

$$y = F(x_1,x_2,x_3,x_4) = x_1(x_2 \dotplus x_3 \dotplus x_4) \dotplus x_2x_3x_4$$

The corresponding dual function is

$$\begin{aligned} y^d = F^d(x_1,x_2,x_3,x_4) &= (x_1 \dotplus x_2x_3x_4)(x_2 \dotplus x_3 \dotplus x_4) \\ &= x_1(x_2 \dotplus x_3 \dotplus x_4) \dotplus x_2x_3x_4 = y \end{aligned}$$

Therefore F is a self-dual function.

The concepts of self-dual, major, and minor functions have an interesting interpretation in terms of points in the switching space. Consider first a self-dual function Φ. From the definition of duality, it follows that for each point p

$$\Phi^d(p) = \overline{\Phi}(\bar{p})$$

and because Φ is self-dual,

$$\Phi(p) = \Phi^d(p) = \overline{\Phi}(\bar{p})$$

Thus to each point p such that

$$\Phi(p) = \Phi^d(p) = 1$$

there corresponds the complement point $\bar{p}$ such that

$$\Phi(\bar{p}) = \Phi^d(\bar{p}) = 0$$

Hence, for a self-dual function, exactly one-half of the points are 1 points and the other half are 0 points, where the 0 points are complements of the 1 points and conversely. This is indicated diagramatically in Fig. 7.1*a*.

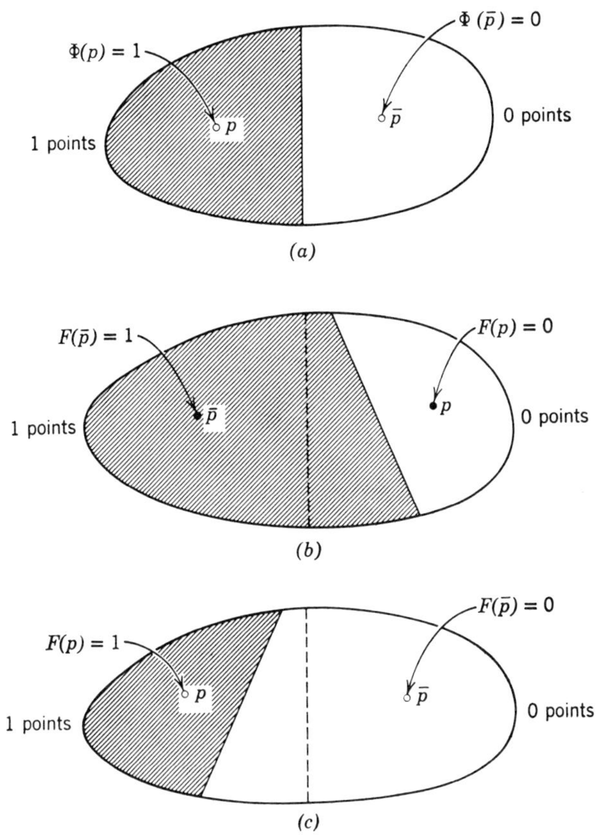

Fig. 7.1. Subsets of the switching space points for dual comparable functions. (*a*) For self-dual functions. (*b*) For major functions. (*c*) For minor functions.

Now let F be a major function. Then from the definition of *duality*,

$$F^d(p) = \bar{F}(\bar{p})$$

for each p and since F is major,†

$$F(p) \geq F^d(p) = \bar{F}(\bar{p})$$

Thus to each point p for which

$$F(p) = 0$$

there corresponds the complement point $\bar{p}$ for which

$$F(\bar{p}) = 1$$

† By the notation $F(p) \geq F^d(p)$ we mean $F \geq F^d$.

If F is not self-dual and the above containments are therefore proper, then there are other points p such that

$$F(p) = F(\bar{p}) = 1$$

Hence, for a (nonself-dual) major function, more than one-half of the points are 1 points and less than one-half are 0 points, where to each 0 point, p there corresponds its complement point $\bar{p}$, which is a 1 point. This is indicated in Fig. 7.1*b*.

By duality, then, for a (nonself-dual) minor function, less than one-half of the points are 1 points, and to each 1 point p there corresponds its complement point $\bar{p}$, which is a 0 point. This is indicated in Fig. 7.1*c*.

In essence, therefore, a major function can be represented as the sum of a self-dual function and a minor function. Stated another way, a major function is obtained from a self-dual function by enlarging the subset of points for which the self-dual function has unit value. A minor function is obtained from a self-dual function by reducing the unit subset. This is indicated diagrammatically in Fig. 7.1 and considered more precisely in the next section.

7.2 PROPERTIES OF DUAL COMPARABLE AND SELF-DUAL FUNCTIONS

In this section we derive a number of properties of dual comparable and self-dual functions. Our interest in these classes of functions stems from the fact that they have some highly restricted Boolean representations which lead to simple realizations. For example, consider a switching function F factored so that

$$y = F(x_1,x_2,\ldots,x_n) = x_1F_1(x_2,x_3,\ldots,x_n) + \bar{x}_1F_0(x_2,x_3,\ldots,x_n)$$

The dual of F can be written as

$$y^d = F^d(\,) = [x_1 + F_1{}^d(\,)][\bar{x}_1 + F_0{}^d(\,)]$$

which can be simplified to

$$F^d(\,) = x_1F_0{}^d(\,) + \bar{x}_1F_1{}^d(\,) + F_1{}^d(\,)F_0{}^d(\,)$$

Since the last term is contained in the other two, it follows that

$$F^d(\,) = x_1F_0{}^d(\,) + \bar{x}_1F_1{}^d(\,)$$

If F and F^d are comparable, then there are some corresponding restrictions on the comparability of the reduced functions F_1 and F_0. In particular, if F is self-dual, we have the following theorem.

Theorem 7.1. A switching function of the form

$$y = F(x_1,x_2,\ldots,x_n) = x_nF_1(x_1,x_2,\ldots,x_{n-1}) + \bar{x}_nF_0(x_1,x_2,\ldots,x_{n-1})$$

is self-dual if and only if $F_0 = F_1{}^d$.

This is the first of the special Boolean representations that can be used for all self-dual functions. Three such representations will be derived in this section and then used in later sections as a basis for synthesis procedures.

Example 7.2. The following switching functions are self-dual:

$$y_1 = x_1x_2x_3 + (x_1 + x_2 + x_3)x_4$$
$$y_2 = x_1(x_2x_3 + \bar{x}_2\bar{x}_3) + \bar{x}_1(x_2\bar{x}_3 + \bar{x}_2x_3)$$

The first function can be written in the form

$$y_1 = x_1x_2x_3\bar{x}_4 + (x_1 + x_2 + x_3)x_4$$

and since $x_1x_2x_3$ is the dual of $x_1 + x_2 + x_3$, this function is of the form of Theorem 7.1. For the second function $(x_2\bar{x}_3 + \bar{x}_2x_3)$ is the dual of $(x_2x_3 + \bar{x}_2\bar{x}_3)$; hence it too is of the form of Theorem 7.1.

The other special representations are somewhat similar but require more understanding of the properties of dual comparable functions. In particular we shall be interested in logical sums and products of dual comparable functions, and the following two simple properties will be quite useful.

Theorem 7.2.

A. A switching function F is minor if and only if

$$F(x_1,x_2,\ldots,x_n) = F(x_1,x_2,\ldots,x_n)F^d(x_1,x_2,\ldots,x_n)$$

B. If F is a minor function, then for any decomposition such that

$$F(x_1,x_2,\ldots,x_n) = F_a(x_1,x_2,\ldots,x_n) + F_b(x_1,x_2,\ldots,x_n)$$

F_a and F_b are minor functions.

Proof.

A. If F is minor then $F^d \geq F$ from which the "only-if" result follows. If

$$F(x_1,x_2,\ldots,x_n) = F(x_1,x_2,\ldots,x_n)F^d(x_1,x_2,\ldots,x_n)$$

then it follows immediately that

$$F^d \geq F$$

and that F is therefore minor.

B. If

$$F(x_1,x_2,\ldots,x_n) = F_a(x_1,x_2,\ldots,x_n) + F_b(x_1,x_2,\ldots,x_n)$$

then, since F is minor,

$$F_a(x_1,x_2,\ldots,x_n) + F_b(x_1,x_2,\ldots,x_n) \le F_a{}^d(x_1,x_2,\ldots,x_n)F_b{}^d(x_1,x_2,\ldots,x_n)$$

This immediately implies that

$$F_a(x_1,x_2,\ldots,x_n) \le F_a{}^d(x_1,x_2,\ldots,x_n)F_b{}^d(x_1,x_2,\ldots,x_n) \le F_a{}^d(x_1,x_2,\ldots,x_n)$$

that

$$F_b(x_1,x_2,\ldots,x_n) \le F_a{}^d(x_1,x_2,\ldots,x_n)F_b{}^d(x_1,x_2,\ldots,x_n) \le F_b{}^d(x_1,x_2,\ldots,x_n)$$

and hence that F_a and F_b are minor functions.|

Although our interest is primarily in special Boolean representations for self-dual functions, we are also interested in similar representations for major and minor functions. Theorem 7.3 is concerned with such a representation for major functions. The dual of the theorem gives a corresponding representation for minor functions.

Theorem 7.3. (*Muroga et al., 1*). A switching function F that can be factored so that

$$F(x_1,x_2,\ldots,x_n) = F_a(x_1,x_2,\ldots,x_n) + F_a{}^d(x_1,x_2,\ldots,x_n)\Phi(x_1,x_2,\ldots,x_n)$$

where Φ is a self-dual function, is a major function and conversely.

Proof. If

$$F(x_1,x_2,\ldots,x_n) = F_a(x_1,x_2,\ldots,x_n) + F_a{}^d(x_1,x_2,\ldots,x_n)\Phi(x_1,x_2,\ldots,x_n)$$

then the dual of F is

$$F^d(x_1,x_2,\ldots,x_n) = F_a{}^d(x_1,x_2,\ldots,x_n)[F_a(x_1,x_2,\ldots,x_n) + \Phi(x_1,x_2,\ldots,x_n)]$$

from which it is evident that

$$F \ge F^d$$

Conversely, let F be a major function and let $\Phi(x_1,x_2,\ldots,x_n)$ be any self-dual function. Since F is major, then by the dual of Theorem 7.2A

$$F(x_1,x_2,\ldots,x_n) = F(x_1,x_2,\ldots,x_n) + F^d(x_1,x_2,\ldots,x_n)$$

Hence, since F^d is a minor function

$$F(x_1,x_2,\ldots,x_n) = F(x_1,x_2,\ldots,x_n) + F^d(x_1,x_2,\ldots,x_n)\Phi(x_1,x_2,\ldots,x_n) \quad |$$

Example 7.3. The following are both major functions where the first is also self-dual.

(1) $y_1 = x_1x_2 \dotplus (x_1 \dotplus x_2)x_3$

(2) $y_2 = (x_1x_2 \dotplus \bar{x}_1\bar{x}_2) \dotplus (x_1\bar{x}_2 \dotplus \bar{x}_1x_2)(x_3x_4 \dotplus x_3x_5 \dotplus x_4x_5)$

The first function is of the form

$$y_1 = F_a(x_1,x_2,x_3) \dotplus F_a{}^d(x_1,x_2,x_3)\Phi(x_1,x_2,x_3)$$

where

$$F_a(x_1,x_2,x_3) = x_1x_2 \quad \text{and} \quad \Phi(x_1,x_2,x_3) = x_3$$

The second function is of the same form; that is,

$$y_2 = F_a(x_1,x_2,\ldots,x_5) \dotplus F_a{}^d(x_1,x_2,\ldots,x_5)\Phi(x_1,x_2,\ldots,x_5)$$

where

$$F_a(x_1,x_2,\ldots,x_5) = x_1x_2 \dotplus \bar{x}_1\bar{x}_2$$

and

$$\Phi(x_1,x_2,\ldots,x_5) = x_3x_4 \dotplus x_3x_5 \dotplus x_4x_5$$

Theorem 7.3 gives a Boolean expression that characterizes major functions. Since major functions include the self-dual class, the latter can also be characterized in terms of this same major-function characterization. This is set forth in the following corollary. The validity follows from the proof of Theorem 7.3 and Theorem 7.2A.

Corollary 7.3.1. A Boolean function F of the form

$$F(x_1,x_2,\ldots,x_n) = F_a(x_1,x_2,\ldots,x_n) \dotplus F_a{}^d(x_1,x_2,\ldots,x_n)\Phi(x_1,x_2,\ldots,x_n)$$

where Φ is a self-dual function, is self-dual if and only if F_a is a minor function.

The wording of Corollary 7.3.1 can be changed slightly as indicated in the following corollary.

Corollary 7.3.2. Let F_a be a minor function and let Φ_0 be an arbitrarily selected self-dual function. Then the function Φ, where

$$\Phi(x_1,x_2,\ldots,x_n) = F_a(x_1,x_2,\ldots,x_n) \dotplus F_a{}^d(x_1,x_2,\ldots,x_n)\Phi_0(x_1,x_2,\ldots,x_n)$$

is self-dual and

$$F_a{}^d \geq \Phi \geq F_a$$

This corollary justifies the previous statements: (1) that a major function is obtained from a self-dual function by enlarging the subset of points for which the self-dual function has unit value, and (2) that a minor function is obtained from a self-dual function by reducing the unit set or enlarging the zero set.

There is an even stronger characterization of self-dual functions than that given by Corollary 7.3.1. The following theorem and corollary comprise the most general statement of the second Boolean form in which all self-dual functions can be expressed. (The first is given by Theorem 7.1.)

Theorem 7.4. If a self-dual function Φ is arbitrarily decomposed into the sum

$$\Phi(x_1,x_2,\ldots,x_n) = F_a(x_1,x_2,\ldots,x_n) \dotplus F_b(x_1,x_2,\ldots,x_n)$$

then Φ can be expressed in the form

$$\Phi(x_1,x_2,\ldots,x_n) = F_a(x_1,x_2,\ldots,x_n) \dotplus F_a{}^d(x_1,x_2,\ldots,x_n)F_b(x_1,x_2,\ldots,x_n)$$

Moreover there exists a self-dual function $\Phi_0 \neq \Phi$ (except when $F_a \equiv 0$ or $\Phi = x_i$) such that

$$\Phi(x_1,x_2,\ldots,x_n) = F_a(x_1,x_2,\ldots,x_n) \dotplus F_a{}^d(x_1,x_2,\ldots,x_n)\Phi_0(x_1,x_2,\ldots,x_n)$$

One such Φ_0 is

$$\Phi_0(x_1,x_2,\ldots,x_n) = F_b(x_1,x_2,\ldots,x_n) \dotplus F_b{}^d(x_1,x_2,\ldots,x_n)\Phi_1(x_1,x_2,\ldots,x_n)$$

where Φ_1 is an arbitrary self-dual function.

Proof. According to Theorem 7.2, F_a and F_b are minor functions. Since Φ is self-dual, then according to the proof of Theorem 7.2, $F_a{}^d > F_b$, from which the first part of Theorem 7.4 follows.

For the second part, Φ_0 can be selected in a number of ways; however, one is sufficient to establish validity. According to Theorem 7.2, F_b is a minor function; hence for any self-dual Φ_1,

$$\Phi_0(x_1,x_2,\ldots,x_n) = F_b(x_1,x_2,\ldots,x_n) \dotplus F_b{}^d(x_1,x_2,\ldots,x_n)\Phi_1(x_1,x_2,\ldots,x_n)$$

is a self-dual function according to Corollary 7.3.1. Substituting this into the desired expression for Φ gives

$$\begin{aligned} F_a(\,) \dotplus F_a{}^d(\,)\Phi_0(\,) &= F_a(\,) \dotplus F_a{}^d(\,)[F_b(\,) \dotplus F_b{}^d(\,)\Phi_1(\,)] \\ &= F_a(\,) \dotplus F_a{}^d(\,)F_b(\,) \dotplus F_a{}^d(\,)F_b{}^d(\,)\Phi_1(\,) \\ &= F_a(\,) \dotplus F_b(\,) \dotplus F_a{}^d(\,)F_b{}^d(\,)\Phi_1(\,) \end{aligned}$$

because $F_a{}^d > F_b$. However, since Φ is self-dual, then

$$F_a(\,) \dotplus F_b(\,) = F_a{}^d(\,)F_b{}^d(\,)$$

so that the above expression can be written as

$$F_a(\,) \dotplus F_a{}^d(\,)\Phi_0(\,) = \Phi(\,) \dotplus \Phi(\,)\Phi_1(\,) = \Phi(\,)$$

This completes the proof.▌

Example 7.4. Consider the self-dual function Φ where

$$\begin{aligned}\Phi(x_1,x_2,x_3,x_4) &= x_1(x_2 \dotplus x_3 \dotplus x_4) \dotplus x_2x_3x_4 \\ &= x_1x_2 \dotplus x_1x_3 \dotplus x_1x_4 \dotplus x_2x_3x_4\end{aligned}$$

Let F_a be one of the terms

$$F_a(x_1,x_2,x_3,x_4) = x_1x_2$$

and

$$F_b(x_1,x_2,x_3,x_4) = x_1x_3 \dotplus x_1x_4 \dotplus x_2x_3x_4$$

Then

$$F_a{}^d(x_1,x_2,x_3,x_4) = x_1 \dotplus x_2$$

and

$$F_b{}^d(x_1,x_2,x_3,x_4) = x_1x_2 \dotplus x_1x_3 \dotplus x_1x_4 \dotplus x_3x_4$$

and

$$\begin{aligned}\Phi(\,) = F_a(\,) \dotplus F_a{}^d(\,)F_b(\,) &= x_1x_2 \dotplus (x_1 \dotplus x_2)[x_1x_3 \dotplus x_1x_4 \dotplus x_2x_3x_4] \\ &= x_1x_2 \dotplus x_1x_3 \dotplus x_1x_4 \dotplus x_2x_3x_4\end{aligned}$$

To illustrate the second form, it is necessary to select a suitable function Φ_0. For this, select

$$\begin{aligned}\Phi_0(\,) &= F_b(\,) \dotplus F_b{}^d\Phi_1(\,) \\ &= [x_1x_3 \dotplus x_1x_4 \dotplus x_2x_3x_4] \dotplus [x_1x_2 \dotplus x_1x_3 \dotplus x_1x_4 \dotplus x_3x_4]\Phi_1(\,)\end{aligned}$$

where Φ_1 can be any self-dual function. Since x_1 appears in three terms of $F_b{}^d$, a considerable simplification occurs if we select

$$\Phi_1(x_1,x_2,x_3,x_4) = \bar{x}_1$$

In this case

$$\begin{aligned}\Phi_0(x_1,x_2,x_3,x_4) &= [x_1x_3 \dotplus x_1x_4 \dotplus x_2x_3x_4] \\ &\quad \dotplus [x_1x_2 \dotplus x_1x_3 \dotplus x_1x_4 \dotplus x_3x_4]\bar{x}_1 \\ &= x_1(x_3 \dotplus x_4) \dotplus x_3x_4\end{aligned}$$

when simplified. Note that Φ_0 is a self-dual function according to Corollary 7.3.1. Substituting the above Φ_0 into the expression for Φ gives

$$\Phi(x_1,x_2,x_3,x_4) = x_1x_2 \dotplus (x_1 \dotplus x_2)[x_1(x_3 \dotplus x_4) \dotplus x_3x_4]$$

which is an expression obtained from the second form.

When a self-dual function is decomposed into the sum of two functions, F_a and F_b, then a form of special interest occurs when F_b is expressed as the product of two functions, one of which is self-dual.

Corollary 7.4.1. If the self-dual function Φ is arbitrarily expressed in the form

$$\Phi(x_1,x_2,\ldots,x_n) = F_a(x_1,x_2,\ldots,x_n) \dotplus \Phi_a(x_1,x_2,\ldots,x_n)F_c(x_1,x_2,\ldots,x_n)$$

where Φ_a is self-dual, then Φ can be expressed as

$$\Phi(x_1,x_2,\ldots,x_n) = F_a(x_1,x_2,\ldots,x_n) \dotplus F_a{}^d(x_1,x_2,\ldots,x_n)\Phi_a(x_1,x_2,\ldots,x_n)$$

Proof. According to Theorem 7.4 and its proof, Φ can be expressed as

$$\Phi(x_1,x_2,\ldots,x_n) = F_a(x_1,x_2,\ldots,x_n) \dotplus F_a{}^d(x_1,x_2,\ldots,x_n)\Phi_0(x_1,x_2,\ldots,x_n)$$

where

$$\begin{aligned}\Phi_0(\,) &= [\Phi_a(\,)F_c(\,)] \dotplus [\Phi_a(\,)F_c(\,)]^d\Phi_1(\,)\\ &= [\Phi_a(\,)F_c(\,)] \dotplus [\Phi_a(\,) \dotplus F_c{}^d(\,)]\Phi_1(\,)\end{aligned}$$

and where Φ_1 is any self-dual function.

Letting $\Phi_1 = \Phi_a$ gives

$$\begin{aligned}\Phi_0(\,) &= \Phi_a(\,)F_c(\,) \dotplus [\Phi_a(\,) \dotplus F_c{}^d(\,)]\Phi_a(\,)\\ &= \Phi_a(\,)F_c(\,) \dotplus \Phi_a(\,) \dotplus F_c{}^d(\,)\Phi_a(\,)\\ &= \Phi_a(\,)\end{aligned}$$

and the result is established.|

This completes the discussion of the second Boolean form.

The third special form for self-dual functions is an expansion in terms of three-input majority functions. In particular, any self-dual function can be expressed as a majority of a single variable and two "smaller" functions. Here "smaller" function means a function with fewer terms in the minimum-sum-of-products form. Before considering this form, however, an auxiliary result is needed which is also of independent interest.

Theorem 7.5 (*Akers, 3; Muroga, 2*).

A. In any sum-of-products expansion of a self-dual function Φ, every pair of terms has at least one literal in common.

B. Any function which has a sum-of-products expansion such that every pair of terms has at least one literal in common is a minor function.

Proof

A. Let T_1 be any term in a sum-of-products expansion. Then

$$\Phi(x_1,x_2,\ldots,x_n) = T_1 \dotplus F_b(x_1,x_2,\ldots,x_n)$$

where F_b is the remainder of the terms of Φ.

Since a product term is always a minor function, then by Theorem 7.4 and its proof, it follows that

$$\Phi(x_1,x_2,\ldots,x_n) = T_1 \dotplus T_1{}^dF_b(x_1,x_2,\ldots,x_n)$$

where $T_1{}^d \geq F_b$. Since T_1 is a product of literals, that is,

$$T_1 = x_i x_j \ldots x_k$$

then

$$T_1{}^d = x_i \dotplus x_j \dotplus \cdots \dotplus x_k$$

Since $T_1{}^d \geq F_b$, then each term of F_b must be contained in $T_1{}^d$. However, a function of the form of $T_1{}^d$ can contain a term only if one of the literals of the term is x_i, or x_j, or ... or, x_k.

B. Let

$$F(x_1,x_2,\ldots,x_n) = T_1 \dotplus T_2 \dotplus \cdots \dotplus T_q$$

and let p_o be any point such that, for some j,

$$T_j(p_o) = 1$$

Since T_j has at least one literal in common with every other term, then

$$T_j(\bar{p}_o) = 0 \qquad \text{for} \quad j = 1,2,\ldots,q$$

where $\bar{p}_o$ is the complement of p_o. Thus for all p_o such that

$$F(p_o) = 1$$

it follows that

$$F(\bar{p}_o) = 0$$

and, therefore, that F is a minor function. The function F may or may not be self-dual, depending on whether there exist points p and $\bar{p}$ such that

$$F(p) = F(\bar{p}) = 0 \quad \blacksquare$$

The three-input majority-function representation of a self-dual function follows directly from the above result.

Corollary 7.5.1 *(Akers, 3).* Any self-dual function Φ can be expressed as

$$\Phi(x_1,x_2,\ldots,x_n) = M(y,\Phi_a,\Phi_b)$$

where y is any variable in the sum-of-products expansion of Φ and where Φ_a and Φ_b are self-dual functions.

Proof. Let

$$\Phi(x_1,x_2,\ldots,x_n) = T_1 \dotplus T_2 \dotplus \cdots \dotplus T_{p-1} \dotplus T_p$$

be a sum-of-products representation of Φ where T_1 and T_p are two of the terms that include the variable y. Consider the functions

$$F_a(x_1,x_2,\ldots,x_n) = T_1 \dotplus T_2 \dotplus \cdots \dotplus T_{p-1}$$
$$F_b(x_1,x_2,\ldots,x_n) = T_2 \dotplus T_3 \dotplus \cdots \dotplus T_{p-1} \dotplus T_p$$

Since the sum of F_a and F_b is a self-dual function, each must be a minor function according to Theorem 7.2B. Let Φ_a and Φ_b be arbitrary self-dual functions so that

$$\Phi_a > F_a \qquad \text{and} \qquad \Phi_b > F_b$$

and consider the function $M(y,\Phi_a,\Phi_b)$. Each unit point of Φ that is a unit point of T_2, or T_3, or. . .or T_{p-1} is a unit point of both Φ_a and Φ_b and hence a unit point of $M(y,\Phi_a,\Phi_b)$. Each unit point of T_1 or T_p is a unit point of Φ_a or Φ_b, respectively. Moreover, each such point has $y = 1$; therefore, it is a unit point of $M(y,\Phi_a,\Phi_b)$. This establishes that

$$M(y,\Phi_a,\Phi_b) \geq \Phi(x_1,x_2,\ldots,x_n)$$

However, $M(y,\Phi_a,\Phi_b)$ is a self-dual function of self-dual functions and is therefore a self-dual function of the variables $x_1,x_2,\ldots,x_n$. As a result, exactly one-half of the points of the n cube are unit points for both Φ and $M(y,\Phi_a,\Phi_b)$. Therefore

$$M(y,\Phi_a,\Phi_b) = \Phi(x_1,x_2,\ldots,x_n) \quad |$$

Example 7.5. Consider the switching function

$$\begin{aligned}\Phi(x_1,x_2,\ldots,x_5) &= x_1x_2 \dotplus x_1x_3 \dotplus x_1x_4 \dotplus x_1x_5 \dotplus x_2x_3x_4x_5 \\ &= T_1 \dotplus T_2 \dotplus T_3 \dotplus T_4 \dotplus T_5\end{aligned}$$

Let $y = x_2$, in which case

$$F_a(x_1,x_2,\ldots,x_5) = x_1x_2 \dotplus x_1x_3 \dotplus x_1x_4 \dotplus x_1x_5$$

$$F_b(x_1,x_2,\ldots,x_5) = x_1x_3 \dotplus x_1x_4 \dotplus x_1x_5 \dotplus x_2x_3x_4x_5$$

To determine Φ_a and Φ_b so that

$$\Phi_a > F_a \qquad \text{and} \qquad \Phi_b > F_b$$

we first determine a minor function F_a' where

$$F_a' \geq F_a$$

For this we use Theorem 7.5B and delete from any term of F_a any literal that is unnecessary, so that the resulting function still has a common literal for every pair of terms. Thus x_2,x_3,x_4,x_5 can be deleted from T_1,T_2,T_3,T_4, respectively, of F_a, in which case

$$F_a'(x_1,x_2,\ldots,x_5) = x_1$$

Since x_1 is self-dual, we let

$$\Phi_a(x_1,x_2,\ldots,x_5) = x_1$$

Similarily, to determine F_b', we delete x_2 from the last term of F_b, which gives

$$F_b'(x_1,x_2,\ldots,x_5) = x_1x_3 + x_1x_4 + x_1x_5 + x_3x_4x_5$$

Again F_b' is self-dual; hence we let

$$\Phi_b(x_1,x_2,\ldots,x_5) = F_b'(x_1,x_2,\ldots,x_5)$$

Thus Φ can be expressed as

$$\begin{aligned}\Phi(x_1,x_2,\ldots,x_5) &= M(y,\Phi_a,\Phi_b)\\ &= M(x_2,x_1,[x_1x_3 + x_1x_4 + x_1x_5 + x_3x_4x_5])\end{aligned}$$

In summary, the following three special forms for self-dual functions have been determined:

$$\Phi(x_1,x_2,\ldots,x_n) = x_1F_1(x_2,x_3,\ldots,x_n) + \bar{x}_1F_1{}^d(x_2,x_3,\ldots,x_n)$$
$$\Phi(x_1,x_2,\ldots,x_n) = F_a(x_1,x_2,\ldots,x_n) + F_a{}^d(x_1,x_2,\ldots,x_n)\Phi_1(x_1,x_2,\ldots,x_n)$$
$$\Phi(x_1,x_2,\ldots,x_n) = M(y,\Phi_a(x_1,x_2,\ldots,x_n),\Phi_b(x_1,x_2,\ldots,x_n))$$

Later in the chapter these will be used to develop special synthesis techniques.

7.3 SELF-DUAL REPRESENTATIVE FUNCTIONS

Before considering the synthesis techniques for self-dual switching functions, we shall show that arbitrary functions can be represented by self-dual functions. Specifically, we shall show that for each switching function, with or without don't-care's, there corresponds a *self-dual representative function* whose realization is readily converted to a realization of the original function. As a result, self-dual realization techniques can be used for arbitrary switching functions.

The self-dual representative function Φ for a given switching function F is determined from one of the following procedures.

1. *Complete Self-Dual Function.* If F is self-dual, then $\Phi = F$.

2. *Complete, Nonself-dual Functions* (*Muroga, 1*). If $F(x_1,x_2,\ldots,x_n)$ is the given function, then

$$\Phi(x_1,x_2,\ldots,x_{n+1}) = \bar{x}_{n+1}F(x_1,x_2,\ldots,x_n) + x_{n+1}F^d(x_1,x_2,\ldots,x_n)$$

The realization for F is obtained from the realization for Φ by setting $x_n = 0$ and $\bar{x}_n = 1$. If desired, the constant unit inputs can be removed by Theorem 2.4B.

3. *Incomplete Functions*† *(with Don't-Care's)*. Let F_M and F_m be the maximum and minimum functions that define the incomplete function $\mathscr{F}$ where any complete function F such that

$$F_M \geq F \geq F_m$$

satisfies $\mathscr{F}$ (See Section 4.10). If F_M is major and F_m is minor then it is shown below that there exists a self-dual function Φ such that

$$F_M \geq \Phi \geq F_m$$

If these conditions are not fulfilled then it subsequently is shown that a self-dual function $\Phi(x_1,x_2,\ldots,x_{n+1})$ exists such that

$$F_M(x_1,x_2,\ldots,x_n) \geq \Phi(x_1,x_2,\ldots,x_n,0) \geq F_m(x_1,x_2,\ldots,x_n)$$

F_M is Major and F_m is Minor

Consider the function

$$F_0(x_1,x_2,\ldots,x_n) = F_m(x_1,x_2,\ldots,x_n) \dotplus F_M{}^d(x_1,x_2,\ldots,x_n)$$

and express it in the form

$$F_0(x_1,x_2,\ldots,x_n) = F_1(x_1,x_2,\ldots,x_n) \dotplus \Phi_2(x_1,x_2,\ldots,x_n)F_2(x_1,x_2,\ldots,x_n)$$

where Φ_2 is any self-dual function. This is always possible since, if necessary, Φ_2 can be a single variable.

Theorem 7.6. In terms of the above notation,

$$\Phi(x_1,x_2,\ldots,x_n) = F_0(x_1,x_2,\ldots,x_n) \dotplus \Phi_2(x_1,x_2,\ldots,x_n)F_0{}^d(x_1,x_2,\ldots,x_n)$$

is a self-dual function such that

$$F_M \geq \Phi \geq F_m$$

Moreover Φ can be expressed as

$$\Phi(\,) = F_1(\,) \dotplus \Phi_2(\,)F_1{}^d(\,)$$

Proof. Since

$$F_0(x_1,x_2,\ldots,x_n) = F_m(x_1,x_2,\ldots,x_n) \dotplus F_M{}^d(x_1,x_2,\ldots,x_n)$$

then

$$F_0{}^d(x_1,x_2,\ldots,x_n) = F_m{}^d(x_1,x_2,\ldots,x_n)F_M(x_1,x_2,\ldots,x_n)$$

and

$$F_0(\,)F_0{}^d(\,) = F_m(\,)F_m{}^d(\,)F_M(\,) \dotplus F_M{}^d(\,)F_m{}^d(\,)F_M(\,)$$

But, since $F_M \geq F_m$ implies $F_M{}^d \leq F_m{}^d$ and since $F_M \geq F_M{}^d$ and $F_m \leq F_m{}^d$, it follows that

$$F_0(\,)F_0{}^d(\,) = F_m(\,) \dotplus F_M{}^d(\,) = F_0(\,)$$

† The remainder of this section can be omitted on a first reading.

It follows from Theorem 7.2A that F_0 is a minor function. Moreover, the same theorem establishes F_1 as a minor function, since

$$F_0(\ ) = F_1(\ ) \dotplus \Phi_2(\ )F_2(\ )$$

According to Corollary 7.3.2, the function

$$\Phi(\ ) = F_0(\ ) \dotplus F_0{}^d(\ )\Phi_2(\ )$$

is self-dual where

$$F_0{}^d \geq \Phi \geq F_0$$

Moreover, since

$$F_0 \geq F_m \qquad \text{and} \quad F_0 \geq F_M{}^d$$

then $F_0{}^d \leq F_M$ in which case

$$F_M \geq \Phi \geq F_m$$

Recall, however, that

$$F_0(\ ) = F_1(\ ) \dotplus \Phi_2(\ )F_2(\ )$$

Therefore,

$$\begin{aligned}\Phi(\ ) &= F_0(\ ) \dotplus F_0{}^d(\ )\Phi_2(\ )\\ &= \{F_1(\ ) \dotplus \Phi_2(\ )F_2(\ )\} \dotplus \{F_1{}^d(\ )[\Phi_2(\ ) \dotplus F_2{}^d(\ )]\}\Phi_2(\ )\\ &= F_1(\ ) \dotplus \Phi_2(\ )F_2(\ ) \dotplus F_1{}^d(\ )\Phi_2(\ ) \dotplus F_1{}^d(\ )F_2{}^d(\ )\Phi_2(\ )\\ &= F_1(\ ) \dotplus \Phi_2[F_2(\ ) \dotplus F_1{}^d(\ ) \dotplus F_1{}^d(\ )F_2{}^d(\ )]\end{aligned}$$

According to Corollary 7.4.1, however, Φ can be expressed as

$$\Phi(\ ) = F_1(\ ) \dotplus \Phi_2(\ )F_1{}^d(\ )$$

Therefore the result is established.∎

Example 7.6. Let

$$\begin{aligned}F_m(x_1,x_2,x_3,x_4) &= \bar{x}_1x_2x_3 \dotplus \bar{x}_1x_2x_4\\ F_M(x_1,x_2,x_3,x_4) &= \bar{x}_1x_2 \dotplus x_2x_4 \dotplus x_1x_3\bar{x}_4 \dotplus \bar{x}_1x_3x_4\end{aligned}$$

The corresponding dual functions are

$$\begin{aligned}F_m{}^d(\ ) &= \bar{x}_1 \dotplus x_2 \dotplus x_3x_4\\ F_M{}^d(\ ) &= x_2x_3 \dotplus \bar{x}_1x_2\bar{x}_4 \dotplus \bar{x}_1x_3x_4 \dotplus x_1x_2x_4\end{aligned}$$

from which it follows that F_m is minor and F_M is major. Computing F_0, one obtains

$$\begin{aligned}F_0(\ ) &= F_m(\ ) \dotplus F_M{}^d(\ )\\ &= x_2x_3 \dotplus x_2x_4 \dotplus \bar{x}_1x_2 \dotplus \bar{x}_1x_3x_4\end{aligned}$$

Since x_2 appears in all but one term, we let Φ_2 be x_2, in which case

$$\begin{aligned} F_0(\,) &= F_1(\,) + \Phi_2(\,)F_2(\,) \\ &= \bar{x}_1x_3x_4 + x_2[\bar{x}_1 + x_3 + x_4] \end{aligned}$$

and

$$\begin{aligned} \Phi(\,) &= F_1(\,) + \Phi_2(\,)F_1{}^d(\,) \\ &= \bar{x}_1x_3x_4 + x_2[\bar{x}_1 + x_3 + x_4] \end{aligned}$$

It is readily verified that

$$F_M \geq \Phi \geq F_m$$

$\mathbf{F_M}$ is Not Major or $\mathbf{F_m}$ is Not Minor

Form the function $F_{0'}$ where

$$F_{0'}(x_1,x_2,\ldots,x_{n+1}) = \bar{x}_{n+1}F_m(x_1,x_2,\ldots,x_n) + x_{n+1}F_M{}^d(x_1,x_2,\ldots,x_n)$$

and factor it arbitrarily so that

$$\begin{aligned} F_{0'}(x_1,x_2,\ldots,x_{n+1}) &= F_{1'}(x_1,x_2,\ldots,x_{n+1}) \\ &\quad + \Phi_{2'}(x_1,x_2,\ldots,x_{n+1})F_{2'}(x_1,x_2,\ldots,x_{n+1}) \end{aligned}$$

where $\Phi_{2'}$ is any self-dual function. Again $\Phi_{2'}$ can be a single variable.

Theorem 7.7. In terms of the above notation, the function Φ' where

$$\begin{aligned} \Phi'(x_1,x_2,\ldots,x_{n+1}) &= F_{0'}(x_1,x_2,\ldots,x_{n+1}) \\ &\quad + F_{0'}{}^d(x_1,x_2,\ldots,x_{n+1})\Phi_{2'}(x_1,x_2,\ldots,x_{n+1}) \end{aligned}$$

is a self-dual function such that

$$F_M \geq \Phi'(x_{n+1} = 0) \geq F_m$$

Moreover Φ' can be expressed as

$$\Phi'(\,) = F_{1'}(\,) + F_{1'}{}^d(\,)\Phi_{2'}(\,)$$

Proof. Since

$$F_{0'}(x_1,x_2,\ldots,x_{n+1}) = \bar{x}_{n+1}F_m(x_1,x_2,\ldots,x_n) + x_{n+1}F_M{}^d(x_1,x_2,\ldots,x_n)$$

then

$$F_{0'}{}^d(x_1,x_2,\ldots,x_{n+1}) = \bar{x}_{n+1}F_M(x_1,x_2,\ldots,x_n) + x_{n+1}F_m{}^d(x_1,x_2,\ldots,x_n)$$

and

$$F_{0'}(\,)F_{0'}{}^d(\,) = \bar{x}_{n+1}F_m(\,)F_M(\,) + x_{n+1}F_m{}^d(\,)F_M{}^d(\,)$$

However $F_M \geq F_m$ and $F_m{}^d \geq F_M{}^d$. Therefore

$$F_{0'}(\,)F_{0'}{}^d(\,) = \bar{x}_{n+1}F_m(\,) + x_{n+1}F_M{}^d(\,) = F_{0'}(\,)$$

It follows from Theorem 7.2A, that $F_{0'}$ is a minor function. Moreover, the same theorem establishes $F_{1'}$ as a minor function since

$$F_{0'}(\,) = F_{1'}(\,) \dotplus \Phi_{2'}(\,)F_{2'}(\,)$$

According to Corollary 7.4.2, the function

$$\Phi'(\,) = F_{0'}(\,) \dotplus F_{0'}{}^{d}(\,)\Phi_{2'}(\,)$$

is self-dual where

$$F_{0'}{}^{d} \geq \Phi' \geq F_{0'}$$

or

$$(\bar{x}_{n+1}F_M(\,) \dotplus x_{n+1}F_m{}^{d}(\,)) \geq \Phi' \geq (\bar{x}_{n+1}F_m(\,) \dotplus x_{n+1}F_M{}^{d}(\,))$$

It follows immediately, by setting $x_{n+1} = 0$ and $\bar{x}_{n+1} = 1$, that

$$F_M \geq \Phi'(x_{n+1} = 0) \geq F_m$$

Recall, however, that

$$F_{0'}(\,) = F_{1'}(\,) \dotplus \Phi_{2'}(\,)F_{2'}(\,)$$

Therefore

$$\begin{aligned}\Phi' &= F_{0'}(\,) \dotplus F_{0'}{}^{d}(\,)\Phi_{2'}(\,)\\ &= \{F_{1'}(\,) \dotplus \Phi_{2'}(\,)F_{2'}(\,)\} \dotplus \{F_{1'}{}^{d}(\,)[\Phi_{2'}(\,) \dotplus F_{2'}{}^{d}(\,)]\}\Phi_{2'}(\,)\\ &= F_{1'}(\,) \dotplus \Phi_{2'}(\,)F_{2'}(\,) \dotplus F_{1'}{}^{d}\Phi_{2'}(\,) \dotplus F_{1'}{}^{d}(\,)F_{2'}{}^{d}(\,)\Phi_{2'}(\,)\\ &= F_{1'}(\,) \dotplus \Phi_{2'}(\,)[F_{2'}(\,) \dotplus F_{1'}{}^{d}(\,) \dotplus F_{1'}{}^{d}(\,)F_{2'}{}^{d}(\,)]\end{aligned}$$

According to Corollary 7.4.1, however, Φ' can be expressed as

$$\Phi'(\,) = F_{1'}(\,) \dotplus \Phi_{2}(\,)F_{1'}{}^{d}(\,)$$

This establishes the result.|

Theorem 7.7 provides a means whereby a realization for an incomplete function can be obtained from a realization of a self-dual function Φ by setting all x_{n+1} inputs to 0 and all $\bar{x}_{n+1}$ inputs to 1.

Example 7.7. Let

$$\begin{aligned}F_M(x_1,x_2,x_3,x_4) &= x_1x_2x_3 \dotplus x_1\bar{x}_2x_4 \dotplus \bar{x}_1\bar{x}_3\\ F_m(x_1,x_2,x_3,x_4) &= x_1x_3x_4 \dotplus \bar{x}_1x_2\bar{x}_3 \dotplus \bar{x}_1\bar{x}_3x_4\end{aligned}$$

It is readily verified that $F_M > F_m$. Now form the function $F_{0'}$ where

$$\begin{aligned}F_{0'}(x_1,x_2,\ldots,x_5) &= \bar{x}_5F_m(x_1,x_2,x_3,x_4) \dotplus x_5F_M{}^{d}(x_1,x_2,x_3,x_4)\\ &= \bar{x}_5[x_1x_3x_4 \dotplus \bar{x}_1x_2\bar{x}_3 \dotplus \bar{x}_1\bar{x}_3x_4] \dotplus x_5[x_1\bar{x}_3 \dotplus \bar{x}_1x_2x_4\\ &\qquad \dotplus \bar{x}_1\bar{x}_2x_3]\end{aligned}$$

Since $\bar{x}_1$ appears in all but two terms, let $\Phi_{2'}$ be $\bar{x}_1$. Then

$$\begin{aligned}F_{0'}(\,) &= x_1x_3x_4\bar{x}_5 \dotplus x_1\bar{x}_3x_5 \dotplus \bar{x}_1[x_2x_3\bar{x}_5 \dotplus \bar{x}_3x_4\bar{x}_5 \dotplus x_2x_4x_5 \dotplus \bar{x}_2x_3x_5] \\ &= F_{1'}(\,) \dotplus \Phi_{2'}(\,)F_{2'}(\,)\end{aligned}$$

and

$$\begin{aligned}\Phi'(\,) &= F_{1'}(\,) \dotplus \Phi_{2'}(\,)F_{1'}{}^{d}(\,) \\ &= x_1x_3x_4\bar{x}_5 \dotplus x_1\bar{x}_3x_5 \dotplus \bar{x}_1[x_1 \dotplus x_3x_5 \dotplus \bar{x}_3x_4 \dotplus \bar{x}_3\bar{x}_5] \\ &= x_1[x_3x_4\bar{x}_5 \dotplus \bar{x}_3x_5] \dotplus \bar{x}_1[x_3x_5 \dotplus \bar{x}_3x_4 \dotplus \bar{x}_3\bar{x}_5]\end{aligned}$$

Then

$$\Phi(x_{n+1} = x_5 = 0) = x_1x_3x_4 \dotplus \bar{x}_1\bar{x}_3$$

from which it immediately follows that

$$F_M \geq \Phi(x_5 = 0) \geq F_m$$

In this section a given switching function has been associated with a representative self-dual function. In every case the realization of the self-dual function is easily transformed to a realization of the given function.

7.4 THE TREE-METHOD REALIZATION OF SELF-DUAL FUNCTIONS

Self-dual switching functions can be expressed in each of three forms

$$\begin{aligned}\Phi(x_1,x_2,\ldots,x_n) &= \bar{x}_1F_0(x_2,x_3,\ldots,x_n) \dotplus x_1F_0{}^{d}(x_2,x_3,\ldots,x_n) \\ &= F_a(x_1,x_2,\ldots,x_n) \dotplus F_a{}^{d}(x_1,x_2,\ldots,x_n)\Phi_0(x_1,x_2,\ldots,x_n) \\ &= M(y,\Phi_a,\Phi_b)\end{aligned}$$

To each of these forms there corresponds a special synthesis approach.

The first of these forms

$$\Phi(\,) = \bar{x}_1F_0(\,) \dotplus x_1F_0{}^{d}(\,)$$

suggests the tree method. In this case, however, the realization need only be determined for F_0, since the realization for Φ can then be determined as illustrated by Example 4.6.

The realization of F_0 by the tree method is no different from the one for any other function and requires no further discussion. The synthesis procedures corresponding to the other special forms for self-dual functions are considered in the following sections.

7.5 SELF-DUAL REALIZATIONS BASED ON THE FORM $F_a \dotplus F_a{}^d\Phi_0$

The self-dual functional form

$$\Phi(x_1,x_2,\ldots,x_n) = F_a(x_1,x_2,\ldots,x_n) \dotplus F_a{}^d(x_1,x_2,\ldots,x_n)\Phi_0(x_1,x_2,\ldots,x_n)$$

implies a realization for Φ in terms of those for F_a and Φ_0. Since F_a is a minor function, this expression is equivalent to the form

$$\Phi(x_1,x_2,\ldots,x_n) = F_a(x_1,x_2,\ldots,x_n)\overline{\Phi}_0(x_1,x_2,\ldots,x_n) \dotplus F_a{}^d(x_1,x_2,\ldots,x_n)\Phi_0(x_1,x_2,\ldots,x_n)$$

Therefore, the procedure of Example 4.6 with x_1 replaced by Φ_0 can be used. Each gate in the realization for F_a is provided an additional input with coefficient so as to displace its gap symmetrically about the center of the map when this input has unit value. This input is connected to the output of the network that realizes Φ_0.

The function Φ_0 is another self-dual function which can be expressed in a manner similar to Φ; therefore, the entire procedure can be repeated to determine a realization for Φ_0.

Example 7.8. Consider the self-dual function

$$\Phi(x_1,x_2,\ldots,x_6) = x_1[x_2 \dotplus x_3 \dotplus x_4(x_5 \dotplus x_6) \dotplus x_5x_6] \dotplus x_2x_3[x_4(x_5 \dotplus x_6) \dotplus x_5x_6]$$

which we wish to express in the form

$$\Phi(\,) = F_a(\,) \dotplus F_b(\,) = F_a(\,) \dotplus F_a{}^d\Phi_0(\,)$$

Let

$$F_a(\,) = x_1(x_2 \dotplus x_3)$$

for which a realization is

$$F_a(\,) = \langle 2x_1 + x_2 + x_3\rangle_{3:2}$$

According to Theorem 7.4, Φ_0 can be the function

$$\Phi_0(\,) = F_b(\,) \dotplus F_b{}^d(\,)\Phi_1(\,)$$

where Φ_1 is an arbitrary self-dual function. Therefore

$$\Phi_0(\,) = x_1[x_4(x_5 \dotplus x_6) \dotplus x_5x_6] \dotplus x_2x_3x_4(x_5 \dotplus x_6) \dotplus x_2x_3x_5x_6 \dotplus \{x_1(x_2 \dotplus x_3) \dotplus x_4(x_5 \dotplus x_6) \dotplus x_5x_6\}\Phi_1(\,)$$

Letting $\Phi_1(\,) = \bar{x}_1$ and simplifying gives

$$\Phi_0(\,) = x_4(x_5 \dotplus x_6) \dotplus x_5x_6$$

for which a realization is

$$\Phi_0(\ ) = \langle x_4 + x_5 + x_6 \rangle_{2:1}$$

In terms of the realizations for F_a and Φ_0, the realization for Φ is

$$\begin{aligned}\Phi(\ ) &= F_a(\ ) \dotplus F_a{}^d(\ )\Phi_0(\ ) \\ &= \langle 2x_1 + x_2 + x_3 + \langle x_4 + x_5 + x_6 \rangle_{2:1} \rangle_{3:2}\end{aligned}$$

Although in certain cases this procedure appears attractive, in general it is difficult to determine a suitable function F_a. One very simple choice is to let F_a be a single term of Φ, in which case it can be realized by a simple AND gate. This choice usually does not yield an efficient network, however, and can perhaps best be described as a canonic procedure based on AND/OR realizations. In fact it can be considered a generalization of the two-level AND/OR gate canonic form.

Recall that for one of these canonic forms the given function is first represented as a minimum sum of product terms

$$F(\ ) = T_1 \dotplus T_2 \dotplus \cdots \dotplus T_M$$

where T_j represents one product term. Each T_j is realized by an AND gate, and the function F is the output of an OR gate the inputs of which are the AND gate outputs. Thus the number of gates is one more than the number of product terms.

The other AND/OR canonic form is the dual of this one. The function is expanded as a minimum product of sum terms and realized as a layer of OR gates feeding an AND gate.

The canonic realization procedure for a self-dual function that is based upon the form

$$\Phi(\ ) = F_a(\ ) \dotplus F_a{}^d(\ )\Phi_0(\ )$$

where F_a is a single term, is somewhat similar to the AND/OR realization that is based on the sum of products form of the function. In general, however, the T-gate realization requires fewer gates than are needed for the AND/OR realization.

A Canonic Realization of Self-Dual Functions (*Akers, 3*).

Let Φ_0 be a given self-dual function of the variables $x_1,x_2,\ldots,x_n$ and let z represent a literal that appears in the largest number of terms of a minimum-sum-of-products expansion. Note that z can represent either a complemented or an uncomplemented variable. Express Φ_0 in the form

$$\Phi_0 = T_1 \dotplus T_2 \dotplus \cdots \dotplus T_k \dotplus zF_a$$

where $T_1,T_2,\ldots,T_k$ are product terms that do not include the literal z and where zF_a denotes the remaining terms of Φ_0. Since Φ_0 is self-dual, then according to Corollary 7.4.1, it can be expressed as

$$\Phi_0 = T_1 \dotplus T_2 \dotplus \cdots \dotplus T_k \dotplus z[T_1{}^dT_2{}^d\ldots T_k{}^d]$$

Using this form of Φ_0 and applying Theorem 7.4 with the term T_1 as F_a, the new expression for Φ_0 becomes

$$\begin{aligned}\Phi_0 &= T_1 \dotplus T_1{}^d[T_2 \dotplus T_3 \dotplus \cdots \dotplus T_k \dotplus zT_1{}^dT_2{}^d\ldots T_k{}^d]\\ &= T_1 \dotplus T_1{}^d[T_2 \dotplus T_3 \dotplus \cdots \dotplus T_k \dotplus zT_2{}^dT_3{}^d\ldots T_k{}^d]\end{aligned}$$

According to Corollary 7.3.1, the function within the brackets is self-dual; hence Theorem 7.4 is used on this part to obtain the expression

$$\begin{aligned}\Phi_0 &= T_1 \dotplus T_1{}^d[T_2 \dotplus T_2{}^d(T_3 \dotplus \cdots \dotplus T_k \dotplus zT_2{}^dT_3{}^d\ldots T_k{}^d)]\\ &= T_1 \dotplus T_1{}^d[T_2 \dotplus T_2{}^d(T_3 \dotplus \cdots \dotplus T_k \dotplus zT_3{}^dT_4{}^d\ldots T_k{}^d)]\end{aligned}$$

where the function within the parentheses is again self-dual. Repeating the procedure yields Φ_0 in the form

$$\Phi_0 = T_1 \dotplus T_1{}^d[T_2 \dotplus T_2{}^d(T_3 \dotplus T_3{}^d[\ldots(T_k \dotplus T_k{}^dz)]\ldots]$$

We have shown, therefore, that a self-dual function Φ_0 can be represented recursively as

$$\Phi_{j-1} = T_j \dotplus T_j{}^d\Phi_j$$

where $j = 1,2,\ldots,k$ and where $\Phi_k = z$. A canonic realization for self-dual functions is based upon this recursive representation.

Let T_1 represent a product term, for example,

$$T_1 = x_{a1}x_{a2}\ldots x_{a\alpha}$$

Then

$$\Phi_0 = T_1 \dotplus T_1{}^d\Phi_1 = x_{a1}x_{a2}\ldots x_{a\alpha} \dotplus (x_{a1} \dotplus x_{a2} \dotplus \cdots \dotplus x_{a\alpha})\Phi_1$$

When Φ_1 has value 0, Φ_0 is the AND function of $x_{a1},x_{a2},\ldots,x_{a\alpha}$, while when Φ_1 has value 1, Φ_0 becomes the dual of the AND function or the OR function of $x_{a1},x_{a2},\ldots,x_{a\alpha}$. Therefore Φ_0 can be realized by a single threshold gate with inputs $x_{a1},x_{a2},\ldots,x_{a\alpha}$ and Φ_1 where this gate realizes $x_{a1}x_{a2}\ldots x_{a\alpha}$ when $\Phi_1 = 0$ and realizes the dual or $x_{a1} \dotplus x_{a2} \dotplus \cdots \dotplus x_{a\alpha}$ when $\Phi_1 = 1$. It follows from Theorem 2.11 that a realization of Φ_0 with inputs $x_{a1},x_{a2},\ldots,x_{a\alpha}$ and Φ_1 is

$$\Phi_0 = \langle x_{a1} + x_{a2} + \cdots + x_{a\alpha} + (\alpha - 1)\Phi_1\rangle_{\alpha:(\alpha-1)}$$

as illustrated in Fig. 7.2.

Since Φ_1 is likewise a self-dual function, it can be similarily realized with

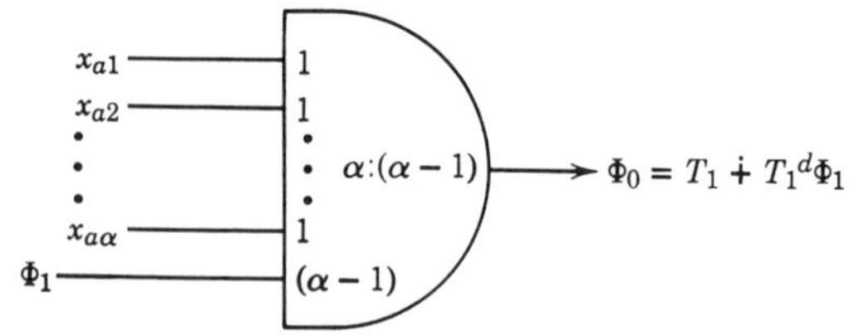

Fig. 7.2. Realization of $\Phi = T_1 \dotplus T_1{}^d\Phi_1$.

inputs $x_{b1},x_{b2},\ldots,x_{b\beta}$ and Φ_2 where Φ_2 can be similarily realized, and so on. The result is the chain network of Fig. 7.3, which realizes Φ_0 with inputs $x_1,x_2,\ldots,x_n$.

In the chain network there is one gate for each of the product terms $T_1,T_2,\ldots,T_k$. Each gate, G_i realizes one of the terms T_i as an AND gate provided the gate G_{i+1} has zero output. When G_{i+1} has unit output, then G_i is the OR gate for the variables of T_i.

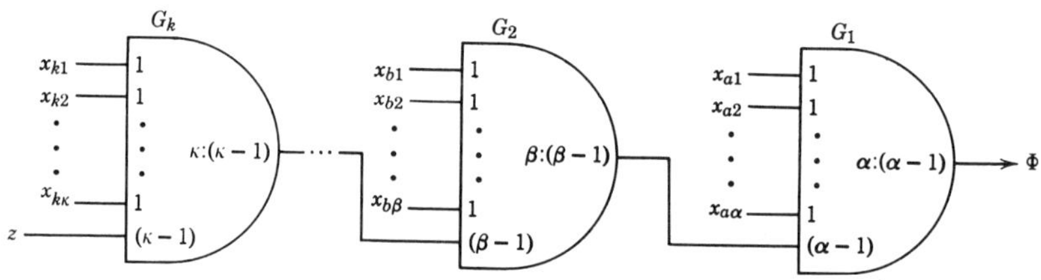

Fig. 7.3. Canonic chain network for self-dual function realization.

Since z was selected as one of the literals that appear most often, k may be considerably less than the number of terms. Hence this canonic realization may yield a considerable saving in gates over the two level AND/OR realization, but at the expense of increasing the number of logic levels.

Example 7.9. Let

$$F(x_1,x_2,x_3,x_4) = x_1x_2 \dotplus x_1x_3 \dotplus x_1x_4 \dotplus \bar{x}_1\bar{x}_2\bar{x}_3$$

The two-level AND/OR realization for F requires five gates.

Form the self-dual representative function

$$\begin{aligned}\Phi(x_1,x_2,\ldots,x_5) &= \bar{x}_5F(x_1,x_2,x_3,x_4) \dotplus x_5F^d(x_1,x_2,x_3,x_4)\\ &= \bar{x}_5[x_1(x_2 \dotplus x_3 \dotplus x_4) \dotplus \bar{x}_1\bar{x}_2\bar{x}_3]\\ &\quad \dotplus x_5[x_1(\bar{x}_2 \dotplus \bar{x}_3) \dotplus \bar{x}_1x_2x_3x_4]\end{aligned}$$

The variable that appears most often is x_1; therefore, let $z = x_1$ and express Φ as

$$\Phi(x_1,x_2,\ldots,x_5) = \bar{x}_1\bar{x}_2\bar{x}_3\bar{x}_5 + \bar{x}_1x_2x_3x_4x_5 + x_1[\bar{x}_5(x_2 + x_3 + x_4) + x_5(\bar{x}_2 + \bar{x}_3)]$$

This can be expressed in the canonic form as

$$\Phi(x_1,x_2,\ldots,x_5) = \bar{x}_1\bar{x}_2\bar{x}_3\bar{x}_5 + (\bar{x}_1 + \bar{x}_2 + \bar{x}_3 + \bar{x}_5)[\bar{x}_1x_2x_3x_4x_5 + (\bar{x}_1 + x_2 + x_3 + x_4 + x_5)x_1]$$

Where the realization is shown in Fig. 7.4*a*.

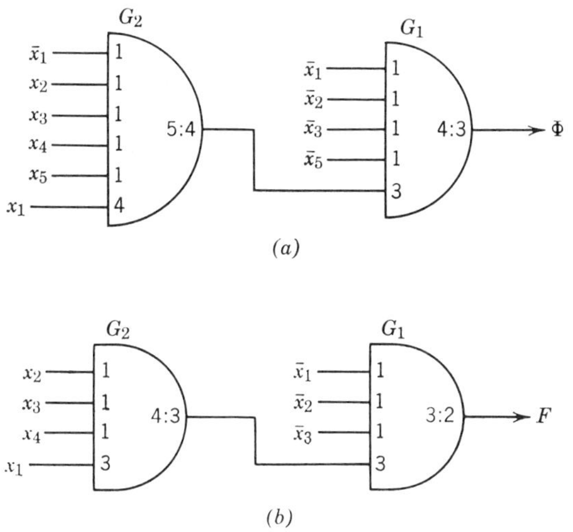

Fig. 7.4. Realizations of Example 7.9. (*a*) Realization of the self-dual representative function. (*b*) Realization of *F*.

To obtain the desired realization for *F*, set $\bar{x}_5 = 1$ and $x_5 = 0$. The resultant constant unit input for G_1 can be removed by changing the gap according to Theorem 2.4B. Gate G_2 can also be simplified by first replacing the x_1 input with coefficient four by two x_1 inputs with coefficients three and one, respectively. The latter is then combined with the $\bar{x}_1$ input to yield a constant input with coefficient one. This is then used to change the gap according to Theorem 2.4B. The final realization of *F* is shown in Fig. 7.4*b*.

Canonic Realization of Arbitrary Functions

The previous realization is for self-dual functions. To use it to realize a function that is not self-dual requires that a new variable be introduced to form a representative self-dual function. This variable is later assigned the constant zero value. One suspects, however, that the conversion to the representative self-dual function would be unnecessary if its purpose were understood.

Consider the two gate realizations of Example 7.9. Since G_1 realizes the term $\bar{x}_1\bar{x}_2\bar{x}_3$, the saving over AND/OR realizations must come from G_2, which realizes

$$x_1x_2 + x_1x_3 + x_1x_4$$

In an AND/OR realization three gates would be required to realize this function. It can be interpreted that the saving occurs because these terms can be expressed as

$$x_1(x_2 + x_3 + x_4) = x_1[x_2x_3x_4]^d$$

where the latter notation means the product of x_1 and the dual of the product term $x_2x_3x_4$. The product term is realized by a single gate, the dual is then obtained by changing the gap, and finally the variable x_1 is multiplied in by adding the appropriately weighted input and changing the gap again.

This situation is denoted by

$$x_1(x_2 + x_3 + x_4) = x_1[(x_2 + x_3 + x_4)^d]_d = x_1[x_2x_3x_4]_d$$

This means that the dual of $x_2 + x_3 + x_4$ is realized (denoted by the superscript d), after which the dual of the realization is obtained by a gap change (denoted by subscript d). The total realization, therefore, is of the form

$$\bar{x}_1\bar{x}_2\bar{x}_3 + x_1[x_2x_3x_4]_d$$

Note that an additional gate is not required for the final OR operation; either one of the T gates can perform this operation.

This illustrates the mechanism by which the self-dual canonic procedure achieves gate savings for non-self-dual functions. Stated concisely—when economic, the procedure realizes the dual of part of the function rather than the part itself and dualizes it.

To verify this in a more general case, let the given function F be written as

$$F(x_1,x_2,\ldots,x_n) = \bar{x}_1F_0(x_2,x_3,\ldots,x_n) + x_1F_1(x_2,x_3,\ldots,x_n)$$

and form the representative self-dual functions

$$\begin{aligned}\Phi(\,) &= \bar{x}_{n+1}F(\,) + x_{n+1}F^d(\,)\\ &= \bar{x}_{n+1}[\bar{x}_1F_0(\,) + x_1F_1(\,)] + x_{n+1}[\bar{x}_1F_1{}^d(\,) + x_1F_0{}^d(\,)]\\ &= \bar{x}_1[\bar{x}_{n+1}F_0(\,) + x_{n+1}F_1{}^d(\,)] + x_1[\bar{x}_{n+1}F_1(\,) + x_{n+1}F_0{}^d(\,)]\end{aligned}$$

In the self-dual canonic procedure, the variable z is selected such that when Φ is expressed as

$$\Phi = T_1 \dotplus T_2 \dotplus \cdots \dotplus T_k \dotplus zT_1{}^dT_2{}^d \ldots T_k{}^d$$

the number of terms in $T_1, T_2, \ldots, T_k$ is a minimum. In our example z can be x_{n+1}, $\bar{x}_{n+1}$, x_1, or $\bar{x}_1$, depending on which of

(1) $\bar{x}_1F_0(\,) \dotplus x_1F_1(\,)$
(2) $\bar{x}_1F_1{}^d(\,) \dotplus x_1F_0{}^d(\,)$
(3) $\bar{x}_{n+1}F_0(\,) \dotplus x_{n+1}F_1{}^d(\,)$
(4) $\bar{x}_{n+1}F_1(\,) \dotplus x_{n+1}F_0{}^d(\,)$

respectively, has the fewest terms. Determining the corresponding self-dual realization for each of these cases, and then setting the corresponding z equal to 0, yields

(1) $F(\,) = x_1F_1(\,) \dotplus \bar{x}_1F_0(\,)$
(2) $F(\,) = [x_1F_0{}^d(\,) \dotplus \bar{x}_1F_1{}^d(\,)]_d = [F^d(\,)]_d$
(3) $F(\,) = \bar{x}_1F_0(\,) \dotplus x_1[F_1{}^d(\,)]_d$
(4) $F(\,) = \bar{x}_1[F_0{}^d(\,)]_d \dotplus x_1F_1(\,)$

Thus, either F_1 or $[F_1{}^d]_d$ is realized and either F_0 or $[F_0{}^d]_d$ is realized depending in each case on which requires the fewer gates.

With this background we consider canonic realizations that achieve gate savings in the same way as for self-dual functions, but without first converting to a representative self-dual function. Actually, this procedure frequently requires fewer gates. Moreover it allows some control of the topology of the network.

One additional notation symbol is useful. When a self-dual function

$$\Phi(\,) = \bar{x}_1F_a(\,) \dotplus x_1F_a{}^d(\,)$$

must be realized, it is often convenient first to realize F_a and then, as in Example 4.6, to realize Φ by adding the variable x_1 to each gate with the appropriate coefficient and gap change. This type of realization is denoted as

$$\Phi = [F_a]_{SD-x_1}$$

which reads: F_a self-dualized with x_1.

Now we describe the canonic realization procedure.

Procedure 7.1

(1) Given a function F, select a variable† x_1 and express F as

† To determine the x_1 that will require the fewest gates, use the procedure for determining z in the canonic self-dual procedure.

$$F(\,) = x_1F_1(\,) \dotplus \bar{x}_1F_0(\,)$$

(2) If $F_1 = F_0{}^d$, then F is self-dual. Express F as

$$F = [F_0]_{SD-x_1} \qquad \text{or} \qquad F = [F_1]_{SD-\bar{x}_1}$$

according to whether F_0 or F_1 has fewer terms, and proceed to Step 5. Otherwise go to Step 3.

(3) Express F as

$$F(\,) = x_1F_a(\,) \dotplus \bar{x}_1F_b(\,) \dotplus F_c(\,)$$

where F_c consists of those terms which do not contain x_1. Compute $F_a{}^d$, $F_b{}^d$, and $F_c{}^d$.

(4) In Step 3 express each F_j, $j = a,b,c$, as

$$F_j \qquad \text{or} \qquad [F_j{}^d]_d$$

according to whether F_j or $F_j{}^d$, respectively, contains the lower number of terms.†

(5) The expression for F obtained in Steps 2 and 4 implies a realization. The functions that remain can be realized as a simple two-level AND/OR canonic form; then the indicated dualizing or self-dualizing operations, if any, are performed; then the multiplying variables, if any, are added. Finally, the OR operation of all terms is achieved by using one of the gates

† Other alternatives to the expression of Step 3 are

$$F(\,) = x_1F_a(\,) \dotplus F_e(\,)$$

where

$$F_e(\,) = \bar{x}_1F_b(\,) \dotplus F_c(\,)$$

$$F(\,) = \bar{x}_1F_b(\,) \dotplus F_f(\,)$$

where

$$F_f(\,) = x_1F_a(\,) \dotplus F_c(\,)$$

and

$$F(\,) = x_1F_1(\,) \dotplus \bar{x}_1F_0(\,)$$

Step 4 is then applied to whichever is selected. In the last form, if $F_1{}^d$ has fewer terms than F_1 and if $F_0{}^d$ has fewer terms than F_0, then instead of writing

$$F(\,) = x_1[F_1{}^d(\,)]_d \dotplus \bar{x}_1[F_0{}^d(\,)]_d$$

as a result of Step 4, one can write

$$F(\,) = [x_1F_0{}^d(\,) \dotplus \bar{x}_1F_1{}^d(\,)]_d = [F^d(\,)]_d$$

Still another alternative if the function is of the form

$$F(\,) = x_1 \dotplus x_2 \dotplus \cdots \dotplus x_q \dotplus F_a(\,)$$

is

$$F(\,) = [x_1x_2 \ldots x_q]_d \dotplus F_a(\,)$$

and then apply Step 4 to F_a.

as the final OR gate. The resulting network will have two or three levels (Examples 7.10 and 7.11, respectively).

Usually an even greater gate savings can be achieved (although sometimes at the expense of additional logic levels) by reapplying the procedure, starting at Step 1, to each of the remaining functions of Step 4 (F_a or $F_a{}^d$, etc.). This will be clarified by the examples.

Example 7.10. Consider the function

$$F(x_1,x_2,\ldots,x_9) = x_1[x_2 \dotplus x_3 \dotplus x_4 \dotplus x_5] \dotplus [x_6(x_7 \dotplus x_8 \dotplus x_9) \dotplus \bar{x}_6\bar{x}_7\bar{x}_8\bar{x}_9]$$

Letting

$$F_a(\,) = x_2 \dotplus x_3 \dotplus x_4 \dotplus x_5$$

and

$$F_c(\,) = x_6(x_7 \dotplus x_8 \dotplus x_9) \dotplus \bar{x}_6\bar{x}_7\bar{x}_8\bar{x}_9$$

we determine that

$$F_a{}^d(\,) = x_2x_3x_4x_5$$
$$F_c{}^d(\,) = x_6(\bar{x}_7 \dotplus \bar{x}_8 \dotplus \bar{x}_9) \dotplus \bar{x}_6x_7x_8x_9$$

Choosing to realize $F_a{}^d$ and F_c, we express F as

$$\begin{aligned} F(\,) &= x_1[F_a{}^d(\,)]_d \dotplus F_c(\,) \\ &= x_1[x_2x_3x_4x_5]_d \dotplus x_6x_7 \dotplus x_6x_8 \dotplus x_6x_9 \dotplus \bar{x}_6\bar{x}_7\bar{x}_8\bar{x}_9 \end{aligned}$$

This corresponds to the two-level realization shown in Fig. 7.5*a*. Each term is realized with a single gate. The gate for the term $x_1[x_2x_3x_4x_5]_d$ is obtained in the following steps:

$$\begin{aligned} x_1[x_2x_3x_4x_5]_d &= x_1[\langle x_2 + x_3 + x_4 + x_5\rangle_{4:3}]_d \\ &= x_1\langle x_2 + x_3 + x_4 + x_5\rangle_{1:0} \\ &= \langle 4x_1 + x_2 + x_3 + x_4 + x_5\rangle_{5:4} \end{aligned}$$

The first step is the dualizing operation given by Corollary 2.10.2, and the second is just the application of Theorem 2.1B.

Each of the other terms is realized by a simple AND gate, and the final OR operation is realized by the gate that realizes x_6x_7 by the application of Theorem 2.1A.

Additional gate savings can be obtained by applying Procedure 7.1 to F_c. Express F_c as

$$F_c(\,) = x_6F_\alpha(\,) \dotplus \bar{x}_6F_\beta(\,)$$

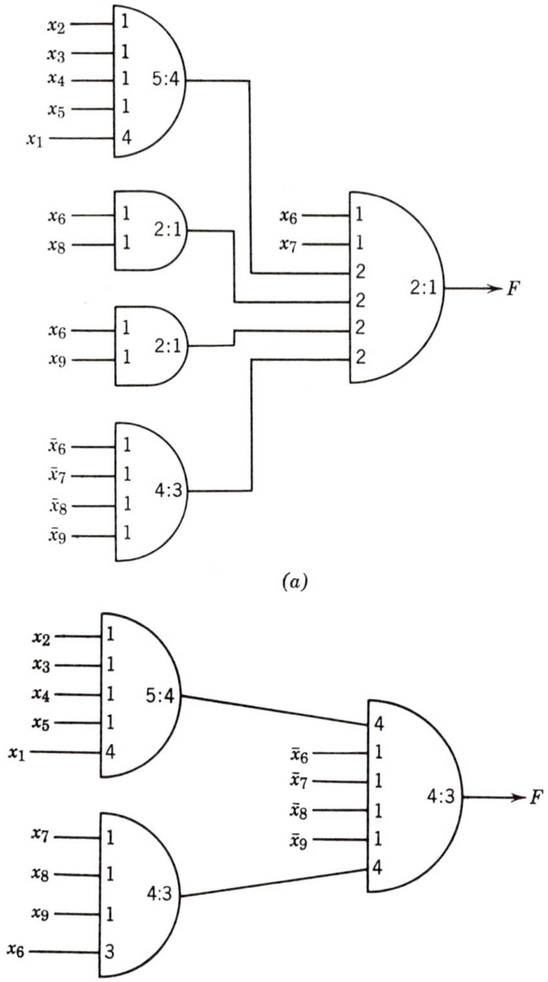

(b)

Fig. 7.5. Realizations of Example 7.10. (*a*) The first realization of *F*. (*b*) The second realization of *F*.

where

$$F_\alpha(\,) = x_7 \dotplus x_8 \dotplus x_9$$
$$F_\beta(\,) = \bar{x}_7\bar{x}_8\bar{x}_9$$

The corresponding dual functions are

$$F_\alpha{}^d(\,) = x_7x_8x_9$$
$$F_\beta{}^d(\,) = \bar{x}_7 \dotplus \bar{x}_8 \dotplus \bar{x}_9$$

Thus F_c can be expressed as

$$F_c(\,) = x_6[x_7x_8x_9]_d \dotplus \bar{x}_6[\bar{x}_7\bar{x}_8\bar{x}_9]$$

and F can be written as

$$F(\,) = x_1[x_2x_3x_4x_5]_d \dotplus x_6[x_7x_8x_9]_d \dotplus \bar{x}_6[\bar{x}_7\bar{x}_8\bar{x}_9]$$

The corresponding realization is shown in Fig. 7.5*b*.

Example 7.11. Consider the function

$$F(x_1,x_2,\ldots,x_5) = x_1(x_2 \dotplus x_3 \dotplus x_4x_5) \dotplus \bar{x}_1(\bar{x}_2\bar{x}_3 \dotplus \bar{x}_2\bar{x}_4 \dotplus \bar{x}_3\bar{x}_4)$$

Let

$$F_a(\,) = x_2 \dotplus x_3 \dotplus x_4x_5$$
$$F_b(\,) = \bar{x}_2(\bar{x}_3 \dotplus \bar{x}_4) \dotplus \bar{x}_3\bar{x}_4 = M_2(\bar{x}_2,\bar{x}_3,\bar{x}_4)$$

Hence

$$F_a{}^d(\,) = x_2x_3(x_4 \dotplus x_5)$$
$$F_b{}^d(\,) = M_2(\bar{x}_2,\bar{x}_3,\bar{x}_4)$$

Then F can be expressed as

$$F(\,) = x_1[F_a{}^d(\,)]_d \dotplus \bar{x}_1F_b(\,)$$
$$= x_1[x_2x_3x_4 \dotplus x_2x_3x_5]_d \dotplus \bar{x}_1[\bar{x}_2\bar{x}_3 \dotplus \bar{x}_2\bar{x}_4 \dotplus \bar{x}_3\bar{x}_4]$$

and this corresponds to the three-level network of Fig. 7.6*a*. The reader should verify the step-by-step derivation of the top two gates using only the product and dualizing rules.

Again, applying Procedure 7.1, this time to $F_a{}^d$ and F_b, gives

$$F(\,) = x_1[x_2x_3[x_4x_5]_d]_d \dotplus \bar{x}_1[\bar{x}_3\bar{x}_4]_{SD-\bar{x}_2}$$

The realization for this representation is obtained as follows:

$$F(\,) = x_1[x_2x_3[\langle x_4 + x_5\rangle_{2:1}]_d]_d \dotplus \bar{x}_1[\langle\bar{x}_3 + \bar{x}_4\rangle_{2:1}]_{SD-\bar{x}_2}$$

Applying Corollary 2.10.2 to the first term, and the self-dualizing procedure of Example 4.6 to the second, gives

$$F(\,) = x_1[x_2x_3\langle x_4 + x_5\rangle_{1:0}]_d \dotplus \bar{x}_1\langle\bar{x}_2 + \bar{x}_3 + \bar{x}_4\rangle_{2:1}$$

Applying Theorem 2.1B to both terms yields

$$F(\,) = x_1[x_2\langle 2x_3 + x_4 + x_5\rangle_{3:2}]_d \dotplus \langle 2\bar{x}_1 + \bar{x}_2 + \bar{x}_3 + \bar{x}_4\rangle_{4:3}$$

and reapplying the same theorem to the first term gives

$$F(\,) = x_1[\langle 2x_2 + 2x_3 + x_4 + x_5\rangle_{5:4}]_d \dotplus \langle 2\bar{x}_1 + \bar{x}_2 + \bar{x}_3 + \bar{x}_4\rangle_{4:3}$$

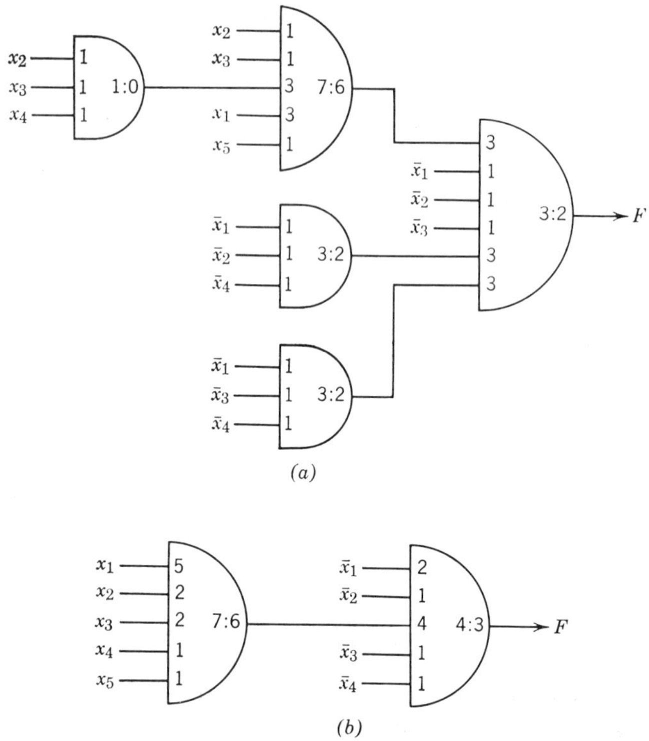

Fig. 7.6. Realizations of Example 7.11. (*a*) Three-level realization. (*b*) Two-level realization.

We then dualize the first term according to Corollary 2.10.2 and then apply Theorem 2.1B so that

$$\begin{aligned} F(\,) &= x_1\langle 2x_2 + 2x_3 + x_4 + x_5\rangle_{2:1} \dotplus \langle 2\bar{x}_1 + \bar{x}_2 + \bar{x}_3 + \bar{x}_4\rangle_{4:3} \\ &= \langle 5x_1 + 2x_2 + 2x_3 + x_4 + x_5\rangle_{7:6} \dotplus \langle 2\bar{x}_1 + \bar{x}_2 + \bar{x}_3 + \bar{x}_4\rangle_{4:3} \end{aligned}$$

Finally the OR operation is obtained according to Theorem 2.1A, so that

$$F(\,) = \langle 4\langle 5x_1 + 2x_2 + 2x_3 + x_4 + x_5\rangle_{7:6} + 2\bar{x}_1 + \bar{x}_2 + \bar{x}_3 + \bar{x}_4\rangle_{4:3}$$

The corresponding network is shown in Fig. 7.6*b*.

7.6 SELF-DUAL REALIZATIONS BASED ON THE FORM $M(\Phi_a,\Phi_b,y)$

The functional form $M(\Phi_a,\Phi_b,y)$ for a self-dual function implies a three-input majority gate realization with input variables Φ_a, Φ_b, and y. The given function Φ, consisting of the sum of p product terms, is expressed as

$$\Phi = M(\Phi_a, \Phi_b, y)$$

(as established by Corollary 7.5.1). The final gate in the realization has inputs Φ_a, Φ_b, and y. But since Φ_a and Φ_b are also self-dual functions, each can be similarly expressed as a three-input majority function of other self-dual functions with still fewer terms, and so on. Thus the process can be iterated until the resulting self-dual functions have only one term and must, therefore, be a single variable.

The procedure is even simpler than it appears, since the following will be shown:

(1) It is not necessary actually to compute Φ_a and Φ_b.
(2) The realization for Φ_a and Φ_b can often share a number of gates.

A Canonic Realization for Self-Dual Functions Using Three-Input Majority Gates (*Akers, 3*)

Let Φ be a given self-dual function (or a self-dual representative function) expressed in minimum sum of products form where p denotes the number of product terms. Arbitrarily number the terms from 1 to p. Let $y_{i,j}$ denote a literal that is common to terms i and j. (If there is more than one variable, use any one of them.)

The canonic realization for Φ is shown in Fig. 7.7. Each circle represents an M_3 gate. All gates except those on the bottom row have two inputs driven by gates on a lower row and one input driven by an independent

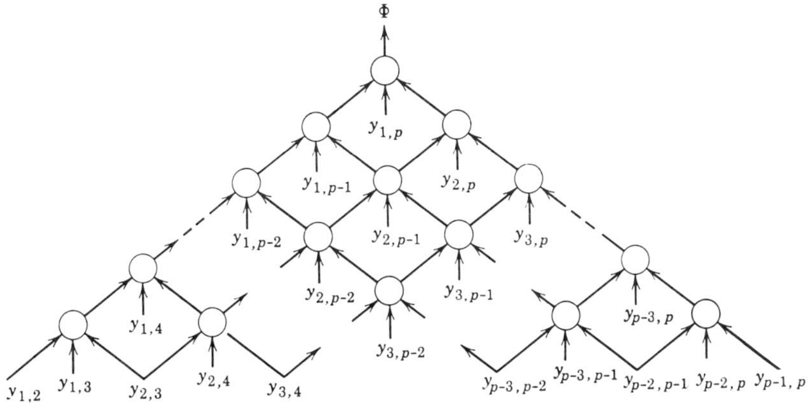

Fig. 7.7. General form of M_3-gate canonic network for Φ.

variable. Note the regularity of numbering of the independent variable inputs along diagonal lines. From this it follows that the number of gates is

$$\binom{p-1}{2} = \frac{1}{2}(p-1)(p-2)$$

The validity of the realization will be established after the following illustrative example.

Example 7.12. Consider the self-dual function of Example 7.5,

$$\Phi(x_1,x_2,\ldots,x_5) = \underset{(1)}{x_1x_2} \dotplus \underset{(2)}{x_1x_3} \dotplus \underset{(3)}{x_1x_4} \dotplus \underset{(4)}{x_1x_5} \dotplus \underset{(5)}{x_2x_3x_4x_5}$$

with the terms numbered as shown. The common literals are

$$\begin{aligned} y_{1,2} &= y_{1,3} = y_{1,4} = y_{2,3} = y_{2,4} = y_{3,4} = x_1 \\ y_{1,5} &= x_2 \\ y_{2,5} &= x_3 \\ y_{3,5} &= x_4 \\ y_{4,5} &= x_5 \end{aligned}$$

The canonic realization is given in Fig. 7.8*a*.

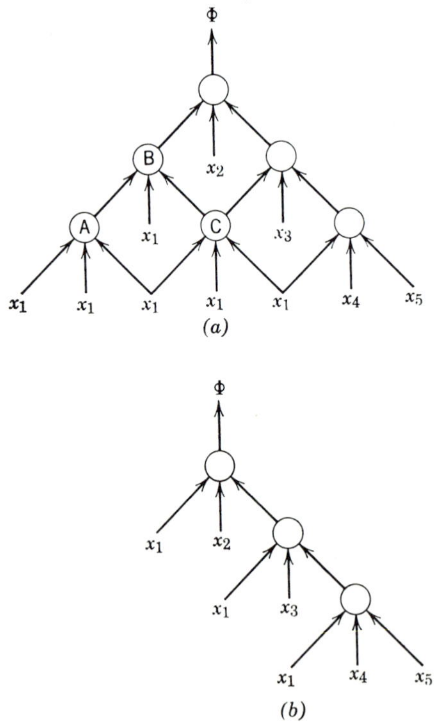

Fig. 7.8. The realization of Example 7.12. (*a*) The canonic realization. (*b*) The simplified realization.

The realization can be simplified by inspection since the three gates in the lower left-hand corner obviously all realize the function x_1. The simplified network is shown in Fig. 7.8*b*. It is readily verified that the realization is valid.

At this point, we establish the validity of the canonic form. For this we derive the following somewhat more general result.

Theorem 7.8. If F is a minor function and if $y_{i,j}$ denotes a literal common to the i and j terms in the sum-of-products expansion, then the canonic network of Fig. 7.7 realizes a self-dual function Φ such that

$$\Phi \geq F$$

Proof. The proof is by induction on the number of terms p. Actually, the procedure is defined only for $p \geq 3$; therefore, we first verify the theorem for $p = 3$.

There are only two essentially different minor functions F with three terms:

$$(1) \quad F(x_1,x_2,\ldots,x_n) = x_1T_1' + x_1T_2' + x_1T_3'$$
$$(2) \quad F(x_1,x_2,\ldots,x_n) = x_1x_2T_1'' + x_1x_3T_2'' + x_2x_3T_3''$$

In Case 1 the same literal is common to all three terms, and in Case 2 different literals are common to each pair of terms. It is easily verified that for Case 1 the output of the canonic network is

$$M(x_1,x_1,x_1) = x_1$$

and for Case 2 the network realizes

$$M(x_1,x_2,x_3) = x_1x_2 + x_1x_3 + x_2x_3$$

Each of these is self-dual and each contains the corresponding minor function F.

Now assume that the realization is valid for $p = m - 1$. We shall show that this implies the validity for $p = m$.

Figure 7.9 shows the realization of the self-dual function Φ that corresponds to the given F. Observe that the network to the left of the dotted line is the canonic realization for the first $m - 1$ terms of Φ. By the induction assumption, the output of its top gate is some self-dual function Φ_a which contains the first $m - 1$ terms. Similarly, from the induction

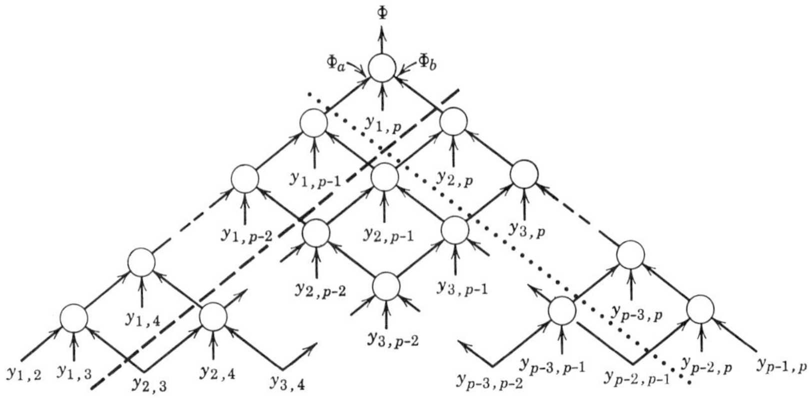

Fig. 7.9. Canonic realizations of Φ, Φ_a, and Φ_b.

assumption, the network to the right of the dashed line realizes some self-dual function Φ_b that contains the last $m - 1$ terms. Thus the output of the entire network is

$$M(y_{1,m},\Phi_a,\Phi_b)$$

It follows from the proof of Corollary 7.5.1 that the output of the network is a self-dual function that contains F. This completes the proof.|

Corollary 7.8.1. If F is self-dual, then the canonic network of Fig. 7.7 realizes F.

Proof. The proof follows from the proof of Corollary 7.5.1.|

The proof of Corollary 7.5.1 implies a number of additional results. The following two results easily follow by the induction argument.

Corollary 7.8.2. In the network of Fig. 7.9, the output of the gate with input $y_{i,j}$ is some self-dual function which contains the terms

$$T_i,T_{i+1},\ldots,T_{j-1},T_j$$

Corollary 7.8.3. If the output of the gate of Fig. 7.9 whose input is $y_{i,j}$ is replaced in the network by *any* self-dual function which contains the terms

$$T,T_{i+1},\ldots,T_{j-1},T_j$$

then the output of the overall network will again be a self-dual function which contains F. Moreover, if F is self-dual, the output will remain unchanged.

The last corollary is often useful in simplifying realizations that are obtained by the canonic procedure. This is demonstrated in the following example.

Example 7.13. Consider the self-dual function of Example 7.12.

$$\Phi(x_1,x_2,\ldots,x_5) = \underset{(1)}{x_1x_2} \dotplus \underset{(2)}{x_1x_3} \dotplus \underset{(3)}{x_1x_4} \dotplus \underset{(4)}{x_1x_5} \dotplus \underset{(5)}{x_2x_3x_4x_5}$$

In Fig. 7.8*a* the function realized by the $y_{1,4}$ gate (Gate B) can be any self-dual function which contains

$$\underset{(1)}{x_1x_2} \dotplus \underset{(2)}{x_1x_3} \dotplus \underset{(3)}{x_1x_4} \dotplus \underset{(4)}{x_1x_5}$$

Referring to Example 7.5, one such function is x_1. Similarly, the function realized by the $y_{2,4}$ gate (Gate C) can be any self-dual function which contains

$$\underset{(2)}{x_1x_3} \dotplus \underset{(3)}{x_1x_4} \dotplus \underset{(4)}{x_1x_5}$$

One such function is again x_1. Thus the realization of Fig. 7.8*b* can be obtained directly.

As should be clear from the last two examples, the particular way in which the terms are numbered may significantly affect the possibilities for making simplifications. One such numbering which frequently leads to a considerable simplification was independently derived by Canaday. Select any variable x_i and factor the given function as

$$\Phi = x_iF_1 \dotplus \bar{x}_iF_0$$

Number the terms of x_iF_1 from 1 to q and those of $\bar{x}_iF_0$ from $q + 1$ to p. Hence x_i is a literal common to the first q terms, and $\bar{x}_i$ is a literal common to the last $p - q$ terms

From the general form of the realization, Fig. 7.7, it follows that the gate with input $y_{1,q}$ and all gates descending from it realize x_i and the gate with input $y_{q+1,p}$ and all gates descending from it realize $\bar{x}_i$. Thus all of these gates can be omitted, and the result is a rectangular, rather than a triangular, array.

The general form of the reduced array is shown in Fig. 7.10. It consists of $q(p - q)$ gates where p is the number of terms in the factored expression for Φ.† This can be compared with

$$\frac{1}{2}(p - 1)(p - 2)$$

gates in the unreduced array. However, it should be noted that in each case p refers to the number of terms in the pertinent expression used for the synthesis and that this number may be larger for the factored than for the

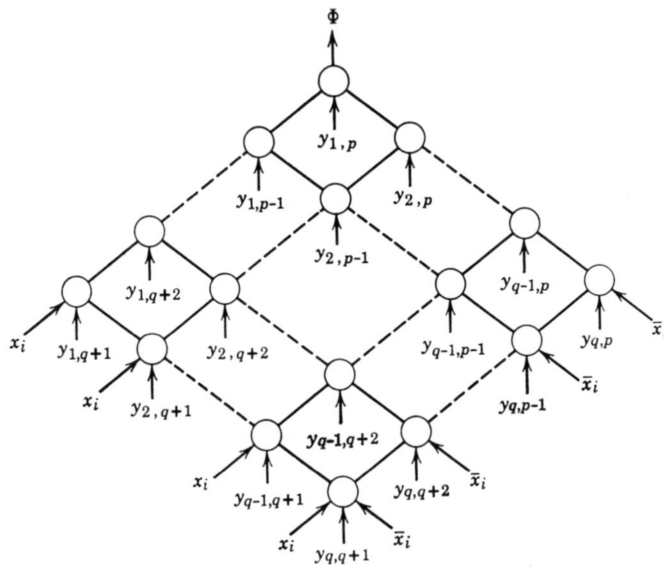

Fig. 7.10. General form of the reduced M_3-gate canonic network for Φ. The terms are numbered according to the factoring $\Phi = x_iF_1 + \bar{x}_iF_0$.

original expression. This situation can occur when Φ has terms that do not contain either x_i or $\bar{x}_i$. In the factored form these terms must be included

† An extension of these ideas, also given by Canaday, is possible when there are two literals x and y such that Φ can be written

$$\Phi = xF_a + xy + yF_b$$

The terms in xF_a are numbered from 1 to q and the terms in yF_b from $q + 2$ to p. Again several gates from the lower left and lower right corners of the general network of Fig. 7.7 can be eliminated, resulting in a rectangular array of $q(p - q)$ gates similar to that of Fig. 7.10. The only difference is that those gates that have inputs of x_i and $\bar{x}_i$ in Fig. 7.10 have inputs of x and y, respectively, in the new array.
Canady also gives some simplification methods for both of his reduced arrays.

in both x_iF_1 and $\bar{x}_iF_0$, and if the resulting expressions do not simplify, then a net increase in the number of terms results. (An example of this situation occurs in Example 7.14.)

Although the canonic procedures are simple, they often yield uneconomical realizations even when the obvious simplifications are made. One reason for this is that unnecessary terms may be realized in both Φ_a and Φ_b. Suppose that y is the variable that is the input of the final gate. Then F can be expressed as

$$F(\,) = yF_a(\,) \dotplus F_b(\,)$$

where F_b can include appearances of $\bar{y}$. Now if F_a is expressed as the sum of two functions,

$$F_a(\,) = F_{a_1}(\,) \dotplus F_{a_2}(\,)$$

so that

$$F(\,) = y[F_{a_1}(\,) \dotplus F_{a_2}(\,)] \dotplus F_b(\,)$$

then it follows that F can be expressed as

$$F(\,) = M(y,[yF_{a_2}(\,) \dotplus F_b(\,)],[yF_{a_1}(\,) \dotplus F_b(\,)])$$

This is analogous to Corollary 7.5.1. The realization procedure under discussion is a special case of this expansion where

$$yF_{a_1}(\,) \dotplus F_b(\,) = T_1 \dotplus T_2 \dotplus \cdots \dotplus T_{p-1}$$
$$yF_{a_2}(\,) \dotplus F_b(\,) = T_2 \dotplus \cdots \dotplus T_{p-1} \dotplus T_p$$

Note that the terms $T_2,T_3,\ldots,T_{p-1}$ are realized in both functions. The following is an example in which this duplication causes unnecessary complexity.

Example 7.14. Consider the self-dual function

$$\Phi(x_1,x_2,\ldots,x_6) = \underset{(1)}{x_1x_3x_5} \dotplus \underset{(2)}{x_1x_5x_6} \dotplus \underset{(3)}{x_3x_5x_6} \dotplus \underset{(4)}{x_4x_5x_6}$$
$$\dotplus \underset{(5)}{x_2x_5x_6} \dotplus \underset{(6)}{x_1x_4x_6} \dotplus \underset{(7)}{x_1x_2x_3x_4} \dotplus \underset{(8)}{x_1x_2x_6}$$
$$\dotplus \underset{(9)}{x_3x_4x_6} \dotplus \underset{(10)}{x_2x_3x_6} \dotplus \underset{(11)}{x_2x_4x_5}$$

Since there are 11 terms, the canonic realization of Fig. 7.7 requires

$$\frac{10 \cdot 9}{2} = 45$$

gates.

To use the reduced canonic realization of Fig. 7.10, we can factor with x_1 (which, as it turns out, is as good a variable as any).

$$\Phi(\;) = x_1(x_3x_5 \dotplus x_5x_6 \dotplus x_4x_6 \dotplus x_2x_3x_4 \dotplus x_2x_6 \dotplus x_2x_4x_5)$$
$$\dotplus \bar{x}_1(x_3x_5x_6 \dotplus x_4x_5x_6 \dotplus x_2x_5x_6 \dotplus x_3x_4x_6 \dotplus x_2x_3x_6 \dotplus x_2x_4x_5)$$

Note that this expression requires 12 terms, since the term $x_2x_4x_5$ must be included with both the x_1F_1 and the $\bar{x}_1F_0$ terms. Thus $p = 12$ and $q = 6$ and the reduced canonic realization requires

$$6(12 - 6) = 36$$

gates, which is a considerable reduction.

Now, however, suppose that we let

$$y_{1,11} = y = x_5$$

$$yF_{a_1}(\;) = \underset{(1)}{x_1x_3x_5} \dotplus \underset{(2)}{x_1x_5x_6} \dotplus \underset{(3)}{x_3x_5x_6}$$

$$yF_{a_2}(\;) = \underset{(4)}{x_4x_5x_6} \dotplus \underset{(5)}{x_2x_5x_6} \dotplus \underset{(11)}{x_2x_4x_5}$$

$$F_b(\;) = \underset{(6)}{x_1x_4x_6} \dotplus \underset{(7)}{x_1x_2x_3x_4} \dotplus \underset{(8)}{x_1x_2x_6} \dotplus \underset{(9)}{x_3x_4x_6} \dotplus \underset{(10)}{x_2x_3x_6}$$

from which it follows that

$$\Phi(\;) = M(y,[yF_{a_1}(\;) \dotplus F_b(\;)],[yF_{a_2}(\;) \dotplus F_b(\;)])$$

and that

$$\Phi(\;) = M(y,\Phi_a,\Phi_b)$$

where Φ_a and Φ_b are any self-dual functions such that

$$\Phi_a \geq yF_{a_1} \dotplus F_b$$
$$\Phi_b \geq yF_{a_2} \dotplus F_b$$

By deleting literals from terms in

$$yF_{a_1}(\;) \dotplus F_b(\;) \qquad \text{and} \quad yF_{a_2}(\;) \dotplus F_b(\;)$$

so that the remaining functions still have the property that each pair of terms has a literal in common, an appropriate Φ_a and Φ_b can be found. Starting with

$$yF_{a_1}(\;) \dotplus F_b(\;) = x_1x_3x_5 \dotplus x_1\enclose{circle}{x_5}x_6 \dotplus x_3\enclose{circle}{x_5}x_6 \dotplus x_1x_4x_6$$
$$\dotplus x_1\enclose{circle}{x_2}x_3\enclose{circle}{x_4} \dotplus x_1x_2x_6 \dotplus x_3x_4x_6 \dotplus x_2x_3x_6$$

and deleting the encircled literals, we obtain after simplification

$$x_1x_3 \dot{+} x_1x_6 \dot{+} x_3x_6$$

which is self-dual and can therefore be Φ_a.

Similarly, starting with

$$\begin{aligned} yF_{a_2}(\,) \dot{+} F_b(\,) = {} & x_4 \textcircled{x_5} x_6 \dot{+} x_2 \textcircled{x_5} x_6 \dot{+} x_2x_4 \textcircled{x_5} \dot{+} x_1x_4x_6 \\ & \dot{+} x_1x_2x_3x_4 \dot{+} x_1x_2x_6 \dot{+} x_3x_4x_6 \dot{+} x_2x_3x_6 \end{aligned}$$

we obtain after simplification

$$x_4x_6 \dot{+} x_2x_6 \dot{+} x_2x_4$$

which is self-dual and can, therefore, be Φ_b

Thus Φ can be expressed as

$$\Phi = M(x_5, M(x_1, x_3, x_6), M(x_2, x_4, x_6))$$

which requires only three gates.

Although the above realization was not obtained using an explicit, formal procedure, it demonstrates the inefficiency of the canonic network, and the economies that can often be achieved by using some of the properties of self-dual functions.

Chapter 8

ADVANCED TREE OPERATIONS— A TEST FOR LINEAR SEPARABILITY

8.1 INTRODUCTION

The function tree displays much of the structure of a function that is important in its realization with T gates. One of the advantages of synthesis methods using the tree is that attention is focused on only a small part of the structure at any one time and, in fact, much of the structure is not used at all. There are situations, however, where additional structure information is needed to improve or extend the realization procedure.

An important case in point is the development of a test for linear separability. If the single-gate tree realization procedure of Chapter 5 fails (that is, if a common range is void) then the question of whether or not the given function is ℓ.s. cannot be answered within the framework of that procedure. As we shall see, however, this question can be answered by additional manipulations on the tree.

Other examples of situations where additional information is needed are: (1) in determining the maximally reliable single-gate realization, that is, the realization with the largest relative gap or the smallest normalized fan-in; (2) in determining the smallest number of gates required to realize a given function. This chapter is concerned with a test for ℓ.s. functions and maximally reliable realizations.

8.2 LINEAR SEPARABILITY TEST: THE BASIC IDEA

The single-gate realization procedure of Chapter 5 fails to yield a realization when on some k-level the common range on a_k is void (see Section 5.5). It is always true that each k-level function pair defines a non-void range on a_k. If the ranges defined by the various function pairs do not have a common intersection, however, then there is no value of a_k that is suitable for all function pairs and the procedure cannot be continued.

It was shown in Example 5.4 that the given function may be ℓ.s. even when the procedure fails. The reason for this was discussed in Section 5.5 and in Chapter 4. The *common range* computed in the single-gate procedure is the range on a_k for going up one level of the tree. The procedure does not compute the *actual range* (the range on a_k for getting all the way to the top of the tree), since the actual range involves the intersection of all of the k-level gaps suitably displaced and its computation requires the solution of a set of simultaneous inequalities.

The actual range must lie entirely within the common range, but it may be smaller in length. Thus if on any k-level of the tree the value of a_k is selected inside the common range but outside the actual range, then the procedure will fail on some higher level of the tree even though the given function is ℓ.s. It is on these concepts that the following basic idea of the ℓ.s. test is based.

Assuming that the single-gate procedure has failed on the k-level of the tree, the void range is traced down the tree. On each lower level it is determined whether a different value for the coefficient selected on that level would allow a nonvoid k-level common range for a_k. If such values exist, a narrower range is defined for that coefficient. A value is selected within that range, and the procedure is continued up the tree in the usual way. If no such value exists, then the void range is traced to the next lower level and the procedure is repeated. If it is necessary to trace the void range to the bottom level of the tree and there are no values of the coefficients that will allow a nonvoid k-level range, then it is established that the given function is not ℓ.s. These ideas are clarified in the following two examples.

Example 8.1. Figure 8.1 shows the first four levels of a function tree. The gaps shown on the 3-level have been determined from the lower levels of the tree. These yield a common range for a_3 which is

$$\overline{R(a_3)} = 1\tfrac{1}{4}{*}1$$

According to the coefficient selection rule, a_3 is chosen as $1\frac{1}{8}$, and this yields the 2-level gaps shown in Fig. 8.1. These yield a void 2-level common range for a_2.

The 2-level cause of the void common range is indicated by the two underlined 2-level range boundaries. One of these is an upper range boundary with value $1\frac{3}{8}$ while the other is a lower range boundary with the same value. Denote these as $\mathscr{U}(a_2)$ and $\mathscr{L}(a_2)$, respectively. In order that the common range for a_2 be nonvoid, it is necessary that

$$\mathscr{U}(a_2) > \mathscr{L}(a_2)$$

a condition which is not satisfied at present.

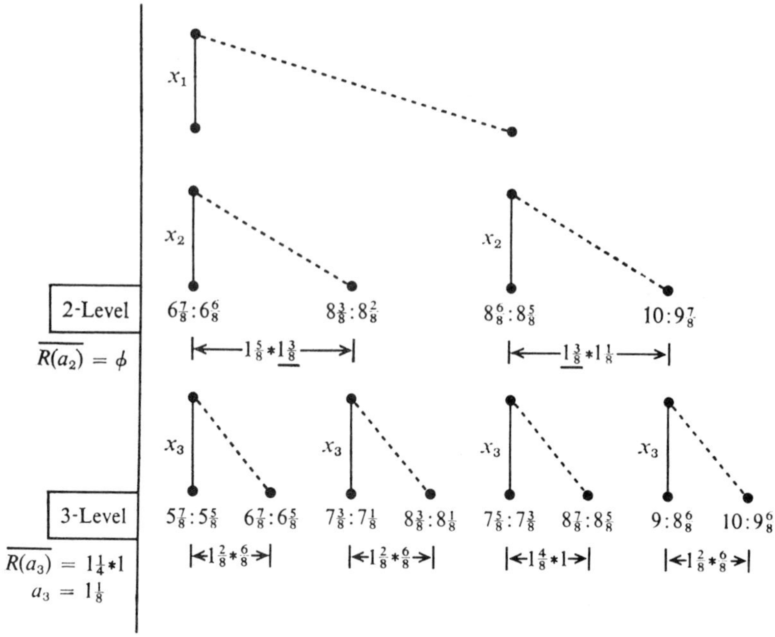

Fig. 8.1. Function tree for Example 8.1 where $\overline{R(a_2)} = \phi$.

We now wish to determine whether a different value for a_3 would allow the above condition to be satisfied. To this end we express $\mathscr{U}(a_2)$ and $\mathscr{L}(a_2)$ as functions of a_3.

Considering $\mathscr{L}(a_2)$ first, it is expressed in terms of the 2-level gap boundaries as†

$$\mathscr{L}(a_2) = [8\tfrac{2}{8}] - [6\tfrac{7}{8}]$$

which is the way in which the lower range boundary was computed. Now each of these 2-level gap boundaries can be expressed in terms of displaced 3-level gap boundaries, some of which may be a function of the choice of a_3.

Consider first the 2-level gap boundary $[8\tfrac{2}{8}]$. It is the maximum of the 3-level displaced boundaries

$$[8\tfrac{1}{8}] \quad \text{and} \quad [7\tfrac{1}{8}] + a_3$$

which for $a_3 = 1\tfrac{1}{8}$ is such that

$$[8\tfrac{2}{8}] = [7\tfrac{1}{8}] + a_3$$

In the expression, a_3 is expressed as a variable.

† Square brackets around a quantity denote it as a gap boundary

The other 2-level gap boundary in the expression for $\mathscr{L}(a_2)$ is $[6\frac{7}{8}]$. It is the minimum of the 3-level displaced gap boundaries

$$[6\tfrac{7}{8}] \qquad \text{and} \quad [5\tfrac{7}{8}] + a_3$$

which for $a_3 = 1\frac{1}{8}$ is such that

$$[6\tfrac{7}{8}] = [6\tfrac{7}{8}]$$

Substituting these values in the expression for $\mathscr{L}(a_2)$ gives

$$\begin{aligned} \mathscr{L}(a_2) &= [8\tfrac{2}{8}] - [6\tfrac{7}{8}] \\ &= [7\tfrac{1}{8}] + a_3 - [6\tfrac{7}{8}] \\ &= \tfrac{1}{4} + a_3 \end{aligned} \tag{8.1}$$

Applying the same procedure to $\mathscr{U}(a_2)$, we first express $\mathscr{U}(a_2)$ in terms of the 2-level gap boundaries:

$$\mathscr{U}(a_2) = [10] - [8\tfrac{5}{8}]$$

Both of these 2-level gap boundaries are caused by the 3-level boundaries that are not displaced by a_3 (when $a_3 = 1\frac{1}{8}$). Therefore, when $\mathscr{U}(a_2)$ is expressed in terms of the appropriate 3-level gap boundaries, it is not a function of a_3; that is,

$$\begin{aligned} \mathscr{U}(a_2) &= [10] - [8\tfrac{5}{8}] \\ &= 1\tfrac{3}{8} \end{aligned} \tag{8.2}$$

The condition that must be satisfied if $\overline{R(a_2)}$ is to be nonvoid is that

$$\mathscr{U}(a_2) > \mathscr{L}(a_2)$$

Substituting Eqs. 8.1 and 8.2 into this inequality gives

$$1\tfrac{3}{8} > \tfrac{1}{4} + a_3$$

which implies

$$\begin{aligned} 1\tfrac{3}{8} - \tfrac{1}{4} &> a_3 \\ 1\tfrac{1}{8} &> a_3 \end{aligned}$$

Thus we have determined a new constraint on the value of a_3. It is a constraint that was violated by our previous choice and one that must be satisfied in order that $\overline{R(a_2)}$ be nonvoid.

Combining this constraint with the previously determined common range

$$R(a_3) = 1\tfrac{1}{4}*1$$

gives a new range on a_3

$$1\tfrac{1}{8}*1$$

within which the value of a_3 must lie.

We now proceed with the usual single-gate procedure using this new range. Choosing $a_3 = 1\frac{1}{16}$, the procedure continues to the top of the tree without further difficulty, as shown in Fig. 8.2.

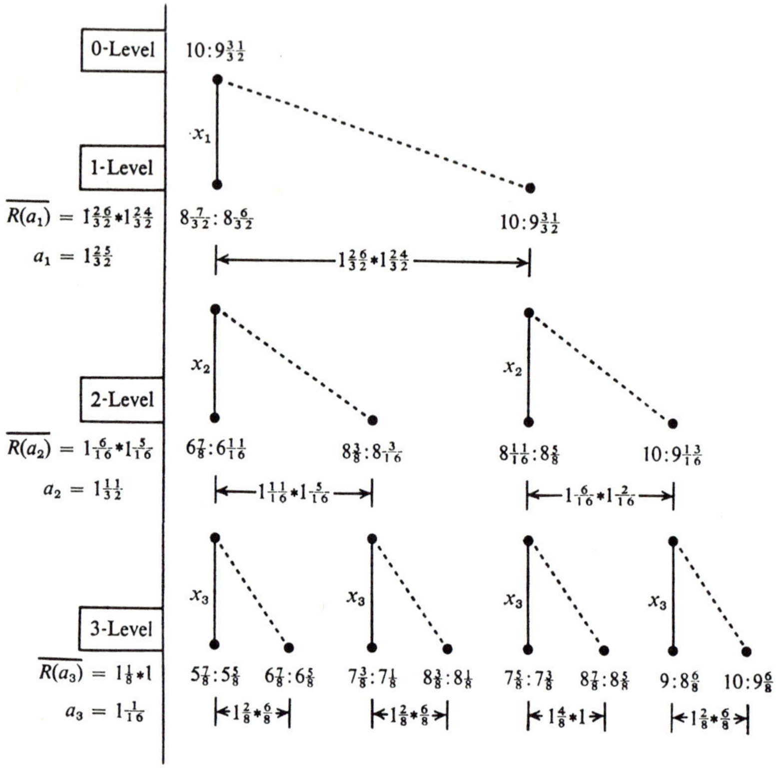

Fig. 8.2. Function tree for Example 8.1 when the realization is completed.

This example illustrates the basic idea of the ℓ.s. test. When a void range is obtained, the desired condition is expressed as a function of one of the coefficients previously selected to determine whether a different choice for that coefficient would yield the desired condition. We shall show later how the procedure is handled when it must be carried beyond one lower level. We remark here, however, that it is never necessary to solve any inequalities simultaneously.

Before describing the general procedure, we consider one more example; this time the function is not linearly separable.

Example 8.2 (*Moore's Function*). In Example 5.3 we attempted a realization of Moore's function using the reduced symmetric tree. The procedure failed on the 1-level, as shown in Fig. 8.3, because of the underlined range boundaries.

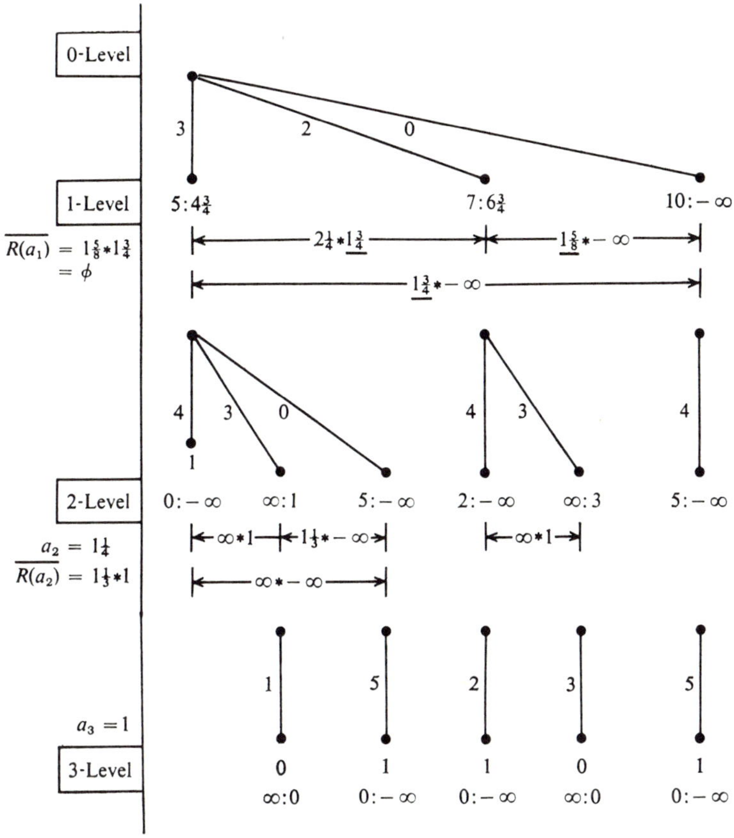

Fig. 8.3. The tree for Moore's function: Example 8.2.

There are two upper range boundaries, $1\frac{3}{4}$ and $1\frac{5}{8}$, both of which do not exceed the lower range boundary $1\frac{3}{4}$. Let $\mathscr{U}(a_1)$ denote the set of these two upper range boundaries,

$$\mathscr{U}(a_1) = \begin{Bmatrix} 1\frac{3}{4} \\ 1\frac{5}{8} \end{Bmatrix}$$

and let $\mathscr{L}(a_1)$ denote the lower range boundary

$$\mathscr{L}(a_1) = 1\frac{3}{4}$$

The void common range for a_1 results from a violation of the necessary condition

$$\mathscr{U}(a_1) > \mathscr{L}(a_1)$$

where the inequality means that each member of the set $\mathscr{U}(a_1)$ must be larger than each member of the set $\mathscr{L}(a_1)$.

The range boundaries of $\mathscr{U}(a_1)$ and $\mathscr{L}(a_1)$ are then expressed in terms of the appropriate 1-level gap boundaries as

$$\mathscr{U}(a_1) = \begin{Bmatrix} 1\tfrac{3}{4} = \tfrac{1}{3}([10] - [4\tfrac{3}{4}]) \\ 1\tfrac{5}{8} = \tfrac{1}{2}([10] - [6\tfrac{3}{4}]) \end{Bmatrix}$$

$$\mathscr{L}(a_1) = 1\tfrac{3}{4} = [6\tfrac{3}{4}] - [5]$$

Now each of the 1-level gap boundaries is expressed in terms of the corresponding 2-level gap boundaries appropriately displaced by a_2. Considering $\mathscr{U}(a_1)$ first, we get

$$\mathscr{U}(a_1) = \begin{Bmatrix} \tfrac{1}{3}(([5] + 4a_2) - ([1] + 3a_2)) \\ \tfrac{1}{2}(([5] + 4a_2) - ([3] + 3a_2)) \end{Bmatrix}$$

and this simplifies to

$$\mathscr{U}(a_1) = \begin{Bmatrix} \tfrac{1}{3}(4 + a_2) \\ \tfrac{1}{2}(2 + a_2) \end{Bmatrix}$$

The $[6\tfrac{3}{4}]$ boundary of $\mathscr{L}(a_1)$ can be similarly expressed as

$$[6\tfrac{3}{4}] = [3] + 3a_2$$

The [5] boundary of $\mathscr{L}(a_1)$ is the result of two 2-level displaced gaps

$$[5] = [5]$$
$$[5] = [0] + 4a_2$$

Thus $\mathscr{L}(a_1)$ consists of the two expressions

$$\mathscr{L}(a_1) = \begin{Bmatrix} ([3] + 3a_2) - [5] \\ ([3] + 3a_2) - 4a_2 \end{Bmatrix}$$

and these simplify to

$$\mathscr{L}(a_1) = \begin{Bmatrix} -2 + 3a_2 \\ 3 - a_2 \end{Bmatrix}$$

The required condition that

$$\mathscr{U}(a_1) > \mathscr{L}(a_1)$$

implies the four inequalities obtained by requiring that each member of $\mathscr{U}(a_1)$ be larger than each member of $\mathscr{L}(a_1)$. Therefore

$$\tfrac{1}{3}(4 + a_2) > -2 + 3a_2$$
$$\tfrac{1}{3}(4 + a_2) > 3 - a_2$$
$$\tfrac{1}{2}(2 + a_2) > -2 + 3a_2$$
$$\tfrac{1}{2}(2 + a_2) > 3 - a_2$$

When simplified, these become

$$1\tfrac{1}{4} > a_2$$
$$a_2 > 1\tfrac{1}{4}$$
$$1\tfrac{1}{5} > a_2$$
$$a_2 > 1\tfrac{1}{3}$$

These four conditions are inconsistent with the previously determined common range on a_2,

$$\overline{R(a_2)} = 1\tfrac{1}{3} * 1$$

and, in fact, are inconsistent among themselves. Since these are the conditions that a_2 must satisfy if the common range on a_1 is to be nonvoid, we can conclude that there is no choice for a_2 which will make $\overline{R(a_1)}$ nonvoid.

It must now be determined whether a different choice for a_3 will allow a nonvoid common range for a_1. The answer in this case is negative, since a_3 is the smallest coefficient and can be arbitrarily selected without adversely affecting the other ranges. Therefore, since there are no choices for a_2 and a_3 which will make the common range for a_1 nonvoid, we conclude that the given function is not ℓ.s.

8.3 LINEAR SEPARABILITY TEST: AN INFORMAL DESCRIPTION

The basic idea of the procedure should now be fairly clear. Although many of the situations that can occur have been illustrated by example, there are points that require additional clarification. This section provides an informal description of the procedure and introduces some additional notation in order to make the description more precise and compact. A step-by-step description and a proof of its validity are given in the following section.

Assume that the single-gate procedure of Chapter 5 has failed on the k-level, in which case

$$\overline{R(a_k)} = \overline{U(a_k)} * \overline{L(a_k)}$$

is void because

$$\overline{U(a_k)} \leq \overline{L(a_k)}$$

Let $\mathscr{U}(a_k)$ be the set of all upper range boundaries of a_k which are equal to or less than $\overline{L(a_k)}$ and let $\mathscr{L}(a_k)$ be the set of all lower range boundaries of a_k that are greater than or equal to $\overline{U(a_k)}$. These are called the *inconsistent upper and lower range boundary sets*, respectively. By the notation

$$\mathscr{U}(a_k) > \mathscr{L}(a_k)$$

we mean the set of inequalities consisting of one member each from $\mathscr{U}(a_k)$ and $\mathscr{L}(a_k)$. This set includes all of the inequalities that are not satisfied because $\overline{R(a_k)}$ is void. Note that this definition corresponds exactly to the $\mathscr{U}(a_k)$ and $\mathscr{L}(a_k)$ used in the two previous examples.

The various k-level range boundaries that appear in the inconsistent range boundary sets are expressed as sums or differences of appropriate k-level gap boundaries. Thereafter, the key step in the procedure is to express these k-level gap boundaries as $(k + 1)$-level gap boundaries appropriately displaced by a_{k+1} where the latter is considered as a variable.

In the single-gate procedure of Chapter 5, each k-level gap boundary is computed as the maximum or minimum of several displaced $(k + 1)$-level gap boundaries. However, not all of the displaced $(k + 1)$-level gap boundaries necessarily achieve the maximum or minimum. For example, a gap boundary may be

$$u = \min \begin{Bmatrix} 4 \\ 4 \\ 5 \end{Bmatrix}$$

where only the two boundaries of 4 actually achieve the minimum. For a particular choice of a_{k+1}, those displaced gap boundaries which actually achieve the maximum or minimum and hence determine the k-level gap boundaries are called the *dominant gap boundaries*. For computational simplicity, when the k-level gaps are expressed in terms of the $(k + 1)$-level gaps, displaced by a_{k+1}, only the $(k + 1)$-level dominant gap boundaries are used; where by dominant we mean those boundaries which were dominant for the value of a_{k+1} previously selected.

After we have expressed $\mathscr{U}(a_k)$ and $\mathscr{L}(a_k)$ as functions of the $(k + 1)$-level dominant gap boundaries and a_{k+1} considered as a variable, the condition

$$\mathscr{U}(a_k) > \mathscr{L}(a_k)$$

implies a set of inequalities with a_{k+1} as the only variable. Some of these inequalities are of the form

$$1 + a_{k+1} > 2$$

or of the form

$$3 - a_{k+1} > 1 + 2a_{k+1}$$

which imply new upper and lower bounds on a_{k+1}. Let $U_I(a_{k+1})$ be the smallest of all of these new upper range boundaries and $L_I(a_{k+1})$ be the largest of the lower boundaries. These in turn define the *implied range* on a_{k+1}, denoted by

$$R_I(a_{k+1}) = U_I(a_{k+1}) * L_I(a_{k+1})$$

The implied range $R_I(a_{k+1})$ is void if

$$U_I(a_{k+1}) \leq L_I(a_{k+1})$$

Otherwise it is defined to be those values of y such that

$$U_I(a_{k+1}) > y > L_I(a_{k+1})$$

The implied range on a_{k+1} specifies those restrictions on the values of a_{k+1} which are implied by the dominant gap boundaries of the inequality set

$$\mathscr{U}(a_k) > \mathscr{L}(a_k)$$

These are necessary conditions on the value of a_{k+1} which were violated by the previous choice for that coefficient.

If the implied range for a_{k+1} is void, or if it is nonvoid but has a void intersection with the common range $\overline{R(a_{k+1})}$, then, clearly, no different choice for a_{k+1} (keeping $a_{k+2}, a_{k+3}, \ldots, a_n$ fixed) will allow a nonvoid common range for a_k.

If the implied range for a_{k+1} is nonvoid and has a nonvoid intersection with the common range $\overline{R(a_{k+1})}$, then there may be a different choice for a_{k+1} which will yield a nonvoid $\overline{R(a_k)}$. However, there still may be inconsistencies of another type.

In the set of inequalities

$$\mathscr{U}(a_k) > \mathscr{L}(a_k)$$

there may be inequalities different from those previously considered that are of the form

$$3 + a_{k+1} > 2 + a_{k+1}$$

or of the form

$$4 > 5$$

Such inequalities do not imply a range on a_{k+1} and are called *residual ranges*. If they are consistent, as in the first of the above cases, they place no restriction on the choice of a_{k+1} and can be ignored. If they are inconsistent as in the second inequality above, then, clearly, no choice for a_{k+1} can make the inequality set defined by

$$\mathscr{U}(a_k) > \mathscr{L}(a_k)$$

consistent.

Determining the residual and implied ranges for a_{k+1} completes one stage or cycle of the procedure. If both ranges are nonvoid, and if the intersection of the implied range and the common range is also nonvoid, then a value of a_{k+1} is selected within this intersection and the single gate procedure is continued up the tree.

If either the residual range or the intersection of the implied and common ranges is void, then it has been proved that there is no choice for a_{k+1} (keeping $a_{k+2}, a_{k+3}, \ldots, a_n$ fixed) that will allow a nonvoid $\overline{R(a_k)}$. In this case, we proceed downward to the $(k + 2)$-level to determine whether a different choice for a_{k+2} will allow a nonvoid common range on a_k. This is accomplished with one or both of the following steps.

(1) If the intersection of the implied range and the common range on a_{k+1} is void, determine the inconsistent boundary sets $\mathscr{U}(a_{k+1})$ and $\mathscr{L}(a_{k+1})$ and repeat the same procedure.

(2) If the residual range is inconsistent, express its gap boundaries as $(k + 2)$-level displaced gaps and proceed as before.

The procedure is continued in the same way. If it reaches the top of the tree, a realization is obtained. If it reaches the bottom of the tree, the function is not linearly separable. A proof of the validity of the procedure is given following the formal description in the next section.

8.4 LINEAR SEPARABILITY TEST: FORMAL DESCRIPTION

The single-gate realization procedure of Chapter 5 (Procedure 5.1) is carried out until, on some k-level of the tree, the common range $\overline{R(a_k)}$ is void. Then proceed as follows:

(1) Determine the inconsistent boundary sets $\mathscr{U}(a_k)$ and $\mathscr{L}(a_k)$.

(2) Express each member of $\mathscr{U}(a_k)$ and $\mathscr{L}(a_k)$ in terms of the k-level gap boundaries.

(3) Express each k-level gap boundary of Step 2 in terms of its $(k + 1)$-level dominant gap boundaries expressed as a function of a_{k+1}.

(4) From the set of inequalities implied by

$$\mathscr{U}(a_k) > \mathscr{L}(a_k)$$

determine the implied range on a_{k+1} and the residual ranges.

(5) Determine the intersection of the implied range and the common range $\overline{R(a_{k+1})}$. Call this intersection the *new common range* on a_{k+1}.

(a) If both the new common range on a_{k+1} and the residual ranges are nonvoid, select a value of a_{k+1} as the mid-value of the new common range and continue the single-gate realization procedure.

(b) If either the new common range or a residual range is void, then proceed as follows:

1. If $k + 2 = n$, the given function is not ℓ.s. since a_n is arbitrary.

2. If $k + 2 < n$, then determine for each of the void ranges the inconsistent boundary sets. Starting with Step 1, repeat this procedure on each of the void ranges, replacing k by $k + 1$.

The complete linear separability test is somewhat iterative in nature. In the first part (Procedure 5.1), we compute certain simple range constraints and make a selection for each coefficient within that range. If it turns out that some constraints not considered by the procedure place restrictions on the coefficients that are not fulfilled by one of the coefficients selected, then Procedure 5.1 will fail to yield a realization. The procedure of Section 8.4 then allows us to trace that inconsistency back down the tree and to determine at least one restriction that the first choice did not fulfill. At this point a somewhat narrower range is determined for some of the coefficients, and a new selection is made within this range. It may be that there are still unfulfilled restrictions that are as yet undiscovered, and if so, the entire procedure of Section 8.4 must be repeated when Procedure 5.1 again fails.

If the procedure terminates at the top of the tree, the function has been proved ℓ.s. and a realization has been obtained. If it terminates at the bottom of the tree, the function has been proved non-ℓ.s. since a fundamental inconsistency has been found among the truth table points (on the bottom level of the tree). We shall now show that the procedure will always terminate at either the top or the bottom of the tree (that is, it will not oscillate indefinitely up and down the tree). From this it follows that the procedure is in fact a necessary and sufficient test.

Assume that the procedure does not terminate but that it oscillates indefinitely, never reaching the top or the bottom of the tree. There must then be a lowest level of the tree to which it returns repeatedly in the course of its oscillations. Call this the ℓ-level. Each time it returns to this level, there is some inconsistency among ℓ-level ranges that can be eliminated by a different choice for a_ℓ. The previous range for a_ℓ was the intersection of the ranges produced by all of the range constraints previously checked. Thus the previous choice for a_ℓ satisfied all of the range constraints previously checked. The fact that this choice for a_ℓ is not adequate, but that there is another choice for a_ℓ which is adequate, implies that there is an ℓ-level range constraint that has not yet been checked. But there are only a finite number of pertinent ℓ-level range constraints. Thus the procedure cannot return repeatedly to the ℓ-level, and the desired property is proved. Thus we have the following theorem.

Theorem 8.1. The procedure consisting of Procedure 5.1 and the procedure of Section 8.4 constitutes a test for linear separability.

We now consider one additional example of the test for linear separability that is somewhat more complex than those previously considered.

Example 8.3. In Example 5.4 an ℓ.s. function was shown for which the single-gate realization procedure failed. The reduced tree of Fig. 5.8*b* is repeated in Fig. 8.4 with the inconsistent ranges on the 2-level underlined.

$\mathscr{U}(a_2)$ has one member; that is,

$$\mathscr{U}(a_2) = 4$$

$\mathscr{L}(a_2)$ has two members; one is the 4, that is underlined, and the other is due to the ordering inequality

$$a_2 > a_3$$

Thus

$$\mathscr{L}(a_2) = \begin{Bmatrix} 4 \\ a_3 = 4 \end{Bmatrix}$$

Expressing $\mathscr{U}(a_2)$ in terms of the 2-level gap boundaries gives

$$\mathscr{U}(a_2) = 4 = [10] - [6]$$

and expressing each of these 2-level gap boundaries as 3-level displaced gap boundaries gives

$$\mathscr{U}(a_2) = ([6] + a_3) - ([2] + a_3) = 4$$

Thus $\mathscr{U}(a_2)$ is independent of a_3 where

$$\mathscr{U}(a_2) = [6] - [2] = 4$$

where [6] and [2] are 3-level gap boundaries.

Expressing the first member of $\mathscr{L}(a_2)$ in terms of the 2-level gap boundaries gives

$$\mathscr{L}(a_2) = 4 = [5] - [1]$$

and expressing these as 3-level displaced gap boundaries gives

$$\mathscr{L}(a_2) = [5] - [1] = 4$$

Hence $\mathscr{L}(a_2)$ can be expressed as

$$\mathscr{L}(a_2) = \begin{Bmatrix} [5] - [1] = 4 \\ a_3 \end{Bmatrix}$$

The inequalities implied by

$$\mathscr{U}(a_2) > \mathscr{L}(a_2)$$

are

$$4 > 4 \qquad \text{residual range}$$

$$4 > a_3 \qquad \text{implied range}$$

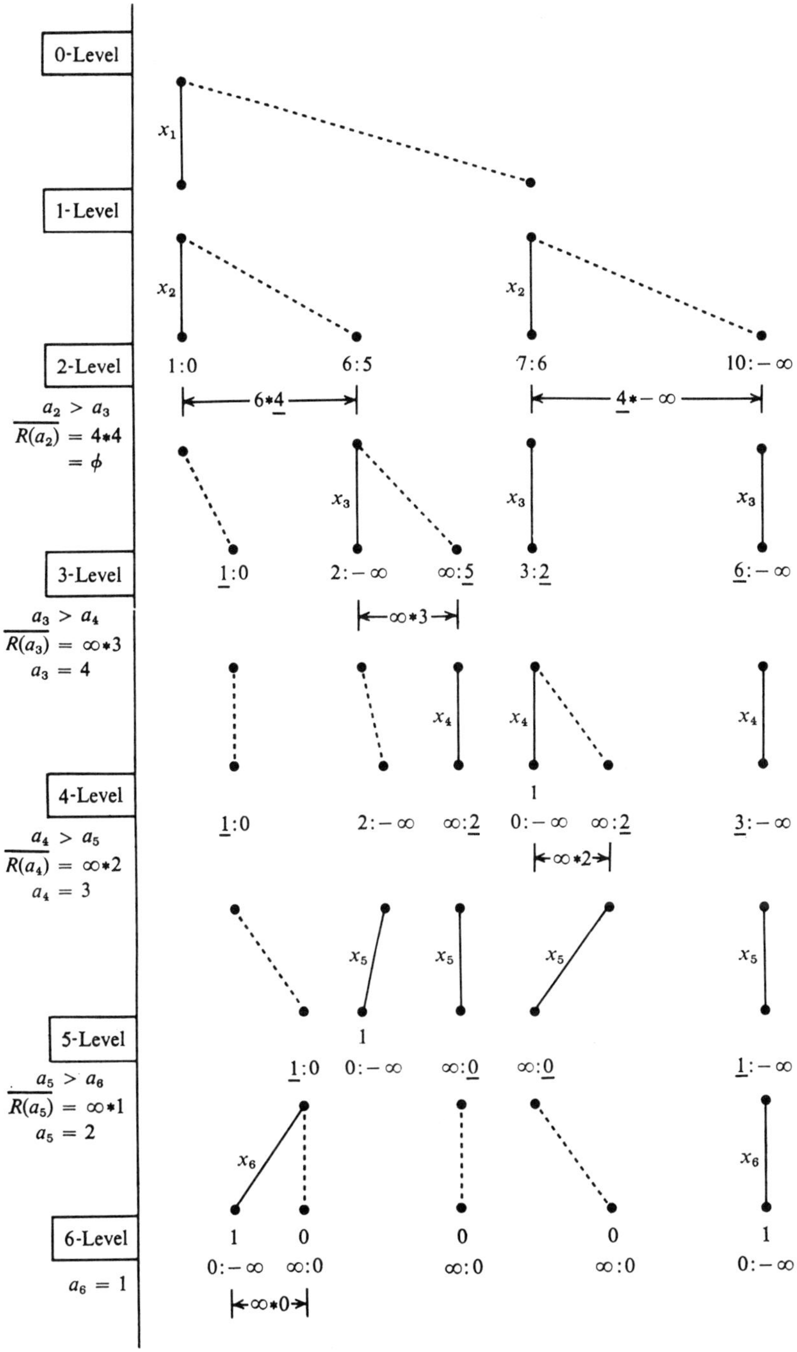

Fig. 8.4. Reduced function tree of Examples 8.3 and 5.4.

The implied range has a nonvoid intersection with the common range on a_3, $(\infty * 3)$; however, the residual range is void. Therefore, the residual range must be taken to the next lower level.

The new inconsistent boundary sets are

$$\mathscr{U}_1 = 4$$
$$\mathscr{L}_1 = 4$$

These have already been expressed as the underlined 3-level gap boundaries

$$\mathscr{U}_1 = 4 = [6] - [2]$$
$$\mathscr{L}_1 = 4 = [5] - [1]$$

Expressing these as 4-level gap boundaries gives

$$\mathscr{U}_1 = ([3] + a_4) - [2]$$
$$\mathscr{L}_1 = ([2] + a_4) - [1]$$

Again the inequality

$$\mathscr{U}_1 > \mathscr{L}_1$$

implies

$$3 + a_4 - 2 > 2 + a_4 - 1$$

or

$$3 - 2 = 1 > 2 - 1 = 1$$

which is again a residual range.

We now define new inconsistent boundary sets that when expressed in terms of 4-level gap boundaries become

$$\mathscr{U}_2 = 1 = [3] - [2]$$
$$\mathscr{L}_2 = 1 = [2] - [1]$$

If we express these in terms of 5-level gap boundaries, they become

$$\mathscr{U}_2 = ([1] + a_5) - ([0] + a_5)$$
$$= [1] - [0] = 1$$

$$\mathscr{L}_2 = ([0] + a_5) - [1]$$
$$= a_5 - 1$$

The inequality implied by

$$\mathscr{U}_2 > \mathscr{L}_2$$

is then

$$1 > a_5 - 1$$

or

$$a_5 < 2$$

which is the implied range for a_5. Intersecting this with the common range on a_5, which is (∞*1), gives the new common range on a_5:

$$\overline{R(a_5)} = 2*1$$

Selecting $a_5 = 1\frac{1}{2}$ and returning to Procedure 5.1, we find that it can be used to obtain a realization as shown in Fig. 8.5. The resultant realization is

$$\langle 3\tfrac{1}{4}\, x_1 + 2\tfrac{3}{4}\, x_2 + 2\tfrac{1}{2}\, x_3 + 2x_4 + 1\tfrac{1}{2}\, x_5 + x_6 \rangle_{7:6\frac{3}{4}}$$

8.5 MAXIMALLY RELIABLE SINGLE-GATE REALIZATIONS

The single-gate realization procedure of Chapter 5, even as augmented in this chapter, does not guarantee realizations which are maximally reliable in any sense such as maximum relative gap or minimum normalized fan-in. Indeed the realization obtained in Example 8.3 has a smaller relative gap and a larger normalized fan-in than the realization of the same function given after Example 5.4.

However, there are certain situations in which the procedure of Chapter 5 can be shown to give realizations that have both maximum relative gap and minimum normalized fan-in. There are other situations in which the realizations obtained in Chapter 5 are not maximally reliable, but additional tree manipulations, similar to those described in Sections 8.1 through 8.4, can be used to obtain realizations which have both maximum relative gap and minimum normalized fan-in.

This section considers these situations in which the tree procedure can be used to obtain maximally reliable realizations. Although the methods to be described are not necessarily applicable to all single-gate realizations, they do appear to apply to a great many "practical" realizations, that is, those with large relative gap and small normalized fan-in. For those situations in which the methods of this section do not apply, the techniques of linear programming as described in Chapter 9 can be used to obtain maximally reliable realizations.

We consider first those situations in which the tree procedure of Chapter 5 can be shown always to give maximally reliable realizations. Suppose that a tree realization in which the smallest coefficient is unity turns out to have a gap length of unity. In other words, it is never necessary to reduce the gap length in the course of the procedure. Suppose further that by tree reduction techniques (Section 4.8) the total number of 1's and 0's in some reduced tree is exactly $n + 1$. Under these conditions we can show that the tree realization as described in Procedure 5.1 gives the realization with maximum relative gap and minimum normalized fan-in.

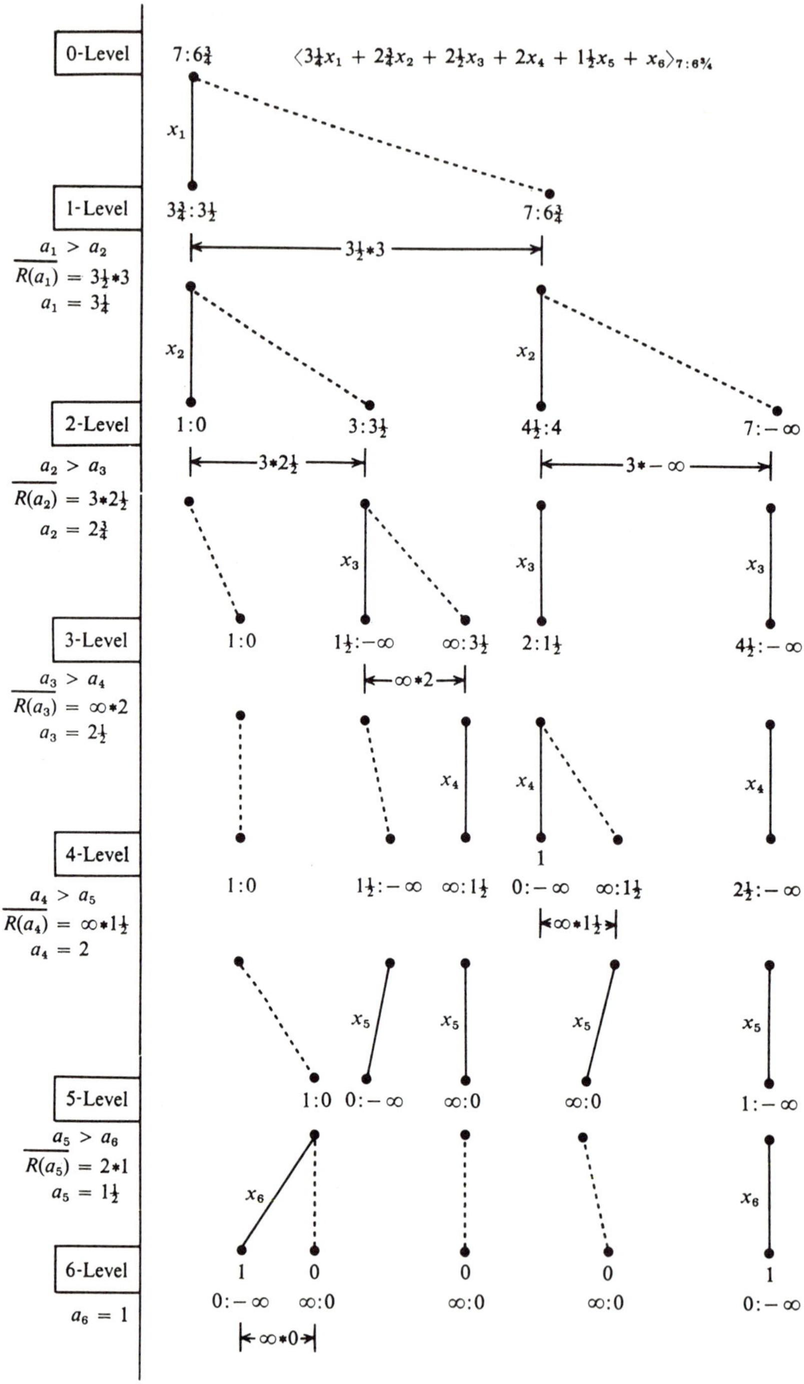

Fig. 8.5. The complete realization of Example 8.3 using the new range for a_5.

Theorem 8.2. If a switching function $F(x_1,x_2,\ldots,x_n)$ has a reduced tree in which the total number of 1's and 0's is exactly† $n + 1$, and if the realization obtained from Procedure 5.1, (using any function tree) has unit gap length when the smallest coefficient is unity, then that realization has maximum relative gap and minimum normalized fan-in.

Proof. Since there are exactly $n + 1$ 1's and 0's in the reduced tree, it follows from Section 4.8 that the realization problem is characterized by $n + 1$ simultaneous inequalities in the $n + 1$ unknowns $a_1,a_2,\ldots,a_n$ and u (assuming that ℓ has been normalized to $u - 1$), together with whatever reducing relations were used to make the reductions.

We now need some concepts from Chapter 9 on simultaneous inequalities. Theorem 9.2 states that, for ℓ.s. functions, the reducing relations can be ignored in obtaining the solution. Thus the realization problem is characterized by just the $n + 1$ inequalities in $n + 1$ unknowns.

Theorem 9.3 states: Given a set of P inequalities in m unknowns ($P \geq m$) in which m of the inequalities are linearly independent, if that set has any finite solution that maximizes or minimizes some linear combination of the variables, then it has a solution in which at least m inequalities are satisfied as equalities.

Both the maximum relative gap and minimum normalized fan-in conditions can be expressed as linear combinations of the variables (maximizing relative gap corresponds to minimizing u when the gap length is normalized to unity). Moreover, both optimization criteria clearly do have solutions with all finite coefficients. Thus they each have a solution in which m inequalities are satisfied as equalities. However, under the assumptions of the theorem,

$$P = m = n + 1$$

and there are only $n + 1$ inequalities. Thus a solution to both optimization problems corresponds to all of the $(n + 1)$ inequalities satisfied as equalities, and hence both optimization problems have the same solution. Moreover, any other optimization problem which has a solution with all finite coefficients also has the same solution.

Now consider the tree realization procedure under the conditions of the theorem. Certainly $a_n = 1$ is the smallest value of a_n under the normalization that the gap length is unity (by the discussion preceding Theorem 3.10). Thus the realization that is obtained by the procedure is a solution to the problem of minimizing a_n under the normalization that $g = 1$. Of course,

† If, because of symmetries certain variables can always have equal coefficients and hence the number of unknowns (including u) is $q + 1$, then the condition is that the number of constant functions in the symmetric tree is $q + 1$.

there may be many solutions to that problem, but they all share the property that $a_n = 1$.

Now consider the problem of obtaining the realization that minimizes $a_{n-1} + a_n$ with $g = 1$. By the previous reasoning, this problem must have at least one solution in common with the problem of minimizing a_n. Thus it must have a solution in which $a_n = 1$. Thus $a_{n-1} + a_n$ can be minimized by selecting the minimum value of a_{n-1} under the assumption that $a_n = 1$. But this is exactly what Procedure 5.1 does; it selects the minimum value of a_{n-1}, under the assumption that $g = 1$ and $a_n = 1$. Thus, under the conditions of the theorem, the realization obtained by Procedure 5.1 also minimizes $a_n + a_{n-1}$ with $g = 1$.

When we repeat this reasoning with each coefficient that is added, it follows that the realization also minimizes

$$a_n + a_{n-1} + \cdots + a_1$$

with $g = 1$, and thus is a minimum normalized fan-in realization.

Now consider the problem of maximizing relative gap. Under the normalization $g = 1$, this is equivalent to minimizing u. The minimum u problem must have one solution in common with all the minimization problems implied by the tree. Since this solution must satisfy the condition that a_n is a minimum, it must have $a_n = 1$. It must also satisfy the condition of minimum $a_{n-1} + a_n$. But since $a_n = 1$, this implies minimum a_{n-1}, which is the value given by the tree procedure. Repeating this reasoning with the remaining coefficients, it follows that the tree realization also minimizes u, with $g = 1$, and hence is a maximum relative gap realization.▌

Example 8.4. Figure 8.6 shows the symmetric tree for the function

$$F = x_1(x_2 \dotplus x_3 \dotplus x_4) \dotplus x_2(x_3 \dotplus x_4x_5) \dotplus x_3x_4x_5$$

Using the reductions due to coefficient ordering (Section 4.8),

$$a_1 > \begin{Bmatrix} a_2 \\ a_3 \end{Bmatrix} > a_4 > a_5 > a_6$$

eliminates all but the encircled constant entries. For example, the leftmost 1 corresponds to the inequality

$$a_1 + a_2 \geq u$$

and is redundant in view of the leftmost encircled entry which corresponds to

$$a_1 + a_4 \geq u$$

and the ordering relation

$$a_2 > a_4$$

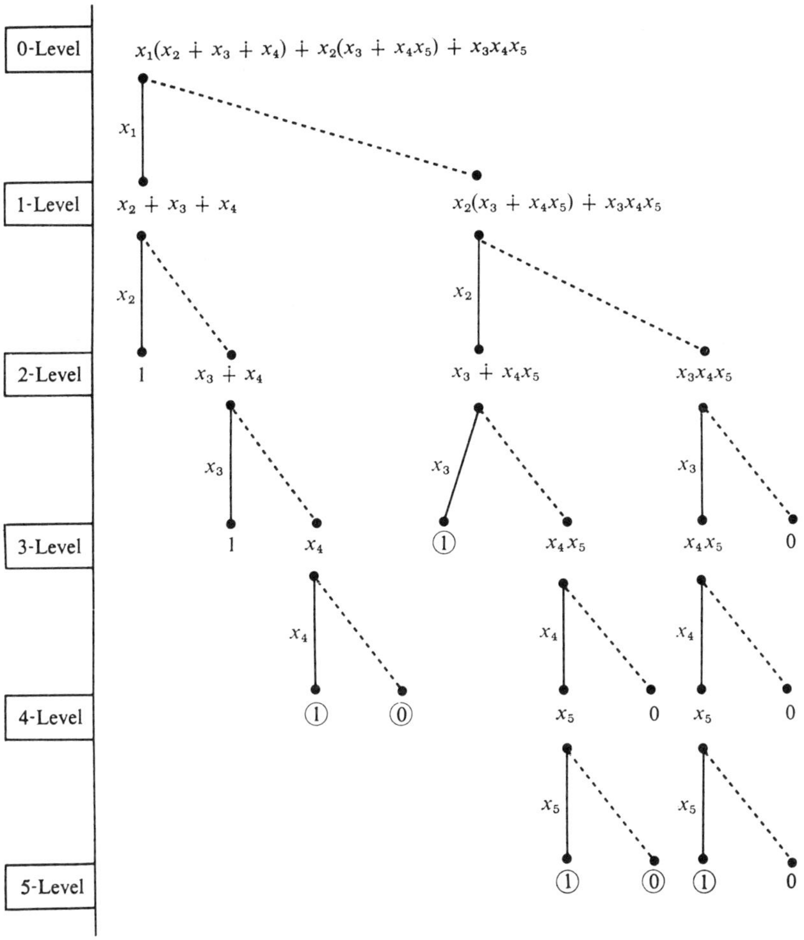

Fig. 8.6. Tree for the function of Example 8.4.

Since there are exactly six encircled entries, the conditions of Theorem 8.2 apply, and the realization obtained from that tree will have maximum relative gap and minimum normalized fan-in. Fig. 8.7 shows that solution to be

$$F = \langle 4x_1 + 3x_2 + 3x_3 + 2x_4 + x_5 \rangle_{6:5}$$

Suppose now that the conditions of Theorem 8.2 are not satisfied. Assume that we are trying to realize a function that has exactly $n + 1$ 1's and 0's in its reduced tree but that now Procedure 5.1 does not yield a realization with unit gap length when $a_n = 1$. That is, the gap length

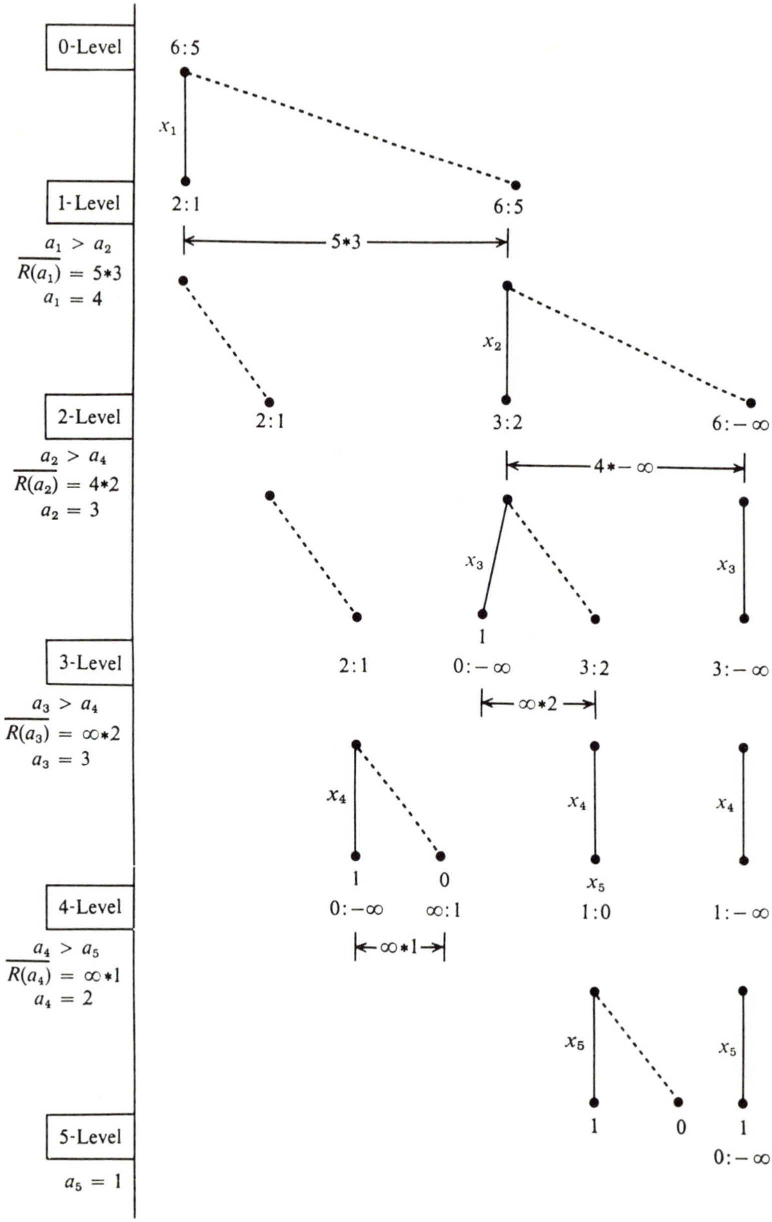

Fig. 8.7. The maximally reliable realization of Example 8.4.

must be reduced in the course of the procedure. Then the reasoning of Theorem 8.2 can no longer be used because each coefficient has not been chosen under the restriction that $g = 1$.

However, there are some additional manipulations that can be performed on the tree so that the reasoning of Theorem 8.2 is again applicable, and hence the resulting realization has both maximum relative gap and minimum normalized fan-in. This procedure was suggested by Coates and Supornpaibul, and is quite useful in obtaining reliable single-gate realizations even in those cases when it cannot be shown to yield the optimum one.

The basic idea is to use the tree procedure to obtain a realization with unit gap length and to select each coefficient with the smallest value that will yield the unit length gap, given the previously selected coefficients. If, after selecting a_n as 1, the gap length never has to be reduced below unity, then this procedure is exactly the same as Procedure 5.1. However, if on any k level, all choices for a_k within its common range would cause a reduction in the minimum gap length on the $(k - 1)$-level, then some additional computations similar to those described in Sections 8.1 through 8.4 are performed. The immediate cause of the potential gap reduction is that the length of the common range on a_k is less than 2 (Section 5.3, p. 178)

$$\overline{U(a_k)} - \overline{L(a_k)} < 2$$

The required condition on the range length for there to be a value of a_k which will give a minimum gap length of unity is

$$\overline{U(a_k)} - \overline{L(a_k)} \geq 2$$

However, this condition is not satisfied by the present choices for

$$a_{k+1}, a_{k+2}, \ldots, a_n$$

This condition is similar to the void common range in the procedure of Section 8.4. It can be traced down the tree in a manner similar to that of Section 8.4 to determine whether a different choice for any coefficient would allow the condition to be satisfied. If the given function is ℓ.s., there will always be such choices, since if necessary the smallest coefficient can be increased in value and the whole realization just renormalized.

When a new range for a coefficient is found which will allow the common range on a_k to be of length at least 2, the smallest value of that coefficient within that range is selected, and the realization is continued going up the tree. If again a common range is found to be of length less than 2, the same procedure is repeated. Eventually the procedure will terminate, giving a realization in which the gap length is unity, and each coefficient has been chosen as the smallest value which will yield the unit gap length, given the previously selected coefficients.

The procedure is exactly the same as that described formally in Section 8.4, except that the inequality to be satisfied is

$$\overline{U(a_k)} \geq \overline{L(a_k)} + 2$$

rather than

$$\overline{U(a_k)} > \overline{L(a_k)}$$

and the inconsistent boundary sets are correspondingly defined.

Since this procedure gives realizations in which each coefficient has the minimum value for unit gap length, given the previously selected coefficients, the reasoning of Theorem 8.2 can be applied.

> ***Corollary 8.2.1.*** If a switching function $F(x_1,x_2,\ldots,x_n)$ has a reduced tree in which the total number of 1's and 0's is exactly $n + 1$, then a realization of F has both maximum relative gap and minimum normalized fan-in if it has unit gap length, and each coefficient a_k has minimum value for the unit gap length, given the choices for $a_{k+1},a_{k+2},\ldots,a_n$, when the coefficients are ordered
>
> $$a_1 \geq a_2 \geq \ldots \geq a_n$$

The same reasoning can be applied in still another situation. Suppose that there are more than $n + 1$ nonredundant 1's and 0's in the tree, but that in the final realization

$$\langle f(p)\rangle_{u:\ell}$$

(obtained by any method), for all of the 1 points, p^1,

$$f(p^1) = u$$

and, for all of the 0 points, p^0,

$$f(p^0) = \ell$$

(in other words, all of the nonredundant 1's and 0's correspond to gap points). Then all of the nonredundant inequalities are satisfied as equalities at the solution point and the same reasoning shows that this solution has both maximum relative gap and minimum normalized fan-in.

Example 8.5 (*Coates and Supornpaibul*). Consider the function

$$F(x_1,x_2,\ldots,x_6) = x_1[x_2 \dotplus x_3 \dotplus x_4(x_5 \dotplus x_6) \dotplus x_5x_6]$$
$$\dotplus\, x_2[x_3(x_4 \dotplus x_5) \dotplus x_4x_5] \dotplus x_3x_4x_5x_6$$

Figure 8.8 shows the R_1 tree for this function with those 1 and 0 functions encircled which are not redundant in view of coefficient ordering,

$$a_1 > a_2 > a_3 > a_4 = a_5 > a_6$$

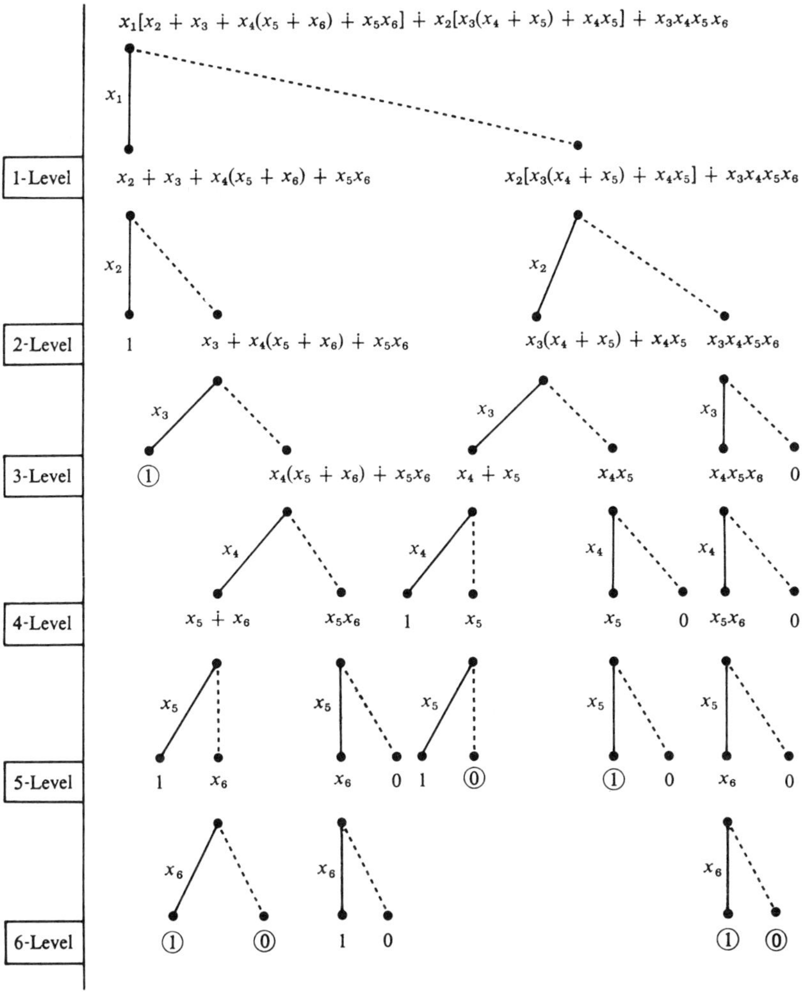

Fig. 8.8. Tree for the function of Example 8.5.

There are 7 encircled points but only 6 unknowns: a_1,a_2,a_3,a_4,a_6 and u. Thus Corollary 8.2.1 cannot be applied. Nevertheless the method of this Section can still be used, although the realization cannot be guaranteed in advance to be maximally reliable. The procedure is shown in Fig. 8.9.

The procedure starts as in the upper realization in the figure and encounters no difficulty until it reaches the 3-level, where the common range

$$R(a_3) = 3*2$$

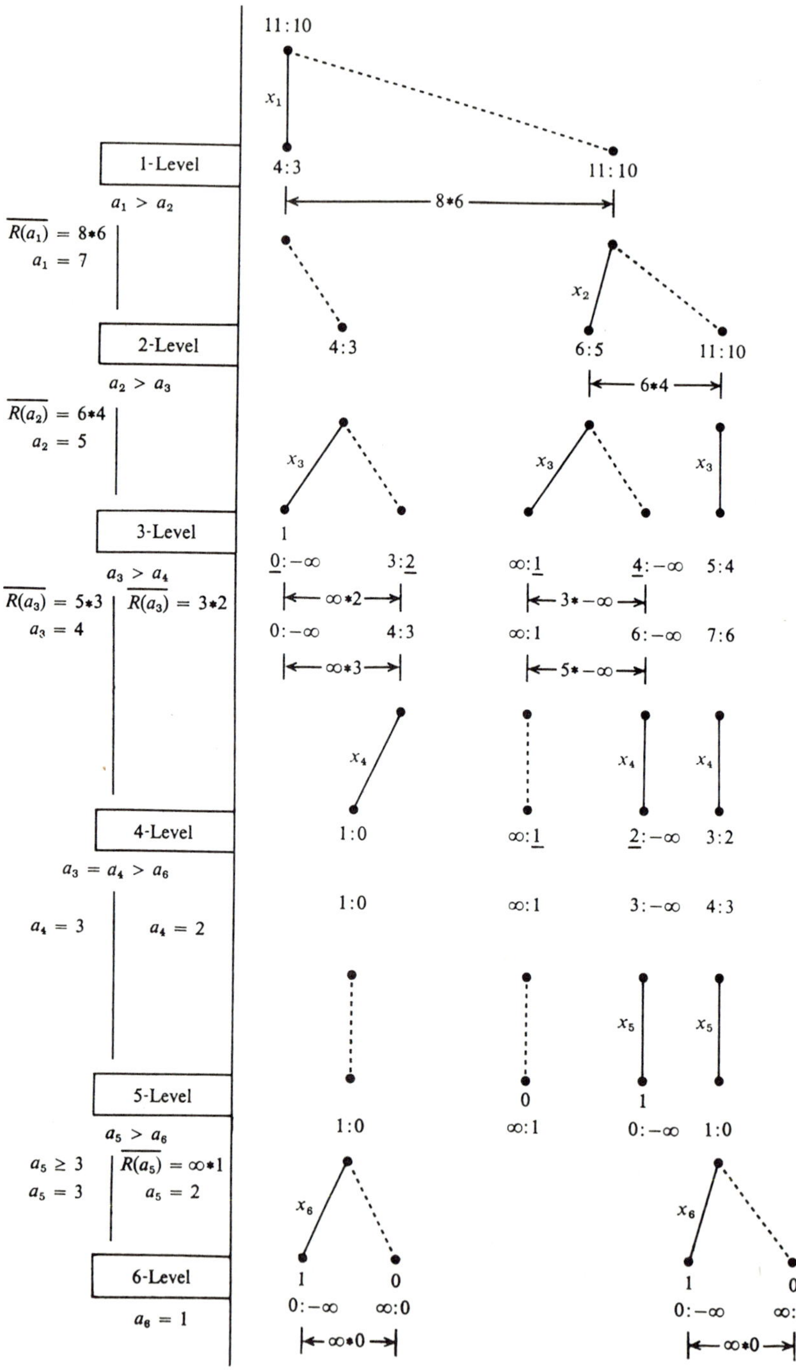

Fig. 8.9. The maximally reliable realization of Example 8.5.

has only unit length. At this point we must trace these range boundaries down the tree. We first determine if there is a different choice for a_4 that will allow the common range on a_3 to have length 2. The condition that must be satisfied by the new realization is

$$\overline{U(a_3)} - \overline{L(a_3)} \geq 2$$

Expressing $\overline{U(a_3)}$ and $\overline{L(a_3)}$ in terms of their dominant 3-level boundaries:

$$\mathscr{U}(a_3) = [4] - [1]$$

$$\mathscr{L}(a_3) = \begin{Bmatrix} [2] - [0] \\ a_4 \end{Bmatrix}$$

where the pertinent 3-level gaps are underlined in Fig. 8.9.

These 3-level dominant boundaries are expressed in terms of 4-level boundaries:

$$\mathscr{U}(a_3) = \{[4] - [1]\} = \{([2] + a_4) - [1]\}$$

$$\mathscr{L}(a_3) = \begin{Bmatrix} [2] \\ a_4 \end{Bmatrix} = \begin{Bmatrix} [0] + a_4 \\ a_4 \end{Bmatrix}$$

The condition

$$\mathscr{U}(a_3) - \mathscr{L}(a_3) \geq 2$$

becomes

$$([2] + a_4 - [1]) - a_4 \geq 2$$

which implies

$$1 \geq 2$$

This residual condition is inconsistent. Thus no choice for a_4 keeping a_5 and a_6 unchanged will allow a unit length gap on the 3-level. To investigate a_5, we trace this residual condition down to the 5-level. Thus

$$1 \geq 2$$

can be expressed in terms of the underlined 4-level boundaries as

$$[2] - [1] \geq [2]$$

and in terms of 5-level boundaries as

$$([0] + a_5) - [1] \geq 2$$

which implies

$$a_5 \geq 3$$

This value for a_5 is within $\overline{R(a_5)}$ and will allow a unit-length 5-level gap. Thus we select

$$a_5 = 3$$

and proceed as in the lower realization. We use our knowledge of coefficient ordering to make a_4 also equal to 3, and no further difficulties are encountered. The final realization is

$$F = \langle 7x_1 + 5x_2 + 4x_3 + 3x_4 + 3x_5 + x_6 \rangle_{11:10}$$

Note that all of the nonredundant 1 and 0 points in the tree of Fig. 8.8 are gap points for this realization. Thus, by the reasoning in the paragraph following Corollary 8.2.1, the realization has maximum relative gap and minimum normalized fan-in.

Example 8.6. We return again to the function of Examples 5.4 and 8.3. Figure 8.5 gives a realization for this function, but the realization given after Example 5.4 has a smaller normalized fan-in and a larger relative gap.

Since the reduced tree in Fig. 8.4 contains exactly seven constant functions for this six-variable function, the methods of this section can be used to obtain the maximally reliable realization.

If the procedure of Example 8.3 is carried out using as the inconsistent boundary set

$$\mathcal{U}(a_2) - \mathcal{L}(a_2) \geq 2$$

rather than

$$\mathcal{U}(a_2) - \mathcal{L}(a_2) > 0$$

then the realization given after Example 5.4 is obtained directly. The details are omitted.

When the conditions of Theorem 8.2 and Corollary 8.2.1 are satisfied (that is, when there are exactly $n + 1$ nonredundant inequalities), then an alternative method of solution is available. Since the reasoning of the proof of Theorem 8.2 shows that one solution is always obtained when the $n + 1$ inequalities are satisfied as equalities, the problem can be set up and solved as $n + 1$ simultaneous equations in $n + 1$ unknowns. Sometimes this is the computationally simplest method.

Example 8.7. Consider the problem of Example 8.6. From the tree of Fig. 8.4, the seven nonredundant simultaneous inequalities are

$$\begin{aligned}
a_1 + a_2 \qquad\qquad\qquad\qquad + a_6 &\geq u \\
a_1 \qquad + a_3 \qquad + a_5 \qquad &\geq u \\
a_2 + a_3 + a_4 \qquad\qquad &\geq u \\
a_3 + a_4 + a_5 + a_6 &\geq u \\
a_1 + a_2 \qquad\qquad\qquad\qquad &\leq \ell \\
a_1 \qquad\qquad + a_4 + a_5 \qquad &\leq \ell \\
a_2 + a_3 \qquad + a_5 \qquad &\leq \ell
\end{aligned}$$

with the additional inequality

$$u > \ell$$

Letting

$$\ell = u - 1$$

and putting the equal sign in all the inequalities gives

$$\begin{aligned}
a_1 + a_2 \qquad\qquad\qquad\qquad + a_6 - u &= 0 \\
a_1 \qquad + a_3 \qquad + a_5 \qquad - u &= 0 \\
a_2 + a_3 + a_4 \qquad\qquad - u &= 0 \\
a_3 + a_4 + a_5 + a_6 - u &= 0 \\
a_1 + a_2 \qquad\qquad\qquad\qquad - u &= -1 \\
a_1 \qquad\qquad + a_4 + a_5 \qquad - u &= -1 \\
a_2 + a_3 \qquad + a_5 \qquad - u &= -1
\end{aligned}$$

It is readily verified that the solution obtained in Example 8.6 satisfies these equations. Whether the solution of these equations is computationally simpler than that of Example 8.6 is open to question.

Thus when there are only $n + 1$ nonredundant inequalities, the methods of this section allow the maximally reliable single-gate realization to be obtained rather easily. However, there are many functions for which it is computationally impractical to determine the minimum number of nonredundant inequalities. Then the method of simultaneous equations cannot be used. However, the method of this section can still be used and, if the number of nonredundant inequalities turns out to be $n + 1$, the optimum realization will be obtained. If not, the method will give a reasonably good realization in any case.

Chapter 9

METHODS FOR SOLVING SIMULTANEOUS INEQUALITIES

9.1 INTRODUCTION

In Chapter 1 the single-gate realization problem was formulated in terms of the consistency and solution of a set of $2^n + 1$ simultaneous linear inequalities in $n + 2$ unknowns. For even moderately large n, this set is too large for the solution to be computationally feasible. In Chapter 3 methods were discussed whereby the number of inequalities can be reduced somewhat by using information about the structure of the given function. However, even with these reductions, the simultaneous-inequality approach does not yield a completely satisfactory solution to the realization problem —particularly for the multigate case.† In fact, one of the primary objectives of this book is the development of a theory that deals directly with the logical function rather than with the simultaneous inequalities. The result is a theory that is useful for both single and multigate synthesis.

Nevertheless, it is important at this point to discuss in some detail the classical methods for solving simultaneous inequalities. These methods may or may not be computationally useful in any particular situation, but an understanding of their theory and application will certainly be useful in a wide variety of situations.

We consider two general methods for solving a set of linear inequalities. The first of these, *successive elimination of variables*, is a simple and straightforward procedure that can be considered in one section. The second, *linear programming*, is considerably more important and computationally simpler. Its explanation, however, requires a somewhat lengthy discussion.

Before considering either method, however, it is necessary to give some

† For some interesting first attempts to use simultaneous inequality concepts for multigate synthesis, see Cameron, 2; Hopcroft, and Mattson; Stabler; and Breuer.

consideration to the detailed formulation of the problem and, in particular, to the method of dealing with reduced sets of inequalities and reducing relations.

9.2 FORMULATION OF THE PROBLEM

The simultaneous-inequality approach to the single-gate test and realization problem was formulated in Section 1.9. Let $F(x_1,x_2,\ldots,x_n)$ be the switching function to be tested and let p be a point of the switching space on which F is defined. The coordinates of p are denoted by

$$(x_1{}^p,x_2{}^p,\ldots,x_n{}^p)$$

Then F can be realized by a single T gate

$$\left\langle\sum_{i=1}^{n} a_i x_i\right\rangle_{u:\ell}$$

if and only if the following set of linear simultaneous inequalities is consistent:

$$\sum_{i=1}^{n} a_i x_i{}^p \geq u$$

for all p such that $F(p) = 1$;

$$\sum_{i=1}^{n} a_i x_i{}^p \leq \ell$$

for all p such that $F(p) = 0$; and

$$u > \ell$$

Furthermore, any solution of these inequalities defines a set of a_i, $i = 1, 2,\ldots,n$, a value for u, and a value for ℓ which constitutes a realization.

There are $2^n + 1$ inequalities in $n + 2$ unknowns. However, since the inequalities are homogeneous, any convenient normalization can be used to reduce the number of unknowns by one. For example, the smallest coefficient can be set equal to 1, or the gap length can be set equal to 1. This second normalization is obtained by setting

$$\ell = u - 1$$

in which case the inequality

$$u > \ell$$

is redundant and the number of inequalities and the number of variables are both reduced by one.

The methods of Chapter 3 can be used to reduce the number of inalitequies in view o` the signs and ordering of the coefficients as well as any other inequalities among the coefficients that are implied by the monotonicities of the given switching function. Before considering the effect of these reductions, however, we present a geometric interpretation of the simultaneous-inequality problem.

Assume that the inequalities have been normalized so that

$$\ell = u - 1$$

in which case they can be expressed as

$$\sum_{i=1}^{n} a_i x_i^p \geq u$$

for all p such that $F(p) = 1$ and

$$\sum_{i=1}^{n} a_i x_i^p \leq u - 1$$

for all p such that $F(p) = 0$. Note that the x_i^p are constants equal to 0 or 1 and the variables are the a_i, $i = 1,2,\ldots,n$ and u. It is convenient for the geometric interpretation to let a_0 represent u so that the inequalities become

$$\sum_{i=1}^{n} a_i x_i^p - a_0 \geq 0$$

for all p such that $F(p) = 1$ and

$$\sum_{i=1}^{n} a_i x_i^p - a_0 \leq -1$$

for all p such that $F(p) = 0$. Each of the inequalities can be represented in a Cartesian coordinate space of $n + 1$ dimensions where each coordinate axis corresponds to one of the variables $a_0,a_1,\ldots,a_n$. (Note that this is a coefficient space and that it is not the same as the switching space. In the switching space the axes correspond to the x_i's; in the coefficient space they correspond to the a_i's.)

Consider first an example in two dimensions. The inequality

$$-a_1 + a_2 \leq 0$$

is represented as shown in Fig. 9.1. The line in the figure is the locus of the equality

$$-a_1 + a_2 = 0$$

The entire half-plane above and including the line is the region in which the inequality is satisfied.

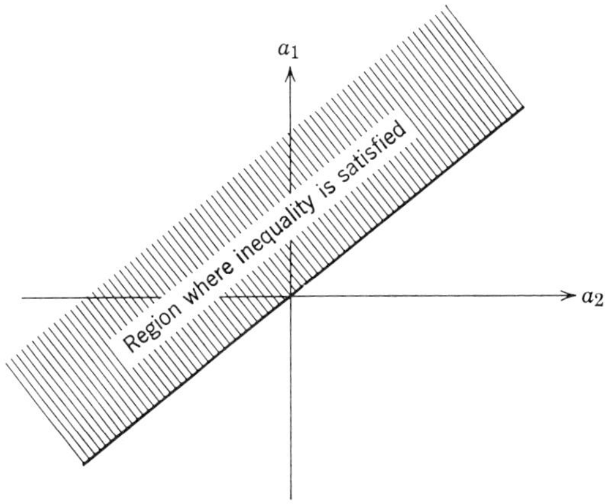

Fig. 9.1. Plot of $-a_1 + a_2 \leq 0$.

Now consider three other inequalities†:

$$a_1 - 2a_2 \leq 0$$
$$2a_1 + a_2 \leq 2$$
$$3a_1 + 2a_2 \leq 6$$

These would imply regions in the coefficient space as shown in Fig. 9.2. The region in which all four inequalities are satisfied is the shaded triangle. This region, including its boundaries, represents all solutions to the four simultaneous inequalities. The boundaries are included because each of the inequalities allows the equality to be included.

The existence of such a shaded area implies that the inequalities are consistent. It is certainly possible, as shown in Fig. 9.3, to have three half-planes that have no common solution area and which, therefore, represent inequalities that are inconsistent.

Notice also that the space of solutions (if any) will always be *convex*. That is, if two points are both solutions, then any point on the line between them is also a solution. In Fig. 9.4 the heavy line encloses a nonconvex region. Observe, however, that the region of solution corresponds not to this nonconvex region but to the shaded region that is convex.

In more than two dimensions, the geometric interpretation is similar but more difficult to illustrate. Each inequality defines a half-space bounded

† These inequalities are for illustrative purposes only and could not occur in an *ℓ.s.* function problem. Since more than two dimensions are required to discuss nontrivial *ℓ.s.* function problems, we defer such a discussion until later.

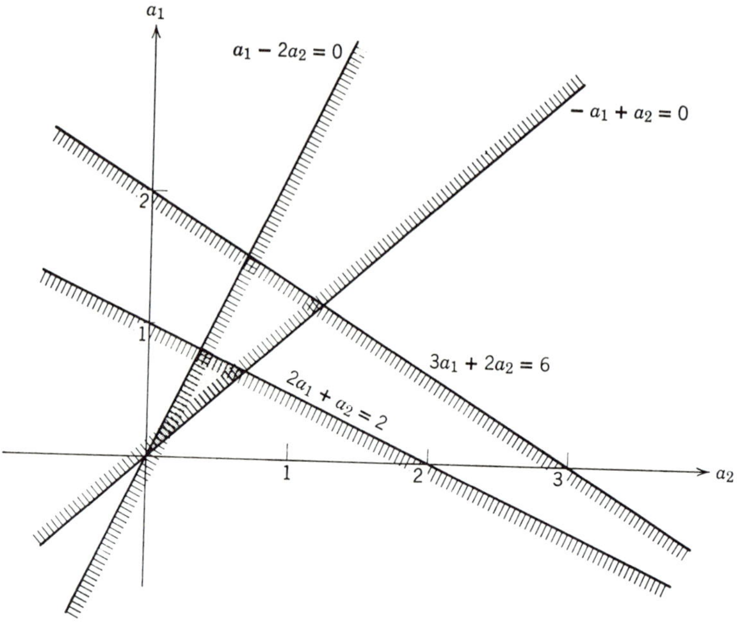

Fig. 9.2. A plot of four simultaneous inequalities.

by the hyperplane implied by its equality. The set of inequalities is consistent if the intersection of all solution half-spaces is nonvoid. In an m-dimension space a nonvoid intersection region will be in the form of an m-dimensional convex polyhedron. This polyhedron, including its boundaries, represents all of the solutions to the set of simultaneous inequalities. That the solution region is convex is established by the following theorem.

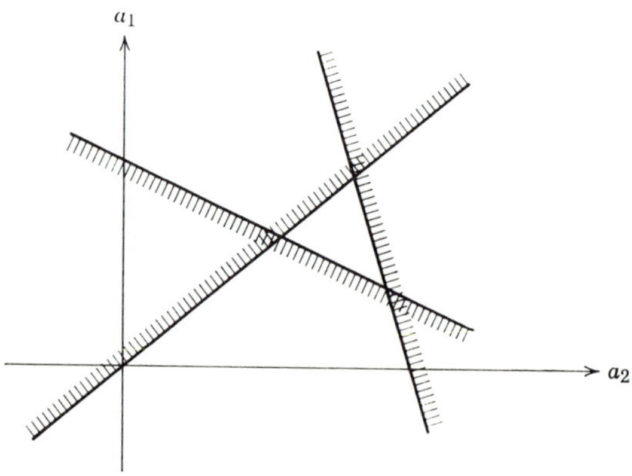

Fig. 9.3. Three loci corresponding to a set of inconsistent inequalities.

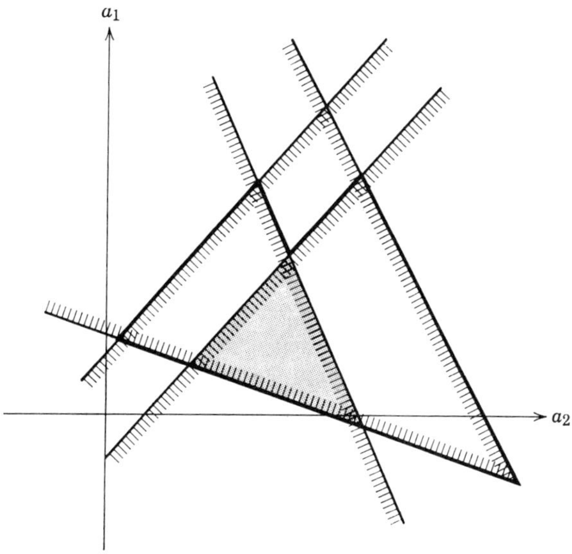

Fig. 9.4. Illustrations of convex and nonconvex spaces.

Theorem 9.1. The solution space of a set of simultaneous linear inequalities is convex.

Proof. Let A_1 and A_2 represent any two points of the coefficient space that are solutions, let the coordinates of each be

$$A_1 = (a_1{}^1, a_2{}^1, \ldots, a_m{}^1)$$
$$A_2 = (a_1{}^2, a_2{}^2, \ldots, a_m{}^2)$$

and let any one of the inequalities be

$$\sum_{i=1}^{m} \alpha_{ij} a_i \leq b$$

where the α_{ij} and b are real numbers. Substituting each of the solution points A_1 and A_2 into this inequality gives

$$\sum_{i=1}^{m} \alpha_{ij} a_i{}^1 = B_1 \leq b$$

$$\sum_{i=1}^{m} \alpha_{ij} a_i{}^2 = B_2 \leq b$$

Now consider any other point, A_3, of the solution space that lies on the line connecting A_1 with A_2. Thus A_3 has coordinate values

$$A_3 = (ca_1{}^1 + (1 - c)a_1{}^2, ca_2{}^1 + (1 - c)a_2{}^2, \ldots, ca_m{}^1 + (1 - c)a_m{}^2)$$

for $0 \le c \le 1$. Substituting A_3 into the inequality gives

$$\sum_{i=1}^{m} \alpha_{ij}(ca_i^{1} + (1 - c)a_i^{2}) = c \sum_{i=1}^{m} \alpha_{ij}a_i^{1} + (1 - c) \sum_{i=1}^{m} \alpha_{ij}a_i^{2}$$
$$= cB_1 + (1 - c)B_2 \le b$$

since $0 \le c \le 1$. Thus A_3 also satisfies all of the inequalities and is therefore a solution.|

Now consider the effect of reducing the number of inequalities by the methods of Chapter 3. We assume that the given function has been altered, if necessary, so that it is a positive unate representative function with all positive coefficients. We further assume that there is some set of r strict inequalities among the coefficients that has been found using reduced function comparability. Denote these by

$$\sum_{i=1}^{n} \beta_{ik}a_i > 0 \qquad k = 1,2,\ldots,r$$

where the β_{ik} are real numbers which may be positive, negative, or zero. (This set, for example, may include the inequality $a_i > 0$.)

Using these strict inequalities as reducing relations, we can delete from the original inequality set those inequalities that are then redundant, thereby yielding a reduced inequality set.

The resulting set of inequalities, consisting of the reduced set plus the reducing relations, is equivalent (with regard to the solution) to the original set. However, the form of the inequalities is somewhat different since the reducing relations are strict inequalities while the original inequalities include the equality sign. This difference is particularly important since one of the methods of solution we wish to consider is linear programming, and it cannot directly handle strict inequalities.

The solution region of the strict inequality

$$-a_1 + a_2 < 0$$

is similar to that of Fig. 9.1, but it must be interpreted differently. The region of solution is the upper half-plane defined by but excluding the line

$$-a_1 + a_2 = 0$$

This is called the *open half-plane* in contrast to the *closed half-plane* which does include the boundary line.

Suppose that we have a set of consistent inequalities, some of which exclude equality. Then the solution region is a polyhedron in the coefficient space. If one of the boundaries of the polyhedron is defined by a strict inequality, then that boundary is excluded from the region of possible solutions. On the other hand, any boundary of the solution space defined by an inequality that includes equality is a part of the solution space.

Now consider the solution space for ℓ.s. functions. The complete set of 2^n inequalities defines a closed-polyhedron† (including all boundaries) solution space. Any reduced inequality set plus the corresponding reducing relations must be solution equivalent to the original inequality set and, therefore, must possess the same closed-polyhedron solution space. But since the solution polyhedron is closed, no strict inequalities (as implied by the reducing relations) can define any of its boundaries. Thus the reduced set of inequalities, not including the reducing relations, defines the complete solution polyhedron, and we have established the following theorem.

Theorem 9.2‡ (*Winder, 1*). For ℓ.s. functions, the solution space defined by any reduced set of inequalities (excluding reducing relations) is the same as that defined by the complete original set.

Thus, for ℓ.s. functions, the reduced inequality set can be solved ignoring the reducing relations, and the solutions will satisfy all of the reducing relations. For non-ℓ.s. functions, this reasoning breaks down. There exist non-ℓ.s. functions that have reduced inequality sets which, excluding the reducing relations, are consistent and have solutions. However, these solutions do not satisfy the reducing relations (since if they did, they would satisfy the original set of inequalities, which by assumption is inconsistent). The following is an example of such a function.

Example 9.1 (*Winder, 1*). Consider the function

$$y = F(x_1,x_2,\ldots,x_6) = x_1x_2x_3 + x_4x_5x_6$$

in which case

$$\bar{y} = \bar{x}_1\bar{x}_4 + \bar{x}_1\bar{x}_5 + \bar{x}_1\bar{x}_6 + \bar{x}_2\bar{x}_4 + \bar{x}_2\bar{x}_5 + \bar{x}_2\bar{x}_6 + \bar{x}_3\bar{x}_4 + \bar{x}_3\bar{x}_5 + \bar{x}_3\bar{x}_6$$

This function is not ℓ.s. since, for example, $y(x_1 = 1,x_4 = 0)$ is not comparable with $y(x_1 = 0,x_4 = 1)$. However, y is a positive unate function of

† A closed polyhedron implies that the boundary planes are included. It does not imply that the polyhedron is necessarily bounded.

‡ The proof presented here is due to R. E. Stearns.

the variables $x_1,x_2,\ldots,x_6$, and the reduced inequality set can be written by inspection.

$$\begin{array}{llllllll}
a_1 + a_2 + a_3 & & & & & - u \geq 0 \\
 & & & a_4 + a_5 + a_6 & & - u \geq 0 \\
 & a_2 + a_3 & & + a_5 + a_6 & & - u \leq -1 \\
 & a_2 + a_3 + a_4 & & + a_6 & & - u \leq -1 \\
 & a_2 + a_3 + a_4 + a_5 & & & & - u \leq -1 \\
a_1 & + a_3 & & + a_5 + a_6 & & - u \leq -1 \\
a_1 & + a_3 + a_4 & & + a_6 & & - u \leq -1 \\
a_1 & + a_3 + a_4 + a_5 & & & & - u \leq -1 \\
a_1 + a_2 & & & + a_5 + a_6 & & - u \leq -1 \\
a_1 + a_2 & & + a_4 & + a_6 & & - u \leq -1 \\
a_1 + a_2 & & + a_4 + a_5 & & & - u \leq -1
\end{array}$$

with the reducing relations

$$a_i > 0 \qquad \text{for} \quad i = 1,2,\ldots,6$$

It is readily verified that the values

$$a_1 = a_2 = a_3 = a_4 = a_5 = a_6 = -1 \qquad \text{and} \quad u = -3$$

satisfy the reduced inequality set although they do not, of course, satisfy the reducing relations.

This example shows that for non-ℓ.s. functions both the reduced inequalities and the reducing relations must be considered. These results suggest the following general procedure for solving reduced sets of inequalities.

(1) By any method solve the reduced set of inequalities ignoring the reducing relations. If the reduced set is inconsistent, the given function is not ℓ.s.

(2) If the reduced set is consistent and a solution has been obtained, then determine, by simple substitution, whether the solution satisfies the reducing relations. If it does not, then the given function is not ℓ.s. If it does, then the function is ℓ.s. and the solution is a valid one.

Throughout the remainder of this chapter we shall assume that this procedure is being followed and shall only consider methods for solving the reduced inequality set without explicitly stating in each case that any solution so obtained must be verified by being substituted into the reducing relations. However, in the examples the complete procedure will be illustrated.

9.3 SUCCESSIVE-ELIMINATION-OF-VARIABLES METHOD (*FOURIER; MOTZKIN*)

The method of solving linear inequalities by successively eliminating variables is similar to the equivalent method for solving linear equations. Starting with a set of simultaneous inequalities in m unknowns, each step of the procedure generates a new set of inequalities in one less unknown. Solutions of this new set, if any, correspond exactly to solutions of the original set. When the procedure has been carried to the point where the set of inequalities contains a single unknown, the values of that unknown are immediately apparent. Then, by successive substitutions, the allowable values of the other unknowns are easily obtained.

The procedure is best described in terms of a specific example.

Example 9.2. Consider the inequalities

$$\begin{aligned} a_1 + 2a_2 &\geq 4 \\ a_1 + a_2 + 2a_3 &\geq 5 \\ a_1 - a_2 + a_3 &\leq 2 \end{aligned}$$

These can be rewritten by rearranging the terms as

$$\begin{aligned} a_1 &\geq 4 - 2a_2 \\ a_1 &\geq 5 - a_2 - 2a_3 \\ 2 + a_2 - a_3 &\geq a_1 \end{aligned}$$

The first and third of these inequalities together imply

$$2 + a_2 - a_3 \geq 4 - 2a_2$$

or

$$3a_2 - a_3 \geq 2$$

and the second and third imply

$$2 + a_2 - a_3 \geq 5 - a_2 - 2a_3$$

or

$$2a_2 + a_3 \geq 3$$

The new set of inequalities contains only a_2 and a_3 as unknowns and is

$$\begin{aligned} 3a_2 - a_3 &\geq 2 \\ 2a_2 + a_3 &\geq 3 \end{aligned}$$

Any values of a_2 and a_3 that satisfy these also satisfy

$$2 + a_2 - a_3 \geq 4 - 2a_2$$
$$2 + a_2 - a_3 \geq 5 - a_2 - 2a_3$$

and, hence, there will always exist a value for a_1 such that

$$a_1 \geq 4 - a_2$$
$$a_1 \geq 5 - a_2 - 2a_3$$
$$2 + a_2 - a_3 \geq a_1$$

Therefore, the inequality set

$$3a_2 - a_3 \geq 2$$
$$2a_2 + a_3 \geq 3$$

constitute the necessary and sufficient restrictions on a_2 and a_3 for which the original set of inequalities has a solution.

One stage of the procedure has now been completed. A new set of inequalities has been obtained that has one less unknown and whose solutions imply solutions for the original set. Note that the new inequality set can be obtained from the original set by first expressing all inequalities so that the $\geq$ signs point the same way.

$$a_1 + 2a_2 \qquad\quad \geq 4$$
$$a_1 + a_2 + 2a_3 \geq 5$$
$$-a_1 + a_2 - a_3 \geq -2$$

By adding the first and third and the second and third (operations that are permissible on such inequalities), the new set of inequalities is obtained directly. The same procedure can be performed on the new set

$$3a_2 - a_3 \geq 2$$
$$2a_2 + a_3 \geq 3$$

to give the inequality

$$5a_2 \geq 5$$

or

$$a_2 \geq 1$$

This is the necessary and sufficient restriction on a_2 for which the inequality set has a solution. It constitutes the "actual range" on a_2 in the sense that any a_2 in this range is guaranteed to yield allowable values of a_3 and a_1.

It is now known that the given set of inequalities is consistent. To determine a solution it is necessary to select a value for a_2 within its range. Selecting

$$a_2 = 2$$

we substitute this value into the second set of inequalities and obtain

$$6 - a_3 \geq 2$$
$$4 + a_3 \geq 3$$

These imply† that

$$-1 \leq a_3 \leq 4$$

This again is the "actual range" on a_3. Selecting

$$a_3 = 0$$

and substituting into the original set of inequalities gives

$$a_1 + 4 \geq 4$$
$$a_1 + 2 + 0 \geq 5$$
$$a_1 - 2 + 0 \leq 2$$

These imply

$$a_1 \geq 0$$
$$a_1 \geq 3$$
$$a_1 \leq 4$$

or

$$3 \leq a_1 \leq 4$$

Thus we can select

$$a_1 = 3$$

and thereby complete a solution to the original set.

The general procedure, as indicated in the following, should now be evident.

Procedure 9.1

(1) Write the inequalities with all inequality signs pointing in the same direction.

(2) Select a variable to be eliminated. This can be any variable that appears with both positive and negative coefficients. (Variables whose coefficients are all of the same sign can be arbitrarily large or arbitrarily small. There will certainly be at least one variable with coefficients of both signs, or the given set of inequalities is trivial.)

† This particular situation would not occur for an ℓ.s. function where an allowable value of 0 for a coefficient implies that the function does not depend upon the corresponding variable.

(3) For each pair of inequalities such that the coefficients of the selected variable are of opposite sign, add the inequalities after an appropriate constant multiplication so that the sum is independent of the selected variable. The totality of all the inequalities generated, plus those inequalities from the original set which do not depend upon the selected variable, constitutes the new set of inequalities.

(4) (a) If at least one inequality of the new set depends upon more than one variable, repeat the procedure on the new set starting with Step 1.
(b) If each inequality of the new set depends upon only one variable, collect all restrictions on those variables. The original set of inequalities is consistent if and only if these restrictions are consistent.

(5) If the restrictions of Step 4 are consistent, select a value within the determined restrictions for each variable that appears in Step 4 and substitute this value into the inequality set that involves two variables. This will result in a consistent set of restrictions on the other variable. Select a value within the restriction, substitute the values of both variables into the inequality set among three variables, and so on.

The procedure is simple to describe and to execute for small sets of inequalities. For larger problems, however, the computational difficulties increase rapidly because of the very large number of inequalities that may be generated by the procedure itself.

If, in the original set of P inequalities, the eliminated variable appears positive in Q and negative in R inequalities, then there will be QR inequalities generated in the first stage of the procedure. This number could be as large as $(P/2)^2$ when

$$Q = R = \frac{P}{2}$$

After a few steps the number of inequalities that must be considered may become astronomically large. As a result, the application of this method is usually limited to situations where the number of original inequalities is small.

Example 9.3. Consider the function

$$F(x_1,x_2,x_3,x_4) = x_1x_2 + x_1x_3x_4 + x_2x_3x_4$$

which has the R_1 reduced set of inequalities†

$$\begin{aligned} 2a_1 \quad &\geq u \\ a_1 + 2a_3 &\geq u \\ 2a_3 &\leq \ell \\ a_1 + a_3 &\leq \ell \\ \\ u &> \ell \\ a_3 &> 0 \\ a_1 &> 0 \end{aligned}$$

together with the additional equalities

$$\begin{aligned} a_1 &= a_2 \\ a_3 &= a_4 \end{aligned}$$

Normalizing by setting

$$\ell = u - 1$$

and multiplying the u inequalities by -1 gives

$$\begin{aligned} -2a_1 \qquad\quad + u &\leq 0 \\ -a_1 - 2a_3 + u &\leq 0 \\ 2a_3 - u &\leq -1 \\ a_1 + \ a_3 - u &\leq -1 \\ \\ a_3 &> 0 \\ a_1 &> 0 \end{aligned}$$

together with

$$\begin{aligned} a_2 &= a_1 \\ a_4 &= a_3 \end{aligned}$$

By Theorem 9.2 and the subsequent discussion, the reducing relations can be temporarily neglected in solving the reduced inequality set:

$$\begin{aligned} -2a_1 \qquad\quad + u &\leq 0 \\ -a_1 - 2a_3 + u &\leq 0 \\ 2a_3 - u &\leq -1 \\ a_1 + \ a_3 - u &\leq -1 \end{aligned}$$

† The R_2 set of inequalities can also be used, in which case the inequality $2a_3 \leq \ell$ is omitted and the reducing relation $a_1 > a_3$ is added.

Eliminating a_1, the new set consists of three inequalities: the third inequality in the original set, the sum of the second and the fourth, and the sum of the first and twice the fourth. These are

$$\begin{aligned} 2a_3 - u &\leq -1 \\ -a_3 \quad &\leq -1 \\ 2a_3 - u &\leq -2 \end{aligned}$$

The first inequality is redundant in view of the third and can be eliminated, yielding

$$\begin{aligned} 2a_3 - u &\leq -2 \\ -a_3 \quad &\leq -1 \end{aligned}$$

This concludes one step of the procedure. Now a_3 can be eliminated from this new set by adding the first inequality to twice the second, yielding

$$u \geq 4$$

This completes the elimination part of the procedure. Since the restrictions on u are consistent, the given function is ℓ.s.

To obtain a solution, we select

$$u = 4$$

(giving $\ell = 3$) and substitute this value into the inequalities in two variables:

$$\begin{aligned} 2a_3 - 4 &\leq -2 \\ -a_3 \quad &\leq -1 \end{aligned}$$

which implies

$$1 \leq a_3 \leq 1$$

Choosing $a_3 = 1$ and substituting the values of u and a_3 into the inequalities among three variables gives

$$\begin{aligned} -2a_1 \qquad + 4 &\leq 0 \\ -a_1 - 2 + 4 &\leq 0 \\ 2 - 4 &\leq -1 \\ a_1 + 1 - 4 &\leq -1 \end{aligned}$$

These imply

$$2 \leq a_1 \leq 2$$

for which the only possible choice is

$$a_1 = 2$$

The complete solution is

$$a_1 = a_2 = 2$$
$$a_3 = a_4 = 1$$
$$u = 4$$
$$\ell = 3$$

Before it is known that this is a valid solution of the nonreduced set of inequalities, it must be verified that the values of the variables satisfy the reducing relations,

$$a_1 > 0$$
$$a_3 > 0$$

which, indeed, they do. Thus the given function is ℓ.s., and a realization has been obtained.

9.4 LINEAR PROGRAMMING†: THE GENERAL IDEA

A consistent set of linear inequalities has a multiplicity of solutions defined by a closed polyhedron. (Unless otherwise indicated, it is assumed that there are no strict inequalities in the set.) It is possible, therefore, to specify additional conditions that the solution must satisfy and thus to reduce the space of possible solutions. The subject of *linear programming* deals with one such additional restriction: namely, that some linear combination of the variables

$$Z = \sum_{i=0}^{n} z_i a_i$$

(where the z_i are real constants) is maximized or minimized. The objective, therefore, is to obtain a solution to the set of linear inequalities that maximizes or minimizes the given linear expression. This type of optimization is particularly useful if Z has some economic or engineering significance.

The linear programming problem can be interpreted geometrically in the a space. Consider, again, the set of consistent inequalities plotted in Fig. 9.2 and suppose that the object is to determine the solution of these that maximizes Z, where

$$Z = a_1 + 3a_2$$

Considering Z as a parametric constant c, the locus of the maximizing expression in the solution space is a straight line, that is,

$$c = a_1 + 3a_2$$

† This is not intended as a complete exposition of the subject of linear programming. For such a presentation see, for example, Dantzig.

As c varies, this line moves parallel to itself as shown in Fig. 9.5a. Note that the triangular solution region from Fig. 9.2a is also shown in Fig. 9.5a.

The optimization problem consists of finding the point or points within the solution triangle for which c has a maximum value. As shown in Fig. 9.5b, these values are defined by the intersection of the optimization line and the solution triangle when the former touches one vertex of the latter. Clearly the solution is the point of intersection.

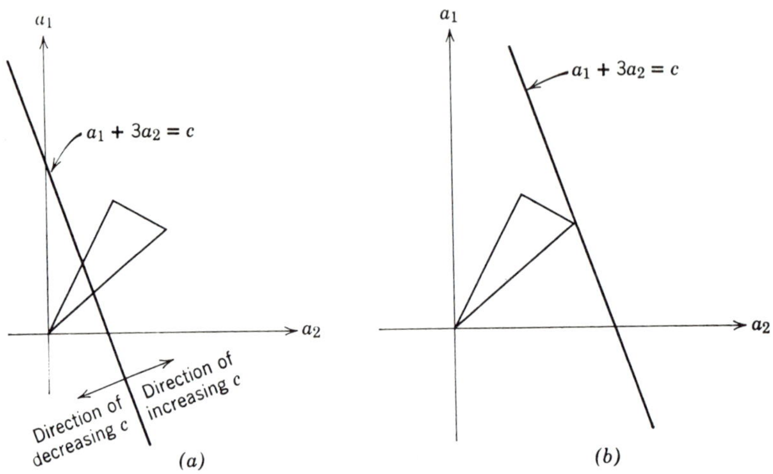

Fig. 9.5. Geometric interpretation of linear optimization. (a) Variation of $c = a_1 + 3a_2$ in the solution space. (b) Position of (max c) $= a_1 + 3a_2$ such that a_1 and a_2 are in the solution triangle.

This point corresponds to two of the inequalities being satisfied as equalities. Reference to Fig. 9.2 shows that the solution is defined by the point so that

$$2a_1 + a_2 = 2$$
$$-a_1 + a_2 = 0$$

The simultaneous solution is

$$a_1 = a_2 = \tfrac{2}{3}$$

and this is the solution which yields the maximum value of c. In this case,

$$c = a_1 + 3a_2 = \tfrac{2}{3} + 3\cdot\tfrac{2}{3} = 2\tfrac{2}{3}$$

From this example it is evident that in a 2-dimensional solution space the optimum solution will always occur at a vertex of the solution polyhedron. (In a degenerate case, the maximizing line may be parallel to one of the boundaries, in which case the entire boundary corresponds to optimum solutions. However, in this case also, the vertices are part of the boundary and nothing is lost by restricting attention to the vertices.)

A vertex of a polygon in 2-space always corresponds to the simultaneous equality of two or more of the inequalities. Thus it can be concluded that in 2-space the solution to a linear programming problem always occurs when at least two of the inequalities are satisfied as equalities.

In an m-dimensional space a similar result holds. The optimal solution always occurs at one of the vertices of the m-dimensional polyhedron, and thus corresponds to at least m of the inequalities being satisfied as equalities. This result is the cornerstone of linear programming and, accordingly, it is presented as a theorem and provided with a formal proof.

Theorem 9.3. Consider a set of P consistent simultaneous linear inequalities in m unknowns ($P \geq m$) in which m of the inequalities are linearly independent. If these inequalities have one finite solution that maximizes or minimizes any linear weighted sum of the variables, then they have such a solution in which at least m of the inequalities are satisfied as equalities.

Proof. Assume that the optimization is one of maximization. Since a minimization problem can be changed to one of maximization by changing the sign of Z, this implies no loss of generality.

Suppose that we are given a purportedly optimum solution that satisfies k of the inequalities as equalities where $k < m$. Another solution will be found which satisfies $k + 1$ of the inequalities as equalities and which yields a value for the function being maximized that is at least as large as the given solution. By iteration, this implies the theorem.

Denote the set of P inequalities as

$$\sum_{i=1}^{m} \alpha_{ij} a_i \leq b_j, \quad j = 1,2,\ldots,P$$

and the maximizing function as

$$Z = \sum_{i=1}^{m} z_i a_i$$

Let the given solution correspond to some point A_1 in the a space, where A_1 has coordinates

$$A_1 = (a_1{}^1, a_2{}^1, \ldots, a_m{}^1)$$

The point A_1 satisfies exactly k of the inequalities as equalities. For convenience, let these be the first k inequalities. The value of the maximizing function for A_1 is

$$Z(A_1) = \sum_{i=1}^{m} z_i a_i{}^1$$

The point A_1 lies in the $(m - k)$-dimensional hyperspace defined by the first k equalities. This is a subspace of the complete m-dimensional a space. Let A_2 be some other point in this subspace where

$$A_2 = (a_1^2, a_2^2, \ldots, a_m^2)$$

In other words, A_2 is some point whose coordinates also satisfy the first k inequalities as equalities.

Now consider the point A_3 defined as the vector sum

$$\begin{aligned} A_3 &= (1 - c)A_1 + cA_2 \\ &= ((1 - c)a_1^1 + ca_1^2, (1 - c)a_2^1 + ca_2^2, \ldots, (1 - c)a_m^1 + ca_m^2) \end{aligned}$$

where c is any real constant positive or negative. Because of the linearity of the inequalities, the coordinates of A_3 also satisfy the first k inequalities as equalities; in other words, A_3 is in the same $(m - k)$-dimensional hyperspace.

Now consider the maximizing function

$$Z(A_3) = \sum_{i=1}^{m} z_i[(1 - c)a_i^1 + ca_i^2]$$

Again, because of linearity,

$$\begin{aligned} Z(A_3) &= (1 - c) \sum_{i=1}^{m} z_i a_i^1 + c \sum_{i=1}^{m} z_i a_i^2 \\ &= (1 - c)Z(A_1) + cZ(A_2) \end{aligned}$$

Thus, if

$$Z(A_2) > Z(A_1)$$

then

$$Z(A_3) > Z(A_1)$$

for any positive c, and if

$$Z(A_2) < Z(A_1)$$

then

$$Z(A_3) > Z(A_1)$$

for any negative c. If $Z(A_2)$ and $Z(A_1)$ are equal, then of course

$$Z(A_3) = Z(A_1)$$

for all c. Thus c can always be selected with the proper sign so that

$$Z(A_3) \geq Z(A_1)$$

We now observe that, for c small enough in magnitude but unequal to 0, A_3 will be a solution to the entire set of P inequalities. The first k inequalities are satisfied with the equal sign, as has already been noted. The remaining $P - k$ inequalities will be satisfied for sufficiently small c, since for c equal to 0 (that is, for A_1) they are satisfied with strict "greater-than" signs, and there is certainly a small enough (but nonzero) value of c for which the nonstrict inequalities will not be violated.

Suppose now that we pick the sign of c so that

$$Z(A_3) \geq Z(A_1)$$

and increase the magnitude of c from 0. We can continue making the magnitude of c larger until one of the remaining $P - k$ inequalities is about to be violated, that is, until it is satisfied with the equal sign. But this value of c corresponds to a solution that satisfies $k + 1$ of the inequalities with equal signs and for which the maximizing function is at least as large as the given one. This is exactly the point we set out to find.

Continuing in this same way, we can eventually find a point that satisfies m inequalities with equal signs and for which the maximizing function is at least as large as the given one.|

This is a very powerful and useful result. Whereas previously the optimum solution could have been any one of the continuum of points in the polyhedron, it has now been shown to be one of a finite number of points at the vertices of the polyhedron or, equivalently, a point for which at least m of the inequalities are satisfied as equalities.

If necessary, the optimum point could be determined by an exhaustive search of the vertices. If there are P inequalities in m unknowns, then the maximum possible number of possibilities is the combinatorial coefficient

$$\binom{P}{m}$$

Some of these may not correspond to vertices of the solution polyhedron. Each selection of m equations, however, can be solved (if possible) and the solutions checked to determine whether they satisfy the remaining inequalities. Those that do are vertices of the polyhedron. For each of these, the function to be maximized can be computed. The optimum solution is the one with the largest maximizing function.

Although this exhaustive procedure will yield the optimum solution, it is computationally forbidding. The Simplex Algorithm, presented in the next section, is a hill-climbing technique that yields the optimum solution with considerably less computation.

9.5 SIMPLEX ALGORITHM: THE BASIC IDEA

For reasons that will become apparent, the application of the Simplex Algorithm to the threshold-gate realization problem is not completely straightforward. Therefore, we first consider the basic idea of the algorithm and then apply it to the T-gate realization problem.

Throughout this discussion, it is assumed that the following conditions are fulfilled:

(1) There are P inequalities in m unknowns where

$$P \geq m$$

(2) All of the inequalities are constrained to be less than or equal to some constant.

(3) All of the variables are nonnegative.

(4) The optimization problem is one of maximization.

First, assume that by some means one vertex of the solution polyhedron has been found. This vertex corresponds to at least m of the inequalities being satisfied as equalities. (In linear programming terminology this is a *basic feasible solution.* Any solution to the given inequalities is called a feasible solution, and any vertex is a basic feasible solution.)

At this point the objective is to determine whether or not the specified vertex is the optimal solution and, if not, to determine another vertex for which the maximizing function is larger. Attention is restricted to the neighboring vertices. The given vertex has m neighboring vertices on the polyhedron, each of which also corresponds to at least m inequalities satisfied as equalities. Moreover, $m - 1$ of these inequalities are the same as for the original vertex.

It will be shown that if the value of the maximizing function Z at the vertex in question is larger than the value at all of the neighboring vertices, then that vertex corresponds to an optimum solution. If the value of Z is equal or larger at one or more neighboring vertices, then one of these is selected as a possible optimum solution and its neighbors investigated, and so on.

To see how this is done, consider first a very special case. Suppose that one of the vertices of the solution polyhedron is at the origin; that is, one possible solution is that all variables have value 0. Suppose, also, that the edges of the polyhedron are the coordinate axes. These are implied by the inequalities

$$a_i \geq 0 \qquad i = 1,2,\ldots,m$$

An example of this special case is shown in Fig. 9.6 for the two variable inequalities

$$\begin{aligned} -a_1 + a_2 &\leq 1 \\ 2a_1 + a_2 &\leq 9 \\ a_1 &\geq 0 \\ a_2 &\geq 0 \\ Z_{\max} &= z_1 a_1 + z_2 a_2 \end{aligned} \tag{9.1}$$

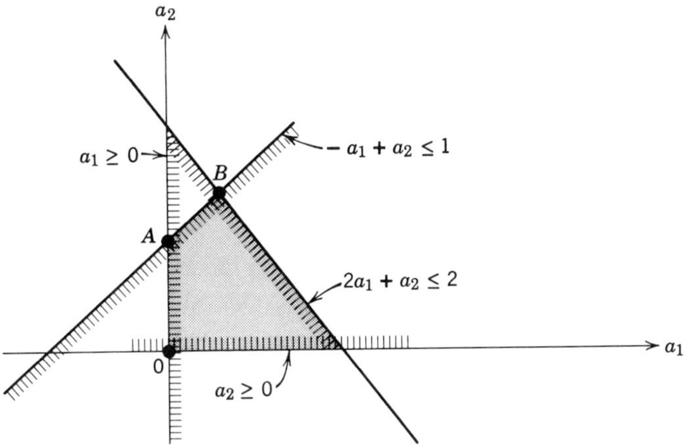

Fig. 9.6. A set of inequalities for which the origin is a basic feasible solution.

In this example, after verifying that the origin is a possible (basic feasible) solution, we wish to determine whether any other vertex corresponds to a higher value of the maximizing function.

Assume first that z_1 and z_2 are both negative, and that

$$Z_{\max} = -a_1 - 2a_2$$

At the origin the value of Z is then zero. Now consider whether there are any nonzero (also necessarily nonnegative) values of a_1 and a_2 for which Z can be larger than zero. Obviously, the answer is no, since if either a_1 or a_2 has any positive value, then Z will be negative. Thus, when both z_1 and z_2 are negative, the origin is the optimal solution.

Now suppose that

$$Z_{\max} = -a_1 + 2a_2$$

Again at the origin, the value of Z is zero. This time, however, the value of Z can be increased by keeping $a_1 = 0$ and increasing a_2 as much as possible. This corresponds to moving along the positive a_2 axis, which is permissible until vertex A of Fig. 9.6 is reached. Thus it is guaranteed that at vertex A the value of the maximizing function is larger than at the origin. If the maximizing function had been of the form

$$Z_{\max} = a_1 + 2a_2$$

then either a_1 or a_2 could be increased, but because of the factor 2, a_2 might be preferable.

This example shows that, for the special case where the origin is a basic feasible solution and where the edges of the polyhedron are the coordinate axes, it is easy to determine whether the origin is an optimum solution and, if not, to find another vertex for which the value of the maximizing function is larger. Now, however, it is necessary to consider how the situation is handled if the origin is not a basic feasible solution or, even if it is, what should be done after the first step, if it is found that there is another vertex for which the maximizing function is larger.

In both cases the answer depends on the property that the optimality of any vertex can be investigated by a transformation of variables, so that it becomes the origin and the edges of the solution polyhedron become the coordinate axes of the new coordinate system. Then the previous reasoning can be used to determine from the signs of the coefficients of the maximizing function (expressed in the new coordinates) whether that vertex is the optimal one. If it is not, then a new vertex with larger Z can be determined. This process can be repeated and is guaranteed to converge to the optimal solution in a finite number of steps. This follows, since (except in certain degenerate cases to be considered later) at each iteration a vertex with a larger Z is found (which assures that no vertex will be selected twice) and there are only a finite number of vertices.

This summarizes the basic idea of the Simplex Algorithm. All that remains is to consider the mechanics of effecting the variable transformations and to investigate some degenerate cases that frequently occur.

9.6 MAKING THE VARIABLE CHANGE—SLACK VARIABLES

In order to make the necessary variable changes, it is convenient to introduce the concept of the slack variable. A *slack variable* is a nonegative variable that is added with unity coefficient to all of the inequalities (except those of the form $a_i \geq 0$) in order to make them equalities.

Example 9.4. The inequality Set 9.1 of the previous example is repeated here:

$$\begin{aligned} -a_1 + a_2 &\leq 1 \\ 2a_1 + a_2 &\leq 2 \\ a_1 &\geq 0 \\ a_2 &\geq 0 \end{aligned} \tag{9.1}$$

$$Z_{\max} = z_1 a_1 + z_2 a_2$$

This set can be altered by the addition of two slack variables a_3 and a_4 so that it becomes

$$\begin{aligned} a_3 \qquad - a_1 + a_2 &= 1 \\ a_4 + 2a_1 + a_2 &= 2 \\ a_1 &\geq 0 \\ a_2 &\geq 0 \\ a_3 &\geq 0 \\ a_4 &\geq 0 \end{aligned} \tag{9.2}$$

$$Z_{max} = z_1 a_1 + z_2 a_2$$

If a particular slack variable is equal to zero, this is the same as if the corresponding inequality is satisfied as an equality. Thus the slack variables identify the vertex being investigated. Moreover, they correspond to the new variables after the variable transformation. This will become clear as we proceed.

Consider the Set 9.1 of inequalities when

$$Z_{max} = -a_1 + 2a_2$$

At the origin where

$$a_1 = a_2 = 0$$

the first two equalities of Set 9.2 become

$$\begin{aligned} a_3 &= 1 \\ a_4 &= 2 \end{aligned}$$

It has already been shown that to increase Z the value of a_2 should be increased. Leaving $a_1 = 0$ and considering a_2 as a variable, the first two inequalities of Set 9.2 become

$$\begin{aligned} a_3 \qquad + a_2 &= 1 \\ a_4 + a_2 &= 2 \end{aligned}$$

Since a_3 and a_4 cannot have negative values, the first equation limits the value of a_2 to 1. This value implies that $a_3 = 0$ and, therefore, that the first inequality is satisfied as an equality. Therefore, the point $a_1 = 0, a_2 = 1$ (vertex A in Fig. 9.6) is a vertex of the solution polyhedron because at least two of the inequalities are satisfied as equalities. To make a change of variable so that vertex A becomes the origin, set

$$a_1 = a_3 = 0$$

We also wish to consider the edges of the solution polyhedron that are incident at vertex A as corresponding to the new coordinate axes. (As a result, the effect on Z of moving away from vertex A along the incident edges can be investigated.) This is accomplished by expressing Z in terms of the variables a_1 and a_3 only. Solving the first equation of Set 9.2 for a_2 gives

$$a_2 = 1 - a_3 + a_1$$

Substituting this value for a_2 into the expression for Z yields

$$\begin{aligned} Z_{\text{max}} &= -a_1 + 2a_2 \\ &= -a_1 + 2(1 - a_3 + a_1) \\ &= 2 - 2a_3 + a_1 \end{aligned}$$

Observe that for

$$a_1 = a_3 = 0$$

which corresponds to vertex A, the value of Z is 2, and this is larger than the value at the origin which was 0. Because the sign of a_1 is positive, the value of Z can be further increased by increasing a_1 and keeping $a_3 = 0$. Since $a_3 = 0$ corresponds to the condition

$$-a_1 + a_2 = 1$$

increasing a_1 while leaving $a_3 = 0$ corresponds in Fig. 9.6 to moving along the edge of the solution polyhedron from vertex A toward vertex B. This will be the next cycle of the procedure. First, however, the change of variable must be completed by substituting the expression for a_2 into the second equation of Set 9.2. Thus

$$a_4 + 2a_1 + a_2 = 2$$

$$a_4 + 2a_1 + (1 - a_3 + a_1) = 2$$

or

$$-a_3 + a_4 + 3a_1 = 1$$

Thus the relations of Set 9.2 can be written as

$$\begin{aligned} a_2 \qquad - a_1 + a_3 &= 1 \\ a_4 + 3a_1 - a_3 &= 1 \\ a_1 &\geq 0 \\ a_2 &\geq 0 \\ a_3 &\geq 0 \\ a_4 &\geq 0 \\ Z_{\text{max}} = 2 - 2a_3 + a_1 & \end{aligned} \tag{9.3}$$

These equations as functions of a_1 and a_3 have the same form as those of Set 9.2 as functions of a_1 and a_2. In this case, the investigation is concerned with the vertex for which

$$a_1 = a_3 = 0$$

The relations of Set 9.2 were for investigations with respect to the vertex for which

$$a_1 = a_2 = 0$$

In each set the equations are expressed so that the maximizing function is expressed in terms of the variables, which are zero at the vertex in question, and each of the other variables appears in only one equation. This completes one cycle of the procedure.

The next cycle is similar to the first. It was previously shown that Z can be increased by making a_1 larger and keeping $a_3 = 0$. From the second equation of Set 9.3, it follows that

$$a_1 = \tfrac{1}{3}(1 + a_3 - a_4)$$

and that a_1 is limited to $\frac{1}{3}$, which occurs when

$$a_3 = a_4 = 0$$

Substituting the above expression for a_1 into the first equation and the maximizing expression of Set 9.3 yields the set.

$$\begin{aligned} a_2 + \tfrac{2}{3}a_3 + \tfrac{1}{3}a_4 &= 1\tfrac{1}{3} \\ a_1 - \tfrac{1}{3}a_3 + \tfrac{1}{3}a_4 &= \tfrac{1}{3} \\ a_1 &\geq 0 \\ a_2 &\geq 0 \\ a_3 &\geq 0 \\ a_4 &\geq 0 \end{aligned} \tag{9.4}$$

$$Z_{\max} = 2\tfrac{1}{3} - 1\tfrac{2}{3}a_3 - \tfrac{1}{3}a_4$$

These equations are in the form for investigations with respect to the vertex

$$a_3 = a_4 = 0$$

Reference to the expression for Z shows that no positive values of a_3 or a_4 can increase Z so that max Z occurs when

$$a_3 = a_4 = 0$$

Substituting these values into the equations of Set 9.4 gives

$$\begin{aligned} a_2 &= 1\tfrac{1}{3} \\ a_1 &= \tfrac{1}{3} \\ Z_{\max} &= 2\tfrac{1}{3} \end{aligned}$$

which are the values associated with vertex B of Fig. 9.6. This completes the solution.

The basic steps in the Simplex Algorithm can now be summarized. This summary depends upon the two following assumptions:

(1) All inequalities are expressed with " $\leq$ " signs and the optimization criterion is maximization.

(2) The set of inequalities

$$a_i \geq 0, \quad i = 1,2,\ldots,m$$

define edges of the solution polyhedron, and one feasible solution is

$$a_1 = a_2 = \cdots = a_m = 0$$

If the second assumption is not satisfied, then some additional work may be necessary. This situation is discussed after the presentation of the step-by-step procedure.

Procedure 9.2 (Simplex Algorithm)

(1) To each inequality that involves more than one variable add a nonnegative slack variable with unity coefficient and express as an equation.

(2) Consider the maximizing expression

$$Z_{\max} = \sum_{i=1}^{m} z_i a_i$$

If all z_i are nonpositive, then the present solution is the optimal one. If there exist positive z_i, select one with the largest value and denote the associated a_i as a_M.

(3) Determine the equation that limits the maximum value of a_M. To accomplish this, compute for each equation that includes a_M the ratio of the constant term to the coefficient of a_M. The equation with the smallest nonnegative quotient is the limiting equation. (In this description it is assumed that the limiting quotient is nonzero. The zero quotient case is considered in the subsequent discussion.) If all quotients are negative, a_M can be increased without bound, and the value of Z has no upper bound. In this case, therefore, the linear programming problem has no solution. This corresponds to the inequalities not forming a finite polyhedron but, instead, including a region that extends to infinity.

(4) Use the limiting equation to eliminate a_M from the remaining equations and from the expression for Z. This can be done by either (a) solving the limiting equation for a_M and substituting this expression into the remaining expressions or (b) subtracting the limiting equation suitably multiplied by a constant from the remaining equations.

(5) The new basic feasible solution is determined by setting equal to zero all of the variables that appear in the expression for Z and "solving" the other equations for the remaining variables. To determine whether this is the optimal solution, return to Step 2.

This completes the description of the procedure. We now discuss what happens if assumption 2 or the assumption of Step 3 is not satisfied.

Consider first assumption 2: that the origin and the coordinate axes are edges of the solution polyhedron. This can be invalid in two ways: (a) the origin is not an allowable solution or (b) the origin is an allowable solution but the inequalities

$$a_i \geq 0, \quad i = 1,2,\ldots,m$$

do not define the edges of the solution polyhedron.

If the origin is not an allowable solution, then there may be some difficulty in finding any feasible solution. There are a number of methods that classical linear programming suggests for finding an initial feasible solution. However, the single T-gate realization problem has its own special technique for this difficulty, and this will be explained in a later section.

The situation in which the origin is a solution but the coordinate axes are not the edges of the solution polyhedron is shown in Fig. 9.7*a*. It occurs when one of the inequalities is of the form

$$\sum_{i=1}^{m} \alpha_i a_i \geq 0$$

where the right-hand side is zero. This is called the *degenerate case*.

Observe that the degenerate case corresponds to a violation of the assumption of Step 3 of Procedure 9.2; that is, the constant term of the equation is zero. Thus, the present discussion covers the violation of both assumptions.

Suppose that it has been determined that Z can possibly be increased by increasing some variable, a_M, from zero and that one of the equations has a zero constant term. Assume first that, when this equation is expressed so that the variable which is nonzero (initially the slack variable) has a positive

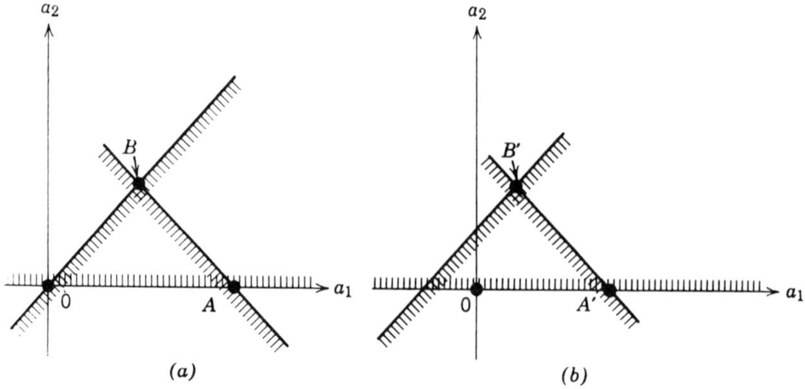

Fig. 9.7. Perturbing the problem when the coordinate axes do not bound the solution polyhedron. (*a*) The coordinate axes do not bound the polyhedron. (*b*) The perturbed problem.

coefficient, the coefficient of a_M is negative. This equation then places no restriction on how large a_M can be, since both a_M and the nonzero variable can be increased together in magnitude so that the equation remains equal to zero. In this situation some other equation constrains the value of a_M, and the degeneracy causes no difficulty.

The difficulty occurs when the coefficient of a_M and that of the nonzero variable are both positive in the equation with zero constant term. This then becomes the controlling equation for the value of a_M, and it is used to eliminate a_M from the remaining equations and from Z. In Fig. 9.7*a* this corresponds to moving to vertex B. So far, the procedure is being carried out in the usual manner. However, since this equation has no constant term, when it is used to eliminate a_M from the expression for Z, the value of Z will not increase, and the convergence argument is no longer valid. It can no longer be guaranteed that the procedure will not go into a cycle and after going to a number of other vertices repeatedly return to the present one and thus never find the optimal solution.

In most instances the possibility of cycling can be ignored and the procedure carried out in the usual way. This is the approach that we shall follow. However, if a cycle ever does occur, then convergence can be guaranteed by using the following perturbation technique.

Any zero constant terms of inequalities with more than one variable are replaced by a small constant, ϵ, giving

$$\sum_{i=1}^{m} \alpha_i a_i = \epsilon$$

This transforms the solution polyhedron as shown in Fig. 9.7*b* and eliminates the difficulty. After a few stages of the procedure, when the further increasing of Z is not dependent upon ϵ, the value of ϵ can be set equal to zero, thereby returning to the original problem.

The next section is concerned with computationally efficient tabular methods for performing the necessary computations. For this, we introduce some additional linear programming terminology.

The *basic variables* are those that are not set equal to zero to define the vertex under investigation. Thus, in the first cycle of the procedure, the slack variables are the basic variables. They are sometimes referred to as being *in the basis*. Going to the next cycle of the procedure where a new vertex is being investigated is called *changing the basis*. One variable is *put into the basis* and another is *removed from the basis*.

The process whereby one equation is used to eliminate a variable from another equation (or to change the basis) is called *pivoting* or *pivoting around that equation*.

9.7 THE TABLEAU

The tableau is a convenient tabular form for keeping track of the computations in the Simplex Algorithm. It is most easily explained with an example.

Example 9.5. Consider again the problem of Example 9.4:

$$
\begin{aligned}
a_3 \qquad - a_1 + a_2 &= 1 \\
a_4 + 2a_1 + a_2 &= 2 \\
a_1 &\geq 0 \\
a_2 &\geq 0 \\
a_3 &\geq 0 \\
a_4 &\geq 0 \\
Z_{\max} &= - a_1 + 2a_2
\end{aligned}
$$

where a_3 and a_4 are the slack variables. This problem is entered into the first tableau as shown in Table 9.1. Notice that, except for the first column which denotes the basic variable in each equation, the rest of the table is nothing more than the coefficients of the variables of each equation (omitting the inequalities $a_i \geq 0$ and zero entries).

Table 9.1. A Set of Tableaux

Basic Variables	Slack Variables		Actual Variables		Constants
	a_3	a_4	a_1	a_2	
A. The Initial Tableau					
a_3	1		−1	①	1
a_4		1	2	1	2
Z_{max}			−1	2	
B. The Second Tableau					
a_2	1		−1	1	1
a_4	−1	1	③		1
Z_{max}	−2		1		−2
C The Third Tableau					
a_2	$\frac{2}{3}$	$\frac{1}{3}$		1	$1\frac{1}{3}$
a_1	$-\frac{1}{3}$	$\frac{1}{3}$	1		$\frac{1}{3}$
Z_{max}	$-1\frac{2}{3}$	$-\frac{1}{3}$			$-2\frac{1}{3}$

The next cycle of the procedure is depicted on the second tableau of Table 9.1*B*. This tableau is constructed as follows:

(1) Find in the Z row of the initial tableau the largest positive number in the variable columns. In this case the number is 2 and it appears in the a_2 column. Thus, a_2 is the variable which will be put into the basis.

(2) For each equation row divide the constant term by the entry in the column selected in (1) (in this case the a_2 column). Encircle the entry in this variable column that corresponds to the smallest nonnegative quotient. In this case the 1 of the first row and in column a_2 is encircled. The corresponding quotient is 1. This corresponds to the restriction on the amount by which a_2 can be increased from 0. After a_2 is increased by this amount, a_3 will have 0 value; therefore a_3 is the variable being taken out of the basis.

(3) The rows in the second tableau (except for their entries in the basis column) are determined as follows:

(a) The row corresponding to the variable being taken out of the basis (in this case the first row) is just the corresponding row in the first table, divided by the value of the encircled coefficient (in this case, 1). In the new tableau the new basis variable is written in the appropriate column of this row.

(b) All other rows including the Z row are obtained by adding or subtracting from the corresponding row in the first tableau, the row being taken out of the basis multiplied by an appropriate constant so as to eliminate the coefficient in the column with encircled entry. Each of these rows keeps its same entry in the basis column.

The second tableau is shown in Table 9.1B. In comparing this with the second cycle of the procedure described in the previous section (that is, with Set 9.3 from Example 9.4) note that the entry in the constant column of the Z row is the *negative* of the value of Z. This negative sign is a consequence of the method of table formation and need cause no difficulty.

The procedure can be repeated once more, yielding the third tableau, shown in Table 9.1C. Because of the negative entries in the Z row, this tableau represents the final solution. The optimum values of the variables are read from the basis and constant columns as

$$a_2 = 1\tfrac{1}{3}$$

$$a_1 = \tfrac{1}{3}$$

and the value of Z is the negative of the constant entry of the Z row.

9.8 SIMPLEX ALGORITHM REALIZATIONS OF T GATES WITH MAXIMUM RELATIVE GAP

In Section 9.1 the single-gate realization problem is formulated as the solution to a reduced set of inequalities of the form

$$\sum_{i=1}^{n} a_i x_i^p \geq u$$

for p so that $F(p) = 1$;

$$\sum_{i=1}^{n} a_i x_i^p \leq \ell$$

for p so that $F(p) = 0$; and

$$u > \ell$$

where the reducing relations have been temporarily neglected as explained in that section. These inequalities can be rewritten as

$$-\sum_{i=1}^{n} a_i x_i^p + u \leq 0$$

for p so that $F(p) = 1$;

$$\sum_{i=1}^{n} a_i x_i^p - \ell \leq 0$$

for p so that $F(p) = 0$; and

$$u > \ell$$

$$a_i \geq 0$$

but this is still not precisely the form required of a linear program to which the Simplex Algorithm can be applied. In this regard the following points must be considered:

(1) Some normalization must be made to eliminate the strict inequality

$$u > \ell$$

(2) Some maximizing or minimizing function must be defined. In this section we consider maximizing relative gap and in the next section minimizing normalized fan-in.

(3) The most crucial difficulty is in finding an initial (basic feasible) solution. The Simplex Algorithm is a procedure for starting with a feasible solution and obtaining a better one according to a given criterion. In the ℓ.s. problem, however, we do not know if a solution exists or how to find one. Indeed, the goal of the ℓ.s. test procedure is to determine whether or not any (feasible) solutions exist. Thus, we appear to be stymied at the very beginning.

When the optimization criterion is maximum relative gap, however, all of these difficulties can be eliminated† in the following way.

(1) The solution is normalized by assigning unit value to ℓ.

(2) The inequality

$$u > \ell$$

is deleted from the set.

(3) The inequality

$$u \geq 0$$

is added to the set.

(4) The maximizing function is selected as

$$Z = u$$

If the optimum solution yields a value of u greater than one (that is, $u > \ell$), then it corresponds to a solution of the original problem. Otherwise, the given function is not ℓ.s.

† This formulation was suggested by L. G. Abraham.

Observe that for ℓ fixed, maximizing u corresponds to maximizing the relative gap:

$$\frac{u - \ell}{u + \ell}$$

It is also evident that there exists a $u > \ell = 1$ if and only if the given function is ℓ.s., and this will be determined by the linear programming problem.

With these changes, the inequalities that characterize the ℓ.s. problem become

$$-\sum_{i=1}^{n} a_i x_i^p + u \leq 0$$

for p such that $F(p) = 1$;

$$\sum_{i=1}^{n} a_i x_i^p - 1 \leq 0$$

for p such that $F(p) = 0$;

$$a_i \geq 0$$

for $i = 1,2,\ldots,n$; and

$$u \geq 0$$
$$Z = u$$

Note that for this new problem, one feasible solution is

$$a_i = 0 \qquad \text{for} \quad i = 1,2,\ldots,n$$
$$u = 0$$

Thus the origin is a basic feasible solution and the Simplex Algorithm can be directly applied.

Example 9.6. Consider the function of Example 9.3,

$$F(x_1,x_2,x_3,x_4) = x_1x_2 \dotplus x_1x_3x_4 \dotplus x_2x_3x_4$$

The reduced inequality set is

$$\begin{aligned} 2a_1 &\geq u \\ a_1 + 2a_3 &\geq u \\ 2a_3 &\leq \ell \\ a_1 + a_3 &\leq \ell \\ u &> \ell \end{aligned}$$

$$\left.\begin{aligned} a_3 &> 0 \\ a_1 &> 0 \\ a_1 &= a_2 \\ a_3 &= a_4 \end{aligned}\right\} \text{reducing relations}$$

Temporarily neglecting the reducing relations, rearranging the remaining inequalities, and making the suggested normalizations yields the set

$$\begin{aligned}
-2a_1 \qquad\quad + u &\leq 0 \\
-a_1 - 2a_3 + u &\leq 0 \\
2a_3 \qquad &\leq 1 \\
a_1 + a_3 \qquad &\leq 1 \\[1em]
a_1 &\geq 0 \\
a_3 &\geq 0 \\
u &\geq 0 \\[1em]
Z &= u
\end{aligned}$$

The first tableau is shown in Table 9.2*A*.

Table 9.2. Tableaux for Example 9.6

Basis Variables	Slack Variables				Actual Variables			Constants
	a_4	a_5	a_6	a_7	a_1	a_3	u	
A. The First Tableau								
a_4	1				−2		1	
a_5		1			−1	−2	①	
a_6			1			2		1
a_7				1	1	1		1
Z_{max}							1	
B. The Second Tableau								
a_4	1	−1			−1	②		
u		1			−1	−2	1	
a_6			1			2		1
a_7				1	1	1		1
Z_{max}		−1			1	2		

C. The Third Tableau

a_3	$\frac{1}{2}$	$-\frac{1}{2}$			$-\frac{1}{2}$	1		
u	1				-2		1	
a_6	-1	1	1		1			1
a_7	$-\frac{1}{2}$	$\frac{1}{2}$		1	$\textcircled{\frac{3}{2}}$			1
$Z_{\max}$	-1				2			

D. The Fourth Tableau

a_3	$\frac{1}{3}$	$-\frac{1}{3}$		$\frac{1}{3}$		1		$\frac{1}{3}$
u	$\frac{1}{3}$	$\frac{2}{3}$		$\frac{4}{3}$			1	$\frac{4}{3}$
a_6	$-\frac{2}{3}$	$\frac{2}{3}$	1	$-\frac{2}{3}$				$\frac{1}{3}$
a_1	$-\frac{1}{3}$	$\frac{1}{3}$		$\frac{2}{3}$	1			$\frac{2}{3}$
$Z_{\max}$	$-\frac{1}{3}$	$-\frac{2}{3}$		$-\frac{4}{3}$				$-\frac{4}{3}$

Because of the 1 in the u column of the Z row, u is the variable to be put into the basis. In the u column, there are only two positive numbers, and in both of the corresponding rows a 0 appears in the constant column; hence, this is a degenerate case. (In fact, this will always be the case in the first cycle of this formulation.)

It is not necessary to add a constant ϵ in the constant column; just select one of the two rows, for instance the encircled one, and proceed. Taking a_5 out of the basis and putting u in yields, the tableau shown in Table 9.2*B*.

The second cycle is also degenerate. We must remove a_4 from the basis and put a_3 into the basis, yielding the tableau shown in Table 9.2*C*.

This is not a degenerate case. We proceed by putting a_1 into the basis and taking a_7 out of the basis. The resulting tableau is shown in Table 9.2*D*.

Since there are no positive numbers in the Z row, the procedure has been completed. The optimum u is $\frac{4}{3}$, which is greater than $\ell = 1$, and thus this solution is a solution to the given inequality problem. The values of the variables are

$$a_3 = \tfrac{1}{3}$$
$$u = \tfrac{4}{3}$$
$$a_6 = \tfrac{1}{3}$$
$$a_1 = \tfrac{2}{3}$$

The fact that one of the slack variables, a_6, is nonzero implies that its inequality, the third, is satisfied with the "greater than" sign and has a "slack" of $\frac{1}{3}$.

Now it must be verified that the solution satisfies the reducing relations. This is found to be the case; hence the coefficients constitute a solution to the given single-gate realization problem. Note that this is exactly the realization obtained in Example 9.3.

This example illustrates that a lower bound can be obtained on the number of cycles required to yield a realization for an ℓ.s. function. This is set forth in the following theorem.

Theorem 9.4. If the reduced inequalities corresponding to an ℓ.s. function contain R variables (excluding ℓ), then at least R cycles are required to obtain a solution using the formulation of this section; that is, at least $R + 1$ tableaux are required.

Proof. Because of the reducing relations, each of the R actual variables must be strictly positive and, as a result, each of the actual variables must be in the basis. Initially, however, none of the actual variables is in the basis and at each cycle only one variable can be put in. Thus, at least R cycles are needed. ∎

In the previous example, there are three variables, a_1, a_3, and u, and exactly three cycles were required, one for each of the variables. In more complex problems, however, it is common that many more than R cycles are needed.

9.9 SIMPLEX ALGORITHM REALIZATIONS OF T GATES WITH MINIMUM NORMALIZED FAN-IN

If any optimization criterion other than maximum relative gap is required, then the problem of finding an initial basic feasible solution may be more difficult. In this section, we show how the results of the maximum relative gap problem can be used as a starting point for the other optimization problems. Section 9.11 describes another approach using the dual formulation of the linear programming problem.

Both approaches are demonstrated using the criterion of minimum normalized fan-in, which is the other important reliability measure for the T gate. However, the same basic ideas apply to any linear optimizing function.

Our approach in this section is to use the procedure of the previous section to obtain an initial basic feasible solution to the minimum normalized fan-in problem so that the usual Simplex Algorithm can be used. It will be shown that the entire procedure can be carried out on a single set of tableaux that are augmented with some additional rows and columns.

Recall that the unnormalized reduced inequality set is

$$-\sum_{i=1}^{n} a_i x_i^p + u \leq 0$$

for p such that $F(p) = 1$;

$$\sum_{i=1}^{n} a_i x_i^p - \ell \leq 0$$

for p such that $F(p) = 0$; and

$$u > \ell$$

For the normalized fan-in formulation, it is convenient to normalize the gap length as

$$\ell = u - a$$

where for the time being a is left as a constant whose value will be determined later. Substituting this expression into the inequality set gives

$$-\sum_{i=1}^{n} a_i x_i^p + u \leq 0$$

for p such that $F(p) = 1$;

$$\sum_{i=1}^{n} a_i x_i^p - u \leq -a$$

for p such that $F(p) = 0$; and

$$a_i \geq 0$$

$$u \geq 0$$

where the redundant inequality $u \geq 0$ has been included. The maximizing condition† can now be added as

$$Z_{\max} = -\sum_{i=1}^{n} a_i$$

(Note that this corresponds to minimizing the sum of the weights.)

The problem has now been expressed as a linear program; however, there is still a difficulty. The origin is not a basic feasible solution; that is,

† Another maximizing condition that is often used is

$$Z'_{\max} = -\sum_{i=1}^{n} a_i - u$$

The condition has engineering implications in those T gate implementations where the threshold is considered as an additional coefficient.

setting all a_i and u equal to zero does not satisfy the inequalities. Thus it is not clear how to start the Simplex Algorithm. We now show how the solution obtained in the previous section can be used as a starting point for the problem of this section.

Although the inequality set of this section looks somewhat different from that of the previous section, they are, of course, simply related because they are different formulations of the same problem. In particular, assume that a solution to the inequality set of the last section has been determined such that

$$u = u^* > 1$$

(Recall that ℓ was normalized to 1.) Now suppose that in the inequality set of this section the value of a is selected as

$$a = u^* - 1$$

which is the gap length of that solution. Now the solution of the first problem becomes a solution to the second.

This is the way that solutions of the maximization problem of the previous section can be used as initial basic feasible solutions to the maximization problem of this section. A tableau is constructed which contains the information about both problems. Initially, the various changes of variable are made so as to maximize u according to the first problem; however, the changes of variable are made on the data of both problems. When the first problem has been completed (if possible) and a value of u^* has been found, then the constant a is selected as

$$a = u^* - 1$$

At this point, it is known which variables should be in the basis to start the second problem. In fact, it will become clear from the example that all of the correct variables of the second problem are already in the basis except for u. Thus the second problem can now be started.

Because of the large amount of overlap between the coefficients of both problems, little extra work is required to keep track of both problems at once. The entire procedure is clarified by the following example.

Example 9.7. Consider again the function of Examples 9.3 and 9.6 for which the reduced inequality set is

$$\begin{aligned}
2a_1 \qquad\qquad &\geq u \\
a_1 + 2a_3 &\geq u \\
2a_3 &\leq \ell \\
a_1 + a_3 &\leq \ell \\
u &> \ell
\end{aligned}$$

Recall that with the normalization of the last section, this becomes

$$\begin{aligned} -2a_1 \qquad\quad + u &\le 0 \\ -a_1 - 2a_3 + u &\le 0 \\ 2a_3 \qquad &\le 1 \\ a_1 + a_3 \qquad &\le 1 \\ Z_1 = u & \end{aligned}$$

(Note the subscript on the Z.)

For the problem of this section, we normalize by setting

$$\ell = u - a$$

and obtain

$$\begin{aligned} -2a_1 \qquad\quad + u &\le 0 \\ -a_1 - 2a_3 + u &\le 0 \\ 2a_3 - u &\le -a \\ a_1 + a_3 - u &\le -a \\ Z_2 = -a_1 - a_3 & \end{aligned}$$

Observe that all coefficients in the inequalities except those in the u column and in the constant column will always be the same in both formulations.

Both of these formulations can be put on the same tableau as shown in Table 9.3*A*. Observe that the a_4, a_5, a_6, a_7, a_1, and a_3 columns are identical and need only be written once. However, there are two u and constant columns, one corresponding to each problem, and there are two Z rows, one for each maximizing function.

Table 9.3. Tableaux for Example 9.7

Basis Variables	Slack Variables				Actual Variables			Constants	Second Problem	
	a_4	a_5	a_6	a_7	a_1	a_3	u		u	c
A. The First Tableau										
a_4	1				-2		1		1	
a_5		1			-1	-2	⓵		1	
a_6			1			2		1	-1	$-a$
a_7				1	1	1		1	-1	$-a$
Z_1							1		//////	//////
Z_2					-1	-1	//////	//////		

B. The Second Tableau

a_4	1	-1			-1	(2)				
u		1			-1	-2	1		1	
a_6			1			2		1	-1	$-a$
a_7				1	1	1		1	-1	$-a$
Z_1		-1			1	2				
Z_2					-1	-1				

C. The Third Tableau

a_3	$\frac{1}{2}$	$-\frac{1}{2}$			$-\frac{1}{2}$	1				
u	1				-2		1		1	
a_6	-1	1	1		1			1	-1	$-a$
a_7	$-\frac{1}{2}$	$\frac{1}{2}$		1	($\frac{3}{2}$)			1	-1	$-a$
Z_1	-1				2					
Z_2	$\frac{1}{2}$	$-\frac{1}{2}$			$-\frac{3}{2}$					

D. The Fourth Tableau

a_3	$\frac{1}{3}$	$-\frac{1}{3}$		$\frac{1}{3}$		1		$\frac{1}{3}$	$-\frac{1}{3}$	$-a/3$
u	$\frac{1}{3}$	$\frac{2}{3}$		$\frac{4}{3}$			1	$\frac{4}{3}$	($-\frac{1}{3}$)	$-4a/3$
a_6	$-\frac{2}{3}$	$\frac{2}{3}$	1	$-\frac{2}{3}$				$\frac{1}{3}$	$-\frac{1}{3}$	$-a/3$
a_1	$-\frac{1}{3}$	$\frac{1}{3}$		$\frac{2}{3}$	1			$\frac{2}{3}$	$-\frac{2}{3}$	$-2a/3$
Z_1	$-\frac{1}{3}$	$-\frac{2}{3}$		$-\frac{4}{3}$				$-\frac{4}{3}$		
Z_2				1					-1	$-a$

E. The Fifth Tableau

a_3		-1		-1		1				$\frac{1}{3}$
u	-1	-2		-4					1	$\frac{4}{3}$
a_6	-1		1	2						$\frac{1}{3}$
a_1	-1	-1		-2	1					$\frac{2}{3}$
Z_2	-1	-2		-3						1

Initially, Z_1 is maximized using the information of the first formulation. Thus, we decide to put u (of the first problem) into the basis and take a_5 out of the basis. The results are placed in the Second Tableau as shown in Table 9.3*B*. Note that the same operations are performed on all columns in the a_4, a_5, a_6, and a_7 rows, even those corresponding to the second problem. Thus the indicated change of variable is made in both problems. The usual operations are also performed in the Z_1 row except that the columns which correspond only to the second problem are ignored.

The Z_2 row is treated in a corresponding way in order to create a zero entry in the column of the variable being put into the basis, and ignoring the columns which correspond only to the first problem. (In this first cycle, there is already a 0 in the entry in the u column so no operation is necessary. Note, however, the second and third cycles.)

The indicated variable change has now been made in both formulations. The procedure continues as before until the first problem is completed in the Fourth Tableau as shown in Table 9.3*D*.

At this point, we know the following about the second problem. If the constant a has a value

$$\begin{aligned} a &= u^* - 1 \\ &= \tfrac{4}{3} - 1 = \tfrac{1}{3} \end{aligned}$$

and if the basis variables are a_3, u (of the second problem), a_6, and a_1, then the solutions will be the same as for the first problem. However, it is clear from the Fourth Tableau that the u of the second problem is not in the basis (its column does not contain a single 1 with the other entries 0), and the u of the first problem is in the basis. It is necessary, therefore, to put the u of the second problem in the basis as shown in the Fifth Tableau. Here a has been set equal to $\frac{1}{3}$, and the columns corresponding to the first problem and the Z_1 row have been omitted since they are no longer needed.

Now we are ready to start the second problem. As a check on the calculations, it can be verified that the initial feasible solution to this second problem is the same as the final solution to the first problem, as, indeed, it is.

Consider now the Z_2 row. Since there are no positive entries, there is no better solution than the one already obtained. The solution is, therefore, complete. For this problem, the solution that maximizes relative gap also minimizes fan-in.

As discussed in Section 8.5, it is no accident that the maximum relative gap and the minimum normalized fan-in realizations for this example are the same. It was shown in Example 9.3 that there are only three nonredundant inequalities for this problem in three unknowns. Thus, by Theorem 9.3, any optimization criterion has, as one of its solutions, the point where these three inequalities are simultaneously satisfied as equalities. Thus all optimization criteria have this one solution in common, and in particular the minimum normalized fan-in and the maximum relative gap solutions are the same.

In general, whenever the number of unequalities corresponding to a switching function of n variables can be reduced to $n + 1$, then, since there are $n + 1$ unknowns (counting u), the same reasoning can be used to show that these two optimization criteria, as well as all others, have the same solution. Moreover, this solution can be obtained using the methods of simultaneous equations (Example 8.7). This situation often occurs for functions of a small number of variables.

9.10 THE DUAL PROBLEM AND RELATED PROPERTIES

To every linear programming problem there corresponds a *dual problem* whose formulation and solution are simply related to those of the original (or *primal*) problem. A consideration of this dual problem not only may decrease the amount of necessary computation but also brings out some additional properties of the solutions.

In order to appreciate better the duality concept, we first introduce a somewhat different tableau on which the Simplex Algorithm can be performed. Consider again the problem of Example 9.6.

$$\begin{aligned}
-2a_1 \qquad\quad + u &\le 0\\
-a_1 - 2a_3 + u &\le 0\\
2a_3 \qquad &\le 1\\
a_1 + a_3 \qquad &\le 1\\
a_1 &\ge 0\\
a_3 &\ge 0\\
u &\ge 0
\end{aligned}$$

$$Z = u$$

Table 9.4. The New Tableaux for the Problem of Example 9.6

A. The Initial Tableau

Basis Variables	a_1	a_3	u	Constant
a_4	-2		1	
a_5	-1	-2	(1)	
a_6		2		1
a_7	1	1		1
Z			1	

B. The Second Tableau

Basis Variables	a_1	a_3	a_5	Constant
a_4	-1	(2)	-1	
u	-1	-2	1	
a_6		2		1
a_7	1	1		1
Z	1	2	-1	

C. The Third Tableau

Basis Variables	a_1	a_4	a_5	Constant
a_3	$-\frac{1}{2}$	$\frac{1}{2}$	$-\frac{1}{2}$	
u	-2	1		
a_6	1	-1	1	1
a_7	$(\frac{3}{2})$	$-\frac{1}{2}$	$\frac{1}{2}$	1
Z	2	-1		

D. The Fourth Tableau

Basis Variables	a_7	a_4	a_5	Constant
a_3	$\frac{1}{3}$	$\frac{1}{3}$	$-\frac{1}{3}$	$\frac{1}{3}$
u	$\frac{4}{3}$	$\frac{1}{3}$	$\frac{2}{3}$	$\frac{4}{3}$
a_6	$-\frac{2}{3}$	$-\frac{2}{3}$	$\frac{2}{3}$	$\frac{1}{3}$
a_1	$\frac{2}{3}$	$-\frac{1}{3}$	$\frac{1}{3}$	$\frac{2}{3}$
Z	$-\frac{4}{3}$	$-\frac{1}{3}$	$-\frac{2}{3}$	$-\frac{4}{3}$

The new tableau is shown in Table 9.4*A*. It is the same as the tableau of Table 9.2 except that the columns corresponding to the slack variables have been omitted. Again, the pivot is around the encircled entry. However, whereas in Table 9.2*B* the u column then became one of the basis columns (with a single 1 entry and the remaining entries 0) and the a_5 column became the nonbasis column, in this new tableau the label on the u column is changed to a_5 and the a_5 column is written in the corresponding space as shown in Table 9.4*B*.

The procedure continues as before as shown in the rest of Table 9.4. This should be compared with Table 9.2. The only difference is in the labeling of the columns and in the omission of columns corresponding to the basis variables. Again the optimum solution can be read directly from the basis variables and the constant columns of the final table as

$$a_3 = \tfrac{1}{3}$$
$$u = \tfrac{4}{3}$$
$$a_6 = \tfrac{1}{3}$$
$$a_1 = \tfrac{2}{3}$$

Besides being somewhat more economical computationally, this new type of tableau displays some interesting relationships among the elements in successive tableaux.

Table 9.5. The General Cycle Using the New Tableau

A. A General Tableau

	a_1	$\cdots$	a_k	$\cdots$	a_m
λ_1	b_{11}	$\cdots$	b_{1k}	$\cdots$	b_{1m}
$\vdots$	$\vdots$		$\vdots$		
λ_j	b_{j1}	$\cdots$	$\boxed{b_{jk}}$ (encircled)	$\cdots$	b_{jm}
$\vdots$	$\vdots$		$\vdots$		
λ_n	b_{n1}	$\cdots$	b_{nk}	$\cdots$	b_{nm}

B. The Next Tableau

	a_1	$\cdots$	λ_j	$\cdots$	a_m
λ_1	$b_{11} - \dfrac{b_{j1}b_{1k}}{b_{jk}}$	$\cdots$	$\dfrac{-b_{1k}}{b_{jk}}$	$\cdots$	$b_{1m} - \dfrac{b_{jm}b_{1k}}{b_{jk}}$
$\vdots$	$\vdots$		$\vdots$		$\vdots$
a_k	b_{j1}/b_{jk}	$\cdots$	$1/b_{jk}$	$\cdots$	b_{jm}/b_{jk}
$\vdots$	$\vdots$		$\vdots$		$\vdots$
λ_n	$b_{n1} - \dfrac{b_{j1}b_{nk}}{b_{jk}}$	$\cdots$	$-b_{nk}/b_{jk}$	$\cdots$	$b_{nm} - \dfrac{b_{jm}b_{nk}}{b_{jk}}$

Consider a general tableau as shown in Table 9.5A where for clarity the basis vectors are shown as λ's. The tableau of the next cycle is shown in Table 9.5B. Observe the general form of the second tableau.

(1) The labels on the encircled row and column have been interchanged.

(2) The encircled entry b_{jk} is replaced by $1/b_{jk}$ and every other entry in that row is divided by b_{jk}. To verify this, observe in Table 9.2 that the entire row with the encircled entry is divided by b_{jk}, and that the a_k column is replaced with the λ_j column, which had an entry of 1 in that row.

(3) To produce the new λ_i row (for any $i \neq j$), the old λ_j row is multiplied by

$$\frac{-b_{ik}}{b_{jk}}$$

and added to the old λ_i row. Thus the entries in the new λ_j column are

$$\frac{-b_{ik}}{b_{jk}}$$

and all of the remaining columns are

$$b_{i\ell} - \frac{b_{j\ell}b_{ik}}{b_{jk}}$$

(for $i \neq j$).

To eliminate some of the difficulty caused by subscripts, Table 9.6 shows the form of the cycle for four entries. The important point to observe is that, except for the minus sign in the encircled column, the operations have a

Table 9.6. A Simple Case of the Next Cycle

A. A Tableau

b	c
(a)	d

B. The Next Tableau

$-\frac{b}{a}$	$c - \frac{bd}{a}$
$\frac{1}{a}$	$\frac{d}{a}$

symmetry with respect to rows and columns. Thus one suspects that there may be a close relationship between the given linear programming problem and another such problem in which the interpretation of the rows and columns are interchanged, including the interpretation of the constant column and the Z row. This, indeed, turns out to be the case.

Example 9.8. Consider again the problem of Example 9.6:

$$\begin{aligned}
-2a_1 \qquad\qquad + u &\le 0 \\
-a_1 - 2a_3 + u &\le 0 \\
2a_3 \qquad &\le 1 \\
a_1 + a_3 \qquad &\le 1 \\
a_1 &\ge 0 \\
a_3 &\ge 0 \\
u &\ge 0
\end{aligned}$$

$$Z_{\max} = u$$

Then consider the problem

$$\begin{aligned}
-2v_1 - v_2 \qquad\qquad + v_4 &\ge 0 \\
-2v_2 + 2v_3 + v_4 &\ge 0 \\
v_1 + v_2 \qquad\qquad\qquad &\ge 1
\end{aligned}$$

$$Z_{\min} = v_3 + v_4$$

$$v_1 \ge 0,\ v_2 \ge 0,\ v_3 \ge 0,\ v_4 \ge 0$$

This is related to the original problem in the following way:

(1) The coefficients of the rows and columns have been interchanged.
(2) The Z row and constant column have been interchanged.
(3) The $\le$ signs in the inequalities have been replaced by $\ge$.
(4) The maximizing problem has been changed to a minimizing problem.

This new problem is called the *dual* of the given one, and the rules of formation of the dual are, in general, given by (1) through (4).

Suppose that we try to solve this dual problem, and further suppose that we solve it without changing it to a maximizing problem and without reversing the inequality signs by multiplying through by -1. When the nonnegative slack variables are added, they would have coefficients of -1; thus

$$\begin{aligned}
-v_5 \qquad\qquad -2v_1 - v_2 \qquad\qquad + v_4 &= 0 \\
-v_6 \qquad\qquad - 2v_2 + 2v_3 + v_4 &= 0 \\
-v_7 + v_1 + \textcircled{v_2} \qquad\qquad &= 1
\end{aligned}$$

$$Z_{\min} = v_3 + v_4$$

$$v_i \geq 0, \quad i = 1,2,\ldots,7$$

Note that the origin is not a feasible solution since setting

$$v_1 = v_2 = v_3 = v_4 = 0$$

would require that

$$v_7 = -1$$

and this violates

$$v_7 \geq 0$$

It is still possible, however, to make a change of variable (that is, to investigate another vertex) in the same way as before. Suppose that the pivot is around the encircled entry. (This corresponds to the encircled entry in the first tableau of the primal problem in Table 9.2*A*.) Making the indicated change of variable gives the equations

$$\begin{aligned}
-v_5 \qquad - v_7 - v_1 \qquad\qquad + v_4 &= 1 \\
-v_6 - 2v_7 + 2v_1 \qquad + 2v_3 + v_4 &= 2 \\
- v_7 + v_1 + v_2 \qquad\qquad &= 1
\end{aligned}$$

$$Z_{\min} = v_3 + v_4$$

To make all of the new basis variables have coefficient -1, the third equation is multiplied by -1, yielding

$$\begin{aligned}
-v_5 \qquad - v_7 - v_1 \qquad\qquad + v_4 &= 1 \\
- v_6 - 2v_7 + 2v_1 \qquad + 2v_3 + v_4 &= 2 \\
v_7 - v_1 - v_2 \qquad\qquad &= -1
\end{aligned}$$

$$Z_{\min} = v_3 + v_4$$

This operation is depicted in the two tableaux of Table 9.7*A* and *B*. Note that the basis variables are written with negative signs, since the slack variables have coefficients of -1.

The origin is still not a feasible solution, but the procedure can be continued as in the remainder of Table 9.7. The final tableau is the solution, since it is feasible (all numbers in the constant column are negative) and

Table 9.7. The Tableaux for the Dual Problem

A. The Initial Tableau

Basis Variables	v_1	v_2	v_3	v_4	Constant
$-v_5$	-2	-1		1	
$-v_6$		-2	2	1	
$-v_7$	1	①			1
$Z_{\min}$			1	1	

B. The Second Tableau

Basis Variables	v_1	v_7	v_3	v_4	Constant
$-v_5$	-1	-1		1	1
$-v_6$	②	-2	2	1	2
$-v_2$	-1	1			-1
$Z_{\min}$			1	1	

C. The Third Tableau

Basis Variables	v_6	v_7	v_3	v_4	Constant
$-v_5$	$-\frac{1}{2}$	-2	1	$\frac{3}{2}$ (circled)	2
$-v_1$	$\frac{1}{2}$	1	-1	$-\frac{1}{2}$	-1
$-v_2$	$-\frac{1}{2}$		1	$\frac{1}{2}$	
$Z_{\min}$			1	1	

D. The Fourth Tableau

Basis Variables	v_6	v_7	v_3	v_2	Constant
$-v_5$	$\frac{1}{3}$	$\frac{4}{3}$	$-\frac{2}{3}$	$\frac{2}{3}$	$-\frac{4}{3}$
$-v_1$	$\frac{1}{3}$	$\frac{1}{3}$	$-\frac{2}{3}$	$-\frac{1}{3}$	$-\frac{1}{3}$
$-v_4$	$-\frac{1}{3}$	$\frac{2}{3}$	$\frac{2}{3}$	$\frac{1}{3}$	$-\frac{2}{3}$
$Z_{\min}$	$\frac{1}{3}$	$\frac{4}{3}$	$\frac{1}{3}$	$\frac{2}{3}$	$-\frac{4}{3}$

cannot be improved (since all coefficients of the minimizing function are positive). The solution is

$$v_5 = \tfrac{4}{3}$$
$$v_1 = \tfrac{1}{3}$$
$$v_4 = \tfrac{2}{3}$$
$$Z_{\min} = \tfrac{4}{3}$$

Comparing these tableaux with those of Table 9.4 shows that all four tableaux are identical except for interchanging and renaming the rows and columns. In fact, Table 9.8 shows how one set of tableaux would have

Table 9.8. The Same Tableaux for the Primal and Dual Problems

A. The Initial Tableau

Dual Problem		$-v_5$	$-v_6$	$-v_7$	$Z_{\min}$
	Primal Problem	a_1	a_3	u	Constant
v_1	a_4	-2		1	
v_2	a_5	-1	-2	①	
v_3	a_6		2		1
v_4	a_7	1	1		1
Constant	$Z_{\max}$			1	

B. The Final Tableau

Dual Problem		$-v_5$	$-v_1$	$-v_4$	$Z_{\min}$
	Primal Problem	a_7	a_4	a_5	Constant
v_6	a_3	$\frac{1}{3}$	$\frac{1}{3}$	$-\frac{1}{3}$	$\frac{1}{3}$
v_7	u	$\frac{4}{3}$	$\frac{1}{3}$	$\frac{2}{3}$	$\frac{4}{3}$
v_3	a_6	$-\frac{2}{3}$	$-\frac{2}{3}$	$\frac{2}{3}$	$\frac{1}{3}$
v_2	a_1	$\frac{2}{3}$	$-\frac{1}{3}$	$\frac{1}{3}$	$\frac{2}{3}$
Constant	$Z_{\max}$	$-\frac{4}{3}$	$-\frac{1}{3}$	$-\frac{2}{3}$	$-\frac{4}{3}$

sufficed for both problems. The first set of row and column labels corresponds to the primal problem; the second set corresponds to the dual problem. It follows from the general form of the tableaux given in Tables 9.5 and 9.6 that this relationship will hold for all primal and dual problems as follows.

Theorem 9.5. Duality Theorem. Given a pair of linear programming problems which are duals, if either the primal or the dual problem has a solution, then both the primal and dual problems have solutions, and

A. The value of the Z for the primal and the dual problems is the same.

B. The tableau which gives a solution to the primal problem also gives a solution to the dual problem, and conversely. The optimum values of the variables for both the primal and the dual problems can be obtained directly from the last row or column of the final tableau.

Proof. Assume that the final tableau such as that of Table 9.8B corresponds to the optimal solution of the primal problem. Then the constant column of the primal problem must have all nonnegative entries, since these are the optimal values of the primal variables which must, by assumption, be nonnegative. But these same entries are the Z_{min} values in the dual problem, and the fact that they are nonnegative is exactly what guarantees that the dual problem is also at the optimal solution.

Similar reasoning, starting with the assumption that the final tableau is an optimal solution of the dual problem, shows that it must also be an optimal solution of the primal problem.|

9.11 THE DUAL SOLUTION TO THE MINIMUM NORMALIZED FAN-IN PROBLEM

Although the primal and dual formulations are quite similar, there are often significant computational advantages in using one formulation rather than the other. These include the following points.

(1) The origin may be an initial feasible solution in one formulation and not in the other; thus starting the procedure may be easier.

(2) Since, in general, different variables are put into and taken out of the basis in each formulation, there may be a significant difference in the number of required iterations.

Thus it is frequently desirable to investigate both formulations of a given problem.

It was first pointed out by Muroga, 1, that in the dual formulation of the minimum fan-in problem, the origin is always a basic feasible solution.

Consider again the unnormalized reduced set of inequalities:

$$-\sum_{i=1}^{n} a_i x_i^p + u \leq 0$$

for p such that $F(p) = 1$;

$$\sum_{i=1}^{n} a_i x_i^p - \ell \leq 0$$

for p such that $F(p) = 0$; and

$$u > \ell$$

Setting

$$\ell = u - 1$$

and adding a maximizing condition† yields

$$-\sum_{i=1}^{n} a_i x_i^p + u \leq 0$$

for p such that $F(p) = 1$; and

$$\sum_{i=1}^{n} a_i x_i^p - u \leq -1$$

for p such that $F(p) = 0$; and

$$Z_{\max} = -\sum_{i=1}^{n} a_i$$

$$a_i \geq 0 \qquad i = 1,2,\ldots,n$$

$$u \geq 0$$

The origin is not a feasible solution to this problem because of the -1's on the right side of the inequalities. The dual problem, however, consists of $\geq$ inequalities with all -1's on the right; therefore, the origin is a basic feasible solution for the dual problem. This is clarified in the next example.

Example 9.9. Consider again the problem of Example 9.7:

$$\begin{aligned} -2a_1 \qquad\quad & + u \leq 0 \\ -a_1 - 2a_3 & + u \leq 0 \\ 2a_3 & - u \leq -1 \\ a_1 + a_3 & - u \leq -1 \end{aligned}$$

$$Z_{\max} = -a_1 - a_3$$

† The methods of this section can also be used to maximize relative gap. After the gap length is normalized at unity, maximizing relative gap corresponds to minimizing u, or maximizing $-u$. Except for this different optimization criterion, the method is unchanged.

for which the origin is not a feasible solution. The dual problem

$$\begin{aligned} -2v_1 - v_2 \qquad + v_4 &\geq -1 \\ -2v_2 + 2v_3 + v_4 &\geq -1 \\ v_1 + v_2 - v_3 - v_4 &\geq 0 \end{aligned}$$

$$Z_{\min} = -v_3 - v_4$$

does have the origin as a feasible solution, since clearly the inequalities are satisfied if all the v's are equal to zero. Thus the minimization can be started immediately, as shown in Table 9.9.

Note that the dual problem solution of Table 9.9 differs slightly from the primal solutions given in Procedure 9.2, Example 9.5, and Tables 9.5 and 9.6. This is because the dual problem is one of minimization rather than maximization and has its inequalities written as $\geq$ rather than $\leq$.

Table 9.9

A. First Tableau

Primal		a_4	a_5	a_6	a_7	$Z_{\max}$
	Dual	v_1	v_2	v_3	v_4	Constant
a_1	$-v_5$	-2	-1		1	-1
a_3	$-v_6$		-2	2	1	-1
u	$-v_7$	1	1	(−1)	-1	
Constant	$Z_{\min}$			-1	-1	

B. Second Tableau

Primal		a_4	a_5	u	a_7	$Z_{\max}$
	Dual	v_1	v_2	v_7	v_4	Constant
a_1	$-v_5$	(−2)	-1		1	-1
a_3	$-v_6$	2		-2	-1	-1
a_6	$-v_3$	1	1	-1	-1	
Constant	$Z_{\min}$	-1	-1	1		

Table 9.9 *(cont.)*

C. Third Tableau

Primal		a_1	a_5	u	a_7	Z_{max}
	Dual	v_5	v_2	v_7	v_4	Constant
a_4	$-v_1$	$-\frac{1}{2}$	$(-\frac{1}{2})$		$\frac{1}{2}$	$-\frac{1}{2}$
a_3	$-v_6$	-1	-1	-2		-2
a_6	$-v_3$	$-\frac{1}{2}$	$\frac{1}{2}$	-1	$-\frac{1}{2}$	$-\frac{1}{2}$
Constant	Z_{min}	$\frac{1}{2}$	$-\frac{1}{2}$	1	$-\frac{1}{2}$	$\frac{1}{2}$

D. Fourth Tableau

Primal		a_1	a_4	u	a_7	Z_{max}
	Dual	v_5	v_1	v_7	v_4	Constant
a_5	$-v_2$	-1	-2		1	-1
a_3	$-v_6$		2	-2	(-1)	-1
a_6	$-v_3$	-1	-1	-1		-1
Constant	Z_{min}	1	1	1	-1	1

E. Fifth Tableau

Primal		a_1	a_4	u	a_3	Z_{max}
	Dual	v_5	v_1	v_7	v_6	Constant
a_5	$-v_2$	-1		-2	-1	-2
a_7	$-v_4$		2	-2	-1	-1
a_6	$-v_3$	-1	(-1)	-1		-1
Constant	Z_{min}	1	-1	3	1	2

(continued)

Table 9.9 (*cont.*)

F. Sixth Tableau

Primal		a_1	a_6	u	a_3	Z_{max}
	Dual	v_5	v_3	v_7	v_6	Constant
a_5	$-v_2$	-1		-2	-1	-2
a_7	$-v_4$	-2	-2	-4	-1	-3
a_4	$-v_1$	-1	-1	-1		-1
Constant	Z_{min}	2	1	4	1	3

It is certainly possible to make the dual problem one of maximization with inequalities written as $\leq$. This is done by multiplying all of its inequalities and its minimizing relation by -1, and is equivalent to multiplying all of the entries in Table 9.9 by -1. If this is done, the previous procedures can be used with no changes. However, rather than do that, Table 9.9 was prepared using the following simple changes in the previously described procedures.

(1) Since this dual problem is one of minimization, it is the *negative* entries in the Z_{min} row that correspond to variables that can be increased from zero. Hence the variable to be put into the basis is the one with the largest negative entry in the Z_{min} row.

(2) Since the inequalities are all written as $\geq$, the signs of the slack variables are all negative, and hence for all feasible solutions the constant terms are all negative. Thus the encircled entry must also be negative. To determine the encircled entry: For each negative entry in the column of the variable to be put into the basis, divide the constant term in its row by the negative entry. Encircle that negative entry with the smallest nonnegative quotient.

(3) Since the slack variables have coefficient -1, the equation (row) with the encircled entry must be normalized so that the entry of the new basis variable is -1.

With these simple changes the procedure is as described previously. The values of the coefficients are, from the last tableau,

$$a_1 = 2$$
$$a_3 = 1$$
$$u = 4$$

Observe, however, that five iterations were required instead of the three of Example 9.7.

An examination of the form of the tableaux of Table 9.9 shows that again, as in Theorem 9.3, there is a bound of at least R cycles on the number of iterations for this formulation. This time the reasoning is the dual of that of Theorem 9.3. All of the variables in the primal problem (which must be strictly positive) correspond to the slack variables in the dual problem. These slack variables have zero entries in the first tableau of the dual problem. However, in the final tableau of the dual problem, the entry in the $Z_{\min}$ row of each of these variables is the value of the corresponding primal variable, which must be strictly positive. Since only one of these entries can be made nonzero in each cycle, there must be at least R cycles.

This bound is not an inherent property of the ℓ.s. realization problem but is only due to this particular formulation. Gabelman, 3, and Sheng, 1, have proposed a formulation for which this bound does not apply, and experience indicates (Winder, 8) that frequently significantly fewer cycles are required. The basic idea is to make a simple change of variable so that the new variables need not be strictly positive, and thus the reasoning of the proof of Theorem 9.3 is no longer applicable.

Suppose that after setting symmetric variables equal, the remaining variables have been ordered as

$$a_1 > a_2 > \cdots > a_R$$

If a_1 must be strictly greater than a_2, then (in the absence of don't-care points†) it must be at least as great as a_2 plus the gap length

$$a_1 \geq a_2 + g$$

In the normalization of this section,

$$g = u - \ell = 1$$

Thus

$$a_1 \geq a_2 + 1$$

Now define

$$\Delta a_1 = a_1 - (a_2 + 1)$$

This new variable, Δa_1 satisfies the inequality

$$\Delta a_1 \geq 0$$

† If there are don't-care points, this condition is not required for the minimum fan-in realization. However, the condition can still be applied and a realization will be obtained, but it may not be the minimum fan-in realization.

and thus does not have to be strictly positive. We can similarly define

$$\Delta a_i = a_i - (a_{i+1} + 1)$$

for $i = 1,2,\ldots,R - 1$, which also satisfy

$$\Delta a_i \geq 0$$

where strict equality is allowed.

We now wish to express the simultaneous inequalities in terms of these new variables. To accomplish this, we write the defining expression for Δa_i as

$$a_i = \Delta a_i + (a_{i+1} + 1)$$

We then substitute into this expression the equivalent expression for a_{i+1} and so on, to obtain

$$a_i = \Delta a_i + \Delta a_{i+1} + \cdots + \Delta a_{R-1} + a_R + (R - i)$$

for $i = 1,2,\ldots,R - 1$.

Now these expressions for a_i are substituted into the set of simultaneous inequalities to obtain a new set in terms of the variables u, a_R, and Δa_i, $i = 1,2,\ldots,R - 1$. Only the first two of these variables need to be strictly positive, and hence the reasoning of Theorem 9.3 implies only that there must be at least two cycles.

Example 9.10. For the function of Example 9.9 with inequality set

$$\begin{aligned}
-2a_1 \qquad\quad + u &\leq 0 \\
-a_1 - 2a_3 + u &\leq 0 \\
2a_3 - u &\leq -1 \\
a_1 + a_3 - u &\leq -1 \\
a_1 &\geq 0 \\
a_3 &\geq 0
\end{aligned}$$

$$Z_{\max} = -a_1 - a_3$$

we know from Example 9.3 that

$$a_1 > a_3$$

Thus we can write

$$\Delta a_1 = a_1 - (a_3 + 1)$$

or

$$a_1 = \Delta a_1 + a_3 + 1$$

Substituting this into the inequality set, we obtain

$$-2\Delta a_1 - 2a_3 + u \le 2$$
$$-\Delta a_1 - 3a_3 + u \le 1$$
$$2a_3 - u \le -1$$
$$\Delta a_1 + 2a_3 - u \le -2$$
$$\Delta a_1 \ge 0$$
$$a_3 \ge 0$$

$$Z_{\max} = -\Delta a_1 - 2a_3 - 1$$

Table 9.10

A. First Tableau

Primal		a_4	a_5	a_6	a_7	$Z_{\max}$
	Dual	v_1	v_2	v_3	v_4	Constant
Δa_1	$-v_5$	-2	-1		1	-1
a_3	$-v_6$	-2	-3	2	2	-2
u	$-v_7$	1	1	-1	(-1)	
Constant	$Z_{\min}$	2	1	-1	-2	1

B. Second Tableau

Primal		a_4	a_5	a_6	u	$Z_{\max}$
	Dual	v_1	v_2	v_3	v_7	Constant
Δa_1	$-v_5$	-1		-1	-1	-1
a_3	$-v_6$		(-1)		-2	-2
a_7	$-v_4$	1	1	-1	-1	
Constant	$Z_{\min}$		-1	1	2	1

(continued)

Table 9.10 (*cont.*)

C. Third Tableau

Primal		a_4	a_3	a_6	u	Z_{max}
	Dual	v_1	v_6	v_3	v_7	Constant
Δa_1	$-v_5$	-1		-1	-1	-1
a_5	$-v_2$		-1		-2	-2
a_2	$-v_4$	1		-1	-3	-2
Constant	Z_{min}		1	1	4	3

The tableaux for this set are shown in Table 9.10. The procedure requires only two cycles. The final solution is

$$\Delta a_1 = 0$$
$$a_3 = 1$$
$$u = 4$$

which implies that

$$a_1 = a_3 + 1 = 2$$

Observe that this savings in number of cycles comes about because Δa_1 is, in fact, zero; that is, the value of a_1 differs from that of a_3 by exactly unity.

Both dual formulations of the minimum fan-in problem always have the origin as a basic feasible solution. Thus the dual problem can always be started. This occurs whether or not the given function is ℓ.s. However, if the given function is not ℓ.s., then certainly the primal problem does not have a solution. In view of the duality theorem (Theorem 9.4), this implies that the dual problem cannot have a solution either. But since the dual problem can always be started, it follows that the value of Z_{min} is not bounded;

$$Z_{min} = -\infty$$

This situation is discussed in Procedure 9.2, Step 3. Stated somewhat differently, one or more of the dual variables is unbounded in value so that the solution polyhedron for the dual problem is not finite and extends to infinity. This condition would manifest itself in a set of tableaux such as

Tables 9.9 or 9.10, by a column appearing with a negative entry in the Z_{min} row (implying that increasing that variable from zero would decrease Z_{min}) and positive entries in all the other rows (implying that there is no limit on how much the variable can be increased).

Thus the test for ℓ.s. realizability in this formulation is that Z_{min} be finite.

9.12 PROPERTIES BASED ON DUALITY*

The relationships between the primal and dual problems lead to a number of interesting and useful properties. The basic form of these relationships is due to the formulation of the dual problem: To each equation in the primal problem there corresponds an (actual) variable in the dual problem, and conversely.

Consider any final tableau that corresponds to an optimum solution as, for example, Table 9.8*B*. The column labels for the primal problem include all of the slack variables that are equal to zero at the solution point; in other words, they denote those inequalities that are satisfied as equalities. The corresponding column labels for the dual problem denote those dual variables that are strictly greater than zero at the solution point. This establishes the following very interesting result.

Theorem 9.6. Consider the solution point of any problem.

A. If an inequality in the primal problem is satisfied as an equality, then the corresponding variable in the dual problem is strictly positive, and conversely.

B. If an inequality in the primal problem is satisfied with some slack, then the corresponding variable in the dual problem is zero, and conversely.

To investigate the situation further, consider the primal and dual problems in matrix form. The primal problem is

$$\begin{bmatrix} \alpha_{11} & \alpha_{12} & \cdots & \alpha_{1n} \\ \alpha_{21} & \alpha_{22} & \cdots & \alpha_{2n} \\ \vdots & & & \\ \alpha_{P1} & \alpha_{P2} & \cdots & \alpha_{Pn} \end{bmatrix} \begin{bmatrix} a_1 \\ a_2 \\ \vdots \\ a_n \end{bmatrix} \leq \begin{bmatrix} c_1 \\ c_2 \\ \vdots \\ c_P \end{bmatrix}$$

$$Z_{max} = [z_1 \quad z_2 \quad \cdots \quad z_n] \begin{bmatrix} a_1 \\ a_2 \\ \vdots \\ a_n \end{bmatrix}$$

* Starred sections can be omitted on a first reading.

or

$$\boldsymbol{\alpha}\mathbf{A} \le \mathbf{C}$$

$$Z_{\max} = \mathbf{ZA}$$

in terms of the obvious matrix notation. The dual problem can be expressed as

$$\boldsymbol{\alpha}^t\mathbf{V} \ge \mathbf{Z}^t$$

$$Z_{\min} = \mathbf{C}^t\mathbf{V}$$

where the superscript t denotes the transpose.

Since it is assumed that

$$P \ge n$$

then at the solution point there are at least n inequalities in the primal problem that are satisfied as equalities. Rearrange the rows of $\boldsymbol{\alpha}$, $\mathbf{A}$ and $\mathbf{C}$ so that these are the first inequalities. Then partition $\boldsymbol{\alpha}$ and $\mathbf{C}$ so that

$$\boldsymbol{\alpha} = \begin{bmatrix} \boldsymbol{\alpha}' \\ \boldsymbol{\alpha}'' \end{bmatrix}$$

$$C = \begin{bmatrix} \mathbf{C}' \\ \mathbf{C}'' \end{bmatrix}$$

where

$$\begin{bmatrix} \boldsymbol{\alpha}' \\ \boldsymbol{\alpha}'' \end{bmatrix} \mathbf{A} \le \begin{bmatrix} \mathbf{C}' \\ \mathbf{C}'' \end{bmatrix}$$

can be expressed as

$$\boldsymbol{\alpha}'\mathbf{A} = \mathbf{C}'$$

$$\boldsymbol{\alpha}''\mathbf{A} < \mathbf{C}''$$

In the dual problem the partitions yield

$$\begin{bmatrix} \boldsymbol{\alpha}' \\ \boldsymbol{\alpha}'' \end{bmatrix}^t \mathbf{V} \ge \mathbf{Z}^t$$

or

$$[\boldsymbol{\alpha}'^t \boldsymbol{\alpha}''^t] \begin{bmatrix} V' \\ V'' \end{bmatrix} \ge \mathbf{Z}^t$$

or

$$\boldsymbol{\alpha}'^t\mathbf{V}' + \boldsymbol{\alpha}''^t\mathbf{V}'' \ge \mathbf{Z}^t$$

According to Theorem 9.5,

$$\mathbf{V}'' = 0$$

at the solution point, since the coordinates of $\mathbf{V}''$ correspond to equations in the primal problem that are satisfied with slack. Thus, at the solution point, the dual equations can be written as

$$\boldsymbol{\alpha}'^t\mathbf{V}' \geq \mathbf{Z}^t$$

Now assume, as in the ℓ.s. problem, that all of the actual variables in the primal problem are strictly positive at the solution point. Then by Theorem 9.5, all of the inequalities in the dual problem are satisfied as equalities, and the dual problem becomes

$$\boldsymbol{\alpha}'^t\mathbf{V}' = \mathbf{Z}^t$$

where each of the components of the matrix $\mathbf{V}'$ is strictly positive. This establishes a part of a following theorem.

Theorem 9.7 (*Willis*). Given any feasible solution to a linear programming problem in which all of the actual variables are strictly positive (as is the case for the ℓ.s. problem), let

$$\boldsymbol{\alpha}'\,\mathbf{A} = \mathbf{C}'$$

represent all the inequalities that are satisfied as equalities for this feasible solution. Also let

$$Z_{\max} = \mathbf{ZA}$$

be the maximizing function. Then the given feasible solution is the optimal solution if and only if there exists a column vector $\mathbf{V}'$ with all components strictly positive, so that

$$\boldsymbol{\alpha}'^t\mathbf{V}' = \mathbf{Z}^t$$

Proof. The "only-if" part was proved by the discussion that preceded the theorem. The "if" part follows immediately by proceeding through the same reasoning in the opposite sequence.|

The theorem implies a procedure for verifying whether a given feasible solution satisfies a given criterion. Suppose that a realization has been obtained for an ℓ.s. function possibly by the tree method or perhaps as a solution to a different linear programming problem. Using the preceding theorem, we can determine whether or not the solution satisfies any given linear optimization criterion.

Example 9.11. Example 9.6 considered the function

$$F(x_1,x_2,x_3,x_4) = x_1x_2 \dotplus x_1x_3x_4 \dotplus x_2x_3x_4$$

and found that the solution with maximum relative gap, when gap length is $\frac{1}{3}$, is

$$a_1 = a_2 = \tfrac{2}{3}$$
$$a_3 = a_4 = \tfrac{1}{3}$$
$$u = \tfrac{4}{3}$$
$$\ell = 1$$

We now use Theorem 9.7 to determine whether this solution also satisfies the criterion of minimum normalized fan-in.

For the minimum fan-in problem the reduced set of inequalities

$$
\begin{aligned}
2a_1 \qquad\quad &\geq u \\
a_1 + 2a_3 &\geq u \\
2a_3 &\leq \ell \\
a_1 + \ a_3 &\leq \ell \\
u &\geq \ell
\end{aligned}
$$

become, after substituting

$$\ell = u - 1$$

and changing the signs of the inequalities to obtain the standard form of a maximizing problem,

$$
\begin{aligned}
-2a_1 \qquad\quad + u &\leq 0 \\
-a_1 - 2a_3 + u &\leq 0 \\
2a_3 - u &\leq -1 \\
a_1 + \ a_3 - u &\leq -1
\end{aligned}
$$

$$Z_{\max} = -a_1 - a_3$$

The inequalities that are satisfied as equalities by the specified solution are now determined. First the given solution is normalized so that it yields a realization with unit gap length. This requires that all coefficients and gap boundaries be multiplied by 3, in which case

$$
\begin{aligned}
a_1 = a_2 &= 2 \\
a_3 = a_4 &= 1 \\
u &= 4 \\
\ell &= 3
\end{aligned}
$$

Substituting these values into the inequalities shows that the first, second, and fourth are satisfied as equalities; thus

$$\boldsymbol{\alpha}' = \begin{bmatrix} -2 & 0 & 1 \\ -1 & -2 & 1 \\ 1 & 1 & -1 \end{bmatrix}$$

The necessary and sufficient condition for the given solution to have minimum normalized fan-in is that

$$\boldsymbol{\alpha}'^t\mathbf{V}' = \begin{bmatrix} -1 \\ -1 \\ 0 \end{bmatrix}$$

or that

$$\begin{aligned} -2v_1 - v_2 + v_3 &= -1 \\ -2v_2 + v_3 &= -1 \\ v_1 + v_2 - v_3 &= 0 \end{aligned}$$

have at least one solution for which v_1, v_2, and v_3 are strictly positive. In this case, it is readily verified that such a solution is

$$\begin{aligned} v_1 &= 1 \\ v_2 &= 2 \\ v_3 &= 3 \end{aligned}$$

and, therefore, that the given values of coefficients and gap boundaries are also optimum with regard to fan-in. Note that, as specified by the theorem, these values of v are the negatives of the entries in the constant column of Table 9.9*F*, which is the final tableau in the dual linear program that minimizes normalized fan-in.

9.13 PROOF OF THEOREM 3.17*

As a final application of the general theory of simultaneous inequalities, Theorem 3.17 from Chapter 3 is proved.

Theorem 3.17. A switching function is ℓ.s. if and only if it is asummable.

Proof. Theorem 3.15 shows that all ℓ.s. functions are asummable; therefore, all that remains is to show that all non-ℓ.s. functions are not asummable. This means that for every non-ℓ.s. function there exists a number k so that the function is k-summable.

Let F be a non-ℓ.s. function, let $p_1,p_2,\ldots,p_q$ be the points for which F has unit value, and let $p_1',p_2',\ldots,p_r'$ be the points for which F is zero. Consider the set of points

$$\{p_{ij}^0\}, \qquad i = 1,2,\ldots,q; \quad j = 1,2,\ldots,r$$

where

$$p_{ij}^0 = p_i - p_j'$$

Moreover, let

$$f(x_1,x_2,\ldots,x_n) = \sum_{i=1}^{n} a_i x_i$$

* Starred sections can be omitted on a first reading.

be any separating function, and consider the set of simultaneous inequalities, qr in number,

$$f(p_{ij}^0) > 0 \qquad i = 1,2,\ldots,q$$
$$j = 1,2,\ldots,r$$

We first show that these inequalities are inconsistent.

Assume the contrary. Then there is some set of a_i, $i = 1,2,\ldots,n$ for which the above set of qr inequalities is simultaneously satisfied. Let f_a denote the separating function with this set of a_i. Let m be the minimum value of $f_a(p_{ij}^0)$. Clearly $m > 0$. Then

$$f_a(p_{ij}^0) = f_a(p_i) - f_a(p_j') \geq m > 0$$

for all $i = 1,2,\ldots,q$ and $j = 1,2,\ldots,r$; or, in other words, f_a separates F. This is a contradiction since F is non-ℓ.s. Hence the set of inequalities

$$f(p_{ij}^0) > 0 \qquad i = 1,2,\ldots,q$$
$$j = 1,2,\ldots,r$$

is inconsistent.

Interpreted geometrically, the consistency of the inequalities would imply the existence of a plane

$$\sum_{i=1}^{n} a_i z_i = 0$$

in the n cube that passes through the origin so that all the points p_{ij}^0 are strictly on one side only. The inconsistency implies that such a plane does not exist.

The nonexistence of such a plane implies that in some sense the points of p_{ij}^0 "surround" the origin. Define the *convex hull* of the points p_{ij}^0 to be the smallest convex set of points which includes the set of p_{ij}^0. Then the nonexistence of the plane implies that the origin is in the convex hull of the points p_{ij}^0. Thus the origin (that is, the point with coordinates $0,0,\ldots,0$) can be expressed as a convex linear combination of the points p_{ij}^0. In other words, there exists a set of nonnegative numbers, λ_{ij}, which sum to unity, so that the vector sum

$$\sum_{i=1}^{q} \sum_{j=1}^{r} \lambda_{ij} p_{ij}^0 = 0 \tag{9.5}$$

where

$$\lambda_{ij} \geq 0$$
$$\sum_{i=1}^{q} \sum_{j=1}^{r} \lambda_{ij} = 1 \tag{9.6}$$

Since the coordinates of the p_{ij}^0 are rational numbers, the λ_{ij} are also rational. Hence there is a positive integer k such that $k\lambda_{ij}$ is an integer for all $i = 1,2,\ldots,q$ and $j = 1,2,\ldots,r$. Multiplying both sides of Eqs. 9.5 and 9.6 by k, and letting $k\lambda_{ij} = N_{ij}$, yields

$$\sum_{i=1}^{q} \sum_{j=1}^{r} N_{ij} p_{ij}^0 = 0 \tag{9.7}$$

where the N_{ij} are integers such that

$$N_{ij} \geq 0$$

$$\sum_{i=1}^{q} \sum_{j=1}^{r} N_{ij} = k > 0$$

Substituting the definition of p_{ij}^0 into Eq. 9.7 gives

$$\sum_{i=1}^{q} \sum_{j=1}^{r} N_{ij} p_i = \sum_{i=1}^{q} \sum_{j=1}^{r} N_{ij} p_j'$$

Thus F is k-summable with

$$k = \sum_{i=1}^{q} \sum_{j=1}^{r} N_{ij}$$

This completes the proof.|

Chapter 10

ADAPTIVE METHODS

10.1 INTRODUCTION

All of the synthesis procedures described in previous chapters might be characterized as *parallel* in nature. That is, the entire switching function is available at any time it is needed by the synthesis procedure. In other words, a computer programmed to perform the procedure must have sufficient memory to store the entire switching function, either as a table of combination or as a Boolean function.

In contrast with this, there are a number of procedures that can be characterized as *serial* or *sequential*. That is, at any instant of time, the procedure is presented with the value of the function for only one input combination, and there is no memory available to store previously presented functional values. The only necessary memory is that required to store the procedure's current estimate of the correct realization (that is, the values of the coefficients and the threshold).

When the functional value for some particular input combination is presented, the procedure is allowed to change the current estimate of the realization, but it can store no other information. The idea is that after each input combination has been presented a sufficient number of times, the procedure's estimate of the realization will have converged to a correct one.

This type of procedure is sometimes called *adaptive*, partly because of its somewhat tenuous relation to certain processes in nerve cells or neurons. We use the adaptive terminology but concentrate our attention on the engineering aspects of the procedures rather than on the anthropomorphic analogies.

10.2 DESCRIPTION OF THE SINGLE-GATE ADAPTIVE PROCEDURE

We shall consider only networks that consist of a single threshold gate with n inputs. The adaptive procedure is a method for determining the coefficients and the threshold of this gate so that it realizes the given switching function F of n variables when the function is linearly separable.

In considering adaptive procedures it is frequently convenient to let the binary variables have values $1,-1$ rather than $1,0$. We let $\zeta_j; j = 1,2,\ldots,n$; denote the binary variables defined on the set $\{1,-1\}$ to distinguish them from those used heretofore, which are defined on the set $\{1,0\}$. Thus the given function F can be expressed as

$$F(\zeta_1,\zeta_2,\ldots,\zeta_n)$$

For Boolean functions this is a trivial change. It simply means that the identity element for the sum operation is represented by the symbol -1 rather than by 0. This is merely a change of symbolism, in which all 0's are replaced by -1's.

With regard to arithmetic functions (such as separating functions, for example) the change is only slightly more significant. The equation

$$\zeta_i = 2x_i - 1$$

is an arithmetic relation between the two variables where x_i denotes the binary variable defined on the set $\{1,0\}$. If the switching function F can be expressed as

$$\left\langle \sum_{i=1}^{n} a_i x_i \right\rangle_t$$

then, since

$$x_i = \tfrac{1}{2}(\zeta_i + 1)$$

it can be expressed as

$$\left\langle \sum_{i=1}^{n} a_i \tfrac{1}{2}(\zeta_i + 1) \right\rangle_t$$

and can be simplified by Theorem 2.4B to

$$\left\langle \sum_{i=1}^{n} a_i \zeta_i \right\rangle_T$$

where

$$T = 2t - \sum_{i=1}^{n} a_i$$

The maps of

$$\left\langle \sum_{i=1}^{n} a_i x_i \right\rangle_t$$

and

$$\left\langle \sum_{i=1}^{n} a_i \zeta_i \right\rangle_T$$

are closely related. In fact, to obtain the latter from the former, multiply each of the functional values of the separating function of the former by 2 and then displace all functional values by

$$-\sum_{i=1}^{n} a_i$$

(add

$$-\sum_{i=1}^{n} a_i$$

to each functional value). Note that the map of

$$\left\langle \sum_{i=1}^{n} a_i \zeta_i \right\rangle_T$$

is symmetric about the functional value zero and that its gap length is twice the gap length of

$$\left\langle \sum_{i=1}^{n} a_i x_i \right\rangle_t$$

The adaptive realization procedure is applicable to functions with don't-care points as well as to complete functions. Therefore, we let m denote the number of points of the n cube for which F is defined, where

$$m \leq 2^n$$

In the course of the adaptive procedure the identity of these m points and the corresponding functional value of F are sequentially available to the procedure in what is called a *training sequence*. The only property required of the training sequence is that if it were allowed to become infinitely long, each point would appear infinitely many times. In other words, the order in which the points appear in the training sequence is immaterial relative to ultimate convergence, provided the number of times a point appears can always be increased by increasing the length of the training sequence.

The m combinations of values of the variables $\zeta_1,\zeta_2,\ldots,\zeta_n$ for which F is defined correspond to points in the n space whose coordinates are $\zeta_1,\zeta_2,\ldots,\zeta_n$. Let $q_1,q_2,\ldots,q_m$ represent these points. Then a training sequence is a sequence of points selected from the set $\{q_1,q_2,\ldots,q_m\}$. Let $q^{(k)}$ represent the kth member of the training sequence; $q^{(k)}$ is a variable defined on the set $\{q_1,q_2,\ldots,q_m\}$. After the kth member of the training sequence has been presented to the procedure, the kth estimate of the coefficients and threshold of the T gate are determined. Let $a_i^{(k)}$ and $T^{(k)}$ denote the kth estimate of a_i and of T respectively. Furthermore let the gap length of the T gate be specified as g. In terms of these definitions, the procedure is described as follows.

Procedure 10.1. Single-Gate Adaptive Procedure

(1) Initially set $T^{(0)} = 0$ and $a_i^{(0)} = 0$; $i = 1,2,\ldots,n$.

(2) At the time the kth input $q^{(k)}$; $k = 1,2,\ldots$ is presented, the current coefficient and threshold estimates are $T^{(k-1)}$ and $a_i^{(k-1)}$; $i = 1,2,\ldots,n$. If the current estimates of these parameters yield the desired T-gate output, then the estimate is not changed. That is, if

$$F(q^{(k)}) = 1$$

and

$$\sum_{i=1}^{n} a_i^{(k-1)}\zeta_i^{(k)} \geq T^{(k-1)} + \frac{g}{2}$$

or if

$$F(q^{(k)}) = -1$$

and

$$\sum_{i=1}^{n} a_i^{(k-1)}\zeta_i^{(k)} \leq T^{(k-1)} - \frac{g}{2}$$

where $\zeta_i^{(k)}$ is the kth value of ζ_i and g is the specified gap length, then

$$a_i^{(k)} = a_i^{(k-1)}; \quad i = 1,2,\ldots,n$$

and

$$T^{(k)} = T^{(k-1)}$$

(3) If for $q^{(k)}$ the functional value of F is specified as 1 but the current estimate of the coefficients and threshold do not give a T-gate output of 1 with the specified gap length, then all coefficients are changed by the same magnitude in such a direction as to make the output of the T gate tend toward 1. In particular, the threshold is decreased, all coefficients with $+1$ inputs are increased, and all coefficients with -1 inputs are decreased. In other words, if

$$F(q^{(k)}) = 1$$

and

$$\sum_{i=1}^{n} a_i^{(k-1)}\zeta_i^{(k)} < T^{(k-1)} + \frac{g}{2}$$

then let

$$T^{(k)} = T^{(k-1)} - \rho$$

and

$$a_i^{(k)} = a_i^{(k-1)} + \rho\zeta_i^{(k)}$$

where ρ is any positive constant. Note that ρ is added or subtracted from $a_i^{(k-1)}$ according to whether $\zeta_i^{(k)}$ is $+1$ or -1.

(4) If for $q^{(k)}$ the functional value of F is specified as -1 but the current estimate of the coefficients and threshold do not give a T-gate output of -1 with the required gap length, then the threshold and all coefficients with -1 inputs are increased and all coefficients with $+1$ inputs are decreased. That is, if

$$F(q^{(k)}) = -1$$

and

$$\sum_{i=1}^{n} a_i^{(k-1)}\zeta_i^{(k)} > T^{(k-1)} - \frac{g}{2}$$

then let

$$T^{(k)} = T^{(k-1)} + \rho$$

and

$$a_i^{(k)} = a_i^{(k-1)} - \rho\zeta_i^{(k)}$$

The procedure and its execution are straightforward. It is illustrated in the following example.

Example 10.1. Let the given switching function be specified by the table of combinations in Table 10.1.

Table 10.1. Function of Example 10.1

$q_j, j =$	ζ_1	ζ_2	ζ_3	$F(\zeta_1,\zeta_2,\zeta_3)$
1	-1	1	-1	-1
2	-1	-1	1	-1
3	1	-1	-1	1
4	-1	1	1	1

By the tossing of a coin the following training sequence was selected:

$$q_1, q_2, q_4, q_3, q_4, q_4, q_4, q_3, q_3, q_2, q_3, q_1, q_2, \ldots$$

and the procedure was carried out as shown in Table 10.2 using

$$\rho = 1$$

$$\frac{g}{2} = 1$$

Table 10.2. Results of Adaptive Procedure of Example 10.1

k	q_j	Point				Estimate Correct?	Coefficients			
	$j =$	ζ_1	ζ_2	ζ_3	$F(\zeta_1,\zeta_2,\zeta_3)$		$a_1^{(k)}$	$a_2^{(k)}$	$a_3^{(k)}$	$T^{(k)}$
0							0	0	0	0
1	1	−1	1	−1	−1	No	1	−1	1	1
2	2	−1	−1	1	−1	No	2	0	0	2
3	4	−1	1	1	1	No	1	1	1	1
4	3	1	−1	−1	1	No	2	0	0	0
5	4	−1	1	1	1	No	1	1	1	−1
6	4	−1	1	1	1	Yes	1	1	1	−1
7	4	−1	1	1	1	Yes	1	1	1	−1
8	3	1	−1	−1	1	No	2	0	0	−2
9	3	1	−1	−1	1	Yes	2	0	0	−2
10	2	−1	−1	1	−1	No	3	1	−1	−1
11	3	1	−1	−1	1	Yes	3	1	−1	−1
12	1	−1	1	−1	−1	No	4	0	0	0
13	2	−1	−1	1	−1	Yes	4	0	0	0
14	4	−1	1	1	1	No	3	1	1	−1
15	1	−1	1	−1	−1	Yes	3	1	1	−1
16	4	−1	1	1	1	No	2	2	2	−2
17	2	−1	−1	1	−1	No	3	3	1	−1
18	1	−1	1	−1	−1	No	4	2	2	0
19	3	1	−1	−1	1	No	5	1	1	−1
20	2	−1	−1	1	−1	Yes	5	1	1	−1
21	4	−1	1	1	1	No	4	2	2	−2

When $k = 21$, the values of the coefficients and threshold give the correct output for all four input points and no further errors occur. In other words, the procedure has converged to a solution.

10.3 ALTERNATIVE DESCRIPTION OF THE SINGLE-GATE ADAPTIVE PROCEDURE

Procedure 10.1 can be shown to converge for all ℓ.s. functions and for all incomplete functions that are realizable with ℓ.s. functions. That is, the estimated values of the coefficients and the threshold converge to values that realize the given function after a finite and bounded number of errors have occurred. To facilitate the proof of convergence, it is convenient to restate the procedure in a slightly different form in terms of the self-dual representation of the given function.

Let the given function be represented as $F(\zeta_1,\zeta_2,\ldots,\zeta_n)$. If the given function has don't-care's, then F represents a complete function that satisfies the given function. According to Section 7.3, F has a self-dual representative function given by

$$\Phi(\zeta_1,\zeta_2,\ldots,\zeta_{n+1}) = \bar{\zeta}_{n+1}F(\zeta_1,\zeta_2,\ldots,\zeta_n) + \zeta_{n+1}F^d(\zeta_1,\zeta_2,\ldots,\zeta_n)$$

We first investigate how the realization of Φ is related to that of F. If F has the realization

$$F(\zeta_1,\zeta_2,\ldots,\zeta_n) = \left\langle \sum_{i=1}^{n} a_i\zeta_i \right\rangle_T$$

then

$$F(x_1,x_2,\ldots,x_n) = \left\langle \sum_{i=1}^{n} a_ix_i \right\rangle_t$$

where

$$T = 2t - \sum_{i=1}^{n} a_i$$

According to Corollary 2.10.1,

$$F^d(x_1,x_2,\ldots,x_n) = \left\langle \sum_{i=1}^{n} a_ix_i \right\rangle_{t'}$$

where

$$t' = \sigma - t = \sum_{i=1}^{n} a_i - t$$

Then

$$F^d(\zeta_1,\zeta_2,\ldots,\zeta_n) = \left\langle \sum_{i=1}^{n} a_i\zeta_i \right\rangle_{T'} = \left\langle \sum_{i=1}^{n} a_i\zeta_i \right\rangle_{-T}$$

since

$$T' = 2t' - \sum_{i=1}^{n} a_i = 2\left(\sum_{i=1}^{n} a_i - t\right) - \sum_{i=1}^{n} a_i = \sum_{i=1}^{n} a_i - 2t = -T$$

According to the Range Theorem,

$$\Phi(x_1,x_2,\ldots,x_{n+1}) = \bar{x}_{n+1}F(x_1,x_2,\ldots,x_n) \dot{+} x_{n+1}F^d(x_1,x_2,\ldots,x_n)$$

$$= \left\langle a_{n+1}x_{n+1} + \sum_{i=1}^{n} a_ix_i \right\rangle_t$$

where

$$a_{n+1} = t - t' = 2t - \sum_{i=1}^{n} a_i = T$$

It follows immediately that

$$\Phi(\zeta_1,\zeta_2,\ldots,\zeta_{n+1}) = \left\langle T\zeta_{n+1} + \sum_{i=1}^{n} a_i\zeta_i \right\rangle_{T''} = \left\langle \sum_{i=1}^{n+1} a_i\zeta_i \right\rangle_0$$

where

$$T'' = 2t - \left(\sum_{i=1}^{n+1} a_i\right) = 2t - \left[\sum_{i=1}^{n} a_i + T\right] = T - T = 0$$

Thus if F as a function of the variables ζ_i, $i = 1,2,\ldots,n$, has a realization with threshold T, then the self-dual representative function of the same variables has threshold zero, and the coefficient of the self-dualizing variable ζ_{n+1} can be the threshold for the realization of $F(\zeta_1,\zeta_2,\ldots,\zeta_n)$. The self-dualization allows the threshold of F to be treated as one of the input variable coefficients and the threshold of Φ to be permanently set to zero. The realization for F can be obtained from that for Φ by setting $\zeta_{n+1} = -1$.

It follows from Corollary 2.10.2 and from the definition of the self-dual representative function as

$$\Phi(q,\zeta_{n+1}) = \bar{\zeta}_{n+1}F(q) \dotplus \zeta_{n+1}F^d(q)$$

that the following relation exists between the functional values of F and Φ.

(1) $F(q) = 1$ implies $\Phi(q,-1) = 1$ and $\Phi(\bar{q},1) = -1$.
(2) $F(q) = -1$ implies $\Phi(\bar{q},1) = 1$ and $\Phi(q,-1) = -1$.

It is convenient to represent the set of points on which Φ is defined as $\{r_j\}$. If the given function is incomplete and is specified at m points where

$$m \leq 2^n$$

then the corresponding self-dual representative function is specified at $2m$ points and

$$2m \leq 2^{n+1}$$

For example, if the given function is defined as in Example 10.1, then the self-dual representative function is defined as shown in Table 10.3. The points below the center line are the complements of the points above.

Table 10.3. The Representative Self-Dual Function Φ Corresponding to the Incomplete Function F of Table 10.1

r_j Points $j =$	ζ_1	ζ_2	ζ_3	ζ_4	$\Phi(\zeta_1,\zeta_2,\zeta_3,\zeta_4)$
a	-1	1	-1	-1	-1
b	-1	-1	1	-1	-1
c	1	-1	-1	-1	1
d	-1	1	1	-1	1
e	1	-1	1	1	1
f	1	1	-1	1	1
g	-1	1	1	1	-1
h	1	-1	-1	1	-1

We now describe the adaptive procedure for the self-dual representative function Φ that corresponds to the given function F. It will be shown that this procedure is computationally equivalent to Procedure 10.1.

The realization to be obtained is of the form

$$\left\langle \sum_{i=1}^{n+1} a_i \zeta_i \right\rangle_0$$

Since Φ is known to be self-dual, the threshold remains at the initial value of zero throughout the entire procedure.

The points in the training sequence for Φ are generated from the training sequence for F. Let $r^{(k)}$ denote the kth point of the training sequence for Φ; then $r^{(k)}$ includes values for ζ_{n+1} where these are not included in $q^{(k)}$. To each $q^{(k)}$ point

$$q^{(k)} = (\zeta_1^{(k)}, \zeta_2^{(k)}, \ldots, \zeta_n^{(k)})$$

there uniquely corresponds two $r^{(k)}$ points

$$(\zeta_1^{(k)}, \zeta_2^{(k)}, \ldots, \zeta_n^{(k)}, -1)$$

and

$$(\bar{\zeta}_1^{(k)}, \bar{\zeta}_2^{(k)}, \ldots, \bar{\zeta}_n^{(k)}, 1)$$

which are complements of each other and hence have complementary values of Φ. Since every current estimate of the final realization for Φ is a self-dual function (because the threshold is permanently set to zero), the functional values are complementary for complementary points. Thus if the realization is correct or incorrect for one of the $r^{(k)}$ points, it is correspondingly correct or incorrect for the associated complement point. For reasons that will be apparent from the procedure, attention can be restricted to only one of the pair of $r^{(k)}$ points. For each $q^{(k)}$, we select the $r^{(k)}$ point such that

$$\Phi(r^{(k)}) = 1$$

The following procedure selects this $r^{(k)}$ for each $q^{(k)}$.

Let

$$q^{(k)} = (\zeta_1^{(k)}, \zeta_2^{(k)}, \ldots, \zeta_n^{(k)})$$

and

$$r^{(k)} = (\xi_1^{(k)}, \xi_2^{(k)}, \ldots, \xi_n^{(k)}, \xi_{n+1}^{(k)})$$

Then

(1) If $F(q^{(k)}) = 1$, select

$$\xi_{n+1}^{(k)} = -1 \qquad \text{and} \quad \xi_i^{(k)} = \zeta_i^{(k)}; \quad i = 1,2,\ldots,n$$

or

(2) If $F(q^{(k)}) = -1$, select

$$\xi_{n+1}^{(k)} = 1 \qquad \text{and} \quad \xi_i^{(k)} = \bar{\zeta}_i^{(k)}; \quad i = 1,2,\ldots,n$$

In each case

$$\Phi(r^{(k)}) = 1$$

If, for example, F is defined as in Table 10.1 and the corresponding Φ is given by Table 10.3, then the r points of Φ that correspond to the q points of F are given by Table 10.4.

Table 10.4. Corresponding Points of F and Φ

q_j Points of F $j =$	r_j Points of Φ $j =$
1	e
2	f
3	c
4	d

Note that when the points of Φ are selected according to this procedure, a functional value of -1 implies erroneous output while a value of 1 implies a correct output.

Procedure 10.2. Alternative Single-Gate Adaptive Procedure

(1) Set $a_i^{(0)} = 0; \quad i = 1,2,\ldots,n + 1$.

(2) Assume the current estimate of the coefficients is $a_i^{(k-1)}$ for $i = 1,2,\ldots,n + 1$, and that the kth member of the training sequence, $r^{(k)}$, has been applied. Let

$$\text{(a)} \quad a_i^{(k)} = a_i^{(k-1)}; \quad i = 1,2,\ldots,n + 1.$$

or

$$\text{(b)} \quad a_i^{(k)} = a_i^{(k-1)} + \rho\xi_i^{(k)}; \quad i = 1,2,\ldots,n + 1$$

according to whether

$$\text{(a)} \quad \sum_{i=1}^{n+1} a_i^{(k-1)}\xi_i^{(k)} \geq \frac{g}{2}$$

or

$$\text{(b)} \quad \sum_{i=1}^{n+1} a_i^{(k-1)}\xi_i^{(k)} < \frac{g}{2}$$

where ρ and g are positive constants.

Observe that the coefficients are altered only when the realized functional value is incorrect. Moreover, the changes in the coefficients tend to correct the erroneous operation.

Procedures 10.1 and 10.2 are computationally equivalent. At every step both procedures give the same estimate of the coefficients, and each involves the same computations. This can be verified by comparing the steps of Example 10.1 with those of Example 10.2.

Example 10.2. Let F be defined by Table 10.1. As in Example 10.1, let

$$\rho = \frac{g}{2} = 1$$

and use the same training sequence. Table 10.5 illustrates the application of Procedure 10.2 to this problem. Compared with Table 10.2, only the following differences occur:

(1) The F column of Table 10.2 is the complement of the ξ_4 column of Table 10.5.

(2) For those points for which the functional value of F is -1, the values of ξ_1,ξ_2,ξ_3 of Table 10.5 are complements of ζ_1,ζ_2,ζ_3 of Table 10.2. Otherwise they are the same.

10.4 PROOF OF CONVERGENCE

The convergence of Procedure 10.2 is established by the following theorem.

> **Theorem 10.1** (*Block; Rosenblatt*). If a switching function F with self-dual representative function Φ is ℓ.s., then Procedure 10.2 yields a set of coefficients $\{a_i\}$ such that
>
> $$\Phi(\zeta_1,\zeta_2,\ldots,\zeta_{n+1}) = \left\langle \sum_{i=1}^{n+1} a_i\zeta_i \right\rangle_0$$
>
> after K modifications of the coefficient vector where
>
> $$K < \left[\frac{g}{\rho} + n + 1\right] \frac{4 \sum_{i=1}^{n+1} \hat{a}_i^2}{g^2}$$
>
> where $\{\hat{a}_i\}$ is a set of minimum coefficients that realizes Φ with gap length g. (Clearly any larger set also bounds K.)

Table 10.5. First 21 Steps Using Procedure 10.2 on Example 10.1

	Training Sequence		Components of r_i				$\langle \sum_{j=1}^{4} a_j^{(k-1)} \xi_j \rangle_0$	Coefficients			
k	q_j $j =$	r_i $i =$	ξ_1	ξ_2	ξ_3	ξ_4	Agrees with F?	$a_1^{(k)}$	$a_2^{(k)}$	$a_3^{(k)}$	$a_4^{(k)}$
0								0	0	0	0
1	1	e	1	−1	1	1	No	1	−1	1	1
2	2	f	1	1	−1	1	No	2	0	0	2
3	4	d	−1	1	1	−1	No	1	1	1	1
4	3	c	1	−1	−1	−1	No	2	0	0	0
5	4	d	−1	1	1	−1	No	1	1	1	−1
6	4	d	−1	1	1	−1	Yes	1	1	1	−1
7	4	d	−1	1	1	−1	Yes	1	1	1	−1
8	3	c	1	−1	−1	−1	No	2	0	0	−2
9	3	c	1	−1	−1	−1	Yes	2	0	0	−2
10	2	f	1	1	−1	1	No	3	1	−1	−1
11	3	c	1	−1	−1	−1	Yes	3	1	−1	−1
12	1	e	1	−1	1	1	No	4	0	0	0
13	2	f	1	1	−1	1	Yes	4	0	0	0
14	4	d	−1	1	1	−1	No	3	1	1	−1
15	1	e	1	−1	1	1	Yes	3	1	1	−1
16	4	d	−1	1	1	−1	No	2	2	2	−2
17	2	f	1	1	−1	1	No	3	3	1	−1
18	1	e	1	−1	1	1	No	4	2	2	0
19	3	c	1	−1	−1	−1	No	5	1	1	−1
20	2	f	1	1	−1	1	Yes	5	1	1	−1
21	4	d	−1	1	1	−1	No	4	2	2	−2

Proof (Due to Novikoff). The proof is in two parts:

A. Let $r^{(\alpha_k)}$ be the member of the training sequence which causes the kth error. According to the rule for changing coefficients, the coefficient vector after the kth error is

$$a^{(\alpha_k)} = a^{(\alpha_{k-1})} + \rho r^{(\alpha_k)} = \rho \sum_{i=1}^{k} r^{(\alpha_i)}$$

since initially the coefficients are zero. Since Φ is ℓ.s. (because F is ℓ.s.) there exists a minimum coefficient vector $\hat{a}$ so that for all r-points

$$r \cdot \hat{a} \geq \frac{g}{2}$$

Consider the expression

$$\hat{a} \cdot a^{(\alpha_k)}$$

Substituting the above expression for $a^{(\alpha_k)}$ gives

$$\hat{a} \cdot a^{(\alpha_k)} = \hat{a} \cdot \left(\rho \sum_{i=1}^{k} r^{(\alpha_i)} \right)$$

$$= \rho \sum_{i=1}^{k} \hat{a} \cdot r^{(\alpha_i)} \geq \frac{1}{2} \rho k g$$

since $\hat{a}$ separates the given switching function. Thus we have shown that

$$\hat{a} \cdot a^{(\alpha_k)} \geq \frac{1}{2} \rho k g$$

or that

$$[\hat{a} \cdot a^{(\alpha_k)}]^2 \geq \frac{1}{4} \rho^2 k^2 g^2$$

By Schwartz's inequality,

$$\hat{a}^2 \cdot a^2 \geq [\hat{a} \cdot a]^2$$

Hence

$$\left(\sum_{i=1}^{n+1} \hat{a}_i^2 \right) \left(\sum_{i=1}^{n+1} (a_i^{(\alpha_k)})^2 \right) \geq \frac{1}{4} \rho^2 k^2 g^2$$

Thus, after k corrections of the coefficients, the sum of the squares of the coefficients is bounded by

$$\sum_{i=1}^{n+1} (a_i^{(\alpha_k)})^2 \geq \frac{\rho^2 k^2 g^2}{4 \sum_{i=1}^{n+1} \hat{a}_i^2}$$

B. Consider now the ith coefficient. The change in its value caused by the jth error is given by

$$a_i^{(\alpha_j)} = a_i^{(\alpha_j - 1)} + \rho \xi_i^{(\alpha_j)}$$

Squaring both sides gives

$$\begin{aligned}(a_i^{(\alpha_j)})^2 &= (a_i^{(\alpha_j - 1)} + \rho \xi_i^{(\alpha_j)})^2 \\ &= (a_i^{(\alpha_j - 1)})^2 + 2\rho a_i^{(\alpha_j - 1)} \xi_i^{(\alpha_j)} + \rho^2 (\xi_i^{(\alpha_j)})^2\end{aligned}$$

Summing over all coefficients gives

$$\sum_{i=1}^{n+1} (a_i^{(\alpha_j)})^2 = \sum_{i=1}^{n+1} (a_i^{(\alpha_j - 1)})^2 + 2\rho \sum_{i=1}^{n+1} a_i^{(\alpha_j - 1)} \xi_i^{(\alpha_j)} + \rho^2 \sum_{i=1}^{n+1} (\xi_i^{(\alpha_j)})^2$$

Every time an error occurs it is because

$$\sum_{i=1}^{n+1} a_i^{(\alpha_j - 1)} \xi^{(\alpha_j)} < \frac{g}{2}$$

Using this in the previous expression gives the bound

$$\sum_{i=1}^{n+1} (a_i^{(\alpha_j)})^2 < \sum_{i=1}^{n+1} (a_i^{(\alpha_j - 1)})^2 + \rho g + \rho^2 \sum_{i=1}^{n+1} (\xi_i^{(\alpha_j)})^2$$

The variables ξ_i are either ± 1; hence

$$(\xi_i^{(\alpha_j)})^2 = 1$$

and the expression becomes

$$\sum_{i=1}^{n+1} (a_i^{(\alpha_j)})^2 < \sum_{i=1}^{n+1} (a_i^{(\alpha_j - 1)})^2 + \rho g + \rho^2 (n + 1)$$

or

$$\sum_{i=1}^{n+1} (a_i^{(\alpha_j)})^2 - \sum_{i=1}^{n+1} (a_i^{(\alpha_j - 1)})^2 < \rho g + \rho^2 (n + 1)$$

This expression gives a bound on the change in the sum of the squares of the coefficients between the jth and the $(j - 1)$th errors. There is a similar expression for all j from 1 to k. Summing all of these and recalling that $a_i^{(\alpha_0)} = 0$, for $i = 1,2,\ldots,n + 1$, gives

$$\sum_{i=1}^{n+1} (a_i^{(\alpha_k)})^2 < k[\rho g + \rho^2 (n + 1)]$$

This is another bound that the sum of the squared coefficients must satisfy after k errors.

Together the two bounds on this sum imply that

$$\frac{\rho^2 k^2 g^2}{4 \sum_{i=1}^{n+1} \hat{a}_i^2} < k[\rho g + \rho^2(n + 1)]$$

which can be expressed as

$$k < \left[\frac{g}{\rho} + n + 1\right] \frac{4 \sum_{i=1}^{n+1} \hat{a}_i^2}{g^2}$$

This completes the proof. Since F and hence Φ are ℓ.s., there does exist a set of coefficients $\{\hat{a}_i\}$ for every g, and hence the number of corrections is finite and convergence is assured.|

Example 10.3. Consider the switching function of Example 10.1. One realization for the function Φ which has a gap length of two is

$$\Phi(\zeta_1,\zeta_2,\zeta_3,\zeta_4) = \langle 2\zeta_1 + 1\zeta_2 + 1\zeta_3 - 1\zeta_4 \rangle_0$$

for which

$$\sum_{i=1}^{n+1} a_i^2 = 4 + 1 + 1 + 1 = 7$$

Thus, when

$$\frac{g}{2} = \rho = 1$$

the bound on the number of corrections before convergence is assured is

$$k < (2 + 4)\frac{7}{1} = 42$$

Observe that the actual number of corrections required, as shown by Table 10.2, is 14.

We can also get a lower bound on the number of corrections using the inequality derived in Part B of the proof:

$$\sum_{i=1}^{n+1} (a_i^{(\alpha_k)})^2 < k[\rho g + \rho^2(n + 1)]$$

The coefficients $\{a_i^{(\alpha_k)}\}$ after the kth correction need not be integers unless ρ is an integer. But since all coefficients must be multiples of ρ, the set $\{a_i^{(\alpha_k)}/\rho\}$ is a set of integer coefficients. Thus we can divide both sides of the above inequality by ρ^2 to get

$$\sum_{i=1}^{n+1} \left(\frac{a_i^{(\alpha_k)}}{\rho}\right)^2 < k\left[\frac{g}{\rho} + n + 1\right]$$

which is an inequality that must be satisfied by the coefficients when they are normalized to integers.

Now for an ℓ.s. function F with self-dual representative function Φ, let $\{\tilde{a}_i\}$ be the *minimal integral square realization*; that is, the set of integer coefficients that realizes Φ so that the sum of their squares, that is,

$$\sum_{i=1}^{n+1} \tilde{a}_i^2$$

is the smallest over all integer weight realizations of Φ. If the adaptive procedure realizes F with k corrections, then certainly

$$\sum_{i=1}^{n+1} \tilde{a}_i^2 \leq \sum_{i=1}^{n+1} \left(\frac{a_i^{(\alpha_k)}}{\rho}\right)^2$$

from which it follows that

$$\sum_{i=1}^{n+1} \tilde{a}_i^2 < k\left[\frac{g}{\rho} + n + 1\right]$$

Thus we have the following corollary.

Corollary 10.1.1. Under the conditions of Theorem 10.1, the number of corrections K is bounded by

$$K > \frac{\sum_{i=1}^{n+1} \tilde{a}_i^2}{\frac{g}{\rho} + n + 1}$$

where $\{\tilde{a}_i\}$ is as defined above.

We can use this result to get a lower bound on the number of corrections required for the function of Examples 10.1 and 10.2. The realization given in Example 10.2 is in fact the minimal integral square realization for Φ. Thus with

$$\frac{g}{2} = \rho = 1$$

the bound is

$$K > \frac{4 + 1 + 1 + 1}{2 + 4} = 1.17$$

Recall that the actual number of corrections was 14.

As another example of the use of this bound, Goto, 3, and Muroga, 9, have independently shown that there exist ℓ.s. functions such that, in all realizations with integer coefficients, the value of each coefficient is at least

$$\tilde{a}_i \sim A2^i \qquad i = 1,2,\ldots,n$$
$$\tilde{a}_{n+1} \sim A2^n \qquad \text{for large } n$$

with $A = 0.1163$. Thus for this class of function

$$\sum_{i=1}^{n+1} \tilde{a}_i^2 \sim A^2[2^{2n} + 2^{2n} + 2^{2(n-1)} + 2^{2(n-2)} + \cdots + 2^2]$$

$$\sim (0.036)\, 2^{2n} \qquad \text{for large } n$$

and the lower bound on the number of corrections is

$$\frac{(0.036)2^{2n}}{\dfrac{g}{\rho} + n + 1} \qquad \text{for large } n$$

Thus there are functions† for which the adaptive procedure requires a number of corrections that increases exponentially with n. This is a very arge number, indeed.

The upper and lower bounds can both be written in terms of the same coefficient set $\{\tilde{a}_i\}$. Since the $\{\tilde{a}_i\}$ are integers and the map of Φ is symmetric around zero, the gap length of the $\{\tilde{a}_i\}$ map must be an even integer. Therefore the coefficient set $\{\tilde{a}_i g/2\}$ must give a realization with gap length greater than or equal to g and hence is an allowable set to use for $\{\hat{a}_i\}$. Using this in the upper bound of Theorem 10.1, we get

Corollary 10.1.2. Under the conditions of Theorem 10.1, the number of corrections is bounded by

$$\frac{\sum_{i=1}^{n+1} \tilde{a}_i^2}{\dfrac{g}{\rho} + n + 1} < K < \left[\frac{g}{\rho} + n + 1\right] \sum_{i=1}^{n+1} \tilde{a}_i^2$$

When $\rho = g/2$, this becomes

$$\frac{\sum_{i=1}^{n+1} \tilde{a}_i^2}{n + 3} < K < [n + 3] \sum_{i=1}^{n+1} \tilde{a}_i^2$$

† The functions given by Goto and Muroga are quite complicated. However, for the simple function

$$y = x_n \dotplus x_{n-1}(x_{n-2} \dotplus x_{n-3}(x_{n-4} \dotplus \cdots \dotplus x_4(x_3 \dotplus x_2(x_1))\ldots)$$

which can be realized using Theorem 2.1, it can be easily seen (Myhill and Kautz) that the integer weights must satisfy

$$\tilde{a}_i = \tilde{a}_{i-1} + \tilde{a}_{i-2}$$

from which it follows by solution of this difference equation that

$$a_i \sim (2.7638)2^{0.6765n} \qquad \text{for large } n$$

which also shows the exponential growth.

Thus the number of corrections is bounded within a factor of about n^2.

For the function class defined by Goto and Muroga, these bounds are

$$\frac{(0.036)2^{2n}}{\dfrac{g}{\rho} + n + 1} < K < \left[\frac{g}{\rho} + n + 1\right](0.036)2^{2n}$$

For large n these upper and lower bounds can be further bounded as

$$\left[\frac{g}{\rho} + n + 1\right](0.036)\,2^{2n} < 0.036\,(2 + \epsilon)^{2n}$$

$$\frac{(0.036)\,2^{2n}}{\dfrac{g}{\rho} + n + 1} > 0.036\,(2 - \epsilon)^{2n}$$

respectively for any positive ϵ. Thus

$$0.036\,(2 - \epsilon)^{2n} < K < 0.036\,(2 + \epsilon)^{2n}$$

and the number of corrections is bounded between two exponentials.

10.5 EXTENSIONS AND MODIFICATIONS OF THE PROCEDURE

There are a number of simple extensions and modifications of Procedure 10.2 that can be established from Theorem 10.1.

Arbitrary Initial Conditions

Procedure 10.2 will converge for any arbitrary initial choices for the coefficients $a_i^{(\alpha_0)}$. The nonzero coefficients will add constant terms to the bounds in the proof of Theorem 10.1, but these constants will eventually become small compared with the terms that depend on k.

Nonconstant $\rho^{(k)}$

When an error is found in Procedure 10.2, the "error point" is added to the coefficients with a constant ρ. However, after this is done, the point in question may still be in error; that is, ρ may not be large enough to correct the error. An alternative procedure which intuitively would seem to converge faster is to use a $\rho^{(k)}$ which depends on k and is large enough to correct the error.

One simple example of this is to select $\rho^{(k)}$ as

$$\rho^{(k)} = p(k)\rho$$

where ρ is a constant and $p(k)$ is the smallest integer such that

$$p(k) \geq \frac{\dfrac{g}{2} - a^{(\alpha_{k-1})} \cdot r^{(\alpha_k)}}{\rho(n + 1)}$$

This is equivalent to repeatedly applying the point $r^{(\alpha_k)}$ to Procedure 10.2 until the output of the T gate is correct for the first time. It is readily verified that $p(k)$ is the number of applications required.

For related procedures, see Agmon; Motzkin and Schoenberg; May, 1; and Ho and Kashyap.

Multiple Points

Assume that at each step in the procedure, more than one point $r_i^{(\alpha_k)}$ is known for which $\Phi = 1$, and the current estimate of the coefficients give an incorrect output. Let $\{r_{\beta_j}^{(\alpha_k)}\}$ be the set of these points, (including $r_i^{(\alpha_k)}$). Then the error-correcting procedure

$$a^{(\alpha_k)} = a^{(\alpha_k - 1)} + \rho \sum_{\beta_j} r_{\beta_j}^{(\alpha_k)}$$

can be used. The proof of convergence is virtually identical with that of Theorem 10.1.

When the set $\{r_{\beta_j}^{(\alpha_k)}\}$ includes all of the points for which the current estimate gives the incorrect output, the procedure is equivalent to that suggested by Kaszerman, 1; and Dertouzos, 1, in a different context.

When the set $\{r_{\beta_j}^{(\alpha_k)}\}$ includes all of those points implied by one point $r^{(\alpha_k)}$ and the ordering of the coefficients (which is assumed known), the procedure is quite similar to that suggested by Gabelman, 1.

Another related procedure is given in Ho and Kashyap.

Nonbinary Variables

The proof of Theorem 10.1 does not depend on the variables being binary. The only place where this restriction was used was in replacing

$$\sum_{i=1}^{n+1} (\xi_i^{(\alpha_j)})^2 = r^{(\alpha_j)} \cdot r^{(\alpha_j)}$$

by $n + 1$. If the variables were not binary, the sum would be bounded by some constant equal to the largest value of the sum of the squared components of any vector r.

Stripped to its bare essentials, there are only three conditions required to prove the theorem.

(1) There must exist a set of m vectors

$$r_1, r_2, \ldots, r_m$$

a vector $\hat{a}$ and a positive number g such that

$$\hat{a} \cdot r_i \geq \frac{g}{2}$$

for $i = 1, 2, \ldots, m$.

(2) The vectors $r_i^{(k)}$ must be selected in an infinite sequence in which each vector appears an infinite number of times.

(3) Vectors $a^{(k)}$ must be selected as

(a) $a^{(0)}$ is arbitrary

(b) If

$$a^{(k-1)} \cdot r^{(k)} \geq \frac{g}{2}$$

then

$$a^{(k)} = a^{(k-1)}$$

(c) If

$$a^{(k-1)} \cdot r^{(k)} < \frac{g}{2}$$

then

$$a^{(k)} = a^{(k-1)} + \rho r^{(k)}$$

where ρ is any positive constant.

These are the only assumptions required for the proof.

10.6 PATTERN RECOGNITION INTERPRETATION

Although the adaptive procedures have been discussed within the framework of switching theory, much of the impetus for work in this area and many of the proposed applications come from the field of pattern recognition.† It is assumed that there is some set of patterns of 1's and −1's that are to be classified into two groups. For example, the patterns may be black and white samples of handprinted letters, to be classified as *A* or not *A*.

In the simplest such recognition system, the patterns would be presented on a two-dimensional array (for instance, of photocells) containing n squares. The output of each square, quantized to 1 or −1, is directly connected to one of the inputs of a T gate as shown in Fig. 10.1.

When the device is in its "learning phase," the patterns to be recognized are sequentially applied to the input array, and when the T gate gives an incorrect output, the coefficients are adjusted as in Procedure 10.1 or 10.2. If the given set of patterns corresponds to a linearly separable function, the procedure will converge, and the patterns will be learned.

More sophisticated recognition systems might contain several T gates connected in two layers as shown in Fig. 10.2. The first layer has its inputs connected to the input array, either in a random manner or accor-

† This is not meant to be a complete exposition of adaptive procedures for recognition problems; see, for example, Nilsson.

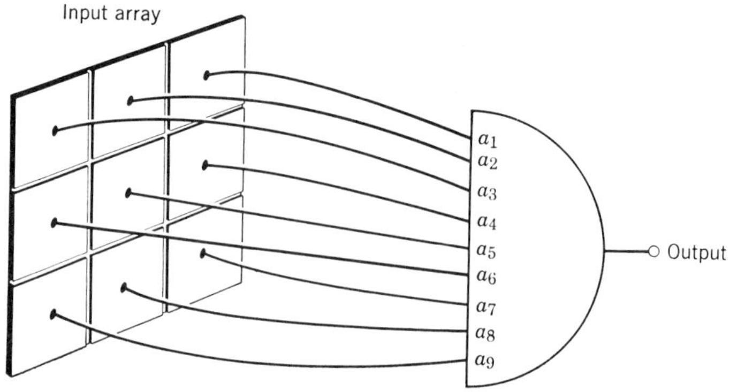

Fig. 10.1. A simple adaptive classifier.

ding to some systematic procedure. The outputs of the first layer are all connected to a single gate which makes the decision.

In one version of this system, the first layer gates are fixed and the weights of the final gate are adjusted, again according to Procedures 10.1 and 10.2. In another version the final gate is fixed as an n input majority gate, and the weights of the second layer gates are adjusted according to a procedure similar to that of Procedures 10.1 and 10.2. Other variations of these systems are possible, but our main interest is in the implications of the single-gate adaptive procedure.

At this point the question naturally arises whether a single T gate is useful as a recognition device. Particularly in the system of Fig. 10.1, what is the justification for believing that a single T gate will be adequate for making

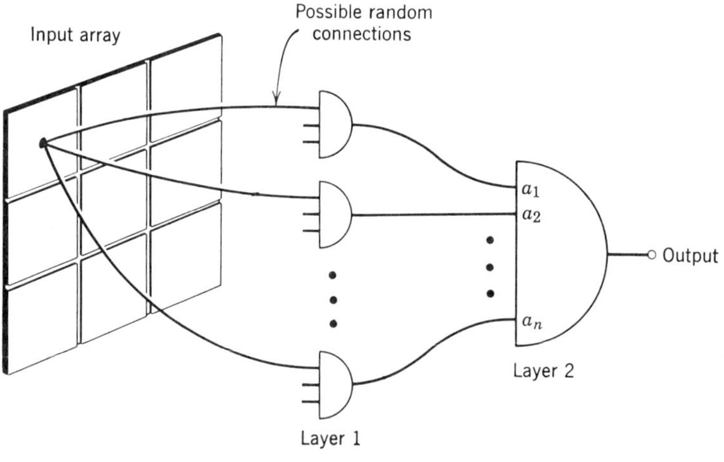

Fig. 10.2. A two-layer recognition system.

the decision? More specifically, what is the reason for believing that the switching function corresponding to a given recognition problem will be linearly separable? After all, in most of our previous work we found that a single gate was not very powerful and that most switching functions required more than one gate. How is the recognition problem different in that respect from the switching theory problem?

There are two somewhat related justifications, one based on counting arguments for don't-care conditions and the other on statistical independence of inputs.

The Don't-Care Justification

Intuitively it is clear that the number of n variable ℓ.s. functions is small compared with the total number of n variable functions. In Appendix B it is shown that B_n, the number of n variable ℓ.s. functions, is bounded by

$$B_n \leq 2\sum_{i=0}^{n}\binom{2^n - 1}{i} < 2\frac{2^{n^2}}{n!} < 2^{n^2}$$

The total number of n variable functions is of course

$$2^{2^n}$$

Looking only at the loosest bound for B_n, it is clear that 2^{n^2} is very small compared to 2^{2^n} and, in fact,

$$\lim_{n \to \infty} \frac{2^{n^2}}{2^{2^n}} = 0$$

For even moderately large n, the number of ℓ.s. functions is a negligibly small fraction of the total number of functions.

However, these bounds are only for completely specified functions. Most recognition problems do not correspond to completely specified switching functions. For example, consider a letter recognition system with a 20-by-20 input array. Here n is 400 and the total number of input points is

$$2^{400}$$

Only a very small fraction of this astronomical number of points will ever be presented to the system. For instance, the learning phase might involve 10 samples of each of the 72 upper and lower case letters, or 720 patterns in all. These are then the only points of interest; the remaining points are all don't-care points.

Whether or not a single threshold gate is adequate for making the decision depends on whether there is an assignment of all of these don't-care points so that the overall function is ℓ.s. Intuitively, one feels that if few enough of the points are specified, it is likely that such an ℓ.s. function is possible.

In Appendix B it is shown that $B_n{}^m$, the number of n variable ℓ.s. functions in which a given m points are specified, is bounded by

$$B_n{}^m \leq 2\sum_{i=0}^{n}\binom{m-1}{i} < 2\frac{m^n}{n!} < m^n$$

The total number of n variable functions in which a given m points are specified is, of course, 2^m.

For m small enough, it is clear that the bounds on $B_n{}^m$ are at least comparable to 2^m for all n. Winder, 6, has shown, for example, that

$$\lim_{n\to\infty}\frac{2\sum_{i=0}^{n}\binom{m-1}{i}}{2^m} = \begin{cases} 1 \text{ for } m < 2n \\ \frac{1}{2} \text{ for } m = 2n \\ 0 \text{ for } m > 2n \end{cases}$$

Thus if the number of specified points is less than twice the number of variables, it appears likely that a single threshold gate will be adequate. (For some experimental verification of this, see Widrow and Angell, 3.) Observe that the example of the letter recognition system with the 20-by-20 input array and 720 input patterns satisfies this condition.

The Statistical Independence Justification

One approach to the recognition problem is through statistical decision theory. Certain assumptions in that theory lead naturally to a threshold gate realization.

Assume that we wish to make a decision as to whether a variable y equals 1 or 0. Suppose that the error criterion is given by Table 10.6.

Table 10.6. A Loss Table

Decision \ Actual	1	0
1	0	ℓ_1
0	ℓ_2	0

Thus if y actually is 0, but the decision is that $y = 1$, then the loss is ℓ_1 units (say dollars).

Assume that the decision is to be based on measurements of n auxiliary quantities which are called characteristics

$$x_1, x_2, \ldots, x_n$$

where the x_i can take on the values 0 or 1.

Suppose that a particular value of that measurement yields

$$x_1 = X_1,\ x_2 = X_2, \ldots,\ x_n = X_n$$

The probabilities of the events $y = 1$ and $y = 0$ are then

$$p(y = 1 | x_1 = X_1, \ldots, x_n = X_n)$$

and

$$p(y = 0 | x_1 = X_1, \ldots, x_n = X_n)$$

where these two probabilities sum to 1.

Referring to the loss table, a decision that $y = 1$ will have an expected loss of

$$L(y = 1) = \ell_1 p(y = 0 | x_1 = X_1, \ldots, x_n = X_n)$$

and a decision that $y = 0$ will have an expected loss of

$$L(y = 0) = \ell_2 p(y = 1 | x_1 = X_1, \ldots, x_n = X_n)$$

The expected loss will be least if y is selected equal to 1 if and only if

$$L(y = 0) \geq L(y = 1)$$

which implies

$$\ell_2 p(y = 1 | x_1 = X_1, \ldots, x_n = X_n) \geq \ell_1 p(y = 0 | x_1 = X_1, \ldots, x_n = X_n)$$

Rearranging this expression, y is selected equal to 1 if and only if

$$\frac{p(y = 1 | x_1 = X_1, \ldots, x_n = X_n)}{p(y = 0 | x_1 = X_1, \ldots, x_n = X_n)} \geq \frac{\ell_1}{\ell_2}$$

Applying Bayes' theorem to the probability ratio yields

$$\frac{p(x_1 = X_1, \ldots, x_n = X_n | y = 1) p(y = 1)}{p(x_1 = X_1, \ldots, x_n = X_n | y = 0) p(y = 0)} \geq \frac{\ell_1}{\ell_2}$$

Rearranging once more gives

$$\frac{p(x_1 = X_1, \ldots, x_n = X_n | y = 1)}{p(x_1 = X_1, \ldots, x_n = X_n | y = 0)} \geq \frac{\ell_1 p(y = 0)}{\ell_2 p(y = 1)}$$

This is the general solution to the decision problem; y is selected equal to 1 if and only if the inequality is satisfied. The quantity on the left is the result of the measurement; the quantity on the right is assumed known beforehand.

Consider the special case where the x_i variables are statistically independent in the above distributions; that is, assume that

$$p(x_1,x_2,\ldots,x_n|y) = \prod_{i=1}^{n} p(x_i|y)$$

Under this assumption the above expression becomes

$$\frac{\prod_{i=1}^{n} p(x_i = X_i|y = 1)}{\prod_{i=1}^{n} p(x_i = X_i|y = 0)} \geq \frac{\ell_1\, p(y = 0)}{\ell_2\, p(y = 1)}$$

which can be written as

$$\prod_{i=1}^{n} \frac{p(x_i = X_i|y = 1)}{p(x_i = X_i|y = 0)} \geq \frac{\ell_1\, p(y = 0)}{\ell_2\, p(y = 1)}$$

Now take logarithms of both sides of this expression. Since the logarithm is a monotonically increasing function of its argument, the inequality is not affected. Thus it is equivalent to select y equal to 1 if and only if

$$\sum_{i=1}^{n} \log \frac{p(x_i = X_i|y = 1)}{p(x_i = X_i|y = 0)} \geq \log \frac{\ell_1\, p(y = 0)}{\ell_1\, p(y = 1)}$$

This is almost in a form suitable for realization with a T gate. A particular term in the sum on the left can be expanded as

$$\log \frac{p(x_i = X_i|y = 1)}{p(x_i = X_i|y = 0)} = x_i\left\{\log \frac{p(x_i = 1|y = 1)p(x_i = 0|y = 0)}{p(x_i = 1|y = 0)p(x_i = 0|y = 1)}\right\} + \log \frac{p(x_i = 0|y = 1)}{p(x_i = 0|y = 0)}$$

which can be verified by substituting 1 and 0 for x_i.

Thus the overall expression can be written

$$\sum_{i=1}^{n} \left\{\log \frac{p(x_i = 1|y = 1)p(x_i = 0|y = 0)}{p(x_i = 1|y = 0)p(x_i = 0|y = 1)}\right\} x_i + \sum_{i=1}^{n} \log \frac{p(x_i = 0|y = 1)}{p(x_i = 0|y = 0)} \geq \log \frac{\ell_1\, p(y = 0)}{\ell_2\, p(y = 1)}$$

which can be rearranged

$$\sum_{i=1}^{n} \left\{\log \frac{p(x_i = 1|y = 1)p(x_i = 0|y = 0)}{p(x_i = 1|y = 0)p(x_i = 0|y = 1)}\right\} x_i \geq \log \frac{\ell_1 p(y = 0)}{\ell_2 p(y = 1)} \prod_{i=1}^{n} \frac{p(x_i = 0|y = 0)}{p(x_i = 0|y = 1)}$$

This is exactly the form of a T gate, in which the coefficients and threshold are functions of the known probabilities and the error criterion. Note that changing the error criterion only implies changing the threshold. Thus we have the following theorem.

Theorem 10.2. Consider a statistical decision problem in which the value of a binary variable y is to be selected as a function of binary variables $x_1,x_2,\ldots,x_n$. If the variables are statistically independent in their effect on the decision, that is, if

$$p(x_1,x_2,\ldots,x_n|y) = \prod_{i=1}^{n} p(x_i|y)$$

then, for any error criterion, the decision can always be made by a T gate.

This is then another justification for using T gates in those situations in which the inputs to the gate are, or can be made to be, statistically independent. For example, one intuitive reason for using the two-layer network of Fig. 10.2 with random connections between the input array and the first layer of gates is that the randomness may produce statistical independence among the inputs to the final gate.

Example 10.4. Consider the following probability distribution relating the binary variable y to three binary variables x_1, x_2, and x_3,

$$p(x_1,x_2,x_3|y) = \prod_{i=1}^{3} p(x_i|y)$$

with

$$p(x_i = 1|y = 1) = \tfrac{2}{3}$$
$$p(x_i = 1|y = 0) = \tfrac{2}{5}$$

for $i = 1,2,3$ and

$$p(y = 1) = \tfrac{1}{2}$$

These numbers completely describe the probability distribution.

Assume that a decision must be made using data based on this distribution with an error criterion

$$\ell_1 = \ell_2 = 1$$

The decision can be made by a threshold gate with weights

$$a_i = \log \frac{p(x_i = 1|y = 1)p(x_i = 0|y = 0)}{p(x_i = 1|y = 0)p(x_i = 0|y = 1)}$$

$$= \log \frac{(\frac{2}{3})(\frac{3}{5})}{(\frac{2}{5})(\frac{1}{3})}$$

$$= \log 3$$

and threshold

$$
\begin{aligned}
t &= \log \frac{\ell_1}{\ell_2} \frac{p(y=0)}{p(y=1)} \prod_{i=1}^{n} \frac{p(x_i = 0 | y = 0)}{p(x_i = 0 | y = 1)} \\
&= \log 1 \cdot 1 \cdot \left(\frac{\frac{3}{5}}{\frac{1}{3}} \right)^3 \\
&= \log \left(\tfrac{9}{5}\right)^3
\end{aligned}
$$

Using base 3 for the logarithms (or equivalently using any base and normalizing by dividing weights and threshold by log 3) yields

$$
\begin{aligned}
a_i &= \log_3 3 = 1 \\
t &= \log_3 \left(\tfrac{9}{5}\right)^3 \\
&\approx 1.60
\end{aligned}
$$

Thus the gate is of the form

$$\langle x_1 + x_2 + x_3 \rangle_{1.60}$$

which is a three-input majority gate.

10.7 EVALUATION OF ADAPTIVE PROCEDURE

Assuming that there are a number of applications for which a single T gate is adequate, we now want to compare the adaptive synthesis procedures with the parallel procedures discussed earlier.

Primarily the adaptive procedure allows computation time to be traded for storage space and complexity of computation. The parallel procedures require enough memory to store a representation of the entire switching function and involve relatively complicated calculations. The adaptive procedure requires essentially no storage space (the points of the function can be generated in any order by some process external to the procedure) and very simple computations. However, this simplicity is counterbalanced by the requirement that each point be presented a large number of times. The adaptive procedure is, in a sense, somewhat inefficient in the utilization of the information presented to it.

In a pattern recognition problem for example, if we are willing to store all of the patterns, then any one of the parallel procedures could be used to "learn" all of the patterns after they have been presented just once. On the other hand, the adaptive procedure requires essentially no storage and very simple calculations but must "see" all of the patterns a large number of times before it can learn them.

One supposed advantage of the adaptive procedure is that, after a set of patterns is learned, if a new pattern is added to the set the procedure can easily adjust the coefficients for this new input. But this advantage is largely illusory. As soon as the coefficients are changed according to the new pattern, there is no guarantee that the gate will give the right answer for any of the patterns it had previously learned. In trying to learn the new pattern it may unlearn all the other patterns. In general, when a new pattern is presented the entire adapting procedure must be repeated, and therefore the same advantages and disadvantages of adaptive versus parallel procedures are applicable.

One possible disadvantage of the adaptive procedure is that it gives no indication when it has converged. If the given function is not ℓ.s., the number of errors will eventually exceed the bound, thus giving the required information. But if the function is ℓ.s. and if the points are generated in some random manner and no external records are kept, there is no way of knowing when the procedure has converged. If it has made no errors for a while, there is no guarantee that it has seen all the points or that there is not some point for which it would give the wrong answer but which has not been generated for a long time. Of course, if the random process is governed by known statistics, some probabilistic assertions can be made about the convergence.

So far sensitivity has not been mentioned. There is no guarantee that the adaptive procedure will converge to a realization with any reasonable sensitivity, even if one exists. Also, if the given problem is at all complicated, any single-gate realization would require sensitivities that are difficult to achieve. In this case, either a multigate realization is necessary or, for some applications, a digital simulation of a T gate can be used.

10.8 POINT COUNTING METHODS

Closely related to the adaptive procedures is a very promising but as yet not very well understood set of procedures based on point counting. We have already seen in Chapter 3 that if, in the set of points for which an ℓ.s. function equals 1, there are more points for which x_1 equals 1 than for which x_2 equals 1, then

$$a_1 > a_2$$

Now we show that the actual number of these points constitutes sufficient information for a test and realization. Although as yet no practical test or realization procedures have been developed, this is one of the more promising areas for research in threshold logic.

Again assume binary variables ζ_i which take on the values 1, -1. A given function F is ℓ.s. if and only if there are a_i and t' such that

$$F(q_j) = 1 \qquad \text{if} \quad f'(q_j) = \sum_{i=1}^{n} a_i \zeta_i^j - t' > 0 \tag{10.1}$$

$$F(q_j) = -1 \qquad \text{if} \quad f'(q_j) = \sum_{i=1}^{n} a_i \zeta_i^j - t' < 0 \tag{10.2}$$

Note that (1) the value f' equals 0 has been excluded, and (2) f' has been defined including the threshold t'. In fact it is convenient to define

$$a_0 = -t'$$

thus:

$$f'(q) = \sum_{i=1}^{n} a_i \zeta_i + a_0$$

Equations 10.1 and 10.2 will be satisfied if and only if

$$F(q_j) f'(q_j) = |f'(q_j)|$$

for all j, as can be verified by direct substitution. Now sum both sides of this equation over all points q_j,

$$\sum_{j=0}^{2^n-1} F(q_j) f'(q_j) = \sum_{j=0}^{2^n-1} |f'(q_j)|$$

This condition is necessary and sufficient for single-gate realizability. Necessity has already been shown. Now suppose that the given function is not ℓ.s. Then for all j, the right side of the equation must be equal to or greater than the left side and, for at least one j, strictly greater; hence the equality cannot hold.

Substituting the expression for $f'(q_j)$ gives

$$\sum_{j=0}^{2^n-1} F(q_j) \left(\sum_{i=1}^{n} a_i \zeta_i^j + a_0 \right) = \sum_{j=0}^{2^n-1} |f'(q_j)|$$

Interchanging summations,

$$\sum_{i=1}^{n} a_i \left(\sum_{j=0}^{2^n-1} \zeta_i^j F(q_j) \right) + a_0 \left(\sum_{j=0}^{2^n-1} F(q_j) \right) = \sum_{j=0}^{2^n-1} |f'(q_j)|$$

Each of the terms inside the round brackets has an interpretation in terms of numbers of truth table points. Let m_1 and m_0 be the number of points for which F equals 1 and -1 respectively. Then

$$\sum_{j=0}^{2^n-1} F(q_j) = m_1 - m_0$$

$$= 2m_1 - 2^n$$

Denote this number by b_0

$$b_0 = m_1 - m_0$$

Let m_{ζ_i} and $m_{\bar{\zeta}_i}$ be the number of points for which F equals 1 and ζ_i is 1 and -1 respectively. Then since $F(\zeta_i = 1)$ and $F(\zeta_i = -1)$ both have 2^{n-1} points, it follows that the number of points for which F equals -1, and ζ_i is 1 and -1 respectively, is

$$2^{n-1} - m_{\zeta_i}$$

$$2^{n-1} - m_{\bar{\zeta}_i}$$

Thus

$$\sum_{j=0}^{2^n-1} \zeta_i^j F(q_j) = m_{\zeta_i} - m_{\bar{\zeta}_i} - (2^{n-1} - m_{\zeta_i} - (2^{n-1} - m_{\bar{\zeta}_i}))$$
$$= 2(m_{\zeta_i} - m_{\bar{\zeta}_i})$$

Denote this number by b_i

$$b_i = 2(m_{\zeta_i} - m_{\bar{\zeta}_i})$$

Thus the necessary and sufficient condition can be written

$$\sum_{i=1}^{n} a_i b_i + a_0 b_0 = \sum_{j=0}^{2^n-1} |f'(q_j)| = \sum_{j=0}^{2^n-1} \left| \sum_{i=1}^{n} a_i \zeta_i^j + a_0 \right|$$

Theorem 10.3. (*Dertouzos 1, 2; Chow, 1*).† A given function F is ℓ.s. if and only if there is a set a_i, $i = 0,1,\ldots,n$ satisfying

$$\sum_{i=0}^{n} a_i b_i = \sum_{j=0}^{2^n-1} \left| \sum_{i=1}^{n} a_i \zeta_i^j + a_0 \right|$$

Note that the only information about F that is contained in this equation is the counting of points implied by b_i, $i = 0,1,\ldots,n$. Furthermore the numbers b_i, $i = 0,1,\ldots,n$ uniquely characterize ℓ.s. functions. Assume, to the contrary, that two different ℓ.s. functions F_1 and F_2 have the same values for b_i, $i = 0,1,\ldots,n$. From the above equation both functions can be realized by the same coefficients and threshold, a_i, $i = 0,1,\ldots,n$. Hence F_1 and F_2 must be the same function. By similar reasoning, it could not be true that F_1 is ℓ.s. and F_2 is not. Thus we have the corollary.

Corollary 10.3.1. The numbers b_i, $i = 0,1,\ldots,n$ (or equivalently m_1 and m_{ζ_i}) constitute a sufficient amount of information for a test and realization procedure.

This is a very significant result. Present ℓ.s. tests require a complete

† This version of the theorem is due to Dertouzos; a slightly weaker version was first proved by Chow.

representation of the function, which may involve a large amount of data. This result shows that only the $n + 1$ numbers $b_0,b_1,\ldots,b_n$ are needed to formulate a test and realization procedure. (In addition these numbers allow a very compact tabular presentation of ℓ.s. function realizations as is given in Appendix D.)

Unfortunately, at present no simple test based on these numbers has been discovered.† For example, a direct solution using the equation in Theorem 10.3 would be very difficult because of the magnitude signs. Iterative methods of solution suggest themselves and have been attempted with some success (Dertouzos, 1).

Even though the equation is difficult to solve, Theorem 10.3 suggests a certain functional form in which a_i might be expressed

$$a_i = \psi_i(b_0,b_1,\ldots,b_i,\ldots,b_n,n)$$

Assuming such a form, the properties of ℓ.s. functions give some additional information. From Procedure 3.1, complementing a variable, x_j, changes the sign of its coefficient, a_j, but leaves the other coefficients unchanged. In the present terminology, complementing x_j corresponds to interchanging m_{ζ_j} and $m_{\bar{\zeta}_j}$ or changing the sign of b_j. Thus changing the sign of b_j changes the sign of a_j but leaves the other coefficients unchanged. Applying this result to the expression for a_i implies that ψ_i is an odd function of b_i and an even function of the remaining b's.

Similar reasoning applies to b_0 and t'. If one further assumes that ψ_i can be expanded in a multidimensional power series, a_i can be expressed in the form

$$a_i = \sum_{p \text{ odd}} C_{ip} b_i^p \quad i = 0,1,\ldots,n$$

where the C_{ip} are functions of n and even functions of $b_0,b_1,\ldots,b_{i-1}, b_{i+1},\ldots,b_n$.

Using a somewhat different approximate geometric approach, Winder, 5, suggests an expression of the form

$$a_i = b_i - \lambda_3 b_i^3 + \lambda_5 b_i^5 - \ldots$$

where λ_p is a function only of n and is independent of the remaining b's. He notes that the first two terms

$$a_i \approx b_i - \lambda_3 b_i^3$$

† Note, however, that the b_i numbers arise naturally in the adaptive procedures discussed earlier. Suppose that the modified procedure of Section 10.5, p. 408 is used with the initial values of the weights set at zero and that all the incorrect points are added in at each step. Then it is readily verified that after the first step the values of the weights are exactly proportional to the b_i. A similar situation occurs in the procedure of Ho and Kashyap.

give very good results for n up to 10, for selected values of $\lambda_3(n)$. No general form for $\lambda_3(n)$ is known at present. Observe that even the first-order approximation

$$a_i \approx b_i$$

will always give the correct sign and ordering of the coefficients as can be verified by the point-counting arguments of Chapter 3.

Example 10.5. Consider

$$F(x_1,x_2,x_3) = x_1 \dotplus x_2x_3$$

for which the truth table is given in Table 10.7.

Table 10.7

x_1	x_2	x_3	F
0	0	0	0
0	0	1	0
0	1	0	0
0	1	1	1
1	0	0	1
1	0	1	1
1	1	0	1
1	1	1	1

The m and b parameters are

$$\begin{aligned} m_1 &= 5 \\ m_0 &= 3 \\ b_0 &= 5 - 3 = 2 \end{aligned}$$

$$\begin{aligned} m_{\zeta_1} &= m_{x_1} = 4 \\ m_{\bar{\zeta}_1} &= m_{\bar{x}_1} = 1 \\ b_1 &= 2(4 - 1) = 6 \end{aligned}$$

$$\begin{aligned} m_{\zeta_2} &= m_{x_2} = 3 \\ m_{\bar{\zeta}_2} &= m_{\bar{x}_2} = 2 \\ b_2 &= 2(3 - 2) = 2 \end{aligned}$$

$$\begin{aligned} m_{\zeta_3} &= m_{x_3} = 3 \\ m_{\bar{\zeta}_3} &= m_{\bar{x}_3} = 2 \\ b_3 &= 2(3 - 2) = 2 \end{aligned}$$

Thus the first approximation

$$a_i \approx b_i$$

gives the realization

$$f' = \sum_{i=1}^{3} a_i\zeta_i + a_0 = 6\zeta_1 + 2\zeta_2 + 2\zeta_3 + 2$$

which corresponds to

$$\langle 6\zeta_1 + 2\zeta_2 + 2\zeta_3 \rangle_{-2}$$

In terms of the x_i variables, this can be written

$$\langle 6x_1 + 2x_2 + 2x_3 \rangle_4 = \langle 3x_1 + x_2 + x_3 \rangle_2$$

which is a satisfactory realization.

Because of their simplicity, point counting methods may very well prove quite useful for test and synthesis purposes. However, at present our knowledge in this area is far from complete.

Appendix A

GENERAL SENSITIVITY ANALYSIS†

The sensitivity analysis of Chapter 1 was confined to definite maps in the zero-one and minus-plus-one models. These cases are particularly simple because, for definite maps, the sensitivity is determined by the gap boundaries; that is, as the weights, logical voltages, and threshold vary, one of the gap boundaries always crosses over the threshold before any of the other map points.

For nondefinite maps, this is no longer true. The sensitivity may be determined by one of the nongap boundary points crossing the threshold. For this reason the sensitivity analysis must consider the entire map, and the results do not come out in a particularly elegant or useful form.

We consider the general case of nonzero-one models with nondefinite maps (see also Wood). We use the notation of Chapter 1.

Consider first the special case when the weights and logical voltages do not change. Since the value of the separating function does not change, its nondefinite or nonzero-one character can have no effect, and the allowable threshold variations are the same as for the definite case. However, since now t can be positive or negative, the notation must be changed somewhat. Let

$$\mathscr{T}_a = \begin{cases} \mathscr{T} & \text{when} \quad t \geq 0 \\ -\mathscr{T} & \text{when} \quad t < 0 \end{cases}$$

Then the allowable threshold variations are governed by the two inequalities

$$t(1 + \mathscr{T}_a) \leq u$$
$$t(1 - \mathscr{T}_a) \geq \ell$$

† For a similar approach see Muroga, 4.

Eliminating t gives

$$\mathcal{T}_a \leq \frac{u - \ell}{u + \ell}$$

or

$$\mathcal{T}_a \leq \left|\frac{u - \ell}{u + \ell}\right|$$

which is the same as the result for definite maps.

Now assume that $\mathcal{T}_a$ is strictly less than this bound. There are then some allowable variations in weights and logical voltages which we can determine by a worst-case analysis.

Let V_1 and V_0 be the voltages corresponding to the logical 1 and the logical 0. Assume that V_1 is positive but V_0 can be positive or negative. Let the threshold be normalized as

$$t = V_1 t'$$

Then the law of operation of the nondefinite, nonzero-one threshold gate is:

$$\text{Output equals 1 if } \sum_{i=1}^{n} a_i V_i \geq V_1 t'$$

$$\text{Output equals 0 if } \sum_{i=1}^{n} a_i V_i < V_1 t'$$

These expressions can be written

$$\sum_{i=1}^{n} a_i(V_1 x_i + V_0 \bar{x}_i) \geq \text{ or } < V_1 t'$$

or

$$V_1 \sum_{i=1}^{n} a_i x_i + V_0 \sum_{i=1}^{n} a_i \bar{x}_i \geq \text{ or } < V_1 t'$$

Define

$$f(p) = \sum_{i=1}^{n} a_i x_i = \sum_{i=1}^{n} a_i x_i^p$$

$$f(\bar{p}) = \sum_{i=1}^{n} a_i \bar{x}_i = \sum_{i=1}^{n} a_i x_i^{\bar{p}} = \sum_{i=1}^{n} a_i \bar{x}_i^p$$

Thus the defining equations can be written

$$V_1 f(p) + V_0 f(\bar{p}) \geq \text{ or } < V_1 t'$$

Divide f into two parts so that

$$f = f_+ - f_-$$

where the terms of f with positive weights are in f_+ and those with negative weights are in f_-.

Then the defining equations can be written

$$V_1 f_+(p) + V_0 f_+(\bar{p}) - [V_1 f_-(p) + V_0 f_-(\bar{p})] \geq \text{ or } < V_1 t'$$

Let p^1 be any point for which the output of the gate should be 1. Let $\mathscr{X}_1$, $\mathscr{X}_0$, and $\mathscr{A}$ represent the allowable percentage tolerances on V_1, V_0, and a_i respectively. The conditions for error-free operation in the worst case are

$$\begin{aligned} &V_1 f_+(p^1)(1 - \mathscr{X}_1)(1 - \mathscr{A}) + V_0 f_+(\bar{p}^1)(1 - \mathscr{X}_0)(1 - \mathscr{A}) \\ &\quad - [V_1 f_-(p^1)(1 + \mathscr{X}_1)(1 + \mathscr{A}) + V_0 f_-(\bar{p}^1)(1 + \mathscr{X}_0)(1 + \mathscr{A})] \\ &\quad \geq t' V_1(1 + \mathscr{T}_a) \end{aligned}$$

Note that the signs of $\mathscr{X}_1$, $\mathscr{X}_0$, and $\mathscr{A}$ depend on the signs of V_0 and V_1. The above equation is written under the assumption that V_1 and V_0 are both positive. To allow V_0 to be negative and to simplify the remaining derivation, we make the following definitions and approximations. Let

$$1 - \mathscr{S}_1 = (1 - \mathscr{X}_1)(1 - \mathscr{A}) \qquad \text{for} \quad V_1 > 0$$

and then make the approximation

$$1 + \mathscr{S}_1 \approx (1 + \mathscr{X}_1)(1 + \mathscr{A}) \qquad \text{for} \quad V_1 > 0$$

and similarly

$$1 - \mathscr{S}_0 = \begin{cases} (1 - \mathscr{X}_0)(1 - \mathscr{A}) & \text{for} \quad V_0 > 0 \\ (1 + \mathscr{X}_0)(1 + \mathscr{A}) & \text{for} \quad V_0 < 0 \end{cases}$$

and

$$1 + \mathscr{S}_0 \approx \begin{cases} (1 + \mathscr{X}_0)(1 + \mathscr{A}) & \text{for } V_0 > 0 \\ (1 - \mathscr{X}_0)(1 - \mathscr{A}) & \text{for } V_0 < 0 \end{cases}$$

These definitions correspond to considering the variations of the logical voltages and the weights as a single variation—for example, in the logical voltages. Making this substitution and also dividing by V_1 and defining

$$\delta = \frac{V_0}{V_1}$$

gives

$$\begin{aligned} f_+(p^1)(1 - \mathscr{S}_1) + \delta f_+(\bar{p}^1)(1 - \mathscr{S}_0) - [f_-(p^1)(1 + \mathscr{S}_1) \\ + \delta f_-(\bar{p}^1)(1 + \mathscr{S}_0)] \\ \geq t'(1 + \mathscr{T}_a) \end{aligned}$$

This is the desired worst-case expression.

Letting p^0 be an arbitrary point for which the output should be 0 and performing a similar computation yields the worst-case expression

$$f_+(p^0)(1 + \mathscr{S}_1) + \delta f_+(\bar{p}^0)(1 + \mathscr{S}_0) - [f_-(p^0)(1 - \mathscr{S}_1) + \delta f_-(\bar{p}^0)(1 - \mathscr{S}_0)] \le t'(1 - \mathscr{T}_a)$$

The rest is algebra. Eliminating t' from these equations and letting

$$K = \frac{1 + \mathscr{T}_a}{1 - \mathscr{T}_a}$$

and

$$S(p) = f_+(p) + f_-(p)$$

yields

$$f(p^1) - Kf(p^0) + \delta(f(\bar{p}^1) - Kf(\bar{p}^0)) \ge \mathscr{S}_1(S(p^1) + KS(p^0)) + \mathscr{S}_0\delta(S(\bar{p}^1) + KS(\bar{p}^0))$$

which is the worst-case restriction on $\mathscr{S}_1$ and $\mathscr{S}_0$. It can be simplified somewhat by letting

$$C = \frac{\mathscr{S}_0}{\mathscr{S}_1}$$

yielding

$$\mathscr{S}_1 \le \frac{f(p^1) - Kf(p^0) + \delta[f(\bar{p}^1) - Kf(\bar{p}^0)]}{S(p^1) + KS(p^0) + C\delta[S(\bar{p}^1) + KS(\bar{p}^0)]}$$

Since this must be true for all points p^1 and p^0, the final restriction can be stated in the following theorem.

Theorem A.1. General Sensitivity. For any $\mathscr{T}_a$ which is

$$\mathscr{T}_a < \frac{u - \ell}{u + \ell}$$

the maximum sensitivity $\mathscr{S}_1$ is

$$\mathscr{S}_1 = \min_{p^1, p^0} \left\{ \frac{f(p^1) - Kf(p^0) + \delta[f(\bar{p}^1) - Kf(\bar{p}^0)]}{S(p^1) + KS(p^0) + C\delta[S(\bar{p}^1) + KS(\bar{p}^0)]} \right\}$$

Note that this minimum might be caused by any point on the map and, therefore, the computation of $\mathscr{S}_1$ may be quite difficult.

For certain special cases the result simplifies. For example, for zero-one (but still nondefinite) models in which

$$V_0 = \delta = 0$$

the result is the corollary.

Corollary A.1.1. General Zero-One Sensitivity. For any $\mathscr{T}_a$ which is

$$\mathscr{T}_a < \frac{u - \ell}{u + \ell}$$

the maximum sensitivity $\mathscr{S}_1$ for a zero-one model is

$$\mathscr{S}_1 = \min_{p^1, p^0} \left\{ \frac{f(p^1) - Kf(p^0)}{S(p^1) + KS(p^0)} \right\}$$

Thus in the general zero-one model also, the sensitivity can be determined only by examining the entire map. For this reason, we do not know in general how to control sensitivity for nondefinite maps.

Once a realization has been obtained, however, it is frequently possible to improve the sensitivity by complementing some of the inputs, thereby changing the sign of their weights.

Consider a separating function

$$\left\langle f_a = \sum_{i=1}^{n} a_i x_i \right\rangle_{u_a : \ell_a}$$

and let $x_1, x_2, \ldots, x_q$ be the variables to be complemented. Then by Theorem 2.4B the equivalent separating function is

$$\left\langle - \sum_{i=1}^{q} a_i \bar{x}_i + \sum_{i=q+1}^{n} a_i x_i \right\rangle_{u_b : \ell_b}$$

where

$$u_b = u_a - \sum_{i=1}^{q} a_i$$

$$\ell_b = \ell_a - \sum_{i=1}^{q} a_i$$

The entire map has been displaced by the sum of a_1 to a_q.

The gates corresponding to different displacements will, in general, have different sensitivities. In fact, two equal displacements obtained by complementing different sets of variables may result in different sensitivities (Example A.1a). It would be of interest to determine which displacement has the best sensitivity. The "best" separating function is not always one of those with all positive or all negative weights (Example A.1b). However, there is no known way to determine this "best" separating function except by trial and error.

One simple result of this type follows from the definite map analysis of Chapter 1:

The positive definite map of a function corresponds to better sensitivity than the negative definite map if and only if the gap of the positive definite map, $u_p:\ell_p$, is less than half way up the map, that is, if and only if

$$u_p + \ell_p < \sum_{i=1}^{n} a_i$$

Example A.1.

a. Equal displacements give different sensitivities.

(1) Consider

$$F = x_1(x_2 + x_3)$$

with a realization

$$\langle 2x_1 + x_2 + x_3 \rangle_{3:2}$$

and

$$\mathscr{T} = 0.1$$

Assuming a general zero-one model and using Corollary A.1.1 gives

$$\mathscr{S}_1 = \frac{3 - 2\left(\frac{1.1}{0.9}\right)}{3 + 2\left(\frac{1.1}{0.9}\right)} = 0.102$$

(2) Complementing x_1 displaces the gap downward by 2 and, using $p^1 = \{1,1,0\}$ and $p^0 = \{1,0,1\}$, gives

$$\mathscr{S}_1 = \frac{1 - 0}{1 + \left(\frac{1.1}{0.9}\right)4} = 0.17$$

(3) Complementing x_2 and x_3 again displaces the gap downward by 2 and, using $p^1 = \{1,1,0\}$ and $p^0 = \{1,0,0\}$, gives

$$\mathscr{S}_1 = \frac{1 - 0}{3 + \left(\frac{1.1}{0.9}\right)4} = 0.127$$

(4) Complementing all variables displaces the gap downward by 4 and yields

$$\mathscr{S}_1 = \frac{(-1) - \left(\frac{0.9}{1.1}\right)(-2)}{1 + \left(\frac{0.9}{1.1}\right)2} = 0.242$$

Note that for this case the threshold is negative and so $\mathscr{T}_a$ changes sign.

In all four cases the sensitivity was implied by the gap boundaries.

b. Mixed map gives best sensitivity.

(1) Consider

$$F = x_1(x_2 \dotplus x_3)$$

with realization

$$\langle 4x_1 + 2x_2 + x_3\rangle_{5:4}$$

Assuming a general zero-one model with $\mathcal{T} = 0$,

$$\mathcal{S}_1 = \frac{5-4}{5+4} = 0.11$$

(2) Complementing all variables displaces the gap downward by 7 and yields

$$\mathcal{S}_1 = \frac{-2-(-3)}{2+(+3)} = 0.20$$

As predicted, the sensitivity of the negative map is better than that of the positive map.

(3) Complementing only x_1 and x_3 displaces the map downward by 5. The sensitivity is now determined not by both of the gap boundaries, but by the point $p^1 = \{1,0,1\}$ and $p^0 = \{0,1,1\}$, which map into 0 and -2 respectively. The sensitivity is then

$$\mathcal{S}_1 = \frac{0-(-2)}{0+6} = 0.33$$

This is better than either definite map and is the best that can be obtained by complementing the variables of the given separating function.

The results of Theorem A.1 can also be specialized to cover nonzero-one models with definite maps. For this case (assuming all positive weights),

$$f_- = 0$$

and the expression for $\mathcal{S}_1$ becomes

$$\mathcal{S}_1 = \min_{p^1,p^0}\left\{\frac{f(p^1) - Kf(p^0) + \delta[f(\bar{p}^1) - Kf(\bar{p}^0)]}{f(p^1) + Kf(p^0) + C\delta[f(\bar{p}^1) + Kf(\bar{p}^0)]}\right\}$$

The sensitivity is now implied by the points p^1 and p^0 which cause the gap. Defining the normalized gap boundaries as

$$u = f(p^1) + \delta f(\bar{p}^1)$$
$$\ell = f(p^0) + \delta f(\bar{p}^0)$$

the fan-in as

$$\begin{aligned}\sigma &= \sum_{i=1}^{n} a_i \\ &= f(p^1) + f(\bar{p}^1) \\ &= f(p^0) + f(\bar{p}^0)\end{aligned}$$

and solving these expressions gives

$$f(p^1) = \frac{u - \delta\sigma}{1 - \delta}$$

$$f(\bar{p}^1) = \frac{\sigma - u}{1 - \delta}$$

$$f(p^0) = \frac{\ell - \delta\sigma}{1 - \delta}$$

$$f(\bar{p}^0) = \frac{\sigma - \ell}{1 - \delta}$$

Substituting these into the expression for $\mathcal{S}_1$, and also substituting the definitions of K and C, yields the corollary that follows.

Corollary A.1.2. General Definite Maps. The sensitivity parameters of any definite map are related by

$$\frac{[\mathcal{T}_a(1 - \delta) + \mathcal{S}_1 - \delta\mathcal{S}_0]\left|\dfrac{u + \ell}{u - \ell}\right| - \delta(\mathcal{S}_1 - \mathcal{S}_0)\left|\dfrac{2\sigma}{u - \ell}\right|}{1 - \delta + \mathcal{T}_a(\mathcal{S}_1 - \delta\mathcal{S}_0)} \leq 1$$

Observe that for the definite zero-one model where δ equals 0, this expression is equivalent to Theorem 1.5 and for the definite minus-plus-one model where

$$\begin{aligned}\delta &= -1 \\ \mathcal{S}_0 &= -\mathcal{S}_1\end{aligned}$$

it is equivalent to Theorem 1.6. In both cases it is necessary to set

$$\begin{aligned}\mathcal{A} &= \mathcal{S}_1 \\ \mathcal{X} &= 0\end{aligned}$$

in Theorems 1.5 and 1.6 to obtain an exact equivalence.

Appendix B

UPPER BOUNDS ON THE NUMBER OF ℓ.s. FUNCTIONS

There are a number of applications, such as those in Chapter 10, for which it is convenient to have bounds on the number of ℓ.s. functions of n variables. In this section we derive a number of upper bounds and give some applications. Lower bounds have also been found (Muroga, 5, 9; Goto, 3) but, although theoretically interesting, they have not been found particularly useful.

We assume a switching function of n variables, defined at m points where

$$m \leq 2^n$$

We define B_n^m to be the number of ℓ.s. functions of n (or fewer) variables defined at m points. The case of m equals 2^n corresponds to a completely specified function.

One very simple bound follows from linear programming. There are m linear inequalities which define an ℓ.s. function. Assume that

$$m > n$$

Then for any optimization criterion, every ℓ.s. function corresponds to at least $n + 1$ inequalities satisfied with the equals sign (Theorem 9.3). Thus an upper bound to the number of ℓ.s. functions is related to the number of ways that $n + 1$ inequalities can be selected out of m inequalities, that is, to

$$\binom{m}{n+1}$$

Since each of the selected inequalities can be assigned as either a 0 or a 1 point, there are 2^{n+1} assignments. The upper bound is then

$$B_n^m \leq 2^{n+1}\binom{m}{n+1}$$

$$= 2^{n+1}\frac{m!}{(n+1)!\,(m-n-1)!}$$

which can be weakened somewhat

$$B_n^m \leq 2^{n+1}\frac{m(m-1)(m-2)(m-n)}{(n+1)!}$$

$$< 2^{n+1}\frac{m^n}{(n+1)!}$$

For

$$m = 2^n$$

this becomes

$$B_n^{2^n} \equiv B_n < \frac{2^{n^2+n+1}}{(n+1)!}$$

Although easy to derive, this bound is not particularly tight. We now give a tighter bound due to Cameron, 1, and Winder, 1.

An ℓ.s. function is defined by the m simultaneous inequalities,

$$\sum_{i=1}^{n} a_i x_i^p - t \geq 0$$

for p such that

$$F(p) = 1$$

$$\sum_{i=1}^{n} a_i x_i^p - t < 0$$

for p such that

$$F(p) = 0$$

Define an $n + 1$ dimensional space in which the coordinate axes correspond to the weights and the threshold. (*Note*: This is not the switching space. There the axes correspond to the x's.)

Consider a particular input combination p. The locus of all values of a_i and t for which

$$\sum_{i=1}^{n} a_i x_i^p - t = 0$$

plot as a hyperplane in this space. The plane goes through the origin since the equation is satisfied by

$$a_i = t = 0$$

All values of a_i and t on one side of this plane, if substituted into the left side of the plane equation, would make it greater than 0, and thus correspond to weight and threshold values which allow

$$F(p) = 1$$

Similarly, those values of a_i and t on the other side of the plane would make the left side of the plane equation less than 0 and correspond to values which allow

$$F(p) = 0$$

Each input combination p implies a similar hyperplane, there being m hyperplanes in all. These hyperplanes partition the $n + 1$ space into a number of regions. All points in any one of these regions correspond to values of a_i and t which imply the same switching function since any two points in the region are on the same side of every hyperplane. Two points in different regions correspond to different switching functions, since they must be on opposite sides of at least one hyperplane.

Thus the number of ℓ.s. functions is exactly equal to the number of regions into which the m planes

$$\sum_{i=1}^{n} a_i x_i^p - t = 0$$

for $p = p_1, p_2, \ldots, p_m$ partition the $n + 1$ space. This number, however, is difficult to compute. We get an upper bound by determining the maximum number of regions into which any m planes passing through the origin can partition an $n + 1$ space.

Since our proof is by induction, let C_j^i be the maximum number of regions into which i hyperplanes passing through the origin can partition a space of j dimensions.

Clearly

$$C_1^i = 2$$

for all i, since a line can be divided into 2 regions by a point at the origin, but additional points at the origin do not increase the number of regions.

Clearly also

$$C_j^1 = 2$$

for all j, since one plane divides any space into exactly two parts.

Now we show that C_j^i satisfies the recursion relation,

$$C_j^i = C_{j-1}^{i-1} + C_j^{i-1}$$

Assume that $i - 1$ planes have already generated C_j^{i-1} regions in j space. We now add the ith plane in such a way as to generate the maximum number of regions. The ith hyperplane can be made to intersect the previous $i - 1$ hyperplanes in, at most, $i - 1$ hyperlines. These $i - 1$ lines can be made to divide the ith plane into, at most,

$$C_{j-1}^{i-1}$$

hyperplane regions since this is the same problem in $j - 1$ dimensions. Since each region into which the ith plane has been partitioned serves as a boundary between two regions in j space, we have added C_{j-1}^{i-1} regions to the C_j^{i-1} regions already implied by the other planes. This proves the recursion relation.

The recursion relation, plus the boundary conditions with i and j equal to 1, defines the function C_j^i. We now show that C_j^i can be expressed as

$$C_j^i = 2\sum_{k=0}^{j-1} \binom{i-1}{k}$$

by demonstrating that the expression satisfies the boundary conditions and the recursion relationship.

First the boundary conditions. For $j = 1$,

$$2\sum_{k=0}^{1-1} \binom{i-1}{k} = 2\binom{i-1}{0} = 2$$

For $i = 1$,

$$2\sum_{k=0}^{j-1} \binom{0}{k} = 2\binom{0}{0} = 2$$

To verify the recursion relationship, we use the well-known and easily verified identity

$$\binom{i-1}{k} = \binom{i-2}{k-1} + \binom{i-2}{k}$$

Thus

$$2\sum_{k=0}^{j-1} \binom{i-1}{k} = 2\sum_{k=0}^{j-1} \binom{i-2}{k-1} + 2\sum_{k=0}^{j-1} \binom{i-2}{k}$$

Consider the first term on the right

$$2\sum_{k=0}^{j-1} \binom{i-2}{k-1}$$

Define $k' = k - 1$. Then

$$2\sum_{k=0}^{j-1}\binom{i-2}{k-1} = 2\sum_{k'=-1}^{j-2}\binom{i-2}{k'}$$

Removing the term for $k' = -1$,

$$= 2\left[\sum_{k'=0}^{j-2}\binom{i-2}{k'} + \binom{i-2}{-1}\right]$$

However, this last term equals 0.

Putting the first term on the right side back into the original equation, and calling the dummy variable k again, yields the desired expression

$$2\sum_{k=0}^{j-1}\binom{i-1}{k} = 2\sum_{k=0}^{j-2}\binom{i-2}{k} + 2\sum_{k=0}^{j-1}\binom{i-2}{k}$$

which is in exactly the form of the recursion relationship. Thus we have shown that

$$C_j^i = 2\sum_{k=0}^{j-1}\binom{i-1}{k}$$

and hence we have the following theorem.

Theorem B.1. (*Cameron, 1; Winder, 1*). The number of ℓ.s. functions of n variables defined on m points is bounded by

$$B_n{}^m \le C_{n+1}^m = 2\sum_{k=0}^{n}\binom{m-1}{k}$$

For completely specified functions this becomes

$$B_n \le 2\sum_{k=0}^{n}\binom{2^n-1}{k}$$

We now give an approximation (Winder, 6) to those bounds valid for

$$m \ge 3n + 2$$
$$n \ge 2$$

We first expand the bound as

$$2\sum_{k=0}^{n}\binom{m-1}{k} = 2\binom{m-1}{n}$$
$$\left[1 + \frac{n}{m-n} + \frac{n(n-1)}{(m-n)(m-n+1)} + \cdots + \frac{n!}{(m-n)\ldots(m-1)}\right]$$

which can be bounded as

$$2\sum_{k=0}^{n}\binom{m-1}{k} < 2\binom{m-1}{n}\left[1+\frac{n}{m-n}+\frac{n^2}{(m-n)^2}+\cdots+\frac{n^n}{(m-n)^n}\right]$$

$$< 2\binom{m-1}{n}\frac{1}{1-\dfrac{n}{m-n}} = 2\binom{m-1}{n}\frac{m-n}{m-2n}$$

Now we expand the binomial coefficient

$$2\binom{m-1}{n}\left[\frac{m-n}{m-2n}\right]$$

$$= \frac{2}{n!}[(m-1)(m-2)\ldots(m-(n-1))(m-n)]\frac{(m-n)}{(m-2n)}$$

Looking only at the rightmost three product terms and the denominator, and using the restriction on m and n,

$$(m-(n-1))\frac{(m-n)(m-n)}{m-2n} = [m-(n-1)]\left[m+\frac{n^2}{m-2n}\right]$$

$$< [m-(n-1)]\left[m+\frac{n^2}{n+2}\right]$$

$$= m^2 - \left[\frac{n^2+n-2-n^2}{n+2}\right]m - \frac{n^2(n-1)}{n+2}$$

$$< m^2$$

Substituting this into the entire expression for the bound, we obtain

$$2\binom{m-1}{n}\left[\frac{m-n}{m-2n}\right] < \frac{2}{n!}[(m-1)(m-2)\ldots(m-(n-2))]m^2$$

which can then be further bounded as

$$< \frac{2}{n!}m^n$$

Thus we have the corollary.

Corollary B.1.1 *(Winder, 6).* For $m \geq 3n+2$ and $n \geq 2$, B_n^m is bounded by

$$B_n^m < 2\frac{m^n}{n!}$$

For completely specified functions the bound becomes

$$B_n < 2\frac{2^{n^2}}{n!}$$

$$= \frac{2^{n^2+1}}{n!}$$

These bounds are considerably tighter than the simple bound derived earlier.

Table B.1 (Cameron, 1) shows the actual values of B_n and the bounds of Theorem B.1 for n from 1 to 5. These can be compared with the total number of functions of n variables, which is 2^{2^n}.

Table B.1. The Number of ℓ.s. Functions Compared with the Bounds of Theorem B.1 and the Total Number of Switching Functions (Cameron, 1)

n	2^{2^n}	B_n	$2\sum_{k=0}^{n}\binom{2^n-1}{k}$	$2\frac{2^{n^2}}{n!}$
1	4	4	4	4
2	16	14	14	16
3	256	104	128	170
4	65,536	1,882	3,882	5,461
5	4.3×10^9	94,572	412,736	559,240

These bounds on single-gate realizability can be used to obtain related bounds for multigate networks.

Corollary B.1.2 (*Winder*, *6*). Consider a two-level network consisting of k T gates, each having n independent variable inputs, and whose outputs connect to an output gate, which then has k inputs. By varying the weights and thresholds of this network, it can realize no more than

$$2^{k+1}\frac{m^{(n+1)k}}{(n!)^k k!}$$

functions with m specified points.

Proof. Each of the k gates can realize at most

$$\frac{2m^n}{n!}$$

functions, and the final gate can realize at most

$$\frac{2m^k}{k!}$$

functions. Therefore, the network can realize no more than

$$\left(\frac{2m^n}{n!}\right)^k \frac{2m^k}{k!}$$

or

$$\frac{2^{k+1}m^{(n+1)k}}{(n!)^k k!}$$

functions.|

Corollary B.1.3 (*Winder*, *6*). Consider a general network of k threshold gates with no feedback. By varying the weights and threshold, this network can realize at most

$$\frac{2^k m^{\left[nk + \frac{k(k-1)}{2}\right]}}{(n!)^k}$$

functions with m specified points.

Proof. Since there is no feedback, the gates can be ordered as first gate, second gate, . . ., kth gate, so that the jth gate only receives inputs from the n independent variables and the $j - 1$ gates labeled 1,2,. . ., $j - 1$. The jth gate can then realize at most

$$\frac{2m^{(n+j-1)}}{(n+j-1)!}$$

functions. The total network can realize at most

$$2^k \frac{m^n}{n!} \frac{m^{n+1}}{(n+1)!} \cdots \frac{m^{n+k-1}}{(n+k-1)!}$$

which is certainly less than

$$\frac{2^k m^n m^{n+1} \ldots m^{n+k-1}}{(n!)^k}$$

This can be simplified to

$$\frac{2^k m^{[n+(n+1)+\cdots+(n+k-1)]}}{(n!)^k}$$

$$= \frac{2^k m^{\left[nk + \frac{k(k-1)}{2}\right]}}{(n!)^k}$$

thus completing the proof.|

Appendix C

SOME EXAMPLES OF LOGICAL DESIGN—THE DONUT COMPUTER

Since this is a book on switching theory rather than logical design, most of the examples in the text have been illustrative of the pertinent material rather than practical logical circuits. In this section, we present some examples of useful logical design using T gates.

C.1. ADDERS

T gates are particularly adapted to adding and other arithmetic circuits. Figure C.1 shows a circuit for the ith stage of a binary adder (Schauer et al.). The inputs x_i and y_i are the ith digits of the numbers to be added and k_{i-1} and $\bar{k}_{i-1}$ are the carries from the preceding stage. The output S_i is the ith digit of the sum and k_i and $\bar{k}_i$ are the carries from the ith stage. The input-output relations of the adder are

$$S_i = k_{i-1}(x_i y_i \dotplus \bar{x}_i \bar{y}_i) \dotplus \bar{k}_{i-1}(x_i \bar{y}_i \dotplus \bar{x}_i y_i) \tag{C.1}$$

$$k_i = x_i y_i \dotplus (x_i \dotplus y_i) k_{i-1} \tag{C.2}$$

The first of these can be rewritten

$$S_i = \bar{k}_i(x_i \dotplus y_i \dotplus k_{i-1}) \dotplus x_i y_i k_{i-1} \tag{C.3}$$

from which follows the realization of Fig. C.1.

This adder requires three T gates per stage with a minimum relative gap of

$$\frac{3-2}{3+2} = 0.20$$

If each gate has a time delay of d seconds, each stage of the adder computes its carry in d seconds and its sum in $2d$ seconds, from which it follows that an N stage adder requires $(N + 1)d$ seconds.

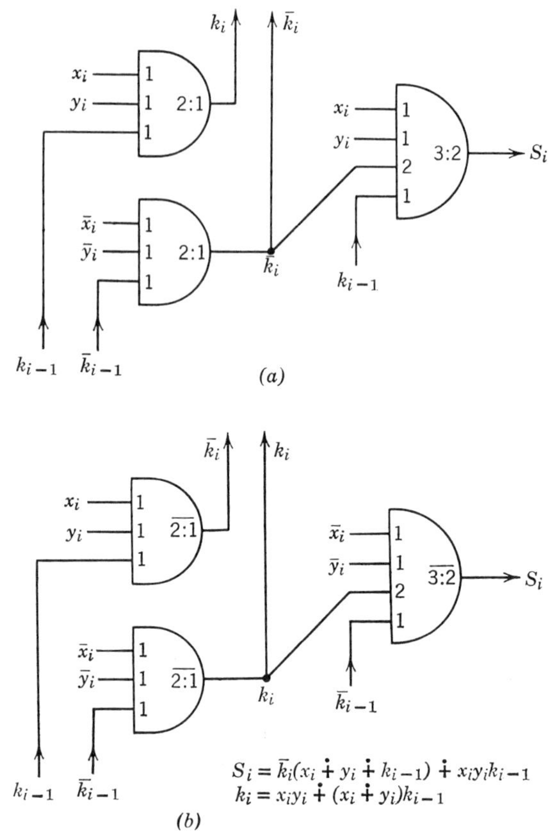

Fig. C.1. Simple adding circuits (Schauer et al.). (*a*) With noncomplementing T gates (*b*) With complementing T gates.

Figure C.2 shows an adder that computes carries two levels at a time and hence operates at twice the speed of those in Fig. C.1. To derive its equation, we substitute the equation for k_{i-1} into the equation for k_i (Eq. C.2) resulting in

$$k_i = x_i y_i \dot{+} (x_i \dot{+} y_i)[x_{i-1} y_{i-1} \dot{+} (x_{i-1} \dot{+} y_{i-1}) k_{i-2}] \tag{C.4}$$

This function is realizable with a single threshold gate, having a relative gap of 0.14. Using this gate to compute the carry and the same gate as in Fig. C.1 to compute the sum, we obtain the adder of Fig. C.2, which requires three gates per stage. Figure C.3 is a similar circuit using both complementing and noncomplementing gates and requires only two gates per stage. All three circuits have a minimum relative gap of 0.14. For each stage, two levels of carries are computed in d seconds and the sum in

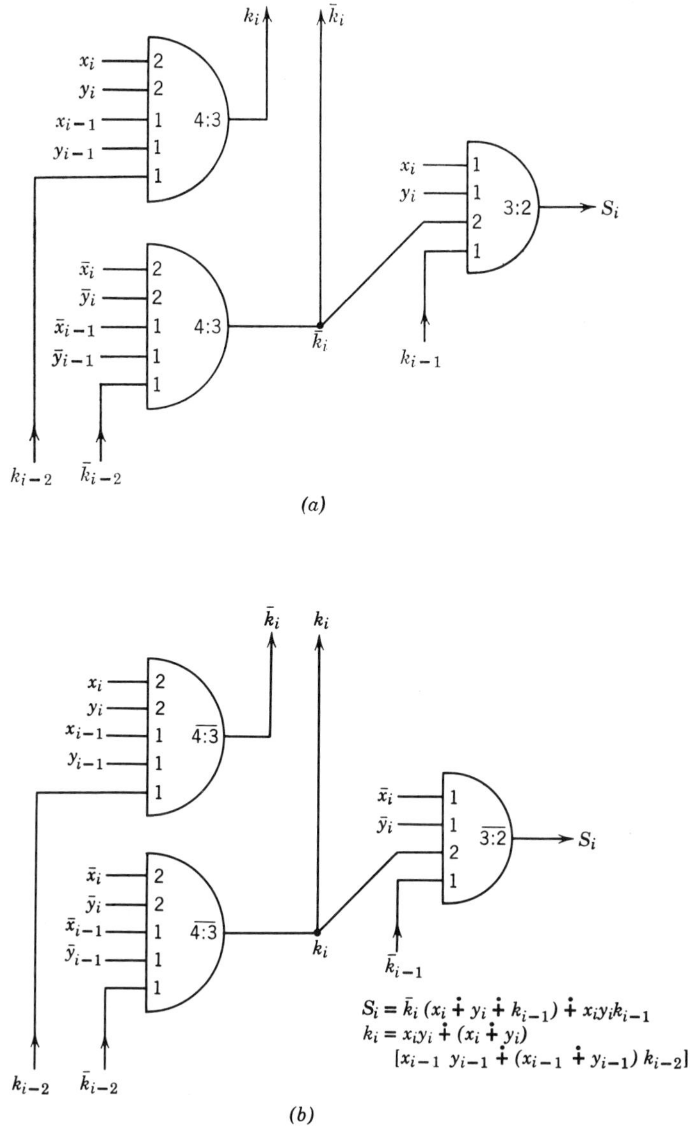

Fig. C.2. An adder with l-bit look-ahead carry. (*a*) With noncomplementing T gates. (*b*) With complementing T gates.

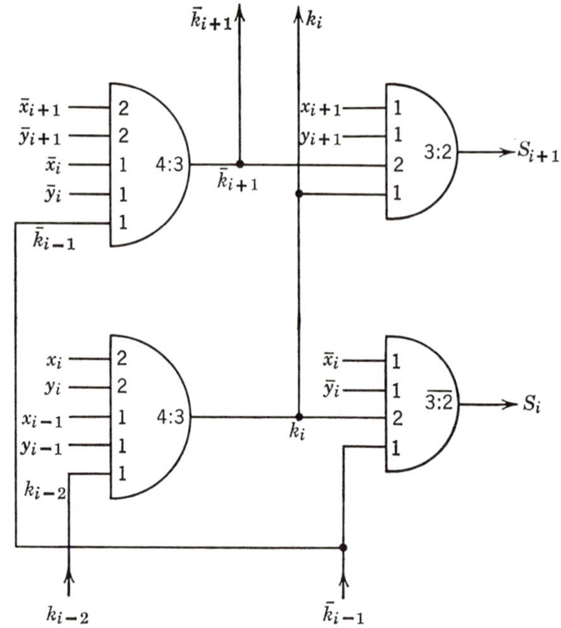

Fig. C.3. Another adder with 1-bit look-ahead carry.

$2d$ seconds, from which it follows that an N stage adder requires $(\lfloor N/2 \rfloor + 1)$ d seconds where $\lfloor N/2 \rfloor$ means the smallest integer larger than or equal to $N/2$.

The procedure whereby the circuit of Fig. C.2 was derived from that of Fig. C.1 can be generalized. Circuits can be designed which require only three threshold gates per stage and which compute carries m stages at a time (where m is any integer) and hence compute at m times the speed of the circuit of Fig. C.1.

Iterating the procedure by which Eq. C.4 was derived from Eq. C.2 results in the expression of k_i in terms of k_{i-m}

$$\begin{aligned} k_i = x_i y_i &\dotplus (x_i \dotplus y_i)\{x_{i-1}y_{i-1} \dotplus (x_{i-1} \dotplus y_{i-1})[x_{i-2}y_{i-2} \dotplus \cdots \\ &\dotplus (x_{i-(m-2)} \dotplus y_{i-(m-2)})[x_{i-(m-1)}y_{i-(m-1)} \\ &\dotplus (x_{i-(m-1)} \dotplus y_{i-(m-1)})k_{i-m}]\ldots\} \end{aligned}$$

This function is realizable by a single T gate:

$$k_i = \left\langle \sum_{k=0}^{m-1} 2^{i-k-1}(x_{i-k} + y_{i-k}) + k_{i-m} \right\rangle_{2^m:(2^m-1)}$$

Using such gates, one can design an N-digit adding network, Fig. C.4,

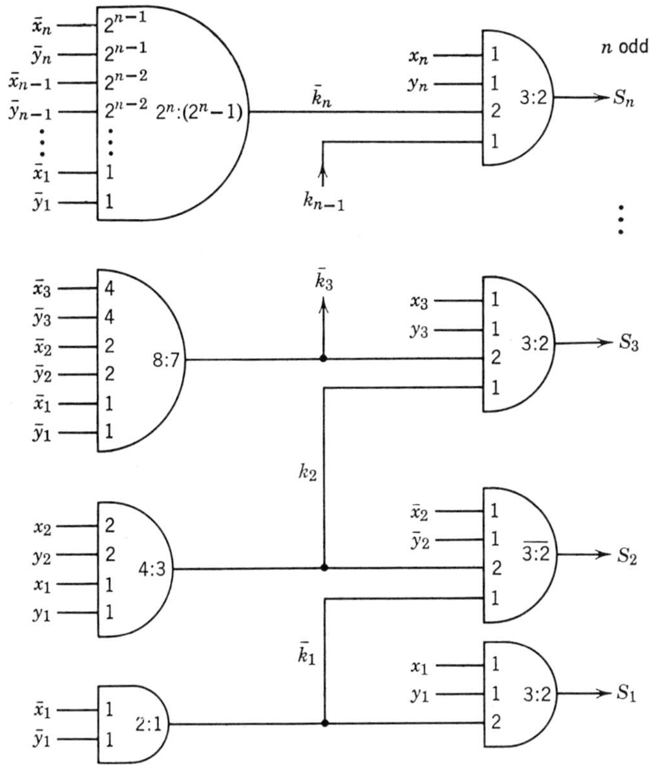

Fig. C.4. An m-bit look-ahead adder.

which requires $2N$ threshold gates and which performs the sum in $2d$ seconds. The relative gap of the mth carry gate (k_m) is

$$\frac{u - \ell}{u + \ell} = \frac{1}{2^{m+1} - 2}$$

which becomes excessively tight for even moderate values of m.

However, if one is willing to use more than one gate to realize this carry function, then circuits exist which have practical tolerances and which still compute m times as fast as the circuit of Fig. C.1 where $m > 2$.

Such a circuit for $m = 4$ is shown in Figs. C.5 and C.6 using noncomplementing and complementing gates, respectively, and requiring five gates per stage. Figure C.7 is a similar circuit using both complementing and noncomplementing gates and requiring four gates per stage. For each stage four levels of carry are computed in d seconds after receiving the carry four levels back. The sum is computed in d additional seconds. There

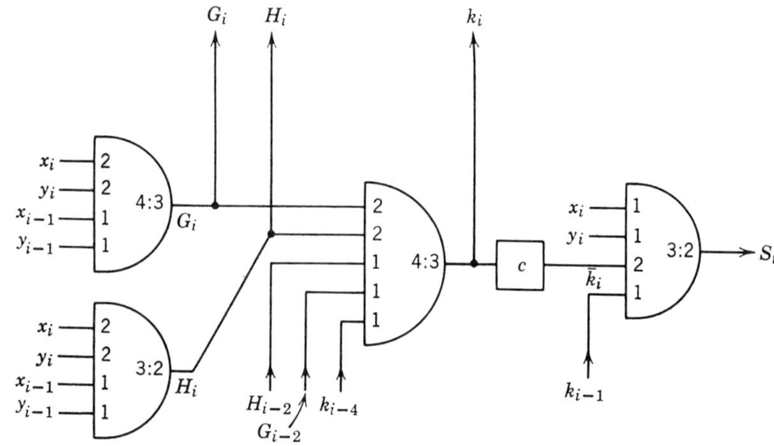

$$S_i = \bar{k}_i(x_i \dotplus y_i \dotplus k_{i-1}) \dotplus x_i y_i k_{i-1}$$
$$k_i = G_iH_i \dotplus (G_i \dotplus H_i)\,[G_{i-2}H_{i-2} \dotplus (G_{i-2} \dotplus H_{i-2})\,k_{i-4}]$$
$$G_i = x_iy_i \dotplus (x_i \dotplus y_i)\,x_{i-1}y_{i-1}$$
$$H_i = x_iy_i \dotplus (x_i \dotplus y_i)\,(x_{i-1} \dotplus y_{i-1})$$

Fig. C.5. A 4-bit look-ahead adder.

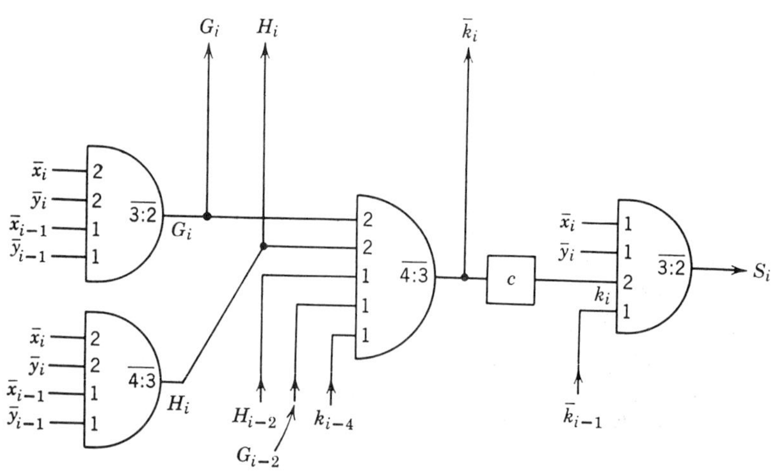

Fig. C.6. The adder of Fig. C.5, but with complementing T gates.

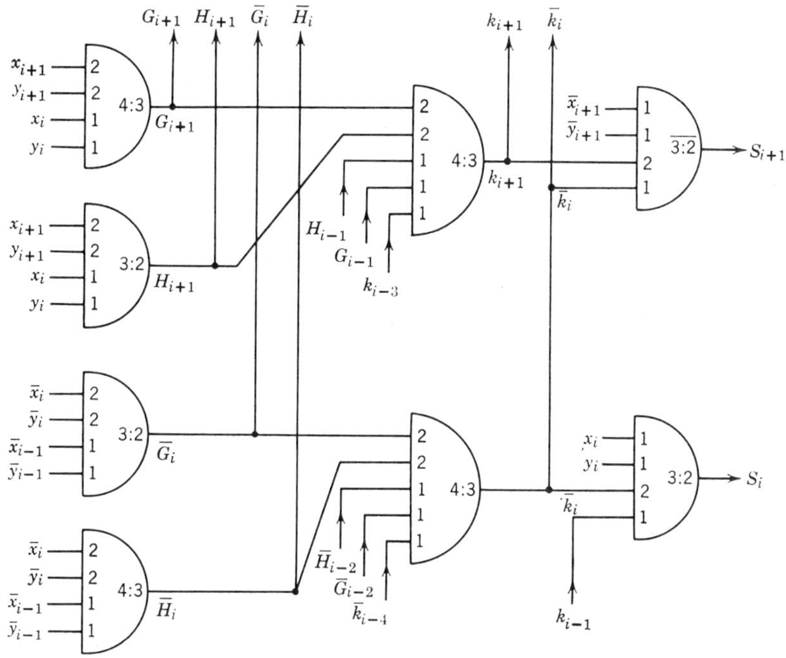

Fig. C.7. Another 4-bit look-ahead adder.

is an initial condition that the first four stages compute their carries in $2d$ seconds. Hence N stages of such an adder require $(\lfloor N/4 \rfloor + 2)d$ seconds.

Figure C.8 is a combination of several previously given circuits which achieves a speed of $(\lfloor N/4 \rfloor + 1)d$ seconds for N stages and which requires three gates per stage, both complementing and noncomplementing with a minimum relative gap of 0.11.

The adder of Fig. C.9 computes carries two stages at a time and alternate sums two stages at a time, thus attaining a speed of $\lfloor (N/2 \rfloor + 1)d$ seconds for N stages and requiring only $\frac{5}{2}$ gates per stage both complementing and noncomplementing with a minimum relative gap of 0.09.

A number of somewhat different adders have been proposed by Fischler and Poe. Figure C.10 shows an octal adder or three stages of a binary adder. Here the carry for the third stage is computed at once (two-bit look-ahead as in Fig. C.4), and its output is used as input to each of the three sum gates; no other carries need be computed for these stages. Note that each sum gate has inputs from all of the digits to be added.

The same design philosophy is used in Fig. C.11, a BCD adder; Fig. C.12 an excess 3 adder; Fig. C.13, a 5–4–2–1 coded decimal adder; and in Fig. C.14 a BCD subtractor. These circuits all require rather tight

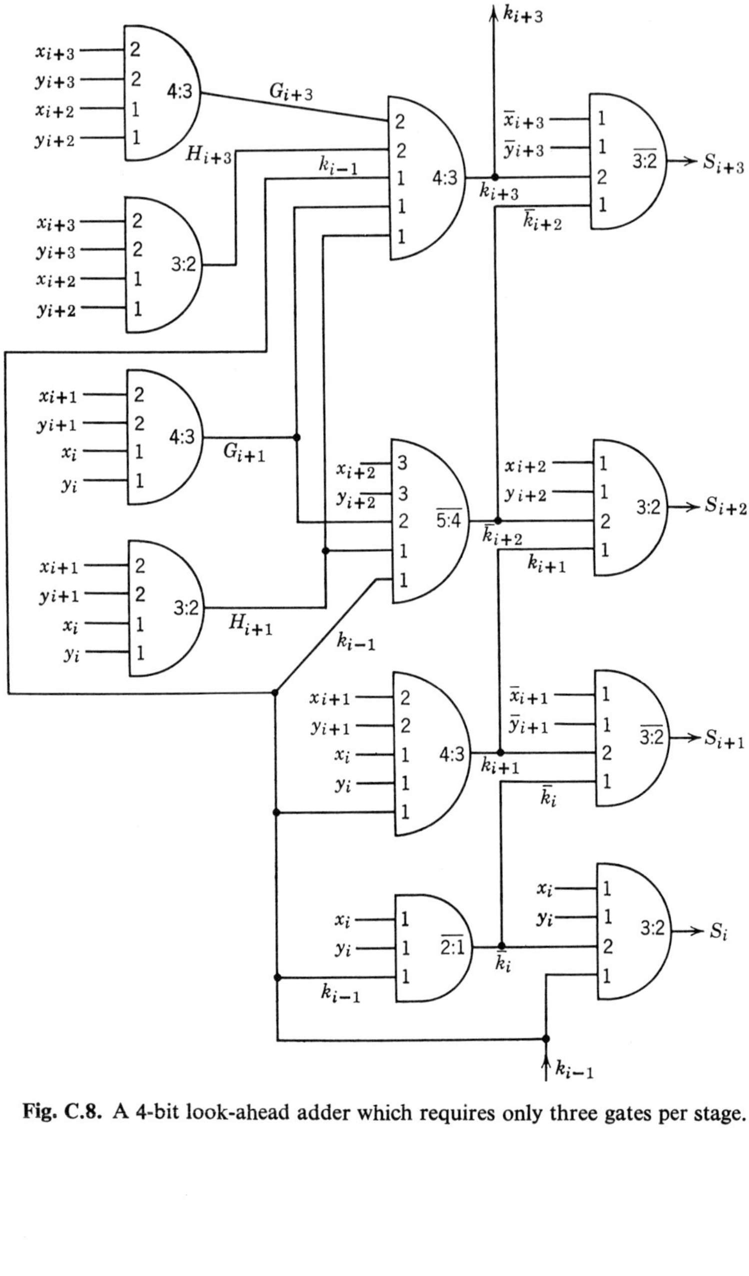

Fig. C.8. A 4-bit look-ahead adder which requires only three gates per stage.

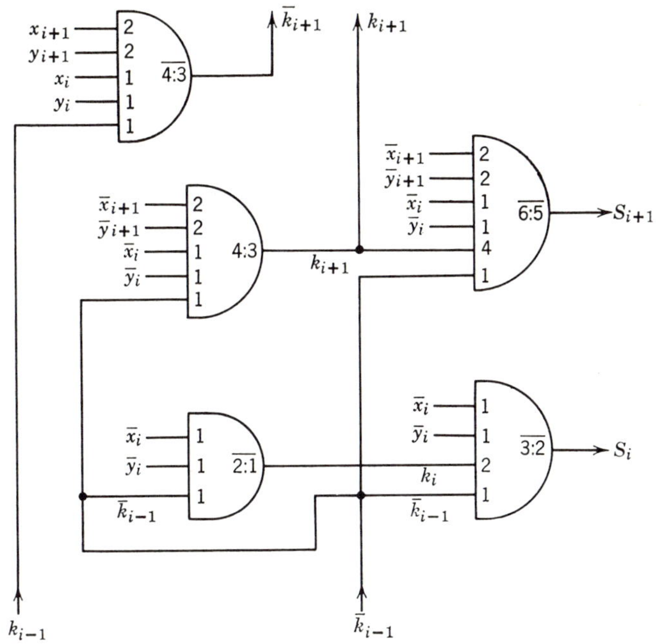

Fig. C.9. An adder with a different S_{i+1} gate.

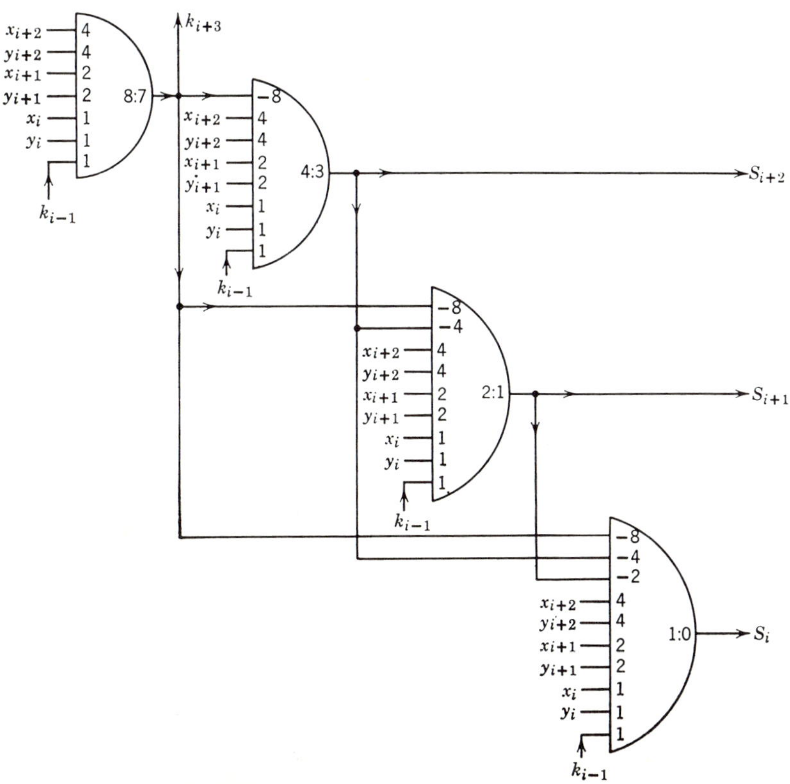

Fig. C.10. An octal adder (Fischler and Poe).

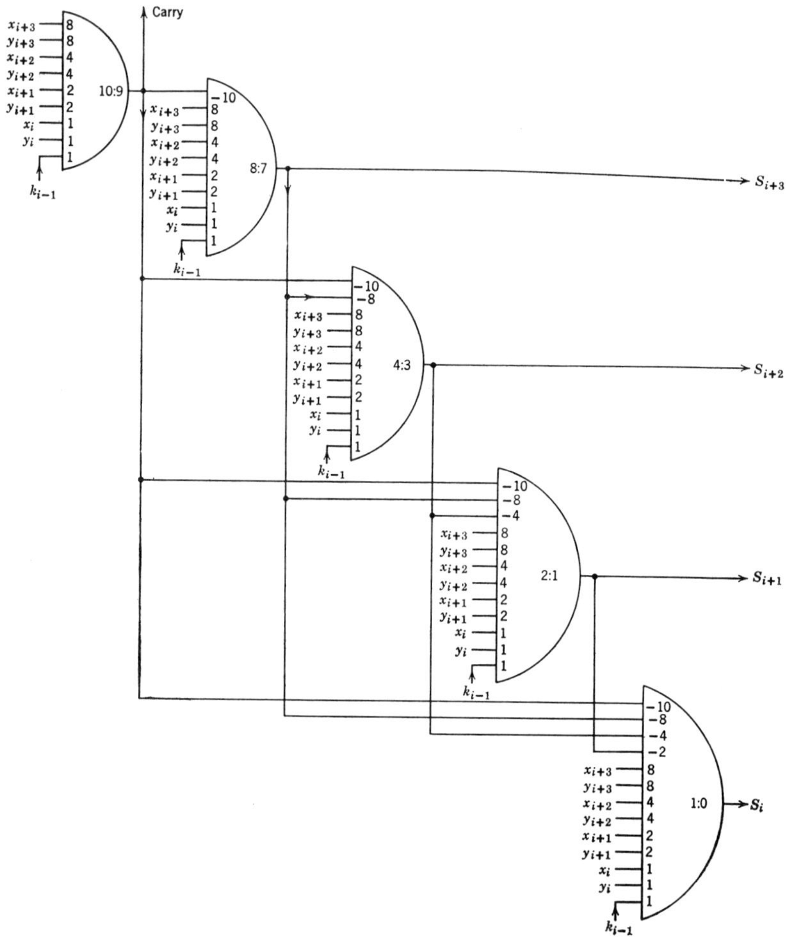

Fig. C.11. A binary coded decimal adder (Fischler and Poe).

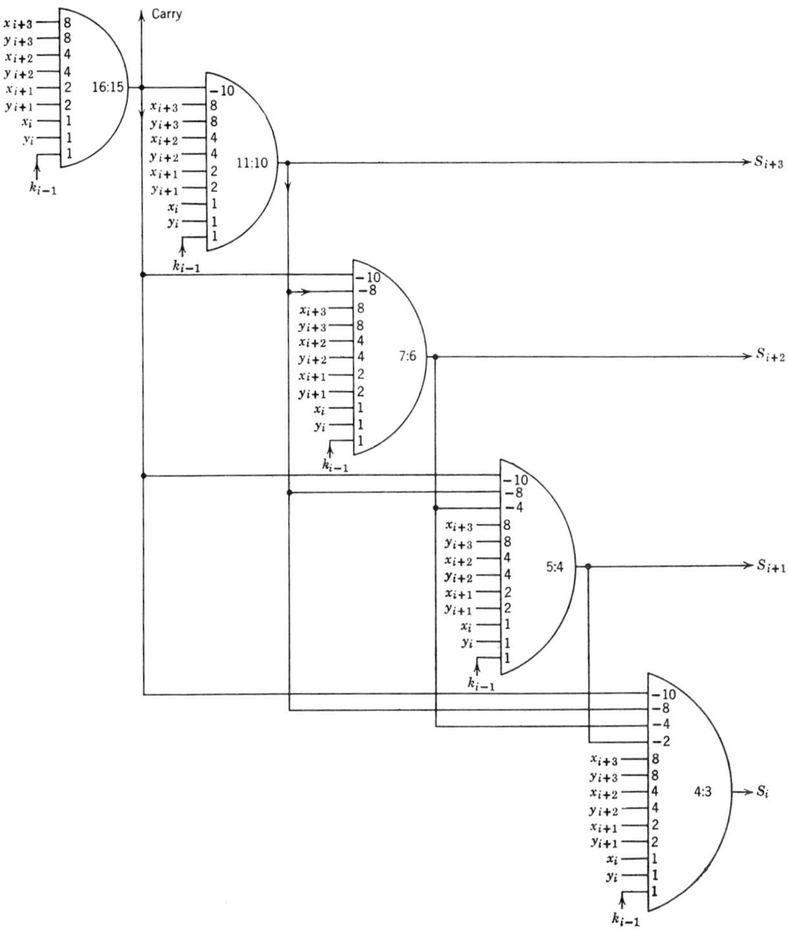

Fig. C.12. An excess-three adder (Fischler and Poe).

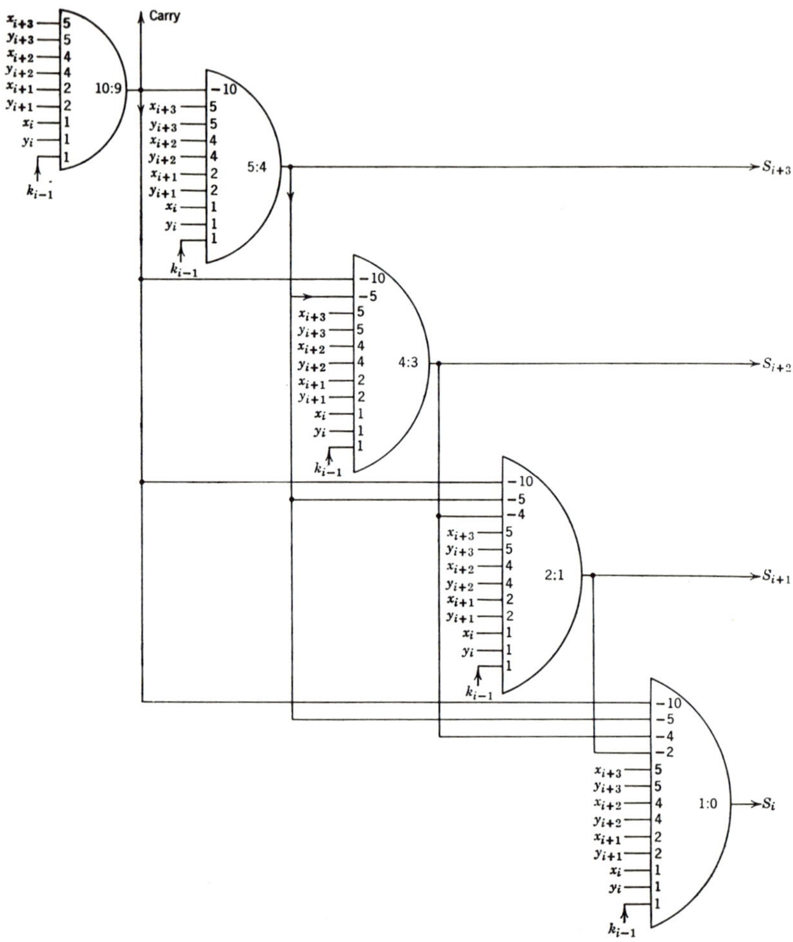

Fig. C.13. A 5-4-2-1 coded decimal adder (Fischler and Poe).

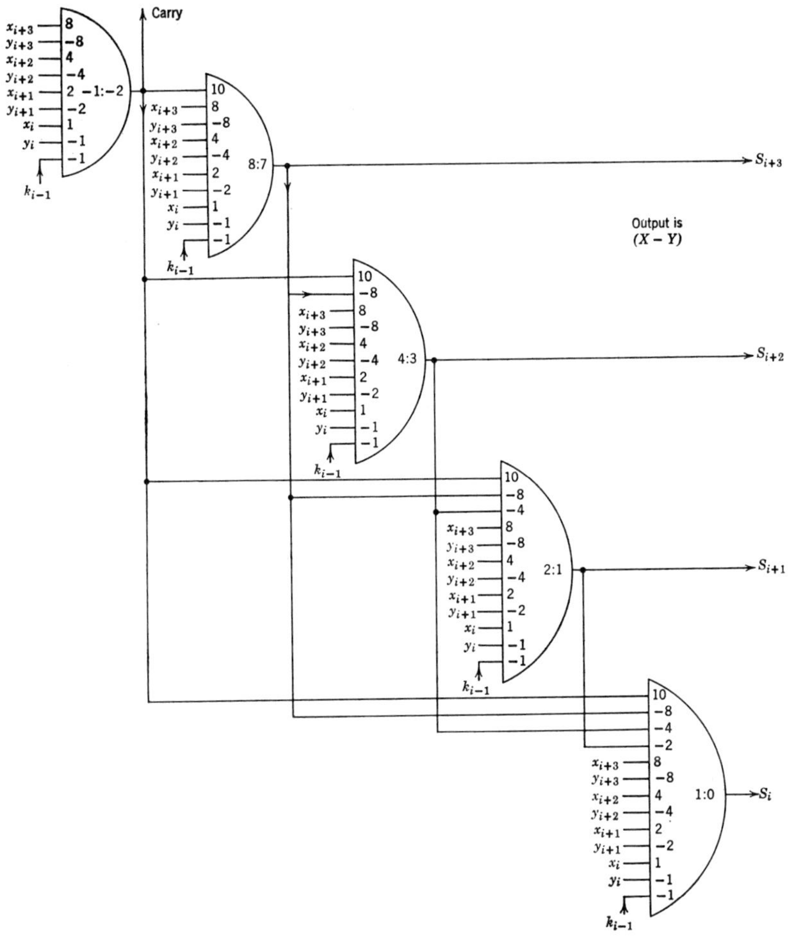

Fig. C.14. A binary coded decimal subtractor (Fischler and Poe).

tolerances. The S_i gate of Fig. C.10 has a relative gap of 0.035 when all coefficients are made positive. Corresponding gates in Fig. C.11 through C.14 have considerably smaller relative gaps.

C.2. FLIP-FLOP STORAGE

Another type of application for which T gates are particularly efficient is in flip-flop storage. Their principal advantage is that they allow threshold logic to be carried out on the inputs to the flip-flop gates in addition to their usual function of storage.

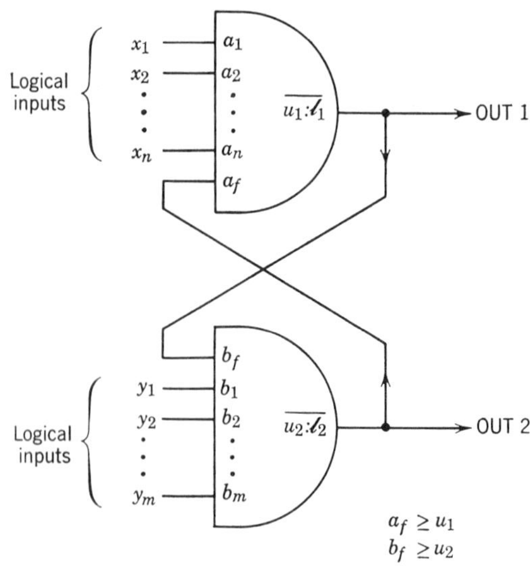

Fig. C.15. A general T-gate flip-flop.

The basic idea of T-gate flip-flops is to connect any two complementing T gates with the same or different gaps in a feedback arrangement as shown in Fig. C.15. Note that the feedback weights ($a_f \geq u_1$ and $b_f \geq u_2$) are such that if the logical inputs of both gates do not exceed the threshold, then if either gate has an output of 1, it will cause the other gate to have an output of 0. Thus the device is bistable and can perform the function of storage.

The flip-flop will change state if the logical inputs on the gate whose output is 1 are such as to exceed its threshold, while the logical inputs on the other gate do not exceed its threshold. Under these conditions, the output of the first gate goes to 0, thus allowing the output of the other gate to go to 1 which in turn will keep the first gate at 0.

Various embodiments of this idea are shown in Figs. C.16, C.17, and C.18.

In Fig. C.16, the inputs IN and $\overline{\text{IN}}$ are data inputs and w (write) is the control input. To store the binary digit x in this register, IN is set equal to x, $\overline{\text{IN}}$ is set equal to $\bar{x}$, and w is set equal to 1. The contents of the flip-flop are not changed if w is equal to 0 or if w is equal to 1 and IN and $\overline{\text{IN}}$ are equal to 0.

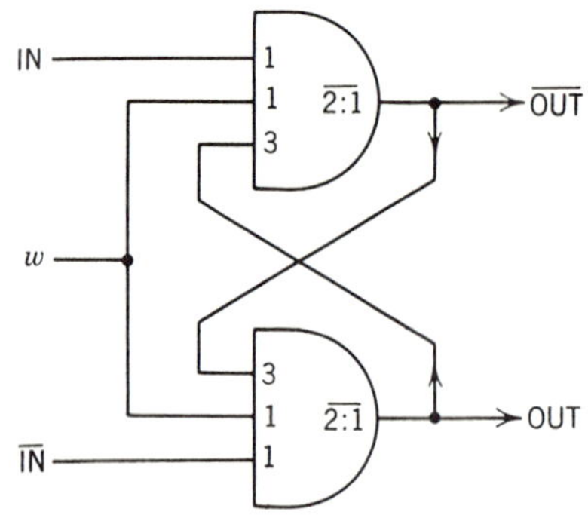

Fig. C.16. A simple T-gate flip-flop.

In Fig. C.17, w is again the control input and IN_1, IN_2, $\overline{\text{IN}}_1$, and $\overline{\text{IN}}_2$ are the data inputs. As long as w equals 0, the data inputs can have any values without setting the flip-flop. Data can be stored in this register either from lines IN_1 and $\overline{\text{IN}}_1$ or from IN_2 and $\overline{\text{IN}}_2$. For example, if IN_1 and $\overline{\text{IN}}_1$ are both equal to 0 and $\text{IN}_2 = x$, $\overline{\text{IN}}_2 = \bar{x}$, and $w = 1$, the binary digit x will be set into the register.

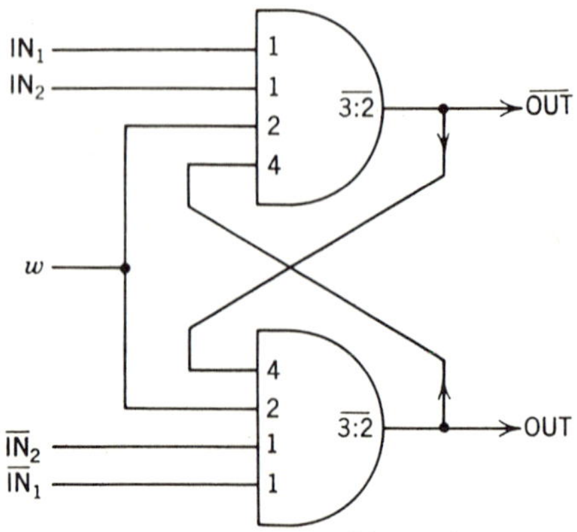

Fig. C.17. A flip-flop with two inputs.

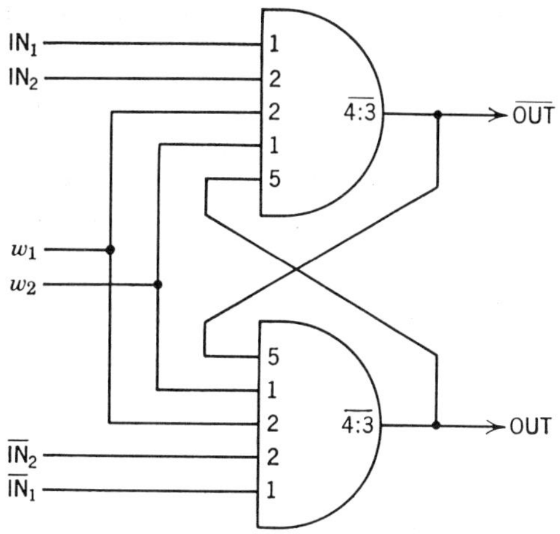

Fig. C.18. A flip-flop with separate control of two inputs.

Figure C.18 shows somewhat more complex threshold logic performed on the input to the flip-flop. Here IN_1, IN_2, $\overline{IN}_1$, and $\overline{IN}_2$ are data inputs and w_1 and w_2 are control inputs. If w_1 and w_2 are both 0, then the data inputs can have any values without setting the flip-flop. Data can be stored from either IN_1 and $\overline{IN}_1$ or from IN_2 and $\overline{IN}_2$. To store data from IN_1 and $\overline{IN}_1$, set $w_1 = 1$, $w_2 = 1$, $IN_2 = 0$, $\overline{IN}_2 = 0$, $IN_1 = x$, and $\overline{IN}_1 = \bar{x}$. To store data from IN_2 and $\overline{IN}_2$, set $w_1 = 1$, $w_2 = 0$, $IN_2 = x$, and $\overline{IN}_2 = \bar{x}$; IN_1 and $\overline{IN}_1$ can have any arbitrary values.

These are just three examples; others can be easily constructed using the same basic idea.

C.3. AN ACCUMULATOR†

One of the logical operations that can be performed on the inputs to a flip-flop gate is that of addition, and thus rather efficient T gate accumulators can be designed. The accumulator to be described performs the operations of CLEAR, ADD (with one-bit "look ahead" carry), COMPLEMENT, SHIFT LEFT, AND, and OR. It consists of only four threshold gates per bit including storage as shown in Fig. C.19.

For each bit the control inputs are labeled D, E, F, G, H, I, J, K, and L. For the ith bit the input data leads are labeled x_i, $\bar{x}_i$, x_{i-1}, $\bar{x}_{i-1}$, k_i, $\bar{k}_i$ k_{i-1}, $\bar{k}_{i-1}$, k_{i-2}, and $\bar{k}_{i-2}$; the output leads are S_i and $\bar{S}_i$.

† See also Muroga et al., 7.

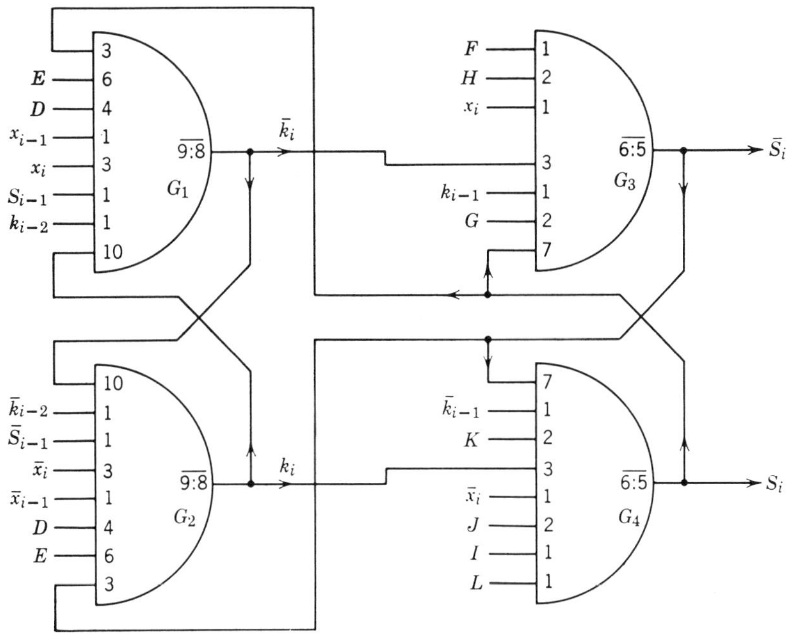

Fig. C.19. An accumulator circuit.

Feedback between G_1 and G_2 forms what is called the "carry flip-flop." The "sum flip-flop" is produced by feedback between G_3 and G_4. The principal storage of the accumulator occurs in the "sum flip-flop" while the storage in the "carry flip-flop" is only temporary. All gates perform threshold logic to control the various operations. A description of how this is accomplished follows. Note that all control inputs are assumed to be zero and all data inputs are assumed arbitrary unless otherwise indicated.

(1) CLEAR. Set $I = J = K = L = 1$.

(2) ADD. This requires two time periods. During the first, the carry is computed and stored in the "carry flip-flop." To accomplish this, set $D = 1$. During the second time period, the sum is computed and stored in the "sum flip-flop." To do this set $H = K = 1$.

(3) COMPLEMENT. This requires two time periods. During the first, S_i and $\bar{S}_i$ are transferred to the "carry flip-flop" to become k_i and $\bar{k}_i$, respectively. To accomplish this, set $E = 1$ and $x_i = \bar{x}_i = 0$ for all i. During the second time period, the outputs of the "carry flip-flop" are transferred to the "sum flip-flop" by setting $H = F = I = K = 1$ and $x_i = \bar{x}_i = 0$ for all i.

(4) SHIFT LEFT. This requires two time periods for a one-bit shift. The first is the same as that for COMPLEMENT; that is set $E = 1$ and $x_i = \bar{x}_i = 0$ for all i. During the second time period, k_i and $\bar{k}_i$ are transferred to the $i + 1$ "sum flip-flop" to become S_{i+1} and $\bar{S}_{i+1}$, respectively. This is accomplished by setting $H = K = 1$ and $x_i = \bar{x}_i = 0$ for all i because the state of the i "sum flip-flop" is changed only if its initial output S_i and $\bar{S}_i$, which now equals k_i and $\bar{k}_i$ (because of the first time period), differs from the initial output of the $i - 1$ "sum flip-flop," S_{i-1} and $\bar{S}_{i-1}$, which now equals k_{i-1} and $\bar{k}_{i-1}$.

(5) AND. This requires one time period and is performed by setting $D = E = I = J = K = 1$. This sets $k_i = \bar{k}_i = 0$ for all i so that the state of each "sum flip-flop" is set according to the dictates of the AND function as computed by G_4.

(6) OR. This requires a single time period and is very similar to AND. It occurs by setting $D = E = H = F = G = 1$.

C.4. DONUT COMPUTER† (*Coates and Lewis, 4*)

To study some of the practical implications of threshold logic, the logical design of a small general purpose computer (the DONUT Computer) was carried out for five different tolerance constraints, and the tightest tolerance version was constructed.

The computer contained a fairly representative spectrum of logical circuits: a fourteen-bit accumulator, about seventy bits of flip-flop registers, and the control logic to perform about fifteen instructions.

The entire logic design of the computer (excluding memory but including registers and accumulator) was carried out using complementing gates with the five limiting values of relative gap and fan-in shown in the Table C.1. In each case the same switching functions were realized but with the different sensitivity constraints. The comparative number of gates (again excluding only memory) is shown in the table.

Table C.1. Gate Requirements for the DONUT Computer

Gap	5:4	3:2		1:0 – NOR Gate	
Fan-In	15	15	5	15	4
Number of Gates	614	882	1052	1702	2127

† Another example of a threshold gate computer is the Musasino, 1, which used parametron majority gates (Muroga et al., 7).

In going from the loosest tolerance NOR gate realization to the tightest tolerance threshold gate realization, there is a savings of a factor of slightly less than 3.5.

The tight tolerance version, which was actually constructed, contained the circuits of Fig. C.2, C.16, C.17, and C.18. In fact, even tighter tolerances were maintained in part of the computer since it was desired to utilize the accumulator circuit of Fig. C.19.

Additional information about detailed circuits and tolerances is included in the reference.

Appendix D

TABLE OF THRESHOLD FUNCTIONS OF SIX OR FEWER VARIABLES

The characterizing numbers derived in Section 10.8 allow a particularly simple tabulation of ℓ.s. functions and their realization. Table D.2 is such a table for functions of six or fewer variables prepared by Dertouzos, 1, 2, 3. A similar table for functions of seven or fewer variables was later prepared by Winder, 7. Other tables have been published by Muroga, 6; Winder, 1; and Liu, 1.

The procedure for using the table will first be explained and illustrated with an example. Then a proof will be given.

Procedure For Using Table D.2

Given a function $F(x_1,x_2,\ldots,x_n)$ where $n \leq 6$:

(1) Find its $n + 1$ characterizing numbers $b_n,b_{n-1},\ldots,b_1,b_0$, where b_0 = (number of points for which $F = 1$) − (number of points for which $F = 0$). For $i \neq 0$, $b_i = 2\{$(number of points for which $F = 1$ and $x_i = 1$) − (number of points for which $F = 1$ and $x_i = 0$)$\}$.

(2) The b_i may be positive or negative. Arrange $|b_i|$ in numerical order according to their value.

(3) Look up the sequence of $|b_i|$ in the table under the $|b_i|$ columns. If it does not appear, F is not ℓ.s. If it does appear, F is ℓ.s. and a minimum sum realization (realization with unit gap length in which

$$\left(\sum_{i=1}^{n} a_i\right) - u$$

is a minimum) is given in the $|a_i|$ columns according to the following rules.

(a) The $|a_i|$ columns are assumed ordered in the same way as the $|b_i|$ columns. For example, if $|b_j|$ was the largest value of $|b_i|$ and consequently the leftmost in the $|b_i|$ sequence, then $|a_j|$ is the leftmost column in the $|a_i|$ columns.

(b) For all a_i, $i \neq 0$, the entry in the $|a_i|$ column is the magnitude of a_i. The sign of a_i is the same as the sign of b_i.

(c) For $i = 0$, let a_0 be the coefficient that would be obtained using Step b above. Then the gap, $u{:}\ell$ (for the zero-one model with independent variables $x_1, x_2, \ldots, x_n$) is given by

$$u{:}\ell = \tfrac{1}{2}(\sigma - a_0 + 1){:}\tfrac{1}{2}(\sigma - a_0 - 1)$$

where

$$\sigma = \sum_{i=1}^{n} a_i$$

Example D.1. Consider the three functions

(1) $F_1 = x_1 \dotplus x_2(x_3 \dotplus x_4)$

(2) $F_2 = x_1 \dotplus \bar{x}_2(x_3 \dotplus x_4)$

(3) $F_3 = x_1(\bar{x}_2 \dotplus x_3 x_4)$

where

$$F_3 = F_2{}^d$$

The table of combinations for these functions is shown in Table D.1

Table D.1 The Functions of Example D.1

x_1	x_2	x_3	x_4	F_1	F_2	F_3
0	0	0	0	0	0	0
0	0	0	1	0	1	0
0	0	1	0	0	1	0
0	0	1	1	0	1	0
0	1	0	0	0	0	0
0	1	0	1	1	0	0
0	1	1	0	1	0	0
0	1	1	1	1	0	0
1	0	0	0	1	1	1
1	0	0	1	1	1	1
1	0	1	0	1	1	1
1	0	1	1	1	1	1
1	1	0	0	1	1	0
1	1	0	1	1	1	0
1	1	1	0	1	1	0
1	1	1	1	1	1	1

The characterizing numbers are:

For F_1

$$\begin{aligned} b_0 &= 11 - 5 = 6 \\ b_1 &= 2(8 - 3) = 10 \\ b_2 &= 2(7 - 4) = 6 \\ b_3 &= 2(6 - 5) = 2 \\ b_4 &= 2(6 - 5) = 2 \end{aligned}$$

For F_2

$$\begin{aligned} b_0 &= 11 - 5 = 6 \\ b_1 &= 2(8 - 3) = 10 \\ b_2 &= 2(4 - 7) = -6 \\ b_3 &= 2(6 - 5) = 2 \\ b_4 &= 2(6 - 5) = 2 \end{aligned}$$

For F_3

$$\begin{aligned} b_0 &= 5 - 11 = -6 \\ b_1 &= 2(5 - 0) = 10 \\ b_2 &= 2(1 - 4) = -6 \\ b_3 &= 2(3 - 2) = 2 \\ b_4 &= 2(3 - 2) = 2 \end{aligned}$$

As expected, in going from F_1 to F_2, complementing x_2 changes the sign of b_2 and, in going from F_2 to F_3, taking the dual of the function changes the sign of b_0.

All three functions have the same $|b_i|$ sequence

$$\begin{matrix} |b_1|, & |b_0|, & |b_2|, & |b_3|, & |b_4| \\ 10, & 6, & 6, & 2, & 2 \end{matrix}$$

This sequence is in Table D.2 for $n \leq 4$ as Number 4. Thus the functions are ℓ.s. The entries in the $|a_i|$ columns are

$$\begin{matrix} |a_1|, & |a_0|, & |a_2|, & |a_3|, & |a_4| \\ 3, & 2, & 2, & 1, & 1 \end{matrix}$$

Table D.2. Threshold Functions of Six or Fewer Variables (Dertouzos)

Number	$\lvert b_i \rvert$						$\lvert a_i \rvert$					
$n \leq 3$												
1	8	0	0	0			1	0	0	0		
2	6	2	2	2			2	1	1	1		
3	4	4	4	0			1	1	1	0		
$n \leq 4$												
1	16	0	0	0	0		1	0	0	0	0	
2	14	2	2	2	2		3	1	1	1	1	
3	12	4	4	4	0		2	1	1	1	0	
4	10	6	6	2	2		3	2	2	1	1	
5	8	8	8	0	0		1	1	1	0	0	
6	8	8	4	4	4		2	2	1	1	1	
7	6	6	6	6	6		1	1	1	1	1	
$n \leq 5$												
1	32	0	0	0	0	0	1	0	0	0	0	0
2	30	2	2	2	2	2	4	1	1	1	1	1
3	28	4	4	4	4	0	3	1	1	1	1	0
4	26	6	6	6	2	2	5	2	2	2	1	1
5	24	8	8	4	4	4	4	2	2	1	1	1
6	24	8	8	8	0	0	2	1	1	1	0	0
7	22	10	10	6	2	2	5	3	3	2	1	1
8	22	10	6	6	6	6	3	2	1	1	1	1
9	20	12	12	4	4	0	3	2	2	1	1	0
10	20	12	8	8	4	4	4	3	2	2	1	1
11	20	8	8	8	8	8	2	1	1	1	1	1
12	18	14	14	2	2	2	4	3	3	1	1	1
13	18	14	10	6	6	2	5	4	3	2	2	1
14	18	10	10	10	6	6	3	2	2	2	1	1
15	16	16	16	0	0	0	1	1	1	0	0	0
16	16	16	12	4	4	4	3	3	2	1	1	1
17	16	16	8	8	8	0	2	2	1	1	1	0
18	16	12	12	8	8	4	4	3	3	2	2	1
19	14	14	14	6	6	6	2	2	2	1	1	1
20	14	14	10	10	10	2	3	3	2	2	2	1
21	12	12	12	12	12	0	1	1	1	1	1	0

(*continued*)

Table D.2. (*continued*)

Number	$\lvert b_i \rvert$							$\lvert a_i \rvert$						
$n \le 6$														
1	64	0	0	0	0	0	0	1	0	0	0	0	0	0
2	62	2	2	2	2	2	2	5	1	1	1	1	1	1
3	60	4	4	4	4	4	0	4	1	1	1	1	1	0
4	58	6	6	6	6	2	2	7	2	2	2	2	1	1
5	56	8	8	8	8	0	0	3	1	1	1	1	0	0
6	56	8	8	8	4	4	4	6	2	2	2	1	1	1
7	54	10	10	10	6	2	2	8	3	3	3	2	1	1
8	54	10	10	6	6	6	6	5	2	2	1	1	1	1
9	52	12	12	12	4	4	0	5	2	2	2	1	1	0
10	52	12	12	8	8	4	4	7	3	3	2	2	1	1
11	52	12	8	8	8	8	8	4	2	1	1	1	1	1
12	50	14	14	14	2	2	2	7	3	3	3	1	1	1
13	50	14	14	10	6	6	2	9	4	4	3	2	2	1
14	50	14	10	10	10	6	6	6	3	2	2	2	1	1
15	50	10	10	10	10	10	10	3	1	1	1	1	1	1
16	48	16	16	16	0	0	0	2	1	1	1	0	0	0
17	48	16	16	12	4	4	4	6	3	3	2	1	1	1
18	48	16	16	8	8	8	0	4	2	2	1	1	1	0
19	48	16	12	12	8	8	4	8	4	3	3	2	2	1
20	48	12	12	12	12	8	8	5	2	2	2	2	1	1
21	46	18	18	14	2	2	2	7	4	4	3	1	1	1
22	46	18	18	10	6	6	2	9	5	5	3	2	2	1
23	46	18	14	14	6	6	6	5	3	2	2	1	1	1
24	46	18	14	10	10	10	2	7	4	3	2	2	2	1
25	46	14	14	14	10	10	6	7	3	3	3	2	2	1
26	44	20	20	12	4	4	0	5	3	3	2	1	1	0
27	44	20	20	8	8	4	4	7	4	4	2	2	1	1
28	44	20	16	16	4	4	4	6	4	3	3	1	1	1
29	44	20	16	12	8	8	4	8	5	4	3	2	2	1
30	44	20	12	12	12	12	0	3	2	1	1	1	1	0
31	44	16	16	16	8	8	8	4	2	2	2	1	1	1
32	44	16	16	12	12	12	4	6	3	3	2	2	2	1
33	42	22	22	10	6	2	2	8	5	5	3	2	1	1
34	42	22	22	6	6	6	6	5	3	3	1	1	1	1
35	42	22	18	14	6	6	2	9	6	5	4	2	2	1
36	42	22	18	10	10	6	6	6	4	3	2	2	1	1
37	42	22	14	14	10	10	2	7	5	3	3	2	2	1
38	42	18	18	18	6	6	6	5	3	3	3	1	1	1
39	42	18	18	14	10	10	6	7	4	4	3	2	2	1
40	42	18	14	14	14	14	2	5	3	2	2	2	2	1
41	40	24	24	8	8	0	0	3	2	2	1	1	0	0

Number	$\|b_i\|$							$\|a_i\|$						
$n \leq 6$														
42	40	24	24	8	4	4	4	6	4	4	2	1	1	1
43	40	24	20	12	8	4	4	7	5	4	3	2	1	1
44	40	24	20	8	8	8	8	4	3	2	1	1	1	1
45	40	24	16	16	8	8	0	4	3	2	2	1	1	0
46	40	24	16	12	12	8	4	8	6	4	3	3	2	1
47	40	20	20	16	8	8	4	8	5	5	4	2	2	1
48	40	20	20	12	12	8	8	5	3	3	2	2	1	1
49	40	20	16	16	12	12	4	6	4	3	3	2	2	1
50	40	16	16	16	16	16	0	2	1	1	1	1	1	0
51	38	26	26	6	6	2	2	7	5	5	2	2	1	1
52	38	26	22	10	10	2	2	8	6	5	3	3	1	1
53	38	26	22	10	6	6	6	5	4	3	2	1	1	1
54	38	26	18	14	10	6	2	9	7	5	4	3	2	1
55	38	26	18	10	10	10	6	6	5	3	2	2	2	1
56	38	26	14	14	14	6	6	5	4	2	2	2	1	1
57	38	22	22	14	10	6	6	6	4	4	3	2	1	1
58	38	22	22	10	10	10	10	3	2	2	1	1	1	1
59	38	22	18	18	10	10	2	7	5	4	4	2	2	1
60	38	22	18	14	14	10	6	7	5	4	3	3	2	1
61	38	18	18	18	14	14	2	5	3	3	3	2	2	1
62	36	28	28	4	4	4	0	4	3	3	1	1	1	0
63	36	28	24	8	8	4	4	6	5	4	2	2	1	1
64	36	28	20	12	12	4	0	5	4	3	2	2	1	0
65	36	28	20	12	8	8	4	7	6	4	3	2	2	1
66	36	28	16	16	12	4	4	6	5	3	3	2	1	1
67	36	28	16	12	12	8	8	8	7	4	3	3	2	2
68	36	24	24	12	12	4	4	7	5	5	3	3	1	1
69	36	24	24	12	8	8	8	4	3	3	2	1	1	1
70	36	24	20	16	12	8	4	8	6	5	4	3	2	1
71	36	24	20	12	12	12	8	5	4	3	2	2	2	1
72	36	24	16	16	16	8	8	4	3	2	2	2	1	1
73	36	20	20	20	12	12	0	3	2	2	2	1	1	0
74	36	20	20	16	16	12	4	6	4	4	3	3	2	1
75	34	30	30	2	2	2	2	5	4	4	1	1	1	1
76	34	30	26	6	6	6	2	7	6	5	2	2	2	1
77	34	30	22	10	10	6	2	8	7	5	3	3	2	1
78	34	30	18	14	14	2	2	7	6	4	3	3	1	1
79	34	30	18	14	10	6	6	9	8	5	4	3	2	2
80	34	30	14	14	10	10	10	7	6	3	3	2	2	2
81	34	26	26	10	10	6	6	5	4	4	2	2	1	1
82	34	26	22	14	14	6	2	9	7	6	4	4	2	1

(*continued*)

Table D.2. (*continued*)

Number	$\lvert b_i \rvert$							$\lvert a_i \rvert$						
$n \leq 6$														
83	34	26	22	14	10	10	6	6	5	4	3	2	2	1
84	34	26	18	18	14	6	6	5	4	3	3	2	1	1
85	34	26	18	14	14	10	10	6	5	4	3	3	2	2
86	34	22	22	18	14	10	2	7	5	5	4	3	2	1
87	34	22	22	14	14	14	6	4	3	3	2	2	2	1
88	34	22	18	18	18	10	6	5	4	3	3	3	2	1
89	32	32	32	0	0	0	0	1	1	1	0	0	0	0
90	32	32	28	4	4	4	4	4	4	3	1	1	1	1
91	32	32	24	8	8	8	0	3	3	2	1	1	1	0
92	32	32	20	12	12	4	4	5	5	3	2	2	1	1
93	32	32	16	16	16	0	0	2	2	1	1	1	0	0
94	32	32	16	16	8	8	8	4	4	2	2	1	1	1
95	32	32	12	12	12	12	12	3	3	1	1	1	1	1
96	32	28	28	8	8	8	4	6	5	5	2	2	2	1
97	32	28	24	12	12	8	4	7	6	5	3	3	2	1
98	32	28	20	16	16	4	4	6	5	4	3	3	1	1
99	32	28	20	16	12	8	8	7	6	5	4	3	2	2
100	32	28	16	16	12	12	12	5	4	3	3	2	2	2
101	32	24	24	16	16	8	0	4	3	3	2	2	1	0
102	32	24	24	16	12	12	4	5	4	4	3	2	2	1
103	32	24	20	20	16	8	4	6	5	4	4	3	2	1
104	32	24	20	16	16	12	8	7	6	5	4	4	3	2
105	32	20	20	20	20	8	8	3	2	2	2	2	1	1
106	30	30	30	6	6	6	6	3	3	3	1	1	1	1
107	30	30	26	10	10	10	2	5	5	4	2	2	2	1
108	30	30	22	14	14	6	6	4	4	3	2	2	1	1
109	30	30	18	18	18	2	2	5	5	3	3	3	1	1
110	30	30	18	18	10	10	10	3	3	2	2	1	1	1
111	30	30	14	14	14	14	14	2	2	1	1	1	1	1
112	30	26	26	14	14	10	2	6	5	5	3	3	2	1
113	30	26	22	18	18	6	2	7	6	5	4	4	2	1
114	30	26	22	18	14	10	6	8	7	6	5	4	3	2
115	30	26	18	18	14	14	10	6	5	4	4	3	3	2
116	30	22	22	22	18	6	6	4	3	3	3	2	1	1
117	30	22	22	18	18	10	10	5	4	4	3	3	2	2
118	28	28	28	12	12	12	0	2	2	2	1	1	1	0
119	28	28	24	16	16	8	4	5	5	4	3	3	2	1
120	28	28	20	20	20	4	0	3	3	2	2	2	1	0
121	28	28	20	20	12	12	8	4	4	3	3	2	2	1
122	28	28	16	16	16	16	12	3	3	2	2	2	2	1
123	28	24	24	20	20	4	4	5	4	4	3	3	1	1

Number	$\|b_i\|$							$\|a_i\|$						
$n \le 6$														
124	28	24	24	20	16	8	8	6	5	5	4	3	2	2
125	28	24	20	20	16	12	12	7	6	5	5	4	3	3
126	26	26	26	18	18	6	6	3	3	3	2	2	1	1
127	26	26	22	22	22	2	2	4	4	3	3	3	1	1
128	26	26	22	22	14	10	10	5	5	4	4	3	2	2
129	26	26	18	18	18	14	14	4	4	3	3	3	2	2
130	26	22	22	22	14	14	14	4	3	3	3	2	2	2
131	24	24	24	24	24	0	0	1	1	1	1	1	0	0
132	24	24	24	24	12	12	12	2	2	2	2	1	1	1
133	24	24	20	20	16	16	16	5	5	4	4	3	3	3
134	22	22	22	18	18	18	18	3	3	3	2	2	2	2
135	20	20	20	20	20	20	20	1	1	1	1	1	1	1

Reprinted by permission, M. L. Dertouzos, "Threshold Element Synthesis," MIT Report ESL–R–200, June, 1964.

Hence the coefficients are

For F_1

$$a_1 = 3, \quad a_2 = 2, \quad a_3 = a_4 = 1$$

and

$$\begin{aligned} u:\ell &= \tfrac{1}{2}(\sigma - a_0 + 1):\tfrac{1}{2}(\sigma - a_0 - 1) \\ &= \tfrac{1}{2}[(3 + 2 + 1 + 1) - 2 + 1]:\tfrac{1}{2}(7 - 2 - 1) \\ &= 3:2 \end{aligned}$$

For F_2

$$a_1 = 3, \quad a_2 = -2, \quad a_3 = a_4 = 1$$

and

$$\begin{aligned} u:\ell &= \tfrac{1}{2}(\sigma - a_0 + 1):\tfrac{1}{2}(\sigma - a_0 - 1) \\ &= \tfrac{1}{2}[(3 - 2 + 1 + 1) - 2 + 1]:\tfrac{1}{2}(3 - 2 - 1) \\ &= 1:0 \end{aligned}$$

For F_3

$$a_1 = 3, \; a_2 = 2, \; a_3 = a_4 = 1$$

and

$$\begin{aligned} u:\ell &= \tfrac{1}{2}(\sigma - a_0 + 1):\tfrac{1}{2}(\sigma - a_0 - 1) \\ &= \tfrac{1}{2}[(3 - 2 + 1 + 1) - (-2) + 1]:\tfrac{1}{2}(3 + 2 - 1) \\ &= 3:2 \end{aligned}$$

Thus the realizations are

$$F_1 = \langle 3x_1 + 2x_2 + x_3 + x_4 \rangle_{3:2}$$
$$F_2 = \langle 3x_1 - 2x_2 + x_3 + x_4 \rangle_{1:0}$$
$$F_3 = \langle 3x_1 - 2x_2 + x_3 + x_4 \rangle_{3:2}$$

Proof of Procedure. Theorem 10.3 shows that the numbers b_i characterize the function for ℓ.s. realizability. It is furthermore clear from the definition of the b_i (Section 10.8) that complementing a variable, x_i, changes the sign of b_i and that taking the dual of a function changes the sign of b_0. It therefore follows that the values of $|a_i|$ can be listed in a table corresponding to the sequence of $|b_i|$. The only part of the procedure requiring proof is the value of the gap $u:\ell$ and the optimum nature of the solution.

Consider a function

$$F(x_1,x_2,\ldots,x_n) = \left\langle \sum_{i=1}^{n} a_i x_i \right\rangle_{u:\ell}$$

with characterizing numbers $b_n,b_{n-1},\ldots,b_0$, and compute its self-dual representative function

$$F'(x_1,x_2,\ldots,x_{n+1}) = \bar{x}_{n+1}F(\,) \dotplus x_{n+1}F^d(\,)$$

From Example 4.6 a realization for $F'(\,)$ is

$$F'(\,) = \left\langle a'_{n+1}x_{n+1} + \sum_{i=1}^{n} a_i' x_i \right\rangle_{u':\ell'}$$

where

$$a_i' = a_i \qquad i = 1,2,\ldots,n$$

and

$$a'_{n+1} = u + \ell - \sigma$$

where

$$\sigma = \sum_{i=1}^{n} a_i$$

It follows from the definition of F' that its characterizing numbers are

$$b_0' = 0 \qquad \text{(since } F' \text{ is self-dual)}$$
$$b_i' = 2b_i \qquad \text{for} \quad 1 \leq i \leq n$$
$$b'_{n+1} = -2b_0$$

Thus for a nonself-dual function such as F, the characterizing numbers, b_i, are simply related to those of its self-dual representative function F'. In particular, the number b_0 is related to the negative of b'_{n+1}.

In the table, the coefficients that are tabulated are those of F' (first proposed by Goto, 2). In other words, one can assume that in the $|b_i|$ columns is tabulated $|b_i'/2|$, $i \neq 0$, and in the $|a_i|$ column is tabulated $|a_i'|$. To relate the primed coefficients to the unprimed ones, we use the realization for F'. Thus, for $i \neq 0$,

$$a_i' = a_i$$

For $i = 0$, the tabulated coefficient a_0 is given by

$$-a_0 = a_{n+1}'$$

because of the negative sign in the relation between b_0 and b_{n+1}'. But from the realization

$$a_{n+1}' = u + \ell - \sigma$$

To proceed further, we use the following auxiliary result.

In the minimum integer sum realization the gap length is always unity. To verify this, assume that the gap length were not unity. Since all the weights are integers, the gap length must also be an integer. But if the gap length were any integer greater than one, for instance, two, then at least one of the weights could be reduced in value by one, which would only reduce the gap length by one. But this is contrary to the assumption of a minimum integer sum realization.

Since the gap length is normalized at unity,

$$u = \ell + 1$$

and the expression for a_{n+1}' becomes

$$a_{n+1}' = 2u - 1 - \sigma$$

Hence the weight a_0 is

$$-a_0 = 2u - 1 - \sigma$$

from which we get

$$u = \tfrac{1}{2}(\sigma - a_0 + 1)$$
$$\ell = \tfrac{1}{2}(\sigma - a_0 - 1)$$

These are the desired expressions.

To investigate the optimality of the tabulated realizations, consider the normalized fan-in of the self-dual representative function F'. Since the gap length is unity, the normalized fan-in is

$$|a_0| + \sum_{i=1}^{n} |a_i| = (\sigma - 2u + 1) + \sigma$$
$$= 2(\sigma - u) + 1$$

Thus if, as is done in the table, the minimum normalized fan-in realization of F' is tabulated, this will correspond to the minimum sum realization for F.▮

BIBLIOGRAPHY (through 1965)

S. Agmon, "The Relaxation Method for Linear Inequalities," *Can. J. Math.*, **6,** 382–392 (1954).

S. B. Akers, (1) "Threshold Logic and Two-Person, Zero-Sum Games," *SCTLD*† (1961), pp. 27–33, September 1961.

——, (2) "On the Algebraic Manipulation of Majority Logic," *IRE Trans. Electron. Computers*, **EC-10, (4),** 779 (December 1961)

——, (3) "Synthesis of Combinational Logic Using Three-Input Majority Gates," *SCTLD* (1962), pp. 149–157, September 1962.

S. Amarel, G. Cooke, and R. Winder, "Majority Gate Networks," *IRE Trans. Electron. Computers*, **EC-13, (9),** 4–13 (February 1964).

H. D. Block, "The Perceptron: A Model for Brain Functioning, I" *Rev. Mod. Phys.* **34, (1),** 123–35 (January 1962).

M.A. Breuer, "Implementation of Threshold Nets by Integer Linear Programming," *IEEE Trans. Electron. Computers*, **EC-14, (6),** 950–952 (December 1965).

A. Butakov, "Synthesis of Threshold Elements on General Purpose Computers," *Proc. Intern. Symp. Relay Systems and Finite Automata*, 1962, pub. by Burroughs Corp., 1964.

S. H. Caldwell, "Switching Circuits and Logical Design," John Wiley & Sons, New York, 1958.

S. H. Cameron, (1) "An Estimate of the Complexity Requisite in a Universal Decision Network," *Bionics Symposium*, WADD Report 60–600, pp. 197–212, December 1960.

——, (2) "The Generation of Minimal Threshold Nets by an Integer Program," *IRE Trans. Electron. Computers* **EC-13, (3),** 299–302 (June 1964).

R. H. Canaday, "Two Dimensional Iterative Logic," Doctoral Dissertation for MIT Electrical Engineering Department; also issued as ESL-R-210.

† *SCTLD* refers to *Proceedings of the Symposium on Switching Circuit Theory and Logical Design*, AIEE, 1960–1963, IEEE 1964–1965.

C. K. Chow, (1) "Boolean Functions Realizable with Single Threshold Devices," *Proc. IRE*, **49,** 370–371 (January 1961).

——, (2) "On the Characterization of Threshold Functions," *SCTLD* (1961), pp. 34–38, September 1961.

——, (3) "Statistical Independence and Threshold Functions," *IEEE Trans. Electron. Computers*, **EC-14,** 1, 66–68 (February 1965).

C. L. Coates, R. B. Kirchner, and P. M. Lewis II, "A Simplified Procedure for the Realization of Linearly Separable Switching Functions," *IRE Trans. Electron. Computers*, **EC-11**, 447–458, (August 1962).

C. L. Coates and P. M. Lewis II, (1) "Linearly Separable Switching Functions," *Journal of the Franklin Institute*, **272,** 360–410 (November 1961).

——, (2) "Realization of Logical Functions by a Network of Threshold Components with Specified Sensitivity," *IRE Trans. Electron. Computers*, **EC-12, (5)**, 443–453 (October 1963).

——, (3) "A Realization Procedure for Threshold Gate Networks," *IRE Trans. Electron. Computers*, **EC-12, (5)**, 454–461 (October 1963). Also *SCTLD* (1962).

——, (4) "DONUT: A Threshold Gate Computer," *IRE Trans. Electron. Computers*, **EC-13, (3)**, 240–247 (June 1964).

C. L. Coates and V. Supornpaipul, "On Maximum Stability Realizations of Linearly Separable Functions," *SCTLD* (1965), pp. 12–24, October 1965.

A. Cobham "The Asummability Condition for Seven Variable Functions" *IBM Research Note*, March 1965.

M. Cohn and R. Lindaman, "Axiomatic Majority-Decision Logic," *IRE Trans. Electron. Computers*, **EC-10**, 17–21 (March 1961).

L. Dadda, "Synthesis of Threshold Logic Combinatorial Networks," *Alta Frequenza*, No. 3, **XXX,** 224–28E–35E–231 (1961).

G. B. Dantzig, "Linear Programming and Extensions," Princeton University Press, Princeton, N.J., 1963.

M. Dertouzos, (1) "Threshold Element Synthesis," Doctoral Dissertation for MIT Electrical Engineering Dept., also issued as ESL–R–200.

——, (2) "An Approach to Single Threshold Element Synthesis," *IRE Trans. Electron. Computers*, **EC-13, (5)**, 519–528. (October 1964).

——, (3) "Threshold Logic: A Synthesis Approach," MIT Press, Cambridge, Mass., 1965.

S. N. Einhorn, "The Use of the Simplex Alogorithm in the Mechanization of Boolean Switching Functions by Means of Magnetic Cores," *IRE Trans. Electron. Computers*, **EC-10, (4)**, 615–622 (December 1961).

C. C. Elgot, "Truth Functions Realizable by Single Threshold Organs," *SCTLD* (1960), pp. 225–245, September 1961, and AIEE Conference Paper 60–1311, October 1960, revised November 1960.

C. C. Elgot and S. Muroga, "Two Open Problems on Threshold Logic," *SCTLD* (1961), pp. 166, September 1961.

K. Fan, "On Systems of Linear Inequalities," *Ann. Math. Stud.* No. 38, W. H. Kuhn and A. W. Tucker, Princeton University Press, Princeton, N.J., 1956.

M. A. Fischler and E. A. Poe, "Threshold Realization of Arithmetic Circuits," *IRE Trans. Electron. Computers*, **EC-11, (2)**, 287–288 (April 1962).

J. B. Fourier, *Solution d'une Question Particulière du Calcul des Inégalités*, 1826.

I. J. Gabelman, (1) "The Functional Behavior of Majority (Threshold) Elements," Doctoral Dissertation for Syracuse University, Electrical Engineering Department.

——, (2) "A Note on the Realization of Boolean Functions Using a Single Threshold Element," *Proc. IRE*, **50, (2)**, 225–226 (February 1962).

——, (3) "The Synthesis of Boolean Functions Using a Single Threshold Element," *IRE Trans. Electron. Computers*, **EC-11, (5)**, 639–642 (October, 1962).

——, (4) "Properties and Transformations of Single Threshold Element Functions," *IEEE Trans. Electron. Computers*, **EC-13, (6)**, 680–684 December 1964).

——. (5) "An Algorithm for Threshold Element Analysis," *IEEE Trans. Electron. Computers*, **EC-14, (5)**, 623–625 (August 1965).

C. A. Gaston, "A Simple Test for Linear Separability," *IRE Trans. Electron. Computers*, **EC-12, (2)**, 134–135. (April 1963).

C. H. Gustafson, D. R. Haring, A. K. Susskind and T. G. Wills-Sanford, "Synthesis of Counters with Threshold Elements," *SCTLD* (1965), pp. 25–35, October 1965.

R. Gonzalez and E. L. Lawler, "Two Level Threshold Minimization," *SCTLD* (1965), pp. 41–44, October 1965.

R. E. Gomory, "Outline of an Algorithm for Integer Solutions to Linear Programs, "*Bull. Amer. Math. Soc.*, **64**, 275–278 (September 1958).

E. Goto, (1) "The Parametron, a Digital Computer Element which Utilizes Parametric Oscillation " *Proc. IRE.* **47**, 1304–1316. (August 1959).

——, H. Takahasi, (2) "Some Theorems Useful in Threshold Logic for Enumerating Boolean Functions," *Proc. IFIP Congress 62*, Munich, August 1962.

——, (3) "Threshold Majority and Bilateral Switching Devices," *Symposium on the Application of Switching Theory in Space Technology*, February 1962, Sunnyvale, Cal.

W. H. Hansen, "Threshold Logic Synthesis by Algebraic Methods," *IRE Trans. Electron. Computers*, **EC-12, (4)**, 401–402 (August 1963).

J. K. Hawkins, "Self Organizing Systems—A Review and Commentary," *Proc. IRE*, **49**, 31–48 (January 1961).

W. H. Highleyman, "A Note on Linear Separation," *IRE Trans. Electron. Computers*, **EC-10, (4)**, 777–778 (December 1961).

Y-C Ho and R. L. Kashyap, "An Algorithm for Linear Inequalities and Its Applications," *IEEE Trans. Electron. Computers*, **EC-14, (5),** 683–688 (October 1965).

J. E. Hopcraft and R. L. Mattson "Synthesis of Minimal Threshold Logic Networks," *IEEE Trans. Electron. Computers*, **EC-14, (5),** 552–560 (August 1965).

O. A. Horna, "Majority Logic Synthesis," *IRE Trans. Electron. Computers*, **EC-12, (2),** 131–132 (April 1963).

G. Hotz, "Digital Filters of Threshold Elements," *Proc. IFIP Congress 62*, Munich, August 1962.

S. T. Hu, "Minimal Linear Decompositions of Switching Functions," *SCTLD* (1965), pp. 36–40, October 1965.

K. R. Kaplan and R. O. Winder, "Chebyshev Approximation and Threshold Functions," *IEEE Trans. Electron. Computers*, **EC-14, (2),** 250–252 (April 1965).

M. Karnaugh, "Pulse-switching Circuits using Magnetic Cores," *Proc. IRE*, **43,** 570–584 (May 1955).

W. H. Kautz, "The Realization of Symmetric Switching Functions with Linear-Input Logical Elements" *IRE Trans. Electron. Computers*, **EC-10, (3),** 371–378 (September 1961).

P. Kaszerman, (1) "A Geometric Test Synthesis Procedure for a Threshold Device," *Information and Control*, **6, (1),** 381–393 (1963).

——, (2) "A 'Region' Concept and Its Application to Threshold Logic," *Information and Control*, **8, (5),** 531–551 (October 1965).

K. Krohn and J. Rhodes, "Nets of Threshold Elements," *Information and Control*, **8, (6),** 579–588, (December 1965).

P. M. Lewis II and C. L. Coates (1), "A Realization Procedure for Threshold Gate Networks," *SCTLD* (1962), pp. 159–168, September 1962.

——, (2) "Threshold Gate Realization of Logical Functions with Don't Cares," *SCTLD* (1963), pp. 41–52, October 1963.

R. Lindaman, (1) "A New Concept in Computing," *Proc IRE*, **48,** 257 (February 1960).

——, (2) "A Theorem for Deriving Majority-Logic Networks within an Augmented Boolean Algebra," *IRE Trans. Electron. Computers*, **EC-9,** 338–342 (September 1960).

C. L. Liu and J. W. S. Liu, "On a Multiplexing Scheme for Threshold Logic Elements," *Information and Control*, **8, (3),** 282–294 (June 1965).

M. T. Liu (1) "The Trisquare Map Method for Realization of Threshold Functions," Doctoral Dissertation, Electrical Engineering Department, University of Pennsylvania, June 1961.

——, (2) "On the Dual Monotonicity of Threshold Functions" *IEEE Trans. Electron. Computers*, **EC-14, (5),** 625–627 (August 1962).

R. McNaughton, "Unate Truth Functions," *IRE Trans. Electron. Computers*, **EC-10,** 1–6 (March 1961), and previously available as

Technical Report No. 4, Applied Mathematics and Statistics Laboratory, Stanford University, October 1957.

H. S. Miiller and R. O. Winder, "Majority Logic Synthesis by Geometric Methods," *IRE Trans. Electron. Computers*, **EC-11, (1)**, 89–90 (February 1962).

R. C. Minnick, (1) "Linear-Input Logic," *IRE Trans. Electron. Computers*, **EC-10**, 6–16 (March 1961).

——, and R. A. Short, (2) "Cellular Linear Input Logic," *SRI* Report AFCRL–64–6, February 1964.

F. Miyata, (1) "Realization of Arbitrary Logical Functions Using Majority Elements," *IRE Trans. Electron. Computers*, **EC-12, (3)**, 183–191 (June 1963).

——, (2) "An Extension of the Method of Cohn and Lindaman," *IEEE Trans. on Electron. Computers*, **EC-13, (5)**, 625–629 (October 1964).

E. F. Moore, "Counter-Example to a Conjecture of McCluskey and Paull," 1957 (unpublished).

T. S. Motzkin, "Beiträge Zur Theorie der Linearen Unglerchungen, 1936, Doctoral Thesis, University of Zurich.

T. S. Motzkin and I. J. Schoenberg, "The Relaxation Method for Linear Inequalities," *Can. J. Math.*, **6**, 393–404 (1954).

C. H. Mays, (1) "Effect of Adaptation Parameters on Convergence Time and Tolerance for Adaptive Threshold Elements," *IEEE Trans. Electron. Computers*, **EC-13**, 465–468 (August 1964). (See also Stanford Electronic Lab. Report SEL–63–027; April 1963.)

——, (2) "The Boundary Matrix of Threshold Functions," *IEEE Trans. Electron. Computers*, **EC-14, (1)**, 65–66 (February 1965).

S. Muroga, S. Toda, and S. Takasu, (1) "Theory of Majority Decision Elements," *Journal of the Franklin Institute*, **271**, 376–418 (May 1961).

S. Muroga, (2) "Functional Forms of Majority Functions and a Necessary and Sufficient Condition for their Realizability," *SCTLD* (1961), pp. 39–46, September 1961.

——, (3) "Restrictions in Synthesis of a Network with Majority Elements," *Proc. IRE*, **49**, 1455 (September 1961).

——, (4) "Majority Logic and Problems of Probabilistic Behavior," *Self Organizing Systems 1962*, pp. 243–281, Spartan Books, 1962 (Proceedings of Conference, May 1962).

——, (5) "Generation of Self-Dual Threshold Functions and Lower Bounds of the Number of Threshold Functions and a Maximum Weight," *SCTLD* (1962), pp. 169–184, September 1962.

——, S. F. Toda and M. Kondo, (6) "Majority Decision Functions of up to Six Variables," *J. Math. of Computers*, October, 1962 (Published in Japanese in 1959 and 1960).

S. Muroga and K. Takoshima (7) "The Parametron Digital Computer MUSASINO–1," *IRE Trans. Electron. Computers*, **EC-8, (3)**, 308–316 (September 1959).

S. Muroga, (8) "Generation and Asymmetry of Self-Dual Threshold Functions," *IEEE Trans. Electron. Computers*, **EC-14, (2)**, 126–136 (April 1965).

——, (9) "Some Bounds on the Number of Threshold Functions and a Maximum Weight," *IEEE Trans. Electron. Computers*, **EC-14, (2)**, 136–148 (April 1965).

——, (10) "Logical Elements on Majority Decision Principle and Complexity of their Circuit," *Proc. Intern. Conf. Information Processing*, Paris, France, June 1959, UNESCO House, Paris, 1960, pp. 400–407.

J. Myhill and W. H. Kautz, "On the Size of Weights Required for Linear-Input Switching Functions," *IRE Trans. Electron. Computers*, **EC-10**, 288–290 (June 1961).

A. E. Negrin, "Synthesis of Practical Three-Input Majority Logic Network," *IRE Trans. Electron. Computers*, **EC-13, (3)**, 296–299 (June 1964).

N. J. Nilsson, *Learning Machines*, McGraw-Hill Book Co., New York, 1965.

A. Novikoff, "On Convergence Proofs for Perceptrons," *Symposium on Mathematical Theory of Automata*, Polytechnic Institute of Brooklyn, April 1962, Proceedings, pp. 615–621.

M. C. Paull and E. J. McCluskey, Jr., "Boolean Functions Realizable with Single Threshold Devices," *Proc. IRE*, **48**, 1335–1337 (July 1960).

W. V. Quine, "Two Theorems about Truth Functions," *Boletin Soc. Matem. Mex.* **Y**, 64–70 (1953).

F. Rosenblatt, "Principles of Neurodynamics," Sparton Publishing Company, Washington, D.C., 1962.

W. D. Rowe, "The Transistor NOR Circuit," *IRE Wescon Convention Record*, Pt. 4, 1957, pp. 231–245.

C. E. Shannon, "A Symbolic Analysis of Relay and Switching Circuits," *Trans. AIEE*, **57**, 713–723, (1938).

R. F. Shauer, R. M. Stewart, A. U. Pohm, and A. A. Read, "Some Applications of Magnetic Film Parametrons as Logical Devices," *IRE Trans. Electron. Computers*, **EC-9, (3)**, 315–323 (September 1960).

C. L. Sheng, (1), "A Method for Testing and Realization of Threshold Functions," *IRE Trans. Electron. Computers*, **EC-13, (3)**, 232–239 (June 1964).

——, (2) "A Graphical Interpretation of Realization of Symmetric Boolean Functions with Threshold Logic Elements," *IEEE Trans. Electron. Computers*, **EC-14, (1)**, 8–18 (February 1965).

——, (3) "Compound Synthesis of Threshold-Logic Networks for the Realization of General Boolean Functions," *IEEE Trans. Electron. Computers*, **EC-14, (6)**, 798–814 (December 1965).

R. C. Singleton, "A Test for Linear Separability as Applied to Self-Organizing Machines," *Self-Organizing Systems, 1962*, Spartan Books, 1962 (Proceedings of the Conference, May 1962), pp. 503–524.

E. B. Stabler, "Threshold Gate Network Synthesis," *SCTLD*, (1965) pp. 5–11, October 1965.

K. Steinbuck and B. Widrow, "A Critical Comparison of Two Kinds of Adaptive Classification Networks," *IEEE Trans. Electron. Computers*, **EC-14, (5)**, 737–740 (October 1965).

O. B. Stram, (1) "Arbitrary Boolean Functions of N Variables Realizable in Terms of Threshold Devices," *Proc. IRE*, **49,** 210–220 (January 1961).

——, (2) "The Profile Technique for the Design of Threshold Device Logic," *SCTLD* (1961), pp. 47–54, September 1961.

Y. Tohma, "Decompositions of Logical Functions Using Majority Decision Elements," *IEEE Trans. Electron. Computers*, **EC-13, (6)**, 698–705 (December 1964).

B. Widrow and M. E. Hoff, (1) "Adaptive Systems in Pattern Recognition," *1960 WESCON Conv. Rev.*, Pt. 4, pp. 96–104 (August 1960).

B. Widrow, (2) "Generalization and Information Storage in Networks of Adaline Neurons," *Self Organizing Systems 1962*, M.C. Yovits, G. T. Jacobi and G. D. Goldstein, ed., Spartan Books, Washington, D.C., 1962.

B. Widrow and J. B. Angell, (3) "Reliable, Trainable Networks for Computing and Control," *Aerospace Eng.* **21,** 78–123 (September 1962).

D. Willis, "Minimum Weights for Threshold Switches," *Switching Theory in Space Techniques*, Stanford University Press, Stanford, Calif., 1963.

R. O. Winder, (1) "Threshold Logic," Doctoral Dissertation for the Mathematics Department of Princeton University, 1962.

——, (2) "Single Stage Threshold Logic," *SCTLD* (1960), pp. 321–332, September 1961, and AIEE Conference Paper 60–1261, October 1960.

——, (3) "More About Threshold Logic," *SCTLD* (1961), pp. 55–64, September 1961.

——, (4) "Some Recent Papers in Threshold Logic," *Proc. IRE*, **49,** 1100 (June 1961).

——, (5) "Threshold Logic in Artificial Intelligence," AIEE 1963 Winter Meeting, published in *Artificial Intelligence*, AIEE Publication, pp. 107–128, January 1963.

——, (6) "Bounds on Threshold Gate Realizability," *IRE Trans. Electron. Computers*, **EC-12, (5)**, 561–4 (October 1963).

——, (7) "Properties of Threshold Functions," *IEEE Trans. Electron. Computers*, **EC-14, (2)**, 252–254 (April 1965).

——, (8) "Enumeration of Seven Argument Threshold Functions," *IEEE Trans. Electron. Computers*, **EC-14, (3)**, 315–325 (June 1965). See also AFCRL Rept. 64–925.

——, (9) Review R62–12U, *IRE Transactions on Electronic Computers*, **EC-11, (6)**, 801 (December 1962).

P. E. Wood, "A Note on Threshold Device Error Analysis," *IRE Trans. Electron. Computers*, **EC-12, (4),** 403–405 (August 1963).

S. Yajima and T. Ibaraki, "A Lower Bound on the Number of Threshold Functions," *IEEE Trans. Electron. Computers*, **EC-14, (6),** 926 (December 1965).

INDEX